Ping Shuai

Understanding Pulsars and Space Navigations

国防工業出版社
National Defense Industry Press

Springer

Ping Shuai
China Academy of Space Technology
Beijing, China

ISSN 2522-0454 ISSN 2522-0462 (electronic)
Navigation: Science and Technology
ISBN 978-981-16-1066-0 ISBN 978-981-16-1067-7 (eBook)
https://doi.org/10.1007/978-981-16-1067-7

Jointly published with National Defense Industry Press
The print edition is not for sale in China (Mainland). Customers from China (Mainland) please order the print book from: National Defense Industry Press.
ISBN of the Chinese edition: 978-7-118-11118-7

This Springer imprint is published by the registered company Springer Nature Singapore Pte Ltd.
The registered company address is: 152 Beach Road, #21-01/04 Gateway East, Singapore 189721, Singapore

Preface

Pulsars are the remnants that result from supernova explosion of the massive stars after evolving very long time and combining gravitational collapse, and belong to high-speed rotation neutron stars, with extremely stable periodicity, thereby known as the most stable clocks in nature. Using the X-ray photons radiated from the pulsars as natural beacons, the process in which the spacecrafts can determine autonomously their orbit, time and attitude parameters to guide them to fly to the target orbits or target celestial body along the designed orbits safely, effectively and just in time is called **X-ray Pulsar-based Navigation (XPNAV)** or **Pulsar Navigation**. Because the X-ray photons are unable to pass through the dense atmosphere of the Earth, they can be detected only in space outside the atmosphere. Apparently, the pulsar navigation can provide the navigation information services for the spacecrafts in the near-earth orbits, deep space and interstellar space, as well as the rovers on the surface of the planets without dense atmosphere, but not directly for the users on the ground. So, the pulsar is also known as the lighthouse for space flight, and hereby the pulsar navigation is substantially a kind of **space navigation**. The pulsar navigation has important engineering value and scientific research significance, and has been a research hotspot in the field of astronautical frontier and disruptive technology in the past 15 years.

Since 2005, my research team and I have been carrying out studies on the theories and methods of the pulsar navigation system. In 2009, the first book on the pulsar navigation co-authored by my colleague and myself, *X-ray Pulsar-based Navigation: Principle and Methodology,* was published by China Astronautic Publishing House. In 2013, the research team made a breakthrough in the studies on key technologies of the pulsar navigation, developed several types of X-ray detector prototypes and products, and constructed a set of ground test system. In 2015, a pulsar navigation test satellite, known as XPNAV-1, was developed by the research team. In November 2016, the world's first dedicated pulsar navigation test satellite XPNAV-1 was successfully launched into the orbit from Jiuquan Satellite Launch Center of China, carrying out the space flight test, to test the performances of the X-ray detectors and demonstrate the feasibility of the pulsar navigation system. At the time, the National Defense Industry Press of China was publishing the second book on the pulsar navigation written by myself, *Pulsars: the Lighthouse for Space Flight*, and then recommended the book to Springer press for publication in English, in September 2017. After reviewing the English abstract and table of contents of the book, the Springer's reviewer considered: "*The book proposal titled "Pulsars: the Lighthouse for Space Flight" is an excellent one, which described the navigation based on the pulsar timing, a universal object. So, the book should be a valuable reference for both scientific community and public. However, some suggestions may be considered before the book is decided for publication...*"

According to the reviewer's suggestions, the author deleted the first chapter, second chapter and some sections of the other four chapters in the original manuscript, and expanded, revised and updated the relevant knowledge and content, especially as a section, adding the spaceflight test results from the XPNAV-1 satellite. In this way, the architecture and content of the original book have undergone great changes, focusing on the concepts, theories and experiments on the pulsars, space flight, space navigation and pulsar navigation, as well as addressing their internal logical relations in detail, which is almost a new creation and writing. Obviously, the original title is no longer appropriate, and hence it is re-entitled *Understanding Pulsars and Space Navigations*. This book is divided into five chapters, in which the basic concepts, development histories, realization approaches, theoretical systems, basic methods and flight tests of the pulsars and space navigations are discussed systematically, deeply and comprehensively. In the Chap. 1, the theoretical prediction of neutron stars as well as the discovery, identification, naming, classification, spatial distribution, physical mechanism and observation characteristics of pulsars are briefly introduced as fundamental knowledge. The Chap. 2 is divided into two parts for describing: one is the development of radio astronomy, principle of radio telescope, modern large radio telescopes and observation of radio pulsars; another is the discovery and physical characteristics of X-rays, X-ray detection mechanism, detector's classification, development of X-ray astronomy and observation of X-ray pulsars. On the basis of summarizing the concepts of space flight, the development history of human space flight is narrated systematically and intensively in the Chap. 3, from the thought sprout and technical development of ancient space flight to the pioneers and founders of modern astronautics, the promoting role of early space research societies, the coming of the age of space flight and the probes flying out of the solar system, and particularly in the end of the Chap. 3, the pulsar's roles in space flights are addressed in detail. In the Chap. 4, firstly, the terms and concepts related to navigation are briefly described and identified; secondly, the development process from ancient south-pointing chariot to modern navigation system is described; thirdly, the basic principles, technical developments and applications including the celestial navigation system, inertial navigation system, radio navigation system and navigation satellite system are discussed, respectively; finally, the ground navigation networks and autonomous navigation technologies for spacecrafts are commented as a key point, which is the problem of space navigation. In the Chap. 5, firstly, the principle and progress on the pulsar navigation are addressed; secondly, the space-time theories under the Newtonian mechanics and general relativity frames are discussed, respectively; thirdly, the fundamental methods of constructing the pulsar timing system and large-scale navigation are presented, and a set of ground test system is designed and built to test the X-ray detectors and quantitatively analyze the performance of pulsar navigation system; finally, the XPNAV-1 satellite and its observation results are addressed in detail, and further the next-generation navigation satellite system is conceived from system architecture.

This book is both a professional reference about pulsar navigation and a high-grade popular science reading material, with the basic characteristics of "five lectures": the first is to lecture the concepts, talking about the concepts of pulsars, spaceflights and navigations; the second is to lecture the histories, talking about the histories of the radio astronomy, X-ray discovery, X-ray astronomy, space flight, navigation system and pulsar navigation; the third is to lecture the approaches, talking about the technical approaches to guide the navigations for movable objects, from ancient south-pointing chariot to modern navigation system; the fourth is to lecture the theory, talking about the basic theories and methods of space-time reference, pulsar timing system and large-scale navigation; the fifth is to lecture the experiments, talking about the ground tests, numerical experiments and spaceflight tests.

The pulsar navigation is a typical interdisciplinary research field, which is complex in technology, difficult to implement, innovative strongly, with the remarkable characteristics of multi-disciplinary, comprehensive and compound development. The purpose of writing this book is to try to build a platform for interdisciplinary research and communication, and cultivate compound talents with innovative potential, so as to better popularize and promote the technical research and system development of pulsar navigation. This book is suitable for scientific and technical personnel engaged in the research on the pulsar navigation system, and it can also be used as a general teaching material or reference book for undergraduate and graduate students who are interested in the pulsar navigation technology and have the relevant professional background knowledge, such as guidance and navigation and control, spacecraft design, theoretical physics, astrophysics, physical electronics, nuclear technology and application, geodesy and surveying engineering, communication and information system, as well as measuring and testing technologies and instruments.

In implementing the pulsar navigation project and writing the manuscript of this book, I would like to express my heartfelt thanks to Mr. Qun Liu who was then Deputy Chief of the R&D department at China Academy of Space Technology (CAST), and Mr. Hu Li who was Chief of the R&D department, for their help and support. I would like to thank the members of the mission design and data processing team of the XPNAV-1 satellite, for their support, understanding and assistance in my work, and they are Dr. Xinyuan Zhang, Dr. Liangwei Huang, Dr. Xiaomin Bei, Dr. Yaojun Wu, Dr. Qingqing Lin, Dr. Jing Meng, Dr. Hengbin Zhang, Senior Engineer Shaolong Chen, Senior Engineer Qiang Chen, Engineer Qian Zhang and Engineer Lingzhong Fu, all of who were then the staffs at the Pulsar Navigation Research Center of Qian-Xuesen Laboratory of the CAST. I wish to thank the members of the engineering team for their technical support in the development and space flight test of the XPNAV-1 satellite, and they are Dr. Lijun Xue at Shenzhen Aerospace Dongfanghong Development Ltd., Dr. Loulou Deng at Beijing Institute of Control Engineering and Dr. Huijun Hu at Shandong Institute of Space Electronic Technology. Finally, I wish to acknowledge Ms. Li You, an Editor at the National Defense Industry Press, who has been following the progress of the manuscript and has provided helpful suggestion in the application of publishing fund, and also to acknowledge the Springer's reviewers who affirmed the publishing value of the manuscript and gave constructive suggestions for revision.

The pulsar navigation technology is with cutting-edge, innovative and interdisciplinary research characteristics. The creation of this book lasted more than 3 years. Particularly, during the final stage of writing the manuscript, just when the Coronavirus Disease 2019 (COVID-19) is rampant over the world, the retina of my left eye detached due to overwork for writing. Fortunately, my vision is gradually recovering after timely surgery. For the writing of this book, I always strive to be correct in concept, clear in historical facts, feasible in method and credible in experiment, and to make it knowledgeable, logical, interesting and readable. Nevertheless, there may be inevitable omissions and defects in the text, and here my apology is extended to every reader who reads this book.

Finally, it is expected that the publication of this book will help researchers to well understand and master the basic concepts and principles, mathematical models and algorithms, flight tests and demonstration of the pulsars and space navigations, to accelerate and haul the leapfrog development of astronautical cutting-edge technology.

Haidian District, Beijing, China
September 2020

Ping Shuai

Introduction

Pulsars are high-speed rotation neutron stars, with extremely stable periodicity, thereby known as the most stable clocks in nature. With the X-ray photons radiated from the pulsars as natural beacons, the spacecrafts can be determined their orbit, time and attitude parameters, so as to achieve autonomous navigation in space. This book is divided into five chapters, in which the basic concepts, development histories, realization approaches, theoretical systems, basic methods and flight tests of the pulsars and space navigations are discussed systematically, deeply and comprehensively. Especially, the basic theories and experimental results of pulsar navigation are addressed in details in this book. The pulsar navigation is a typical interdisciplinary research field, which is complex in technology, difficult to implement, innovative strongly, with the remarkable characteristics of multi-disciplinary, comprehensive and compound development. This book is to try to build a platform for interdisciplinary research and communication, and cultivate compound talents with innovative potential, so as to better popularize and promote the technical research and system development of pulsar navigation. So, This book is both a professional reference about pulsar navigation and a high-grade popular science reading material, suitable for scientific and technical personnel engaged in the research on the pulsar navigation system, and it can also be used as a general teaching material or reference book for undergraduate and graduate students who are interested in the pulsar navigation technology and have the relevant professional background knowledge.

Contents

Acronyms

1G-NAV	First-Generation Navigation Satellite System
2G-NAV	Second-Generation Navigation Satellite System
2SOPS	2nd Space Operation Squadron
3G-NAV	Third-Generation Navigation Satellite System
A/S	Anti-spoofing
ABMA	US Army Ballistic Missile Agency
AFB	US Air Force Base
AFSCN	Air Force Satellite Control Network
AGN	Active Galactic Nuclei
AIS	American Interplanetary Society, or Automatic Identification System
AIT	Assemble, Integration and Test
ANS	Astronomical Netherlands Satellite
AOD	Age of Data
AP	Accretion Powered pulsar
ARGOS	Advanced Research and Global Observation Satellite
ARS	American Rocket Society
AS	Apollo-Saturn combination
ASCA	Advanced Satellite for Cosmology and Astrophysics
ASK	All-Sky Monitor
AX	Anomalous X-ray pulsar
AXAF	Advanced X-ray Astrophysics Facility
BCRS	Barycentric Celestial Reference System
BDCS	BeiDou Coordinate System
BDS	BeiDou Navigation Satellite System
BDT	BeiDou Navigation Satellite System Time
BIH	Bureau International de l'Heure
BIPM	Bureau International des Poids et Measures
BIS	British Interplanetary Society
BJ	Big Joe rocket
BTS	BIH Terrestrial System
CalTech	California Institute of Technology

CAST	China Academy of Space Technology
CCD	Charge-Coupled Device
CDMA	Code Division Multiple Access
CEO	Celestial Ephemeris Origin
CEP	Celestial Ephemeris Pole
CGRO	Compton Gamma-Ray Observatory (or GRO)
CIO	Conventional International Origin/Celestial Intermediate Origin
CIP	Celestial Intermediate Pole
CM	Command Module
CMB	Cosmic Microwave Background
CNES	Centre National d'Etudes Spatiales (French)
CNS	Celestial Navigation System
CPU	Central Processing Unit
CS	Commercial Service
CTP	Conventional Terrestrial Pole
CTS	Conventional Terrestrial System
CV	Cataclysmic Variable star
CXO	Chandra X-ray Observatory
DARPA	Defense Advanced Research Project Agency (US)
DASS	Distress Alerting Satellite System
DME	Distance Measuring Equipment
DNA	Deoxyribonucleic Acid
DNSS	Defense Navigation Satellite System
DoD	Department of Defense (US)
DOF	Degrees of Freedom
DOP	Dilution of Precision
DORIS	Doppler Orbitography and Radiopositioning Integrated by Satellite
DSN	NASA Deep Space Network
ECEF	Earth-Centered Earth-Fixed
ECI	Earth-Centered Inertial
EPTA	European Pulsar Timing Array
ESA	European Space Agency
ET	Ephemeris Time
EU	European Union
EXOSAT	European X-ray Observatory Satellite
FAA	Federal Aviation Administration (US)
FAST	Five-hundred-meter Aperture Spherical Telescope
FCC	Federal Communication Commission (US)
FDMA	Frequency Division Multiple Access
FFT	Fast Fourier Transform
FGST	Fermi Gamma-Ray Space Telescope
FOC	Full Operational Capability
FOV	Field of View
FWHW	Full Width at Half Maximum
GALCIT	Guggenheim Aeronautical Laboratory, Caltech

GAST	Greenwich Apparent Sidereal Time
GCAS	Ground Control Approach System
GCRS	Geocentric Celestial Reference System
GDL	Gas Dynamic Laboratory
GE	General Electronic Company
GEO	Geostationary Orbit
GINS	Gimbaled Inertial Navigation System
GIOVE	Galileo In-Orbit Validation Element
GIRD	Group for the Study of Reactive Motion (the acronym in Russian)
GLONASS	Global Navigation Satellite System
GLST	GLONASS Time
GMST	Greenwich Mean Sidereal Time
GMT	Greenwich Mean Time
GNSS	Global Navigation Satellite System
GPS	Global Positioning System
GPST	GPS Time
GRGS	Grope de Recherche en Géodésie Spatiale (French)
GSFC	Goddard Space Flight Center
GST	Galileo Navigation Satellite System Time
HEAO	High-Energy Astrophysics Observatory
HEXTE	High-Energy X-ray Timing Experiment
HMXB	High-Mass X-ray Binary
HST	Hubble Space Telescope
HTPC	High Time-resolution Photon Counter
HXMT	Hard X-ray Modulation Telescope
IAA	International Academy of Astronautics
IAC	International Astronautical Congress
IAF	International Astronautical Federation
IALA	International Association of Lighthouse Authorities
IAU	International Astronomical Union
IBM	International Business Machines Corporation
ICAO	International Civil Aviation Organization
ICBM	Inter-Continental Ballistic Missiles
ICG	International Committee on Global navigation satellite system
ICRF	International Celestial Reference Frame
ICRS	International Celestial Reference System
IERS	International Earth Rotation and Reference Systems Service
IGEB	Interagency GPS Executive Board
IGN	Institut Géographique National (French)
IGSO	Inclination Geo-Synchronization Orbit
ILS	Instrument Landing System
IMU	Inertial Measurement Units
IND	Interplanetary Network Directorate
INS	Inertial Navigation System
INTEGRAL	International Gamma-Ray Astrophysics Laboratory

IOV	In-Orbit Validation
IPTA	International Pulsar Timing Array
IRNSS	Indian Region Navigation Satellite System
ISS	International Space Station
IT	Italian National Institute of Metrological Research
ITRF	International Terrestrial Reference Frame
ITRS	International Terrestrial Reference System
ITS	Instantaneous Terrestrial System
ITT	International Telephone and Telegraph Corporation
ITU	International Telecommunication Union
IUGG	International Union of Geodesy and Geophysics
IURS	International Radio Science Association
JATO	Jet-Assisted Takeoff
JAXA	Japan Aerospace Exploration Agency
JPL	Jet Propulsion Laboratory
JPO	Joint Program Office
KIK	Command and Measurement Complex (the acronym in Russian)
LEO	Low Earth Orbit
LGM	Little Green Man
LJ	Little Joe Rockets
LLR	Lunar Laser Ranging
LM	Lunar Module
LMC	Large Magellanic Cloud
LMXB	Low-Mass X-ray Binary
LORAN	Long-Range Navigation
MADAN	Multi-mission Attitude Determination and Autonomous Navigation
MANS	Microcosm Autonomous Navigation System
MEMS	Micro-Electro-Mechanical System
MEO	Medium Earth Orbit
MIT	Massachusetts Institute of Technology
MJD	Modified Julian Day
MLS	Microwave Landing System
MOST	Molonglo Observatory Synthesis Telescope
NACA	National Advisory Committee on Aeronautics
NANOGrav	North American Nanohertz Observatory for Gravitational Waves
NASA	National Astronautics and Space Administration (US)
NAVSTAR	Navigation Satellite Time and Ranging
NCST	Naval Center for Space Technology (US)
NGA	U.S. National Geospatial-Intelligence Agency
NICER	Neutron-star Interior Composition Explorer
NII	Scientific Research Institute (the acronym in Russian)
NIP	Scientific Measurement Stations (the acronym in Russian)
NIST	National Institute of Standards and Technology (US)
NNSS	Navy Navigation Satellite System
NPL	British National Physical Laboratory

NRAO	National Radio Astronomy Observatory (US)
NRII	Jet Propulsion Research Institute (the acronym in Russian)
NRL	Naval Research Laboratory (US)
NS	Neutron Star
NTS	Navigation Technology Satellite
NTSC	National Time Service Center, China
OAO	Orbiting Astronomical Observatory
OCX	Next-Generation Operational Control System
OP	French National Metrology and Testing Laboratory
OS	Open Service
OSIRIS-REx	Origins, Spectral Interpretation, Resource Identification, Security-Regolith Explorer
OSO	Orbiting Solar Observatory
PCA	Proportional Counter Array
PNT	Positioning, Navigation and Timing
PPS	Precise Positioning Service
PPTA	Parkes Pulsar Timing Array
PRS	Public Regulated Service
PSR	Pulsating Source of Radio
PT	Pulsar Time
PTA	Pulsar Timing Array
PTB	Federal Physical Institute of Germany
QZSS	Quasi Zenith Satellite System
RANT-600	Radio Astronomical Telescope Academy Nauk 600
RF	Radio Frequency
RMS	Root Mean Square
ROSAT	Röntgen Satellite
RP	Rotation Powered pulsar
RXTE	Rossi X-ray Timing Explorer
SA	Selective Availability
SAR	Search and Rescue
SCC	System Control Center
SDD	Silicon Drift Detector
SECOR	Sequential Collation of Range
SEXTANT	Station Experiment for X-ray Timing and Navigation Technology
SGR	Soft Gamma-Ray Repeater
SI	International System of Units
SINEX	Solution Independent Exchange
SINS	Strap-down Inertial Navigation System
SLBM	Submarine-Launched Ballistic Missiles
SLR	Satellite Laser Ranging
SM	Service Module
SMC	Small Magellanic Cloud
SoL	Safety of Life
SPS	Standard Positioning Service

SS-ANARS	Space Sextant—Autonomous Navigation Attitude Reference System
SSB	Solar System's Barycenter
SST	Spitzer Space Telescope
STEP	Space Test Experiments Platform
TACAN	Tactical Air Navigation System
TAI	Temps Atomique International (International Atomic Time)
TAOS	Technology for Autonomous Operational Survivability
TASS	Telegraph Agency of the Soviet Union
TCB	Barycentric Coordinate Time
TCG	Geocentric Coordinate Time
TDB	Barycentric Dynamical Time
TDMA	Time Division Multiple Access
TDRS	Tracking and Data Relay Satellite
TDT	Terrestrial Dynamical Time
TEO	Terrestrial Ephemeris Origin
TIO	Terrestrial Intermediate Origin
TOA	Time of Arrivals
TRW	Thompson Ramo Wooldridge Inc.
TSXS	Time-resolved Soft X-ray Spectrometer
TT	Terrestrial Time
TT&C	Telemetry, Tracking and Control
TWSTFT	Two-Way Satellite Time and Frequency Transfer
UK	United Kingdom
UNESCO	United Nation Educational, Scientific and Cultural Organization
URE	User Ranging Error
US	United States
USA	Unconventional Stellar Aspect experiment or United States of America
USNO	United States Naval Observatory
USRA	U.S. Universities Space Research Association
USSR	Union of Soviet Socialist Republics
UT	Universal Time
UTC	Coordinated Universal Time
VfR	German Society for Space Flight (Acronym in German)
VHF	Very High Frequency
VLA	Very Large Array
VLBI	Very Long Baseline Interferometer
VOR	Very High-Frequency Omni-Directional Range
WGRF	Work Group of Reference Frame
WGS	World Geodetic System
XB	X-ray Binary Stars
XMM	X-ray Multi-mirror Mission
XNAV	X-ray Source-Based Navigation for Autonomous Position Determination

XPNAV	X-ray Pulsar-Based Navigation
XTI	X-ray Timing Instrument
XTIM	X-ray Timing Program

Chapter 1
Introduction to the Pulsars

In 1967, Professor Antony Hewish and his student Jocelyn Bell at the University of Cambridge coincidently discovered a flickering radio pulse source with a repetition period of 1.3373 s, when they were investigating interplanetary scintillation. This pulsating source of radio was then named **pulsar**. It was shown by the detailed investigation that the pulsar was just possibly the neutron star predicted hypothetically by Lev D. Landau, a famous physicist from the Soviet Union, in 1932. The neutron stars result from the supernova explosion of the massive stars after evolving very long time and combining gravitational collapse, with a kind of extremely physical environment, including an extremely high temperature, density, pressure, magnetic field and gravitational field. Typically, they have a radius on the order of 10 km, a mass between 1.44 and 3.2 solar masses and a core density up to 10^{12} kg/cm^3. The pulsars belong to the neutron stars with very rapid rotation and have extremely stable periodicity. Especially, the periodic change rates of millisecond pulsars can get to $10^{-19}-10^{-21}$ s/s, and hereby the millisecond pulsars are known as the most stable clocks in nature. The pulsars are not only very significant for the studies on astronomy, astrophysics and cosmology but also have the potential application values for high-energy physics, time–frequency science and space science. The X-ray signals radiated from the pulsars can provide high-precision autonomous navigation information for spacecrafts. So, the pulsars are able to become the lighthouses of space flights for human beings.

1.1 Theoretically Predicted Neutron Stars

As everyone knows, the end of stellar evolution will produce finally compact stars, including white dwarfs, neutron stars and black holes. The white dwarf was discovered empirically when scientists could not understand what are the celestial bodies or the radiating sources, the neutron star was accidentally found on the basis of theoretical prediction, and the black hole has been investigating theoretically up to now.

P. Shuai, *Understanding Pulsars and Space Navigations*, Navigation: Science and Technology 11, https://doi.org/10.1007/978-981-16-1067-7_1

As fundamental knowledge of understanding pulsars, the history from the discovery of neutrons to the prediction of neutron stars is introduced briefly in this section.

1.1.1 A Brief Story of Discovering Neutron

Ernest Rutherford (1871−1937), a New Zealand-born British physicist, is known as the father of nuclear physics. In 1907, he moved to the Victoria University of Manchester (today the University of Manchester) in the United Kingdom and became Director of the Department of Physics. He proved with Thomas Royds (1884−1955) that alpha radiation is helium nuclei. In 1911, he thereby pioneered the Rutherford model of the atom: the mass of the whole atom almost concentrates on the very small core area that is called **atomic nucleus** ; the electrons orbit around the atomic nucleus; the atomic nucleus is with positive charges and the electrons with negative charges. However, the stability of in-orbit electrons could not be interpreted by Rutherford's model. At the time, Niels Henrik David Bohr (1885−1962) just received his doctorate degree from the University of Copenhagen in Denmark. And then Bohr was invited from Rutherford to conduct post-doctoral work at Victoria University of Manchester. Bohr adopted Rutherford's nuclear structure to Max Planck's quantum theory and hence created the Bohr model of the atom, in which he proposed that energy levels of electrons are discrete and that the electrons revolve in stable orbits around the atomic nucleus but can jump from one energy level (or orbit) to another. In 1913, Bohr published a paper on the structure of atoms and molecules, in which the model of the atomic structure was demonstrated completely. The issue on the stability of the atomic structure could be interpreted perfectly by Bohr's model, and thus the theory of the atomic structure had been developed and enriched further. According to Rutherford's and Bohr's models, it was commonly known at that time that the mass of the atomic nucleus should equal the number of protons with positive charges. In fact, it was shown by the experimental investigations that both are not equal unexpectedly. In other words, the atomic nuclei should still contain other particles besides the protons with positive charges. What were exactly "other particles"? It was a really puzzling problem of physics at that time. The problem was then resolved by James Chadwick (1891−1974) who had ever earned his master's degree and doctorate degree under Rutherford's supervision, and had been Rutherford's assistant director of research at the Cavendish Laboratory of Cambridge University for over a decade. Chadwick discovered that the "other particles" were exactly neutrons.

In 1930, two nuclear physicists from Germany, Walther Bothe and his student Herbert Becker, found that an unusual penetrating radiation (or ray) was produced if the energetic alpha particles emitted from polonium fell on certain light elements, specifically like beryllium, boron or lithium. **Alpha particle** is identically a helium-4 nucleus composed of two protons and two neutrons, together emitted from the decay of some radioactive elements. Since the observed radiation was not influenced by an electric field, it was thought to be gamma ray at that time. Consequently, it was not investigated further by Bothe and Becker. Two years later, Irène Joliot-Curie (1897–1956) and Frédéric Joliot (1900–1958) in Paris showed their new experimental results: if this unknown radiation fell on paraffin wax or any other hydrogen-containing compound, it ejected protons of very high energy. Paraffin wax is a hydrocarbon high in hydrogen content, hence offers a target dense with protons.

On hearing of the Joliot–Curie couples' results, Chadwick at the Cavendish Laboratory did not believe the gamma-ray hypothesis. Chadwick and his colleagues had already conducted experiments on disintegrating light elements using alpha particles emitted from polonium, and also developed more accurate and efficient methods for detecting, counting and recording the ejected protons. Chadwick quickly performed repeatedly the Joliot–Curie couples' experiment: the hydrogen and helium atoms were respectively disintegrated by using the ray produced from alpha particles falling on the beryllium atoms, and as a result, the hydrogen and helium nuclei were ejected. Since a gamma ray had to have impossibly high energy and momentum to scatter the massive protons, the gamma-ray hypothesis was untenable. If it was assumed that the new rays were the neutral particles without charges and their mass almost equal to the mass of protons, the experimental results could be interpreted reasonably. Chadwick performed a series of experiments to measure the range of these protons, the velocity of hydrogen and helium nuclei, and the mass of the new particles. Meanwhile, he also measured how the new particles impacted the atoms of various gasses. He finally concluded that the new particles are not composed of gamma rays, but uncharged particles with about the same mass as the protons. So, these uncharged particles were named **neutrons** . In 1932, Chadwick published a classic paper in *Nature*, "Possible Existence of a Neutron", and thereby declared the neutrons were discovered [1]. Therefore, Chadwick won the Nobel Prize in Physics for the discovery in 1935.

Currently, it is measured more accurately by the experiments that the neutron's mass is $1.674927211 \times 10^{-27}$ kg, the proton's mass $1.672621637 \times 10^{-27}$ kg and the electron's mass $9.10938215 \times 10^{-31}$ kg. Obviously, neutrons and protons have almost equal mass, and the mass of electrons is only 1/1836 of the protons. The atomic diameter is about the order of 10^{-10} m, and the atomic nucleus only has 10^{-4} of the atomic size. Nevertheless, the mass of the entire atom almost concentrates on the atomic nucleus, and hence its density is very high, up to the order of 10^{12} kg/cm^3.

1.1.2 Neutron Stars in the Universe

From the atomic structure size, it can be shown that the vast majority of the interior of the atom is empty. Hydrogen is the lightest element on the periodic table and has one proton and no neutron. The elements except for hydrogen on the periodic table contain the neutrons. For example, helium-4 located in the second period has two protons and two neutrons. If the electrons and protons inside the atoms can be combined to become neutrons, the substance consists of neutrons. And thus, the atomic sizes would be compressed into very small nuclei. It needs enough high pressure which the electrons are compressed into the atomic nuclei and bounded with the protons. Of course, the high pressure could not be obtained on the ground, but there are possibly the celestial bodies with the extreme physical conditions in the universe. Therefore, it would certainly be imagined from the neutrons inside the atomic nuclei that there are possibly the neutron stars in the universe.

After Chadwick just declared his discovery, the Soviet physicist Lev Davidovich Landau (1908–1968) theoretically proposed the predictions: there is possibly a stable celestial body consisted of the neutrons in the universe, neutron star; it has a very small size, high density and faint radiation. In 1934, the German astronomer Walter Baade (1893–1960) and the Swiss astronomer Fritz Zwicky (1898–1974) proposed the existence of a new form of the neutron star, which would be the endpoint of stellar evolution. They wrote in the papers [2, 3]:

> ...with all reserve we advance the view that a supernova represents the transition of an ordinary star into a neutron star, consisting mainly of neutrons. Such a star may possess a very small radius and an extremely high density.

In 1939, the American theoretical physicist Julius Robert Oppenheimer (1904–1967) and his student George Volkoff (1914–2000) assumed that the neutrons in a neutron star formed a degenerate cold Fermi gas, and first worked out the mass, density and diameter of the neutron star using a simple equation of state. Thereby, the model of the neutron star was created theoretically. The model needed to be verified by actual observations. And thus, this made astronomers so warm and high that they searched prospectively the neutron stars using optical telescopes. Unfortunately, the candidate objects like white dwarfs were only observed. It was shown from analyzing the observation data that the pressure inside the white dwarfs could not furthest reach the required order of neutron stars. Except for the white dwarfs, other candidate objects were also found no longer. It is well known that the luminosity of a star is proportional to its surface area. Typically, the radius of a neutron star is only 10 km. And thereby, its surface area is so small that its luminosity is only a part of billions compared to an ordinary star. At that time, optical telescopes were the main equipment for astronomical observations, and hence the neutron stars predicted in the universe could not be discovered naturally. Neutron stars were thought to be too faint to be detectable and then little work was done on them.

Until 1967, just before the pulsar discovery, the Italian astrophysicist Franco Pacini (1939–2012) published a paper titled "Energy Emission of a Neutron Star" in *Nature* [4]. He pointed out in this paper: if neutron stars were spinning and had large magnetic fields, then electromagnetic waves would be emitted; the rapid rotation of highly magnetic neutron stars might be the source of energy in the Crab Nebula. Unbeknown to him, the radio astronomer Antony Hewish and his research assistant Jocelyn Bell at the University of Cambridge were shortly to detect radio pulses from the stars that are now believed to be highly magnetized, rapidly spinning neutron stars, known as pulsars.

In the past 35 years since the neutrons were discovered, the neutron stars had not been found. Meanwhile, the density of the neutron stars was so extremely high from the theoretical model that it was surprising. Therefore, the hypothesis that there were neutron stars in the universe had generally been doubted for a long time. There are two reasons for the undiscovered neutron stars: one is that the physical features of neutron stars, such as the rapid rotation, radio radiation, pulsing emission and so on, are all unknown completely; another is that both the exposure time and the energy integration time, for the optical observation and the radio astronomy, are required so longer that the shorter periodic signals are usually smoothed out.

1.2 The Discovery and Identification

1.2.1 Observations of Interplanetary Scintillations and Quasars

When we look carefully at the clear night sky in summer, it can be seen that all visible stars are scintillating. It is also commonly known as "twinkling of stars". In simple terms, the twinkling of stars is a view phenomenon resulting from the variation of the light intensity, while the lights from the distant stars pass through different layers of a turbulent atmosphere. Its root cause is due to the irregularity and disturbance of the air density in the atmosphere of the Earth. Similarly, there are also scintillations in radio astronomy. It is usually called "radio twinkling", but occurs in the radio band rather than in the visible band of the electromagnetic spectrum. The scintillations of radio waves from celestial sources also occur due to random refraction in the terrestrial ionosphere, the ionized interplanetary gas of the solar system and the ionized interstellar gas of the Milky Way Galaxy. In all three regions, the radio waves from distant celestial sources traverse a medium with fluctuations of refractive index sufficient to deviate radio signals into paths which cross before they reach the radio telescope, giving rise to interference and hence to variations in the radio signal strength.

The scintillation of radio waves due to the terrestrial ionosphere was observed as early as 1951 by Antony Hewish (1924–), a radio astronomer at the University of Cambridge in the United Kingdom. And then, Hewish focused on investigating the intensity fluctuation from extragalactic radio sources. In 1964, while observing several bright sources from celestial radio waves, Hewish's research team discovered the unusual fluctuations of intensity in a few of the sources. Analyzing the observational data in detail, they believed that the fluctuation resulted from irregularities in the density of the plasma associated with the solar wind, which is called **interplanetary scintillation**. The **solar wind** is a stream of charged particles with supersonic speed released from the upper atmosphere of the Sun, also called solar corona. The solar wind is essentially the plasma composed mostly of electrons, protons and alpha particles, with thermal energy between 1.5 and 10 keV. Its speed can reach the range from 200 to 800 km/s, and continue to affect the whole solar system. As the radio waves emitted from celestial sources cross through the solar wind, the phenomenon of radio scintillation will appear.

Generally, a radio source with a large angular diameter can be regarded as a complex consisted of many sources with a small angular diameter. For the same observation time, since the intensities of many sources with small angular diameters vary randomly, the fluctuating variation of average observation has not discrepancy in overall effect. Therefore, the angular sizes of radio sources can be measured by observing the interplanetary scintillation. From Hewish's investigation results, it is shown that the interplanetary scintillation is related to the radio wavelength; only if the angular diameters of radio sources are less than 1 arc-second in the meter band, the interplanetary scintillations appear obviously.

A quasi-stellar radio source with no corresponding visible object was first discovered in the late 1950s and then named "quasar" by a Chinese-American astrophysicist Hong-Yee Chiu (1932–) in 1964. The **quasar** is an active galactic nucleus with very high luminosity and has the most distant and highest energy. Moreover, the angular diameters of the quasars are usually less than 1 arc-second. The observation of the interplanetary scintillation became an effective approach to investigating the quasars in the late 1950s and the early 1960s. With the discovery of the first quasar, a craze for searching the quasars in all-sky radio surveys was set off widely in radio astronomy at that time.

Just under this background, in order to improve further the capability of observing the interplanetary scintillations and searching the faint quasars, Hewish's research team at the University of Cambridge decided to build a large array radio telescope at the Mullard Radio Astronomy Observatory in 1965, known as **Interplanetary Scintillation Array**. The radio telescope consists of 1 central receiver and a 2048-dipole antenna array. The antenna array is rectangular arrangement over almost five acres (approximately 20,235 m^2) on the ground, the west–east 470 m long and the north–south 45 m wide. It includes 16 rows and has 128 dipoles in each row. The radio telescope is operated at a wavelength of 3.7 m (corresponding frequency 81.2 MHz), and at a time resolution of about 0.1 s. This high time resolution set it apart from many other radio telescopes of the time, and it amounts to a parabolic antenna radio telescope with an aperture of 165 m and has the capability recording rapid fluctuation signals. Since astronomers did not expect emission from a radio source to feature such rapid variation, the ordinary receivers only got the average signals with an integration time of several minutes to smooth out the random fluctuation noises, and thereby did not detect the rapid variation signals, with periods of less than 1 s. The radio telescope designed by Hewish's research team was intended originally to survey the interplanetary scintillation in the sky. Considering the frequency of interplanetary scintillation is as higher as the order of 1 Hz, the high time resolution of 0.1 s for the radio telescope is demanded to accurately record the signal intensity variation. So, it creates an advantage condition to capture the short-period signals from radio pulsars. Just as Andrew Lyne and Francis Graham-Smith said in their monograph, *Pulsar Astronomy* [5], "it was an investigation of interplanetary scintillation that led to the discovery of pulsars, even though the discovery was a by-product rather than the purpose of the investigation".

1.2.2 *Discovery of Pulsating Radio Sources*

In 1965, Professor Hewish recruited a female doctoral graduate student, Susan Jocelyn Bell (now Professor Bell-Burnell, 1943–). Jocelyn Bell was born in Lurgan town in County Armagh, Northern Ireland. Her father was an architect who had helped design the Armagh Planetarium, and during visits, she was encouraged by the staff to pursue astronomy professionally. Young Jocelyn also found her father's books on astronomy. She graduated from the University of Glasgow in 1965, with a Bachelor of Science Degree in natural philosophy (Physics). To achieve her astronomic dream, Jocelyn Bell was intending to go to the Jodrell Bank Observatory at the University of Manchester. Unfortunately, her application was lost by the staff of the Jodrell Bank Observatory. And thus, she came to the University of Cambridge and became Hewish's postgraduate student. At Cambridge, she attended New Hall (now Murray Edwards College) and worked together with Hewish's team to construct the Interplanetary Scintillation Array. This radio telescope was completed finally in July 1967 and soon carried out the observations to the sky survey. At that time, Jocelyn Bell's research direction was on the observation of interplanetary scintillation, and her doctoral dissertation title is "*The Measurement of Radio Source Diameters Using a Diffraction Method*". According to the observation plan, Jocelyn Bell would sequentially observe a batch of radio sources, she repeated observation once each week, and the recording paper tape of every day was 30 m long. The radio telescope was so sensitive that both radio source signals and interference ones from the background were recorded together. It was one of Jocelyn Bell's tasks to identify the radio sources from the interference signals every day. In repeating observation once every week, the radio source signals always appeared in the same direction every time, while the interference signals were random. Obviously, the same area in the sky was searched and surveyed so repeatedly that the probability to discover the pulsar was increased correspondingly.

In the solar system, the non-homogeneity of interstellar medium is resulted from the solar wind. At night, since the observatory is always on the back to the Sun, it is little impacted from the Sun, and hence the stronger interplanetary scintillation is not observed generally. However, Jocelyn Bell surprisedly found that a series of periodic "scintillation sources" with different intensities occurred in the direction known non-existent radio sources, when she investigated the observation data at night on August 6, 1967. During one-month observation, these scintillation signals were always flickering appearance and all occurred at the same sidereal time. In other words, the signals go through the meridian circle once every 23 h and 56 min, which is just the period of the Earth's rotation. After excluding the possibility of being affected by the Sun and the interference from the Earth, it was shown that the signal might come from a certain celestial object outside the solar system. And then, Jocelyn Bell reported the observed phenomenon to her supervisor Professor Hewish.

In order to further observe the unusual scintillation source, Hewish's research team installed again a dedicated receiver with higher time resolution on November 28, 1967. It was shown from the observation results that the radio signals emitted from the scintillation source were very regular and once pulsating signal could be recorded every 1.3373 s. The pulse is itself about 20 ms wide, and it was also shown that the radio source's size would not be greater than the Earth and must have come from outside the solar system. Meanwhile, by carefully studying a lot of recorded observation data, Jocelyn Bell found that there were similar phenomena and still three scintillation sources with similar regularity. At that time, a science fiction novel was being popular in the United Kingdom. The extraterrestrial intelligence is described as the Little Green Man (LGM) in this novel. At first, the research team guessed that these unusual signals came possibly from an extraterrestrial civilization, and hence the four scintillation sources discovered first were catalogued respectively as LGM-1, LGM-2, LGM-3 and LGM-4.

Apparently, any extraterrestrial intelligence should live on the planets orbiting around a star, and the signals from the planets would have the Doppler frequency shifts. However, the pulsating signals obtained by the research team only have the Doppler frequency shifts resulted from the Earth orbiting around the Sun, while there are not the phenomena of the radio sources moving. Moreover, the four discovered pulsating signals come respectively from different directions in the universe, and thereby it is impossible that the extraterrestrial intelligence lived in the different locations would emit coincidentally the same kind of signals. Finally, it was confirmed by the research team that the radio sources were the pulsating signals emitted from the natural celestial objects outside the solar system, and were named **pulsars**. Furthermore, the first discovered pulsar was also called CP 1919, where "C" denotes the Cambridge, "P" denotes the pulsar and "1919" denotes the right ascension expressed by hour and minute, also known as 19h19min.

1.2.3 The Pulsars Identified as Neutron Stars

In February 1968, Hewish and Bell, et al. published their classic paper in the *Nature*, entitled "Observation of a Rapidly Pulsating Radio Source" [6, 7]. In this paper, they declared: an unusual radio source located in outside the solar system was discovered; it emitted a short and regular radio pulse, known as pulsar; it was pointed out that the density of the pulsar was very high, and possibly the neutron star ever predicted theoretically more than 30 years ago. Moreover, it was also thought in this paper that the pulsating periodicity was resulted from the oscillation of the neutron star along its radial direction. Apparently, it has been confirmed presently that the deduction of the radial oscillation is wrong. In 1974, Antony Hewish and his colleague, another radio astronomers Martin Ryle (1918–1984), were together awarded the Nobel Prize in physics by the Royal Swedish Academy of Science for their pioneering research in radio astrophysics: Ryle for his observation and inventions, in particular of the aperture synthesis technique, and Hewish for his decisive role in the discovery of pulsars.

A few days before the *Nature* letter appeared, the discovery of the pulsar was discussed at a colloquium at the University of Cambridge. The news spread rapidly and radio astronomers immediately turned their attentions to confirming the remarkable results. At that time, the world's large radio telescopes almost pointed together to the pulsar located in the constellation of Vulpecula, now known as PSR B1919+21, where PSR stands for Pulsating Source of Radio, the letter "B" stands for the 1950 system of coordinates, and the numbers "1919+21" stands for its position, right ascension and declination. The tide of pulsar observation was set off widely in a short time, and many papers on the observation report and theoretical analysis of the pulsars were published in various kinds of astronomical journals. Only a fortnight separated from the first paper, and another paper from the Jodrell Bank Observatory gave some remarkable extra details of the radio pulses from the first pulsar in the *Nature*. By the end of 1968, total papers related to the pulsars were published more than 100, and the number of pulsars discovered got to 23 [8].

Particularly, two significant pulsars, Vela Pulsar and Crab Pulsar, were found at the time. The Vela Pulsar is associated with the Vela supernova remnant in the constellation of Vela, which the supernova explosion is estimated about 11,000–12,300 years ago. The association of the Vela pulsar with the Vela supernova remnant, which was made by astronomers at the University of Sydney in 1968, was the first direct observational proof that the neutron stars came from the supernova explosions [9]. The Vela Pulsar has the emitting signals in the radio, visible light, X-ray and gamma-ray energy -bands, in which the gamma ray is the brightest in the high-energy gamma-ray sky and the visible light is the third brightness of all known pulsars. Its pulse period and distance are respectively about 89.33 ms and 959 light-years. The Crab Pulsar is the central star in the Crab Nebula , and a remnant of the supernova explosion which was observed by astronomers of the Song Dynasty in China in 1054 [10]. The Crab Pulsar is a relatively young neutron star, and its pulse period and distance are respectively about 33.7 ms and 7100 light-years. The Crab Nebula is often used as a calibration source in X-ray astronomy, which is very bright in X-rays, and its flux density and spectrum are known to be constant, of course, with the exception of the pulsar itself. The pulsar provides a strong periodic signal that is used to check the timing of the X-ray detectors.

After the pulsar was discovered in the Crab Nebula, could Pacini's prediction in 1967 be verified? It was a fundamental issue on the pulsars identified as the neutron stars, and also a problem which most astronomers and astrophysicists would endeavor to solve at that time. In 1968, an Australian-born astrophysicist Thomas Gold (1920–2004) independently presented that the pulsars are the rapidly rotating neutron stars and the magnetic flux density on their surface about 10^8 T [11]. Gold's neutron star model, similar to Pacini's model published in 1967 [4], explicitly argued that this model could explain the pulsating radiation observed by Jocelyn Bell and Hewish. Many observation features, especially the stability of pulse period, can be explained using the models of rotating neutron stars. The pulse period will increase a few as the pulsar loses gradually its spinning energy. In 1969, Gold also pointed out that the energy loss of the Crab Pulsar was roughly equal to the energy consumption of the Crab Nebula. And thus, it is basically accepted that the pulsars are the rapidly rotating neutron stars. In fact, the observational evidence that the pulsars are identified as neutron stars mainly come from the following three points:

(1) The typical pulse periods of pulsars are from 1.6 ms to 8.5 s.
(2) The pulse periods are always increased slowly, and the contrary cases have not been observed. Of course, it is an exception which pulsars happen accidentally the glitches of periods.
(3) The pulsars are the astronomical clocks with extremely stable periodicity, and for the decimal fraction of second of their pulsating periods, the 13th–21st significant digits have been observed.

During a pulse period of 1.6 ms, transmission distance of light is 480 km. The value is the upper limit of size of the source emitting the pulsating signals. Moreover, the pulsars are the astronomical clocks with extremely stable periodicity, and hence the relation between the emitting source and its whole body source must have been inter-coupling. So, the candidates of the pulsars are only possible the compact stars, including white dwarf, neutron star and black hole.

If a compact star in the universe behaves as an astronomical clock with extremely stable periodicity, there are only three possible mechanisms: pulsation, rotation and binary system. According to Eddington's stellar pulsation equation, the relation between the pulsating period and the celestial average density can be expressed as

$$P = \frac{C}{\sqrt{\overline{\rho}}}, \tag{1.1}$$

where P is the stellar pulsation period, $\overline{\rho}$ is the stellar average density, and C is the constant.

According to expression (1.1), if the density of the white dwarf is used to calculate the stellar pulsation period, the period is the order of 10 s and it is impossible for the period to be less than 1 s. If the density of the neutron star is used, the periodic range is 1–10 ms. Comparing to the pulse periods for most of the discovered pulsars, the calculating periods of the white dwarfs are not accordant with the observational facts, while those of the neutron stars are too short indeed. Therefore, the pulsation is firstly eliminated in three mechanisms.

Generally, the rotational angular velocities of stars will not be greater than a certain critical value, which is decided by the balance conditions between the gravitational and centrifugal forces on the equatorial matters. For a rotating star with dynamics stability, the lower limit of its rotating period is

$$P \geqslant \sqrt{\frac{3\pi}{G\overline{\rho}}}, \tag{1.2}$$

where P is the stellar rotational period, $\overline{\rho}$ is the stellar average density and G is the gravitational constant.

According to inequality (1.2), the lower limit of the rotational period is calculated as 1 s for the white dwarfs, but the pulse periods of many pulsars are less than 1 s; for the neutron stars, their lower limit is 1 ms, despite being close to the critical period, but up to now, the periods of all discovered pulsars have not been beyond the lower limit value.

For the binary system of three mechanisms, if the stellar radius is replaced as the orbital radius of the binary system in estimating the average density, a longer period can be calculated similarly using inequality (1.2). Since there are no periodic radiations for the black holes and anything including the electromagnetic waves can not get away from their event horizons, it is impossible to receive any periodic signals.

In summary, it can be shown from the very regular, stable and short periodicities of the pulsars that the rapidly rotational neutron stars are uniquely possible candidates. Certainly, the white dwarfs as the candidates can not be excluded from the pulsars with the longer rational periods. Currently, it is a well-known conclusion that the pulsars are the neutron stars.

1.3 The Name, Classification and Distribution

1.3.1 Naming Rules

After the discovery of the pulsar, the naming rule for the pulsars was initially expressed as "NP rrrr", where the capital letter "N" will be replaced by initials of the observatory's name discovering the pulsar, the capital letter "P" stands for the word "Pulsar" and "rrrr" is four digits denoting the right ascension of the pulsar in the 1950.0 coordinate system, in which the first two digits are expressed in hours and the last two digits in minutes. The names and their capital letters of the observatories early discovering the pulsars are listed in Table 1.1, where "CP 1919" stands for the first pulsar discovered by Hewish's research team at Cambridge and its right ascension is 19h19min, and "NP 0531" stands for the Crab Pulsar discovered by the National Radio Astronomy Observatory (NRAO) in the United States and its right ascension is 5h31min. Moreover, it should be noted that the Crab Pulsar was also often called "NP 0532" in early references as its position was uncertain.

In order to avoid the trouble of using the identifiers with the initial letters of observatories, "PSR" is directly used to denote the Pulsating Source of Radio (or Pulsar). And thus, the first discovered pulsar is named "PSR 1919" again. With discovering more and more population of the pulsars, the four-digit number will be repeated sometimes if the right ascension is only used to differentiate various pulsars. And hence, both right ascension and declination are used together in the naming rules. According to astronomical practice, the positive sign "+" is put on the front of the northern declination and the negative sign "−" on the front of the southern declination. Moreover, the significant digits of the declination may be extended sometimes to 0.1 ° or 0.01 ° to distinguish with adjacent pulsars. As same as marking the locations of stars, the reference epoch of pulsars' right ascension and declination is also considered. In the past, the beginning of the Besselian year with the interval

Table 1.1 The names and their initials of observatories early discovering pulsars

Number	Name of observatory	Identifier with initial letter	Notes
1	Cambridge Observatory, UK	CP	CP 1919
2	Harvard College Observatory, USA	HP	
3	National Radio Astronomy Observatory, USA	NP	NP 0531
4	Arecibo Observatory, USA	AP	Managed by Cornell University
5	Jodrell Bank Observatory, UK	JP	
6	Molonglo Observatory Synthesis Telescope (MOST), Australia	MP	Operated by the University of Sydney

of 25 calendar years was used as an epoch in most of the stellar catalogues, and expressed as the capital letter "B", such as epoch B1900.0, B1925.0, B1950.0 and so on. Consequently, the naming rule was revised to be "PSR Brrrrsdddd", where "rrrr" denotes four digits of the right ascension, in which the first two digits are still expressed with hours and the last two digits with minutes; "s" denotes the positive sign "+" or the negative one "−"; "dddd" denotes four digits of the declination, in which the first two digits are expressed with degrees, the third and fourth digits are extended respectively to 0.1 ° and 0.01 °. In some cases, the declination is denoted only by the first two digits. Supposed that the right ascensions of two pulsars are 19h13min, while their declinations are respectively 16°02′ and 16°40′, the former is named PSR B1913+16, and the latter is named PSR B1913+167 due to its declination conversed into 16.7 °. Sometimes, the reference epoch B1950.0 is regarded as a default, and the letter "B" is also neglected in the past some references. And thus, the names of the two pulsars are respectively written into PSR 1913+16 and PSR 1913+167. Furthermore, the discrepancies of the pulsar's angular positions in some globular clusters are so small that they can not be distinguished distinctly. And thereby a capital litter is usually appended behind the declination digits to distinguish the different positions, such as PSR 0021–72C, PSR 0021–72D and so on.

At present, according to the International Astronomical Union (IAU) regulation, the naming rule of pulsars is the standard epoch J2000.0 has been used uniformly in stellar catalogues since 1984; the declinations are denoted by four digits, in which the first two digits are still expressed with angular degrees and the latter two digits with arc-minutes rather than 0.1 ° or 0.01 °. For the pulsars discovered before 1993, two coordinate systems B1950.0 and J2000.0 may be used, and distinguished respectively by the letter "B" and "J"; for the pulsars discovered after 1993, it is regulated that only J2000.0 can be used to name the pulsars. In other words, the naming rule of the pulsars discovered after 1993 is "PSR Jrrrrsdddd", where the last two digits "dd" are expressed in arc-minutes, and other signs are as same as the above. For example, the first pulsar discovered in 1968 can be named PSR B1919+21 using epoch B1950.0, and also PSR J1921+2153 using epoch J2000.0; it is worth noting that PSR J1921+2153 stands for the right ascension of 19h21min and the declination of $+21°53'$. In 1993, a nearby binary millisecond pulsar was discovered in the Parkes 70 m survey [12] and named PSR J0437–4715, where only epoch J2000.0 can be used, the right ascension expressed as 4h37min and the declination as $-47°15'$. PSR J0437–4715 is the brightest and most stable in all known millisecond pulsars.

The naming of X-ray sources or X-ray pulsars is not normalized like the above radio pulsars. In 1948, X-rays radiated from the Sun was first discovered by the sounding rocket; in 1962, the X-ray source outside the solar system was successfully detected by the sounding rocket with an altitude of 230 km; and then, dozens of X-ray sources were also detected successively by the sounding rockets and balloons outside the dense atmosphere. For early discovered X-ray sources, they were named by their constellations as well as their discovery sequence in the same constellation. The first X-ray source outside the solar system came from the constellation of Scorpius, and thereby it was named Scorpius X-1 (or simply Sco X-1). Moreover, other X-ray sources are also named similarly, such as Cygnus X-1, Cygnus X-2, Hercules X-1, Centaurus X-3, LMC X-4 and SMC X-3, where LMC and SMC are acronyms, and respectively Large Magellanic Cloud and Small Magellanic Cloud.

The first dedicated X-ray astronomy satellite, Uhuru, was launched successfully from the Italian San Marco launch platform near Mombasa in Kenya on December 12, 1970. "Uhuru" is the Swahili word for "Freedom". The satellite had been operated by the U.S. National Astronautics and Space Administration (NASA), and X-ray sources had been explored using it on the Earth orbit from 1970 to 1973. After launching Uhuru, X-ray sources have often been named by initials of the satellite's names first discovering them, as well as their position parameters in which the right ascension is expressed with hours and the declination with degrees. For example, "A" stands for the satellite Aril-5, "S" for SAS-3, "E" for Einstein, "EXO" for EXOSAT, "V" for Vela, as well as "U" for Uhuru in which 3U and 4U respectively denote the third and fourth X-ray source catalogues from the satellite Uhuru.

1.3.2 Fundamental Classification

Currently, the signals emitted from pulsars have been observed in the radio, infrared, visible, ultraviolet, X-ray and gamma-ray bands of the electromagnetic spectrum. However, it is shown from the actual observations that most of the pulsars radiate the signals only in the radio and X-ray bands, while a few pulsars in the infrared, visible, ultraviolet and gamma-ray bands. According to the different frequency band ranges of the signals radiated from pulsars, the pulsars can be classified into radio, optical,

X-ray and gamma-ray pulsars, among which X-rays and gamma rays are high-energy photons, and most of radiating energy from pulsars is concentrated on these high-energy bands. The X-rays and gamma rays can be easily detected only by using the miniaturized equipment, but the equipment must be mounted on the vehicles in outer space to carry out the detections as the X-rays and gamma rays can not cross through the dense atmosphere of the Earth. Therefore, spacecrafts can navigate autonomously by using soft X-ray signals (with an energy band of 1–15 keV) from pulsars as natural beacons. Generally, it should be noted that hard X-rays (15–200 keV) and gamma rays (more than 200 keV) from pulsars can not be used to autonomously navigate spacecrafts as their photon fluxes are too low.

Certainly, the same pulsar may radiate the signals simultaneously in several frequency bands. For example, the signals radiated from the Crab pulsar (PSR B0531+21) have been observed in the radio, infrared, visible, ultraviolet, X-ray and gamma-ray bands. Moreover, the infrared and X-ray radiations from PSR B1509–58 have been also detected. According to different sources of the pulsars' radiant energy, the pulsars can be classified into three categories: Rotation Powered pulsar (RP), Accretion Powered pulsar (AP) and Anomalous X-ray pulsar (AX).

The RP pulsar is a type of neutron stars' outward radiating energy at the expense of their own rotational kinetic energies, and many X-ray pulsars belong to the type. It is shown from the theoretical analysis that the pulse periods are the spinning periods of neutron stars. Moreover, on the basis of the observational facts that the pulse periods of radio pulsars are increasing uniformly, it is shown clearly that the rotations of neutron stars are slowing down and their rotational kinetic energy also decreasing gradually. For a spherical neutron star with a rotational angular velocity of ω, its rotational kinetic energy E_k can be expressed as

$$E_k = \frac{1}{2} J\omega^2, \tag{1.3}$$

where J is the moment of inertia of neutron star.

For a sphere with uniform mass distribution, there is

$$J = \frac{2}{5} mR^2, \tag{1.4}$$

where m and R are respectively the mass and radius of the neutron star.

And then, by finding the time derivative on the right and left sides of formula (1.3), the loss rate of the rotational kinetic energy can be expressed as

$$\dot{E}_k = J\omega\dot{\omega} = -\frac{4\pi^2 J\dot{P}}{P^2}, \tag{1.5}$$

where P is the spinning period of neutron star, and $\omega = \frac{2\pi}{P}$.

For the Crab Pulsar, its energy loss rate is about 4.7×10^{31} W, while the radiating power of the Crab Nebula is 1.0×10^{31} W. Apparently, it is shown that the losing energy of the Crab Pulsar can enough provide the radiation of the Crab Nebula and itself. Generally, it is speculated that the radiating energy of the single pulsar is possibly transformed from the rotational energy of itself.

The AP pulsar is a type of neutron star which obtains radiant energy by accreting material from the companion star in a binary system consisting of a neutron star and its companion star. Particularly, X-ray binary system is a typical representative of the type of pulsar. It is different from ordinary radio pulsar that the AP pulsar's rotation does not gradually slow down with radiant energy loss other than more and more rapid. The orbital period in the X-ray binary system is very short, which shows that the distance between two sub-stars is very near. The pulsar accretes abundant materials from its companion star, and the materials gathered around the pulsar rapidly falling onto two poles of the pulsar along the magnetic field lines. And thus, high-energy radiations including X-rays are generated as the stream of materials continuously impacts the pulsar's surface. According to the mass and gravity of the companion stars, the AP pulsars are also subdivided into two types, High-Mass X-ray Binary (HMXB) and Low-Mass X-ray Binary (LMXB).

The HMXB is a binary system consisting of a neutron star and a high-mass companion star, where the neutron star produces X-ray radiation and orbits the companion star. The companion stars are typically 10–30 solar masses in size. They are obviously immense in comparison to the neutron stars only with a mass of 1.44–3.2 solar masses and a typical radius of 10 km. It is similar to the mechanism of forming the solar wind, and an extensive high-energy particle stream can be produced by the immense companion star. The companion star has much immense mass and high temperature that the intensity of its stellar wind is far greater than that of the solar wind. The abundant high-energy material streams carried with the stellar wind are absorbed and captured as they travel through the gravitational field of the neutron star, and thus an "atmosphere" around the neutron star will be formed; the materials are accelerated along the magnetic field lines to impact on the solid crust of the neutron star; finally, the releasing energy from the impaction are transformed into X-rays radiating. In the process of accreting materials, an accretion disk will generally be produced around the neutron star by using the materials from the high-mass companion star.

Alternatively, the LMXB is a binary system with a low-mass companion star, even less than one solar mass, where a neutron star can inhabit. For the LMXB system, the intensity of stellar wind from the low-mass companion is so weak that the stellar wind can not be used to capture the matter from the companion star as a kind of main approach. Since mass of the neutron star is almost equal to that of its companion star, the gravitational field around the binary system is controlled by two components of the system together, and thus an equipotential surface of gravity similar to the Arabic number "8" shape is produced, which is also called **Roche lobes**. The two sub-stars of the LMXB system are located in the centers of two Roche lobes respectively. The Roche lobes are subdivided into three types: detached binary, semidetached binary and contact binary. Considered that the neutron star has extremely high density, it can be regarded as a point source of the Roche lobes, and the whole Roche lobes will be surrounded by the atmosphere from its companion star. Therefore, the matter from the companion star easily gets into the Roche lobe of the neutron star through intersection, and then it is controlled and captured by the neutron star. Using the abundant materials from the companion star, an accretion disk can also be built around the neutron star. Similar to the high-mass companion stars, the materials within the accretion disk will fall onto the magnetic poles along the magnetic field lines, and continuously impact on the solid crust of the neutron star nearby the poles, so that high-energy particle streams including X-rays are brought up ultimately.

In X-ray binary systems, pulsars continuously accrete the matter from the companion stars, and thereby their rotations are accelerated and their pulse periods gradually shortened from a typical 10 s order to dozens of milliseconds. In this way, the ordinary pulsars can slowly be evolved into millisecond pulsars. Furthermore, due to the shielding effect of the accretion disk, the magnetic field strength of the millisecond pulsars is usually four magnitudes lower than that of the ordinary pulsars. So, many observational features that the millisecond pulsars have, such as the older age, shorter period and weaker magnetic field, all may be interpreted reasonably.

It should be noted that although pulsars within binary systems can produce significant amounts of X-ray photons, the pulsations of the type of pulsars are more complex. The complexity results from the combined effects of both the rotating neutron star and the orbit of the star around its companion. Moreover, along the line-of-sight to an observer, the neutron star is probably eclipsed by its companion, so that its signals can not be received. The eclipsing binary systems introduce additional signal processing complexity.

In addition to the above X-ray binaries, the first radio binary pulsar, PSR B1913+16, was discovered by Russell A. Hulse and his supervisor Joseph H. Taylor using a 305-m radio telescope at Arecibo Observatory in 1974. PSR B1913+16 together with another neutron star are in elliptical orbits around a common center of mass and they form a binary star system. The period of the orbital motion is 7.75 h, and the two neutron stars are believed to be nearly equal in mass, about 1.4 solar masses. Up to now, radio emissions have been observed from only one of the two neutron stars. Most of radio binaries consist of a neutron star and a white dwarf, and the masses of companion stars are typically 0.08−0.87 solar masses and their orbital periods longer. For example, the companion star of PSR B0820+02 is a white dwarf and their orbital period is about 1232 days; another pulsar PSR J2051–0827 with its companion both are neutron stars and their orbital period is only 0.099 days.

The AX pulsars are the neutron stars powered by the decay of their extremely powerful magnetic fields, and their magnetic field strengths get to $10^{13}-10^{15}$ Gauss, while that of the Sun is only about 50 Gauss. The AX pulsars are widely believed to be a kind of magnetars which are young, isolated and highly magnetized neutron stars, and characterized by soft high-energy bands, faint luminosities and slow rotation periods of about 10 s. In general, the rotational energy loss of pulsars can be determined by observing their rotational period changes. It is not sufficient from the rotational energy loss of the AX pulsars to provide their X-ray radiation energy.

The powered sources by decaying the magnetic fields are only one of theoretical interpretations. Besides, another interpretation is the accretion model, similar to the AP pulsars, but accreting matter can not provide X-ray radiation energy so sufficient that a part of the radiant energy is still obtained by rotating energy loss. At present, a population of the discovered AX pulsars is dozens, but their counterparts in the infrared, visible or gamma-ray bands have never been observed.

Furthermore, according to different pulse periods, pulsars are also divided into two types, ordinary pulsars and millisecond pulsars, in which the former period range is typically 0.05–8.5 s and the latter mostly 1–10 ms. The millisecond pulsars have been detected in radio, X-ray and gamma-ray portions of the electromagnetic spectrum. In 1982, the American astrophysicist Donald C. Backer (1943–2010), et al. discovered the first millisecond pulsar, PSR B1937+21, with a spinning period of 1.558 ms [13]. The rotating period of PSR B1937+21 is only about 1/20 of the Crab Pulsar, while its age gets to 400 million years and is more five orders than the Crab Pulsar; its magnetic field strength is smaller and about 10^8 Gauss; its periodic change rate gets to 1.049×10^{-19} s/s. The PSR B1937+21 is a X-ray binary system and the best probe to explore the globular cluster. Its discovery started a pioneer to search the planetary system outside the solar system.

It is well known that most of neutron stars are older stars with an age of about one billion years. Why can these older stars rapidly rotate yet? The leading theory for the origin of millisecond pulsars is that they are old, rapidly rotating neutron stars which have been spun up through accretion of matter from a companion star in the close binary system. For this reason, it is considered that the millisecond pulsars are related to the LMXB systems. In the binary systems, the companion stars are usually too faint to be observed. The matter accreted from the companion stars continuously impact on the accretion disks and produce an impact force; the kinetic energy will be transferred onto the surface of neutron stars, and hence the neutron stars are accelerated to rotate. However, there has been recent evidence that the standard evolutionary model fails to explain the evolution of all millisecond pulsars, especially young millisecond pulsars with relatively high magnetic fields, like PSR B1937+21. It is shown that different millisecond pulsars must form by at least two distinct processes, while the nature of other process remains a mystery [14]. The millisecond pulsars have extremely stable periodicity and their periodic change rates get to 10^{-19}–10^{-21} s/s. In other words, the millisecond pulsars, which can be used to time with high precision, are better clocks than atomic clocks [15]. Currently, there are approximately 200 millisecond-pulsars that have been discovered and catalogued.

Particularly, it is worth noting that a Polish astronomer Aleksander Wolszczan (1946–) discovered a millisecond pulsar with a rotating period of 6.22 ms, PSR B1257+12, by using Arecibo radio telescope on February 9, 1990. It was found to have anomalies in the pulse period, which led to investigations as to the cause of the irregular pulses. And thus, Wolszczan and Dale Frail published a paper on the first confirmed discovery of the planet outside the solar system. By using refined methods, more planets were found again orbiting this pulsar in 1994. The millisecond pulsar is located in the constellation of Virgo, and 2300 light years away from the Sun. It has a planetary system with three known extra-solar planets respectively named

"Draugr" (PSR B1257+12A), "Poltergeist" (PSR B1257+12B) and "Phobetor" (PSR B1257+12C), in which Draugr is the lowest mass planet, with a mass of 0.02 earth-masses, Poltergeist's mass is 4.3 earth-masses, and Phobetor's mass is 3.9 earth-masses. It is shown from theoretic analysis that these planets were formed by the neutron star capturing matter from the remnants of the destroyed companion star. Significance of PSR B1257+12 is not only to first discover pulsar's planets, but also the first planet system outside the solar system.

The millisecond pulsar, PSR B1937+21, is rotating roughly 641 times per second, but another millisecond pulsar PSR J1748–2446ab, discovered in 2005, rotating 716 times a second. According to theories of neutron star structure and evolution, the pulsars' rotational speed limits can be estimated as follows:

(1) The spinning rates of pulsars would not be greater than 1500 cycles per second. Otherwise, the pulsars would break apart [16].
(2) Pulsars would radiate the gravitational wave to further restrain their accelerated rotations before they spin at an extremely high speed. It is shown from actual observational results that the spinning rates of pulsars seem to have been controlled within 1000 cycles per second, corresponding a period of 1 ms.

By analyzing data from two spacecrafts, Rossi X-ray Timing Explorer (RXTE) and International Gamma-Ray Astrophysics Laboratory (INTEGRAL), it was believed in early 2007 that neutron star XTE J1739–285 was spinning at a rate of 1122 cycles per second. However, this result is not statistically significant, only with a level of 3 sigma significance, and has not also been obtained by other astronomers re-analyzing these data. Therefore, it is an interesting candidate for further observations, but current results are inconclusive. Meanwhile, it is also believed that gravitational radiation plays an import role in slowing down the rate of rotation. Additionally, the accreting millisecond X-ray pulsar that spins at a rate of 599 cycles per second, IGR J00291+5934 [17], will be a prime candidate for helping to detect the gravitational waves in the future. Generally, most of such X-ray pulsars only spin at a rate of around 300 cycles per second.

1.3.3 Spatial Distribution

With successfully launching a series of astronomical satellites in 1970s, the observational means of pulsars have been extended from the radio to infrared, visible, ultraviolet, X-ray and gamma-ray bands of electromagnetic spectrum. As of 2020, a population of the catalogued pulsars has gotten to 2811, in which most of pulsars are the radio pulsars and concentrate on around the Galactic plane, and there are only dozens of pulsars in the globular cluster. Although the pulsars located in out of the Milky Way system have always been searched for a long time, there are only about 45 pulsars discovered in the LMC and SMC up to now [18]. Wherever pulsars are in the Milky Way system or out, population of the discovered pulsars is far less than that of the theoretically predicted ones. It is shown from theoretical estimations that there are 350 thousand radio pulsars whose radiation fluxes are more than 0.3 mJy, in which if the radiating beams of 20% sweep across through the Earth, there would be 70 thousand pulsars to be found, while the total of pulsars in the Milky Way system be more than 200 thousand. Certainly, if their radiating beams do not sweep across nearby the Earth, the pulsars are unable to be discovered.

There are 759 X-ray sources catalogued in the Galactic coordinate system, where population of the LMXB is 290, that of the HMXB is 152, that of Cataclysmic Variable stars (CV) is 141, that of X-ray Neutron Stars (NS) is 95, that of Active Galactic Nuclei (AGN) is 6 and that of other X-ray sources is 75 [19]. The **Galactic coordinate system** is a celestial coordinate system in spherical coordinates, with the Sun as its center, the primary direction aligned with the approximate center of the Milky Way system, and the fundamental plane parallel to an approximation of the Galactic plane but offset to its north. It uses the right-handed convention, meaning that coordinates are positive toward the north and toward the east in the fundamental plane. The Galactic longitude measures the angular distance of an object eastward along the Galactic equator from the Galactic center. The Galactic latitude measures the angle of an object north or south of the Galactic equator as viewed from the Earth, and is positive to the north and negative to the south. Both the Galactic longitude and latitude are measured in degrees. The X-ray sources exist throughout the X-ray sky, as is evident from their space distributions, although the clustering near the Galactic plane is clear. The 95 neutron stars include 68 RP pulsars, 11 AX pulsars, 3 Soft Gamma-ray Repeaters (SGR) and 13 other neutron stars, which are uniform distribution relative to the Galactic plane. Most of the pulsars are located nearby in the Galactic plane, and a few of the pulsars located in the region with high Galactic-latitude and relatively distant distance. In the international celestial reference system at epoch J2000.0, distributions of the RP pulsars mostly concentrated on the region with the declination from -60 to $+60°$.

The statistical results of the pulsars with remarkable features are listed in Table 1.2, where SAX denotes an Italian-Dutch X-ray astronomy satellite launched in 1996, named BeppoSAX. The names, distances and characteristic ages of the pulsars with a distance of less than 300 parsecs are listed in Table 1.3. It can be seen from Tables 1.2 and 1.3 that these catalogued pulsars are mostly located in the Galactic System, their distance are from hundreds to tens of thousands of light years, and their characteristic ages from tens of thousands to billions of years. Among the discovered and catalogued pulsars, population of the pulsars with the emitting signals in the optical or gamma-ray frequency bands is relatively few. There are 6 optical pulsars, including PSR B0531+21 (Crab Pulsar), PSR J0835–4510 (Vela Pulsar), PSR B0540–69 (LMC), PSR B0656+14, PSR B0633+17 (Geminga) and PSR B1509-58. There are around 100 gamma-ray pulsars, mostly discovered by the Fermi Gamma-ray Space Telescope (FGST).

Table 1.2 Statistics of the pulsars with remarkable features

Number	Name	Remarkable feature	Period (ms)	Distance (pc*)	Discovered year
1	PSR B1919+21	The first radio pulsar	1337.30	700	1967
2	PSR B1913+16	The first radio binary system	59.03	6438	1974
3	PSR B1937+21	The first radio millisecond pulsar	1.56	3600	1982
4	PSR J1121−60ab	The first X-ray pulsar	4840	5700	1971
5	SAX J1808.4−3658	The first X-ray binaries	2.49		1996
6	PSR B1257+12	The first pulsar with planets	6.22	705	1990
7	PSR J0738−4042	The first pulsar observed to have been affected by asteroids	374.92		2013
8	PSR J0737−3039	The first double pulsar binary system	22.70	1150	2003
9	PSR J1311−3430	The first millisecond pulsar discovered via gamma-ray pulsations	2.50		2012
10	PSR J2144−3933	The longest period radio pulsar	8510	180	1998
11	PSR J1841−0456	The longest period anomalous X-ray pulsar with a magnetic field strength of 8×10^{14} Gauss	11766		1997
12	PSR J1748−2446ad	The shortest period pulsar	1.40	5518	2004
13	PSR J0437−4715	The closest and brightest millisecond pulsar with the most stable period	5.75	156	1993

*Note: "pc" indicates a parsec, the unit of distance, which is equal to 3.262 light years

1.4 The Physical Mechanisms and Characteristics

1.4.1 Lighthouse Effect

In Sect. 1.2.3, it has been identified that pulsars are neutron stars. In that way, why can the neutron stars emit stably the pulsating signals? It is shown by many observations and investigations that the durations of the pulsating signals are always shorter than those of the whole pulse cycles. And thus, it would be imagined that pulsars are

Table 1.3 Pulsars with a distance of less than 300 parsecs

Number	Name	Distance (pc)	Characteristic ages (million years)
1	J0835–4510	290	0.0113
2	J0659+1414	290	0.111
3	J0633+1746	156	0.342
4	J1741–2054	250	0.387
5	J1856–3754	161	3.76
6	J0453+0755	260	17.5
7	J0108–1431	238	166
8	J2144–3933	165	272
9	J0437–4715	156	1590
10	J1045–4509	300	6710
11	J0030+0451	244	7580

like the lighthouses for navigating ships on the sea. As early as 1968, two famous astrophysicists Picini and Gold had presented a magnetic dipolar cap radiation model of neutron stars. They believed that pulsars are rapidly rotating neutron stars and have extremely strong magnetic fields; the electrons and protons in the magnetic fields form a radiation beam at every magnetic pole by curvature radiation or synchrotron radiation mechanisms; there is an angle between the rotational axis and magnetic axis of neutron star; therefore, if the radiation beams of rotating neutron stars sweep across the detection equipment mounted on the ground stations or spacecrafts, the equipment will receive the pulsating signals. It is called the **lighthouse effect** (or **lighthouse model**) of neutron star, as shown in Fig. 1.1.

Obviously, the radiating signals do not come from the whole surfaces of neutron stars, but from a concentrating radiation region; the pulse periods always equal to the spinning periods of neutron stars; the wider are the radiation beams from the magnetic poles, the longer the durations of the observed pulses. It is shown from the lighthouse model that the discovered and catalogued pulsars are the spinning neutron stars whose radiation beams coincidently sweep across nearby the Earth. Certainly, many neutron stars can not be observed as their radiation beams never point to around the Earth. Furthermore, for the neutron stars far away from the Earth, their emission pulse signals detected near the Earth are very faint. In addition, since the radio waves are affected by the dispersion of abundant interstellar free electrons, the observed pulsation features will be very indistinct. Consequently, it is very difficult to observe these neutron stars near the Earth. In these reasons, the population of the currently known pulsars is far less than the theoretically estimated value. Considered that

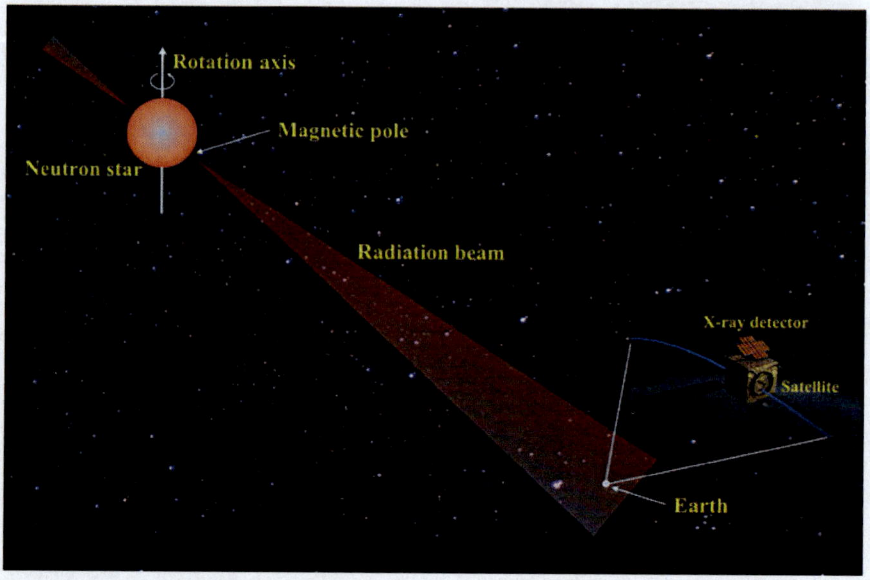

Fig. 1.1 Lighthouse model of the neutron star

pulsars are very far away from the solar system, their conical radiation beams can cover the whole solar system, and thus the signals emitted from the pulsars are often regarded as plane waves.

Pulsars are rapidly rotating neutron stars and their rotational rates get to hundreds of cycles per second. The rotational rates are enough to produce the pulsating signals. Why can the neutron stars rotate so rapidly? It is the law of conservation of angular momentum, which all rotating objects must be followed unconditionally. Generally, the angular momentum can be decided by three factors, the object's mass, extensional size and rotational rate. For a rotating object with constant mass, if its extensional size is increased, its rotational rate must be decreased to keep the conservation of angular momentum; on the contrary, if its extensional size is decreased, its rotational rate must be increased. It is well known that neutron stars result from the supernova explosion of massive stars after undergoing a very long evolution with combining gravitational collapse, and thus are condensed so tightly that their densities are extremely high. The typical radii of neutron stars are about 10 km,while their masses are equal to 1.44–3.2 solar masses. Therefore, the neutron stars can rotate very rapidly and stably.

1.4.2 *Radiation Mechanisms*

It is known by the above that the pulsars' radiation energy come possibly from their rotational power, gravitational power of accreting matter or extremely strong magnetic power. However, what are the mechanisms used to transform all kinds of powers into the radiant energy and how are the pulsating signals also brought up? It is a fundamental issue on the radiation mechanisms of pulsars. It is shown by theoretic studies that there are mainly five possible radiation mechanisms for pulsars, the curvature radiation from magnetic poles, synchrotron radiation from magnetic equators, high-energy radiation from hot spots, accreting matter radiation and cyclotron radiation. Whatever is any kind of these radiation mechanisms, they will produce the wave beams with continuous radiation spectrum at a certain region on the surfaces of pulsars, and the beams point to various directions.

1.4.2.1 Curvature Radiation

In the extremely strong magnetic field of the pulsar's magnetosphere, a relativistic electron may be constrained to follow the path of a magnetic field line very closely, with pitch angle nearly zero; the field line will generally be curved, so that the electron will be accelerated transversely to its motion and then radiated; the radiation is usually called **curvature radiation**. The so-called **relativistic electron** refers to a type of electron whose velocity is approximately to the light velocity, and the electron whose velocity is less than the light velocity is called **non-relativistic electron**.

Why do the signals emitted from pulsars always appear to a beam shape? It results from the extremely strong magnetic fields of pulsars. A highly rapid rotation and magnetization neutron star radiate its energy in the way of electromagnetic waves at the same rotation frequency. It is shown by studying the Crab Pulsar that only if the magnetic field strength on the surface of a pulsar gets to an order of 10^{12} Gauss, the radiant energy of electromagnetic wave will possibly equal to the rotational energy of the pulsar losing. It seems hard to believe that there is so strong magnetic field on the surface of a pulsar. In fact, it can easily be interpreted according to the theory that neutron stars come from massive star evolutions. Supposed that a star with a radius of 10^6 km evolves to the late term and collapses into a neutron star, its radius is only about 10 km. Since all atoms inside the star are ionized and magnetic induction lines frozen in the plasmas, the magnetic flux of the collapsed star is constant. That is,

$$C = 4\pi R^2 B, \tag{1.6}$$

where C is the magnetic flux of the collapsed star, R is the radius of the star, and B is the magnetic field strength on the surface of the star.

It is clearly known from formula (1.6) that the magnetic field strength B will increase 10^{10} times, if the magnetic flux C keeps a constant and the radius R is decreased to 10^{-5} of its original value. On the basis of the observation data of the rotation slowdown rates of radio pulsars, it is obtained by the calculation that the magnetic field strengths of most pulsars are $10^{11}-10^{13}$ Gauss, while those of few pulsars are weak, only 10^8-10^{10} Gauss.

Obviously, most of pulsars have extremely strong magnetic fields. The charged particles will obtain very high speeds as they motion along the magnetic field lines in the extremely strong magnetic fields. Since the curvatures are usually very great around the magnetic poles in the extremely strong magnetic fields, the charged particles with high speeds are accelerated again, and thus the curvature radiations produced. The very strong electromagnetic waves will be emitted by the curvature radiation mechanism, where the electromagnetic waves in each frequency band, including the radio and X-ray bands, will together form a radiation cone with special directionality near the magnetic poles and radiate energy toward space.

1.4.2.2 Synchrotron Radiation

Synchrotron radiation is the electromagnetic radiation emitted when the relativistic electrons are accelerated radially by the Lorentz force in magnetic fields. In other words, the electrons are subject to an acceleration perpendicular to their velocities. The synchrotron radiation was first observed from a General Electric synchrotron accelerator built in 1946 and announced in 1947, and then also named sync-radiation. The electrons from the sync-radiation generally make circular or spiral motions.

It is well known that a rotating magnetic field can produce an electric field. The rotating neutron stars with ultra-strong magnetic fields can produce the electric field strengths of above 10^{14} V/m on their surfaces. For so strong electric fields, it is enough to overcome the force of gravity, and thus all electrons and protons will be pulled out to form the plasmas around the neutron stars. The plasmas are controlled by the ultra-strong magnetic fields, and they revolve together with magnetic induction lines around the rotation axes of neutron stars. The rotational linear velocities are proportional to the distances away from the neutron stars. Supposed that the rotation velocities of the charged particles get to the light velocity c at distance r_c, then a cylinder with the magnetic axis of neutron star as the central axis and with r_c as the radius will be formed, and usually called **light-velocity cylinder**. For the Crab Pulsar, the radius of its light-velocity cylinder is about 1600 km.

Outside the light-velocity cylinder, the charged particles are constrained no longer by the magnetic field of neutron star; they will escape rapidly from the neutron star and become a possible source of cosmic rays. In the region where the light-velocity cylinder intersects with the magnetic equator of neutron star, the highly rapid electrons make the spiral orbit motions around the magnetic field lines, and thus a beam of sync-radiation will be emitted in their motion direction. If the sync-radiation beams sweep across nearby the Earth, the pulsating signals from the neutron stars will be produced. Whether the sync-radiation beams are emitted along the magnetic axes or along the magnetic equators at the light-velocity cylinders, it is an important prerequisite to suppose that the magnetic axes of neutron stars are not parallel to their rotation axes. In other words, only if there is an angle between the magnet axis of neutron star and its rotation axis, the pulsating signals will be produced and the lighthouse effect formed.

1.4.2.3 High-Energy Radiation

A high-energy radiation is believed to come from hot spot effects. The mechanism producing the hot spot effects is similar to the solar flare, a sudden flash of increased Sun's brightness near its surface. Pulsars are the remnants after the supernova explosions of massive stars. The stars have extremely high temperature as they explode. With ending the supernova explosions, original neutron stars are produced in their interior cores. Meanwhile, the temperatures on the surfaces of the neutron stars without thermonuclear reaction will cool gradually. In the neutron stars evolving and cooling, there are discrepancies of declining rates of their surface temperatures. Therefore, it will occur that local temperatures are higher than around ones on the surfaces of the stars. The higher temperature regions are called the hot spots, and their particles are very active and have higher speeds. When the high-speed particles impact the around-low-speed ones, X-ray radiations will be produced. Obviously, the radiations from the pulsars occur possibly in the hot spots on their surfaces rather than near the magnetic poles.

1.4.2.4 Accreting Matter Radiation

In the binary system, the accretion disk can be built around the pulsar by absorbing the matter from the companion star. The matter is accelerated along the magnetic field lines under a strong magnetic field, and thus gravitational potential energies are transformed into kinetic ones so that the falling velocities of the matter can get to 10^5 km/s. Along the magnetic field lines, the materials impact continuously at such high velocity on the solid crusts of neutron stars, and then the releasing energies will be transformed into the high-energy particle radiations, such as X-rays, near the magnetic poles.

1.4.2.5 Cyclotron Radiation

Cyclotron radiation is an electromagnetic radiation emitted by accelerating non-relativistic electrons deflected by a magnetic field. The Lorentz force on the electrons acts perpendicular to both the magnetic field lines and the electrons' motions through them, creating accelerations of the electrons that cause them to emit radiation as a result of the accelerations they undergo, as they spiral around the lines of the magnetic field. The polarization of the radiation may be understood from a description of the electron acceleration as seen by an observer. For the simplest case of the observer in the plane of gyration, with no streaming motion, the polarization is linear. An observer above or below the plane will see a circular component, giving elliptical radiation whose ellipticity diminishes to zero at edge of the beam where the polarization approaches pure circular. A cross-section across the cone of radiation from an electron with a large component of velocity along the field line shows similar polarization characteristics. The cyclotron radiation from plasma in the interstellar medium or around black holes and other astronomical phenomena is an important source of information about distant magnetic fields.

It should be pointed out that the radiation mechanism of pulsar is still a difficult research topic. Although a lot of observational data has been obtained, it is not clear how to determine the actual position of radiation source on the surface of neutron star and to describe the motion process of charged particles in the magnetic field yet. So, the theoretical models of pulsar radiations are still being studied intensively.

1.4.3 Internal Structures

The structure of a typical model neutron star is shown in Fig. 1.2, from which it can be seen that the density inside neutron star gradually increases with the depth increasing, and the star consists of five segments, from the surface to interior, sequentially the atmosphere, outer crust layer, inner crust layer, neutron matter region and core region.

For the neutron star with stable state, there should be the atmosphere to maintain the interstellar pressure balance. The atmosphere is a transitional zone from high pressure inside the star to zero pressure of outer space. According to the hydrostatic equilibrium equation, the higher is fluid density, the rapider the change of the pressure relying on radius. For the neutron star with a mass of 1.4 solar masses and radius

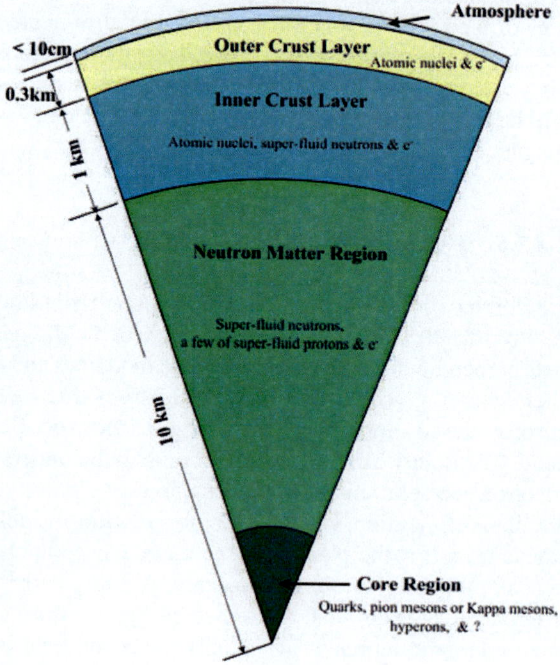

Fig. 1.2 Model of typical neutron star structure

of 15 km, it is computed easily that its density is 2×10^{11} kg/cm^3 and its surface acceleration 8.3×10^8 km/s^2. Apparently, the atmospheric pressure of neutron star is decreased so rapidly that its atmospheric thickness is only 0.1–10 cm and average density 0.01–100 g/cm^3. As same as an ordinary star, the thermal radiation of neutron star comes from its atmosphere, and thus the characteristics of X-ray radiation will be decided by the atmosphere. So, the basic components and characteristics of the neutron star's atmosphere can be detected by observing its thermal radiation.

Below the atmosphere of neutron star, the matter in the zone with a density of less than 4.3×10^8 kg/cm^3 consists of nuclei and electrons, and thereby the zone is called **outer crust layer**, with a thickness of about 0.3 km. If the temperature of the outer crust layer is lower than the dissolution temperature, a solid state will occur. In the outer crust layer, the Coulomb lattice of heavy atomic nucleus is always balanced with relativistic-degenerate electron gas each other.

With the density increasing, the Fermi energy of electrons will rise, and the number of protons inside atomic nucleus also increases. When the density of matter is higher than 4.0×10^8 kg/cm^3, with the neutron number increasing, a part of neutrons will almost drip out from the atomic nuclei. This density is usually referred to as the **neutron drip density**. When the density further increases and approaches to the density of atomic nuclei, all protons inside the atomic nuclei are almost neutronized and neutrons free out of the nuclei. Finally, the atomic nuclei themselves will disappear. The inner crust layer is a zone composed of the nuclear lattice with neutron rich, super-fluid neutron gas and electron gas. The thickness of the zone is about 1 km, and its density is between the neutron drip density and the atomic nucleus one.

The density of the inner crust layer has continuously gone up to that of the atomic nuclei. In the zone where the density has been higher than the density of the atomic nuclei, there are not any atomic nuclei; most parts of the zone are composed of anisotropic super-fluid neutrons, and there are a small amount of super-fluid protons and positrons. The zone is usually called **neutron matter region**, with a thickness of about thousands of meters. Neutrons are main matter in the region, and the population of neutrons is 8 times that of protons in an extreme case.

With the depth increasing, the density inside neutron stars also continuously rises. When the density is more than 2–3 times the density of atomic nuclei, energy per unit baryon will be very low, and thus bring up various subatomic particles, such as quark matter, pi meson or kappa meson condensation, hyperon matter, and so on. Under so high density, neutrons may be squeezed to form mesons or kaons and further a solid core. This zone is usually referred to as core region, with a thickness of about 1 km.

The quarks would naturally be "up" and "down", but they might change into the "strange" variety. Currently, there are two main views on the quark matter of possibly occurring inside neutron stars. For one view, there are the first-order phase transitions in the core regions of neutron stars and the phase transitions between nucleon matter and quark matter are separated by a cross-section of discontinuous density. The types of these neutron stars are usually called **mixed stars**. For another view, it is possible from macro-scale that both the nucleon matter and the quark matter inside neutron stars can exist together, and thus the density from the nucleon matter to the quark matter will change continuously. Generally, the types of these neutron stars are referred to as **hybrid stars.**

1.4.4 *Physical Conditions*

Pulsars are the neutron stars with highly rapid rotating, and have the extreme physical conditions, such as extremely strong gravitational field, extremely high density, extremely high temperature and pressure, as well as extremely strong magnetic field and electric field.

A typical radius of pulsar is only about 10 km and it has a mass of 1.44–3.2 solar masses. Therefore, the gravity on the pulsar's surface is so strong that it is very difficult for any matter to escape the gravitational constraints, and even light is able to break away from the constraints only by a way of parabolic curve. Most free electrons are squeezed into the atomic nuclei by the extremely strong gravity of pulsars, and then the electrons are forced to combine with the protons to form the neutrons.

It is known from the above that the internal structure of typical pulsar is divided into five segments from the outer to the inner. The solid outer crust layer below the atmosphere takes on a crystalline lattice structure, where the matter is composed of neutron nuclei and electrons, and the density increased rapidly and up to an order of 10^3–10^8 kg/cm^3. Below the outer crust layer, the density of the inner crust layer has been more than 4.0×10^8 kg/cm^3. The neutron matter region is just below the inner crust layer and its matter consists of the degenerated neutron fluid as well as a few of protons and electrons. In the region, the neutrons appear on super-fluid states and the atoms do on super-conduction ones, and the density is more than an order of 10^{11} kg/cm^3. The core region of neutron star is a solid core with a range of about 1 km, and its density is 2–3 times that of the atomic nuclei. There is no other place in the universe where the physics of so high density matter can be investigated.

The temperature on the surfaces of neutron stars are about 10^7 K, and the core temperature inside neutron stars up to 6×10^9 K. The core pressures of neutron stars are up to an order of 10^{28} atm, which is 3×10^{16} times that of the Sun's core. Therefore, neutron stars are much hotter than the Sun, and have extremely high temperature and pressure environments.

The interiors of neutron stars are filled with plasmas. During the star evolving, the magnetic fields have been reserved by the plasmas. It is shown by theoretical calculations that if the magnetic field strength of a main sequence star is 10^{-2} T (1T = 10^4 Gauss), the magnetic field of the neutron star formed by the star evolving will get to 10^8 T. It is seen from the actual observations that the magnetic fields of ordinary pulsars are typically 10^6–10^8 T, while those of the AX pulsars can reach 10^{14} T. The magnetic field on the surface of a neutron star is 2×10^{13} times that of the Sun and 10^{16} times that of the Earth. It is well known that a rapidly rotating magnetic field will produce the electric field. So, the electric fields on the surfaces of pulsars are also extremely strong and up to 10^{14} V/m.

1.4.5 *Formation Processes*

On the formation processes of pulsars, it is generally accepted that the solid supernova remnants are produced by the supernova explosions of massive stars, and thus the proto neutron stars will be formed in the centers of these remnants. The proto neutron stars bring up gradually the theoretically predicted neutron stars after they undergo the gravitational collapses and the phase transitions. However, by comparing the discovered more than 200 supernova remnants with the catalogued more than 2800 pulsars, it is shown that there are only tens of the counterparts each other. And thus, the above interpretation on the formation processes of pulsars will be suspected inevitably.

At present, most scholars still insist that pulsars result from the supernova explosions. Their reasons are: on the one hand, the pulsars corresponding to the supernova remnants can not be observed as their radiation beams do not sweep through the near-Earth space; on the other hand, all remnants of the supernova explosions are spread so rapidly out space that their spatial densities are too low to have been observed, and thereby the pulsars whose counterparts of the supernova explosion remnants is not found, are actually the counterparts of the diffused remnants.

Meanwhile, some scholars also believe that the supernova explosion is not unique approach to bring up neutron stars. For example, when the mass of a white dwarf by accreting matter is more than the Chandrasekhar limit, the white dwarf will gradually form a neutron star by the gravitational collapse and the phase transition. Obviously, the formation processes of pulsars still need to be studied further.

1.5 The Observational Features

1.5.1 *Pulse Profiles*

Pulse profile is a periodic change curve that the intensity of the signals emitted from the pulsar varies with time, also called **light curve**. The profile of each pulsar is unique identifier, a representation of the features of pulsars, which is determined by the internal mechanism of pulsars and the external space environment. The pulse profiles vary in terms of shape, size, cycle's length, and intensity. Some basic physical parameters can be obtained by the pulse profiles, such as pulse period, pulse amplitude, total number of pulse peaks, and stability of pulse period. Most of the discovered pulsars emit the periodic pulsation signals in the wider bands of radio frequencies; the pulse widths are about one thirtieth of the entire periods; the pulse shapes in millisecond timescale are very complicated. A typical pulse profile usually includes one main pulse signal, as well as one or more sub-pulse signals. The structures of the sub-pulse signals are more complex and their timescales even get to an order of microseconds. In other words, the single pulse radiation of pulsars has very high polarization characteristics; the pulse intensities and shapes are different from one another, and vary confusedly and disorderly. Nevertheless, an accumulated pulse profile, which can be obtained by folding hundreds or more of the pulse series together, is very stable. The folded pulse profile will clearly show an output shape with one, two or more peaks.

100 pulse-sequences of PSR B1133+16 observed at a central frequency of 600 MHz are plotted in Fig. 1.3, where the top folded profile results from synchronously averaging 500 pulse-sequences. It is seen from Fig. 1.3 that although the single pulse-sequence is very instable, and their intensities are also different and sometimes even disappear, the average pulse from folding hundreds of the pulses has very stable periodicity. It is shown from actual observation that Time-Of-Arrival (TOA) of the average pulse from the RP pulsars is very stable and its long-term stability can be comparable to the stability of the atomic clock. In order to get the precise TOA observables, it is necessary to fold and synchronously average hundreds and even thousands of pulses.

The radio pulsars have been observed for more than 40 years and the abundant observation data is accumulated. By comparison, the observational data of X-ray pulsars is very few. Because the gas molecules in the Earth's atmosphere, like H_2O and CO_2, intensively absorb X-rays, it is impossible on the ground to receive the X-ray signals emitted from pulsars. Generally, the X-rays from pulsars can be detected only in space outside the atmosphere. Only by launching the dedicated astronomy satellites

Fig. 1.3 The folded pulse profile of PSR B1133+16

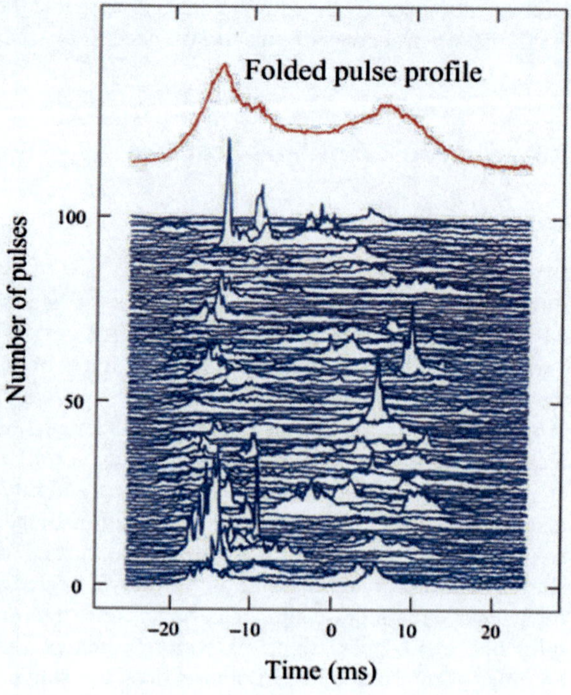

or the spacecrafts carrying X-ray detection devices, X-ray pulsars can be observed in outer space for a long time. Apparently, cost of observing the X-ray pulsars relative to the radio ones is high. With X-ray pulsar-based navigation (XPNAV) technology developing, more and more dedicated X-ray pulsar observation satellites will be launched, and a large amount of X-ray data from the pulsars obtained and used to build the database for the XPNAV.

The pulse profiles of the pulsars in the radio, visible, X-ray and gamma-ray bands are different from one another and some pulsars have periodic pulse radiations only in individual spectral bands. For example, The pulse profiles of the Crab Pulsar (PSR B0531+21) and Vela Pulsar (PSR B0833–45) have been observed in the radio, visible, X-ray and gamma-ray bands, but the shapes of pulse profiles have discrepancies in the deferent frequency bands for each pulsar. In addition, the pulse profiles of PSR B1909–58 and PSR B1055–52 have been observed in the radio, X-ray and gamma-ray bands, those of PSR B0633+17 (the Geminga Pulsar) in the visible, X-ray and gamma-ray bands, and those of PSR B1706–44 and PSR B1951+32 in the radio and gamma-ray bands. The X-ray and gamma-ray signals emitted from the Geminga are first observed respectively by the High Energy Astrophysics Observatory-2 (HEAO-2, also referred to as Einstein Observatory) and German X-ray astronomy satellite ROSAT (short for Röntgensatellit). And then, the optical counterpart of the Geminga is also discovered, and the Geminga is a blue spectrum star with the luminosity of 25.5 magnitudes.

Because there are discrepancies of the spatial dispersions of electromagnetic wave signals from different frequency bands, their radiation mechanisms are also different. For the pulse radiation with different energies, it is also possible that there is a special radiation cone on the surface of pulsar, and thus there are discrepancies of the pulse profiles among different frequency bands. Although the pulse profiles in different frequency bands have discrepancies, their pulse periods are same for a pulsar. It has been well known that the signals from pulsars are produced with their rotations. The radiation cones in whatever shapes or with different locations, all will sweep through detection devices with the pulsars rotating. So, the pulse periods from different frequency bands should be same. Moreover, there is the phase difference between the pulse profiles in the radio and X-ray bands. If the phase difference between both can be determined accurately, the relation between the pulse profiles in the two frequency bands is established definitely, and the timing models from the X-ray pulsars will be made more accurately by using the radio observation data. And thus, a pulsar-time reference system based on space-ground integration will be built.

1.5.2 Periodic Features

Pulsars are highly rapid rotation neutron stars, and their periodic ranges are typically 1.4 ms–8.5 s, while the periods of a few of pulsars get to tens of seconds. Generally, the periodic change rates of most pulsars are positive, and thus it is shown that the rotating velocities of the pulsars have been slowing down gradually. The periodic change rates of ordinary pulsars are 10^{-13}–10^{-15} s/s, while the millisecond pulsars have very stable periodicities and their rates can usually get to 10^{-19}–10^{-21} s/s. Therefore, the millisecond pulsars are generally credited as the most stable clocks in nature. The pulsar timing and navigation are studied by utilizing the very stable periodicity of the pulsars.

Seven millisecond-pulsars, including PSR B1937+21, PSR J1713+0747, PSR J0621+1002, PSR J0751+1807, PSR B1855+09, PSR J2033+1734 and PSR J2322+2057, have been observed and investigated in the Arecibo Observatory since 1982, and their periodic change rates are respectively 1.051193×10^{-19} s/s, 8.5314×10^{-21} s/s, 4.732×10^{-20} s/s, 7.7848×10^{-21} s/s, 1.78354×10^{-20} s/s, 1.1112×10^{-20} s/s and 9.650×10^{-21} s/s [20]. Obviously, the long-term stabilities of the millisecond pulsars are higher than those of the current hydrogen atomic clocks in laboratories.

Generally, the upper limits of pulsars' ages are defined as the ratios of the pulse periods to the periodic change rates, and also referred to as **characteristic ages** of pulsars. For example, the pulse period and its rate of the Crab Pulsar are respectively 0.03372069749 s and 4.2277×10^{-13} s/s on May 1, 2016, and thus it is easily calculated that its characteristic age is 2529 years. In fact, the supernova explosion of the Crab nebula was first observed and recorded by astronomers of the Song Dynasty in China in 1054, and thereby calculating until 2016, its age should be 962 years. Apparently, there is a large discrepancy between the two ages. According to the magnetic dipole model of pulsars, a modified formula of the characteristic age is derived as follows.

$$\tau = \frac{P}{(n-1)\dot{P}}, \tag{1.7}$$

where τ is the characteristic age of pulsars, P and $\dot{P}$ are respectively the pulse period and its change rate, and n is a modified coefficient and usually is valued as 3.

Consequently, it is calculated by using formula (1.7) that the characteristic age of the Crab Pulsar is 1264 years and close to the above result of 962 years. In general, the periodic change rates of the ordinary pulsars with short periods are great and correspondingly their characteristic ages small. It is also shown clearly that the short period pulsars are younger. The pulse periods gradually increase with time, the periodic change rates slow down, and the characteristic ages of the ordinary pulsars will be older and older.

However, no all pulsars have stable rotation periods. The periods of some pulsars are suddenly faster under the background of steady growth for a long time, and then rapidly slow down until recovering the previous period-increasing-rate. The phenomena, which the periods of pulsars suddenly jump, are usually called **glitch**. For the younger pulsars, the glitches happen easily and the relative change rates of periods are usually $10^{-6} - 10^{-10}$. For example, the glitch of the Vela Pulsar happened from February 24 to March 3, 1969; the relative change rate of the period got to 2×10^{-7} and recovered to normal level after the following several weeks. Thereafter,

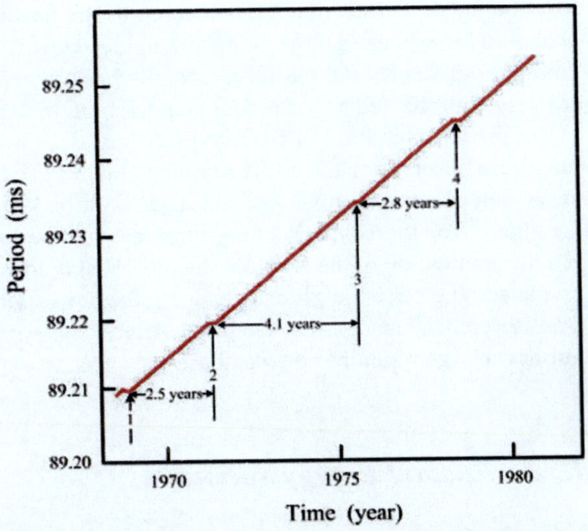

Fig. 1.4 Glitch of Vela Pulsar from 1968 to 1980

more than 10 glitches from the Vela Pulsar had again been observed from 1971 to 1994, as shown in Fig. 1.4. In general case, the periodic change rate of the Vela Pulsar is about 3.53×10^{-17} s/s, far less than the rate during the glitch occurring, and the change direction between the two rates is just contrary. In addition, seven glitches from the Crab Pulsars had also been observed from 1969 to 1996 and the large glitch got to 10^{-8} orders, and in November 2017, the Crab pulsar suffered from the largest glitch in the observed records, with 10^{-7} orders. And At present, the glitches from tens of pulsars have been observed. The glitches occur at intervals of a few years for the young pulsars, and the older are the pulsars, the longer the intervals of the glitches, so that the majority of pulsars have not been observed to glitches even during several decades of observation.

The mechanism of the glitch is one of the research hotspots in the field of radio astronomy. One view is that the glitches result from the "star-quakes" of neutron stars. The centrifugal force produced by rapidly rotating neutron star deforms the crust of neutron star and its equator partly rises; with the rotating power gradually transformed into radiant energy, the rotating rate of the neutron star slows down and the centrifugal force gradually decreases; and thus, a part of the equator of the star's crust will fracture and collapse with the centrifugal force decreasing, so that the matter inside neutron star redistributes and the rotating rate increases. It is shown from the calculated results that if the crust of neutron star collapses to 1 mm, the glitch will occur. The star-quake is not only "crust-quake" occurring on the surface of neutron star, but also may be "core-quake" occurring in the solid core. The structures and characteristics inside neutron stars can be understood by the star-quakes. Hence, studies on the glitches and their recovering processions become a kind of effective probes to explore interior compositions of the neutron stars. Although it can be interpreted by the star-quake model that the glitches occur randomly and irregularly, why the pulse periods may be recovered to the normal level before the glitches is unable to be explained convincingly. Another view is that the glitches result from a kind of complex orbital motions. The glitches are not irregular, but the regularities are very complicated and can not be understood by the human beings yet.

No matter what kind of the interpretations, it is necessary to verify them by the long-term observational data. The typical glitches occur at average interval of several years. The stable periodicity of pulsar can not be kept during several months after the glitch. And thereby, it is a long-term task to monitor and investigate the glitches. For the application of the XPNAV, the pulsars should continuously be monitored and predicted, and once the glitches happen, the corresponding pulsars will be isolated from the pulsar navigation database. After recovering to the normal periods, the pulsars are again put into the database.

1.5.3 Radiant Energy Spectrum

The radiant energy spectrum of pulsars is a non-thermal radiation with a power-law spectrum, and can be expressed as

$$E = E_0 \nu^{\alpha}, \tag{1.8}$$

where E is the radiant energy, ν is the radiant frequency, and α is the power exponent, with a typical value of -1.5.

The power-law spectrums of radiant energies are typical characteristics of the synchrotron and curvature radiations from high-energy electrons. Moreover, the radiant features of high polarizations are also observed. Many pulsars have very high linear polarizations whose intensities are far greater than circular polarizations. Certainly, the power-law spectrums are also necessary results of the radiations from highly rapid rotation neutron stars with extremely strong magnetic fields. The charged particles (mainly containing electrons) can be accelerated by rapidly rotating strong magnetic fields and they obtain higher energies; when the electrons make spiral motion around the magnetic field lines or accelerated motion along them, the synchrotron and curvature radiations with the power-law spectrum will be produced. It is generally considered that the magnetic field strengths at two magnetic poles are strong, and thus the radiations mainly concentrate on the magnetic axis and form two beams with contrary direction. Furthermore, it is also considered that the magnetic axis and the rotation axis do not coincide, and thereby the radiant beams emitted along the magnetic axis will form the lighthouse effects. Obviously, Issue on the power-law spectrum distribution of the radiant energy can be well explained by utilizing the magnetic dipole model of pulsars. However, it is difficult that the radiations with the power-law spectrum are reasonably explained by using the pulsating model of pulsars. So, the pulsating model is also eliminated from the theoretic models of pulsars.

1.5.4 Distance Measurement

In astronomy, the interstellar medium is the matter and radiation that exist in space between the star systems in a galaxy. The interstellar medium is generally regarded as a kind of rare plasmas. When the electromagnetic waves emitted from pulsars travel through the interstellar medium, a phenomenon referred to as dispersion will occur. The **dispersion** is a basic property that the refractive index of the interstellar medium is decreased with the incident electromagnetic wave frequency decreasing. In other words, there are discrepancies among the propagation velocities of the electromagnetic waves with different frequencies as they travel through the interstellar medium.

For the radio signals of wider frequency bands emitted from pulsars, the time which the low-frequency components arrive in the Earth is later than that the high-frequency components. In general, the dispersion measure (*DM*) can be expressed as

$$DM = \int_0^L n_e dl = \overline{n}_e L, \tag{1.9}$$

where L is the distance of pulsar, n_e and $\overline{n}_e$ are respectively the actual and average electron densities, and l is the path length along the line-of-sight of pulsar.

For the XPNAV, because X-rays are high-energy photons with a higher frequency of above 10^9 GHz, the measurement errors of TOA resulting from the dispersion effects are usually less than 0.1 ns. And thus, it is unnecessary in the application of the XPNAV to make de-dispersion, and the dispersion will be ignored.

If the average density of free electron is known in interstellar space, the distances of radio pulsars can be determined approximately by comparing pulse arrival times at different radio frequencies. Supposed that the plasmas in the interstellar medium are enough rare and the interstellar magnetic fields very weak, the group velocity of the electromagnetic wave with a frequency of ω in the interstellar medium can be expressed as

$$v_\omega = c\left(1 - \frac{\omega_p^2}{\omega^2}\right)^{\frac{1}{2}}, \tag{1.10}$$

where c is the velocity of light in vacuum, and ω_p is the frequency of plasma and $\omega_p^2 \equiv \frac{4\pi n_e e^2}{m}$, in which m is the mass of plasma and e is the charge quantity of plasma.

It is known from formula (1.10) that the group velocities from radio pulsars are slightly less than the velocity of light in vacuum and associated with the frequencies of the electromagnetic waves. The lower are the frequencies of electromagnetic waves, the smaller the group velocities in the interstellar medium.

From formula (1.9) and (1.10), the arrival time of the pulse signal after crossing the distance (L) can be expressed as

$$t_\omega = \int_0^L \frac{dl}{v_\omega} = \frac{1}{c}\int_0^L \left(1 + \frac{\omega_p^2}{2\omega^2}\right) dl = \frac{L}{c} + \frac{2\pi e^2}{mc\omega^2} DM. \tag{1.11}$$

Supposed that the arrival times of the same pulse signal are observed by respectively using high-frequency (ω_H) and low-frequency (ω_L), it can be gotten from formula (1.11) that

$$DM = \frac{mc\omega^3}{4\pi e^2}\frac{\Delta t_\omega}{\Delta\omega}, \tag{1.12}$$

where $\omega = \frac{1}{2}(\omega_H + \omega_L)$; $\Delta\omega = \omega_H - \omega_L$; Δt_ω is an arrival-time difference between the high-frequency and low-frequency pulse signals, which is an observable measurement; $\Delta t_\omega = t_{\omega L} - t_{\omega H}$.

The dispersion measure (DM) can be solved by using formula (1.12). Moreover, the average density ($\overline{n}_e$) of the free electrons in the interstellar space is also gotten by other ways. The distance of pulsar (L) is finally determined by using formula (1.9). Certainly, the $\overline{n}_e$ may be gotten by comparing with the pulsars whose distances are known. For example, the distance of the Crab Pulsar is determined by using optical ways, and further the $\overline{n}_e$ can be calculated with the dispersion measurement. The dispersion measurement is an effective method to estimate the distances of pulsars. By utilizing the method, the range of measuring the pulsar's distance is about $0.1-18$ kpc. The $\overline{n}_e$ stands for a real electron content of different directions in the Galaxy, and it is very difficult to accurately determine the $\overline{n}_e$. Furthermore, an assumed condition using the dispersion measurement is that the free electrons are uniformly distributed in interstellar space. So, it is still very difficult to accurately determine the distances of pulsars at present.

1.5.5 Pulsar Polarization

The electromagnetic wave is a transverse wave, and thus its electric vectors will oscillate on the plane perpendicular to the direction of the wave propagation. **Polarization** is a phenomenon that the spatial distribution of the electric vector oscillations has lost symmetry relative to the direction of the electromagnetic wave propagation.

In general, the polarization is divided into linear polarization and circular polarization. The former is that all electric vectors are parallel one other and their direction constant. The latter is that the directions of all electric vectors will be rotated with the radiation frequencies. The radiations of many radio pulsars have very strong linear polarizations, whose strengths are far higher than the circular polarizations. In some cases, the linear polarization rates even get to 100 percent for the radio pulsars. It is a key feature for the synchrotron and curvature radiation mechanisms of pulsars. By utilizing the polarizations of the radiations from pulsars, the interstellar magnetic field strengths can be determined.

The signals radiated from pulsars have traveled through a very long distance before they arrive at the observers, and carry the interstellar medium with weak magnetic fields. And thus, the polarization plane will bring up the **Faraday rotation**, which is that when the electromagnetic waves of the linear polarization cross through the electromagnetic fields, a rotation of the polarization planes relative to the incident waves will be produced under the action of the electromagnetic fields. The Faraday rotation is a magneto-optical phenomenon, discovered by a British physicist Michael Faraday (1791–1867) in 1845, and also called **Faraday effects**.

The rotational angle of the polarization plane is associated with the wavelength of electromagnetic wave, electron density, length of the propagation path and interstellar magnetic field strength, and can approximately be expressed as

$$\alpha = 0.81\lambda^2 n_e L B_p, \tag{1.13}$$

where α is the rotation angle of polarization plane (at a unit of radians), λ is the wavelength of electromagnetic wave (at a unit of meter), n_e is the electron density (at a unit of the electron counts per cubic centimeter), L is the distance of pulsar (at a unit of parsec), and B_p is the component of the interstellar magnetic field strength parallel to the line-of-sight (at a unit of micro Gauss).

Supposed that L and n_e are known, B_p can be calculated by using formula (1.13), comparing two rotational angles (α) obtained from different wavelengths. And thereby, the interstellar magnetic field strengths with an order of micro Gauss can be obtained by using the method to process the observational data from pulsars. Of course, it is also so far the most effective method to measure the interstellar magnetic field strength. Therefore, an important approach to studying the interstellar medium characteristics is provided by observing pulsars.

1.5.6 Spatial Motion

For the observers on the Earth, all observed pulsars are moving relative to the Earth. By correcting the revolution of the Earth around the Sun and its self rotation, the movements of the pulsars relative to the Solar System's Barycenter (SSB) can be gotten, which is referred to as **spatial motions**. For the majority of pulsars, the spatial motions are not obvious.

Generally, the spatial motions are divided into two components, tangential velocity and radial velocity. In practice, the right ascension (α) and declination (δ) of pulsars can be measured directly on the celestial sphere, and their rates are respectively expressed as

$$\left.\begin{aligned} \mu_\alpha &= n\frac{d\alpha}{dt} \\ \mu_\delta &= n\frac{d\delta}{dt} \end{aligned}\right\}, \tag{1.14}$$

where n is the total seconds per year (3.156×10^7 s /year), μ_α is the proper motion of right ascension, and μ_δ is the proper motion of declination.

Obviously, the angular displacement of pulsar perpendicular to the celestial equator on the celestial sphere is μ_δ, while the angular displacement parallel to the celestial equator should be $\mu_\alpha \cos\delta$, and thus total proper motion per year of pulsar is

$$\mu^2 = (\mu_\alpha \cos\delta)^2 + \mu_\delta^2, \tag{1.15}$$

where μ is always positive, μ_α is positive toward the east and negative toward the west, and μ_δ is positive toward the north and negative toward the south.

The orientation of the proper motion of pulsar is expressed by the azimuth (ϕ), that is

$$\tan\phi = \frac{\mu_\alpha \cos\delta}{\mu_\delta}, \tag{1.16}$$

where ϕ is measured along clockwise from the direction of the north celestial pole and its quadrant is decided by the positive sign or negative sign of μ_α and μ_δ.

The relation between the proper motion velocity and tangential velocity of pulsars is actually that between the arc-length and the central angle. And thus, the tangential velocity can be expressed as

$$V_t = \frac{\pi}{180 \times 3600} \cdot \frac{r\mu}{n}, \tag{1.17}$$

where r is the distance of pulsar, and μ is the total proper motion of pulsar (at a unit of radian).

References

1. Chadwick J (1932) Possible existence of a neutron. Nature 129(3252):312
2. Baade W, Zwicky F (1934) On super-novae. Proc Natl Acad Sci USA 254–259
3. Baade W, Zwicky F (1934) Remarks on super-novae and cosmic rays. Phys Rev 46(1):76–77
4. Pacini F (1967) Energy emission from a neutron star. Nature 216(5115):567
5. Lyne A, Graham-Smith F (2012) Pulsar astronomy, 4th ed. Cambridge University Press, Cambridge
6. Hewish A, Bell SJ et al (1968) Observation of a rapidly pulsating radio source. Nature 217(5130):709–713
7. Bell-Burnell SJ (1977) Petit four—after dinner speech published in the annals of the New York Academy of Science Dec 1977. Annals NY Acad Sci 302:685–689
8. Wu XJ, Wen XS (2005) Fifteen lectures on modern astronomy. Beijing University Press, Beijing (Chinese)
9. Large MI, Vaughan AE, Mills BY (1968) A pulsar supernova association. Nature 220(5165):340–341
10. He XT (2009) Crab Nebula and her bright pearl. Hunan Science & Technology Press, Changsha (Chinese)
11. Gold T (1968) Rotating neutron stars as the origin of the pulsating radio sources. Nature 218(5143):731–732
12. Johnston S, Lorimer DR et al (1993) Discovery of a very bright, nearby binary millisecond pulsar. Nature 361(6413):613–615
13. Backer DC, Kulkarni SR et al (1982) A millisecond pulsar. Nature 300(5893):615–618
14. Kızıltan B, Thorsett SE (2009) Constraints on pulsar evolution: the joint period-spin-down distribution of millisecond pulsars. Astrophys J Lett 693(2):109–112
15. Matsakis DN, Taylor JH, Eubanks TM (1997) A statistic for describing pulsar and clock stabilities. Astron Astrophys 326:924–928
16. Cook GB, Shapiro SL, Teukolsky SA (1994) Recycling pulsars to millisecond periods in general relativity. Astrophys J Lett 423:117–120
17. Hartman JM, Galloway DK, Chakrabarty D (2011) A double outburst from IGR J00291+5934: implications for accretion disk instability theory. Astrophys J 726:26–40
18. Majid WA, Lamb RC, Macomb DJ (2004) X-ray pulsars in the small magellanic cloud. Astrophys J 609(6):133–143
19. Sheikh SI (2005) The use of variable celestial X-ray sources for spacecraft navigation. PhD dissertation, University of Maryland, USA
20. Kaspi VM, Taylor JH, Ryba MF (1994) High-precision timing of millisecond pulsars 3: long-term monitoring of PSRs B1855+09 And B1937+21. Astrophys J 428(2):713–728

Chapter 2
Observations of Radio and X-ray Pulsars

From Sect. 1.3.2, it is known that the pulsars radiate the signals in the radio, infrared, visible, ultraviolet, X-ray and gamma-ray bands of the electromagnetic spectrum, but the atmosphere of the Earth only opens "two observable windows" for human beings: one is radio window and another is optical window. In other words, only can the radiation signals in the radio and visible bands freely reach the ground, while those in the infrared, ultraviolet, X-ray and gamma-ray bands are obstructed outside the atmosphere and observed only by using the detection devices carried with the spacecrafts in outer space. Currently, most of the catalogued pulsars radiated the signals only in the radio and X-ray bands, and a few of the pulsars in the infrared, visible, ultraviolet and gamma-ray bands. In the applications of constructing the pulsar timing system and space navigation, the pulsars emitting the radio and X-ray signals at the same time are usually selected as the catalogued objects in the database. Therefore, it has important engineering application value and scientific research significance to intensively discuss the observations of the radio and X-ray pulsars. In this chapter, the contents related to the radio and X-ray pulsars are introduced from two aspects: one is to describe the technical developments and principles, and another is to discuss the observation approaches and devices.

2.1 A Brief History of Radio Astronomy

Since ancient times, people have seen stars in the visible-light window. Although the radio window is always opened, nobody knows all about it and the radio radiations from celestial bodies. In 1865, a Scottish physicist and mathematician James Clerk Maxwell (1831–1897) published a classic paper, "A Dynamical Theory of the Electromagnetic Field", in which it is demonstrated that the electric and magnetic fields travel through space as waves moving at the velocity of light. In 1873, Maxwell published his classic work, *A Treatise on Electricity and Magnetism*, in which a set of partial differential equations to describe the characteristics of electromagnetic

P. Shuai, *Understanding Pulsars and Space Navigations*, Navigation: Science and Technology 11, https://doi.org/10.1007/978-981-16-1067-7_2

fields and their interaction relations was presented and later referred to as **Maxwell's equations**. The equations for electromagnetism have always been called the second great unification in physics after the first one realized by Isaac Newton (1643–1727). Maxwell's equations brought first together electricity, magnetism and light as different manifestations of the same phenomenon and predicted that the electromagnetic waves from stars radiate at any wavelength. As a result, many astronomers started to investigate the electromagnetic radiations from celestial bodies with the ways similar to optical observations. However, the radio waves from celestial bodies have never been received in the low-frequency bands because of the limit of the technical conditions at that time. Moreover, according to the theory of blackbody radiation, the radiations from ordinary celestial bodies are not obvious in the radio-frequency bands. So, it had been believed for a long time that it was vain to investigate the universe in the radio-frequency bands.

In the early 1930s, an engineer at Bell Telephone Laboratories, Karl Guthe Jansky (1905–1950), built a directional antenna with a diameter of about 30 m and a height of 6 m, mounted on a turntable that slowed it to be rotated in any direction, in order to investigate the static interference of short wave transatlantic voice transmissions. The antenna was designed to receive radio waves at a frequency of 20.5 MHz (wavelength about 14.6 m). Jansky noticed that his analog pen-and-paper recording system kept recording a repeating signal of unknown origin. Since the signal peaked about every 24 h, Jansky initially suspected that the interference might come from the Sun. And then, it was shown by further analyzing that the signal peaks repeated exactly with a period of 23 h and 56 min. That is to say, the time interval between the signal peaks was the exact length of a sidereal day, and the source of the interference should be certain celestial object outside the solar system. By comparing his observations with optical astronomical maps, Jansky eventually concluded that the source of the interference came from the center of the Milky Way, and the radiation source was strongest when the antenna was pointed to the constellation of Sagittarius. In 1933, his discovery was widely publicized by the *New York Times* and his paper "Electrical Disturbances Apparently of Extraterrestrial Origin" was also published, but he found little support from astronomers. On the one hand, two kinds of different technical approaches are applied, respectively, to the optical and radio telescopes while few astronomers were familiar with radio observation technology at that time. On the other hand, the resolution of Jansky's radio antenna was so low that the radio radiations from celestial bodies were unable to be really observed. Jansky wanted to investigate the radio waves from the Milky Way in further detail, but he could not be supported by Bell Telephone Laboratories or astronomers, and thus did no further work in the field of astronomy after 1935. Nevertheless, his classic paper has been regarded as a mark of the birth of radio astronomy. To commemorate his historical merits in the field of radio astronomy, the fundamental unit of radio flux density was named after Karl G. Jansky (Jy) in 1966.

In 1937, an American radio amateur, Grote Reber (1911–2002), was inspired by Jansky's work and built a parabolic radio telescope with a diameter of 9 m in his backyard. After experiencing two detection failures when his receiver was operated, respectively, at the frequencies of 3300 MHz and 900 MHz, Reber successfully detected the radio signals from outer space in his third attempt at 160 MHz. He began to repeat Jansky's observations, and then plotted first a radio source map of the Milky Way by conducting the sky survey. His observation data revealed the existence of radio sources like Cygnus A and Cassiopeia A for the first time. In 1940, the radio source map was published in *the Astrophysical Journal*, and widely attended by international astronomical community. For nearly a decade from 1937, only Reber had been engaging dedicatedly in radio astronomy research all over the world. So, Reber really became a pioneer of radio astronomy.

During the Second World War, the radar stations of the British Army had ever received the interference signals from the Sun in the same day. During the daytime, the direction of the interference signals moved with the Sun, and the interference signals would disappear after sunset; the interference signals occurred repeatedly in the second day, and until the third day gradually declined and eventually disappeared. It was confirmed by astronomical observations later that a large sunspot and solar flare was passing through central line of the solar disk at that time and thus the interference signals detected by the radar stations were exactly the radio signals from the Sun. After the end of the Second World War, some members from the radar team of the British Army, such as Martin Ryle, Antony Hewish and Bernard Lovell, entered in the research field of radio astronomy, and received the radio signals emitted from celestial objects by sufficiently utilizing idle radars on the battleground. Soon after, a series of amazing discoveries was gotten by radar astronomical observations and thereby opened up an age of radio astronomical research.

1950s is a key age of rapidly developing the radio astronomy and almost every carefully designed observation can get important findings. At the University of Cambridge, Ryle, Hewish and their colleagues used the Cambridge interferometer to map the radio sky, and published the Second Cambridge Catalogue of Radio Sources (2C) and the Third Cambridge Catalogue of Radio Sources (3C). Meanwhile, astronomers sufficiently utilized the extensive radio observation data to investigate the Sun's activity, search the supernova remnants, measure the velocity of meteors, discover the anagalactic nebulae and plot the map of hydrogen distribution of the Milky Way. The "four great discoveries" in the field of modern astronomy in 1960s, including the quasar, pulsar, Cosmic Microwave Background (CMB) and interstellar molecules, were closely related to the rapid development of radio astronomy. The view of astronomical research is expanded by radio astronomical observations, and furthermore a new field of astronomy also created.

2.2 Principles of Radio Telescopes

According to the radar detection principle, the radio telescopes receive the radio signals at single wavelength and display a curve of denoting signal intensity changes in the receiver. The electromagnetic radiations in the radio-frequency band can cross through the interstellar gas and dust which the visible light is unable to penetrate, and thus the radio telescopes can observe deeper into the universe than the optical telescopes. Moreover, the radio observations are not affected from solar diffuse radiations and clouds, and can be conducted in the daytime and at night. Obviously, the radio telescopes are really a kind of astronomical observation devices at all weather and full time.

2.2.1 Composition of Radio Telescope System

A radio telescope system is usually composed of the antenna, receiver (or amplifier), recorder, as well as data processing and displaying device, as shown in Fig. 2.1. Data collecting, recording, processing and displaying of modern radio telescopes are implemented by the computers. The antennas are usually rotational-parabolic ones, whose advantages are that there are large receiving surfaces to collect radio waves and wide frequency bands, as well as it is easy to mechanically sweep and track. Since the mirror errors of the antennas must be less than one-twentieth of their operating wavelength, the shorter are the wavelengths, the higher the requirements for the machining accuracies of antennas. To observe meter wave signals, the antennas can be made by using metal meshes; to observe centimeter wave signals, it is necessary to make the antennas by using highly smooth metal plates. Generally, for a radio telescope system, the huge antennas are their significant marks and important components.

A basic process of radio observation is: firstly, the radio signals from celestial objects are reflected by the parabolic surface of the antenna to focus on the receiver installed at its focal point; secondly, the signal intensities are amplified 10–1000 times by the power amplifier; thirdly, the amplified signals can be recorded by the recorder; finally, the signals are processed by the computer and the results displayed. In recent years, with the wide application of digital technology, the computers can not only carry out data processing, but also aperture synthesizing and spectrum analyzing.

To estimate the performance of a radio telescope, there are generally two key indicators, sensitivity and resolution. The **sensitivity** is the smallest radiation flux density observed by the radio telescope, that is,

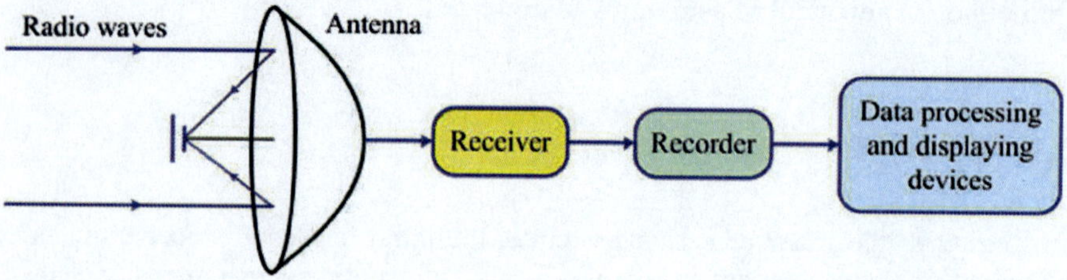

Fig. 2.1 Flowchart of radio telescope system

$$S_{\min} = \frac{2kT_s}{A\sqrt{\tau \cdot \Delta\nu}}, \tag{2.1}$$

where $S_{\min}$ is the smallest observable flux density (at a unit of W/(m^2·Hz)), k is the Boltzmann constant, T_s is the antenna system noise temperature (at a unit of K), A is the effective area of antenna (at a unit of m^2), τ is the integrated time of signals (at a unit of s) and $\Delta\nu$ is the effective noise bandwidth (at a unit of Hz).

It is known from formula (2.1) that the smallest observable flux density is inversely proportional to the effective area of the antenna. Apparently, a significant approach to enhance the sensitivity of radio telescope is to make large aperture antenna. Moreover, there are also other effective ways to improve the sensitivity, such as increase in the effective noise bandwidth and observational time, as well as decline in the antenna's system noise temperature.

In the radio telescope system, the parabolic antenna is often used, and has two functions: one is to collect energy, and another is to control orientation. The larger is the effective area of the antenna, the more the radiation energy collected from celestial bodies. Only if the direction of the radiation signal from celestial bodies is parallel to that of the major axis of the parabolic antenna, the reflected signals will be focused on the focal point. And then, the radio signals are collected by the "feed source" installed at the focal point and transmitted to the receiver. The principle which the radio radiation signals from celestial bodies are focused on by the parabolic antenna is illustrated in Fig. 2.2, where the lengths of light paths ABF, CDF, EGF and

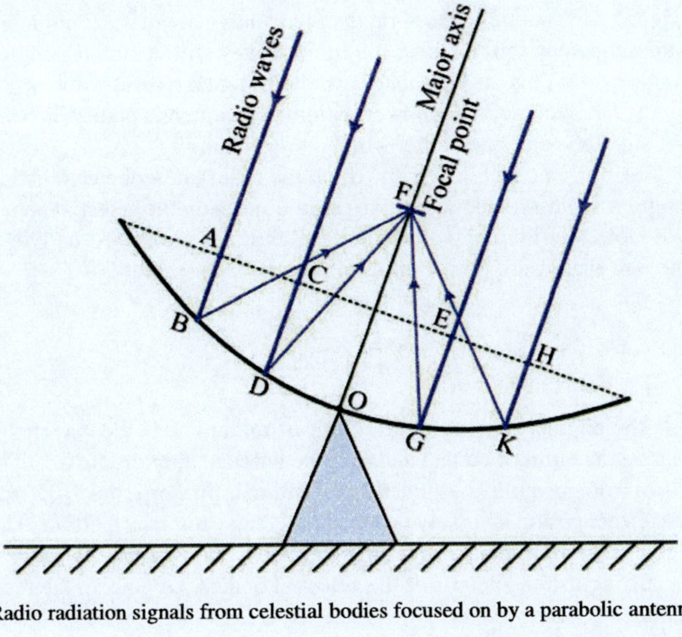

Fig. 2.2 Radio radiation signals from celestial bodies focused on by a parabolic antenna

Fig. 2.3 Antenna's receiving power pattern

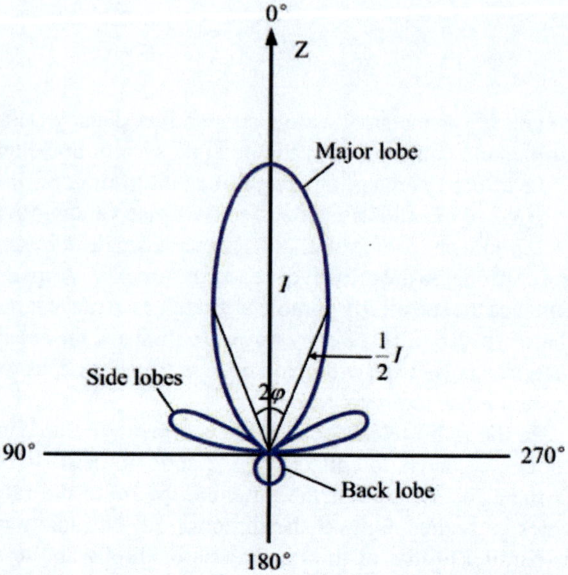

HKF are equal. In other words, there is same phase of the reflected signals of each light path at focal point F, and thus the reflected signals from the parabolic surface are overlain one other and strengthened at the focal point. If there is a big deviation between the direction of the radio radiation signals and that of the major axis, the reflected signals will not be focused on the feed source at the focal point. Therefore, the radio telescope can only receive the radio waves within smaller angle deviated from the major axis. This angle is usually called **angular resolution**, whose size is decided by the antenna receiving power pattern. The antenna pattern is composed of major lobe, side lobe and back lobe, which is shown in Fig. 2.3.

The **resolution** is a capability with which the radio telescope can distinguish the smallest angular separation between two point objects on the celestial sphere, which is usually denoted by the angular resolution, and also expressed as two times the angle at the half maximum power point of the major lobe, that is,

$$2\varphi = 1.22\frac{\lambda}{D}, \tag{2.2}$$

where 2φ is the angular resolution (at a unit of radian), λ is the wavelength of the radio telescope (at a unit of meter) and D is the antenna aperture (at a unit of meter).

It is known from formula (2.2) that the resolution is proportional to the wavelength of the radio telescope and inversely proportional to the antenna aperture. The smaller is the angular resolution, the stronger the capability with which the radio telescope can clearly distinguish the details of the celestial bodies. Of course, the resolution is also the higher.

2.2.2 Two-Element Radio Interferometer

It is well known that the wavelength of radio-frequency band is 10^4-10^7 times that of visible frequency band. It is known from formula (2.2) that if the resolution of a optical telescope with a aperture of 10 cm is achieved by single radio telescope, the antenna aperture of the radio telescope will be required to get to 1–1000 km. Apparently, it is impossible to make the radio telescope with so huge aperture. At present, the world's largest aperture for the rotation parabolic antennas is only 100 m, and that for the Earth-fixed spherical antenna also reaches 500 m. The resolutions of the single radio telescopes are always far less than those of the optical telescopes. Therefore, the radio interferometry is introduced to improve the resolutions of the radio telescopes. It is a powerful technique for aperture synthesis in astronomy.

According to the interferometric principle of electromagnetic waves, the combination of two radio antennas separated from certain distance can be used to improve the resolution of the radio telescope. The combination is usually called **two-element radio interferometer**. Two same radio antennas A_1 and A_2 are installed, respectively, at two ends of baseline D, and connected together to the same receiver by the transmission lines with equal length. A schematic diagram of the two-element radio interferometer is illustrated in Fig. 2.4. Generally, the radiation signals from the same radio source do not arrive simultaneously at the two antennas. Supposed that the radio incident angle is α, the time when the radiation signals arrive at antenna

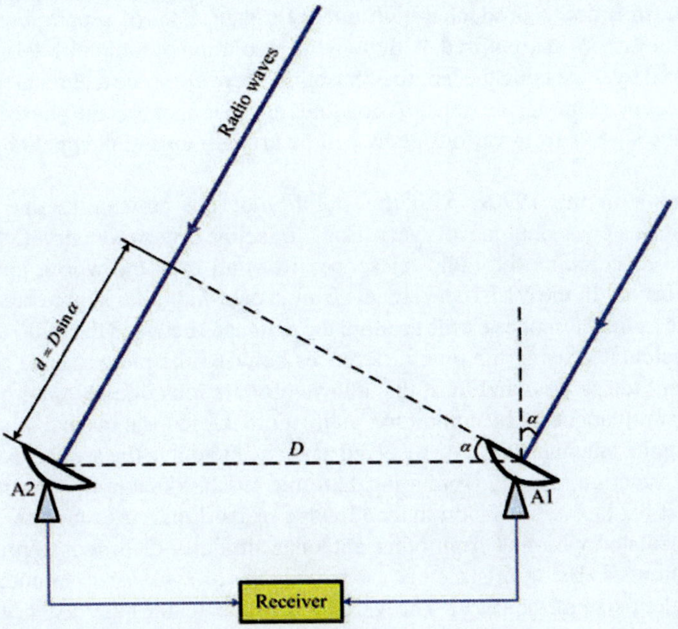

Fig. 2.4 Schematic diagram of two-element radio interferometer

A_1 is earlier than that at antenna A_2, and then the path difference between both can be expressed as

$$d = D \sin \alpha. \tag{2.3}$$

When d is an even multiple of half of the operating wavelength of the radio telescope, the radio waves with the same phases received by the two antennas will be added to each other and further strengthen. When d is an odd multiple of half of the wavelength, the two waves with the opposite phases will cancel each other out and further weaken to reach the minimum. Since the radio source generates diurnal apparent motion with the Earth rotating, α is always varied so continuously that the receiver outputs the periodic superposition signals with strong and weak intervals to form interferometric fringe pattern. The angular resolution of the two-element radio interferometer is still expressed as formula (2.2), but here$D$ no longer denotes the aperture of the single radio antenna other than the distance between the two radio antennas. Obviously, the resolution of the radio interferometer is equivalent to a radio telescope whose aperture is length of the baseline. In other words, the observational effects of the single-antenna radio telescope with large aperture can be achieved by using two small aperture radio telescopes.

Modern radio interferometers consist of widely separated radio telescopes observing the same objects that are connected together using coaxial cables, waveguides, optical fibers or other types of transmission lines. This not only increases the total signal collected, but is also used in an aperture synthesis to vastly improve the resolution. In order to produce a high-quality image, a lot of separations between different telescopes are required. Although the resolution of radio interferometer can be improved by increasing the length of baseline, there are some technical difficulties in practice. For example, the transmission lines are so longer that the phase difference between the signals from various paths will be brought up and the quality of signals affected.

Beginning in the 1970s, with the stability of the radio telescope receivers improving, a new technique of Very Long Baseline Interferometry (VLBI) was introduced to combine the radio telescopes from all over the world, and even in the Earth orbits. In the VLBI, the signals from a celestial radio source are collected at multiple radio telescopes on the ground; the distances between the radio telescopes are then calculated using the time differences between the radio signals arriving at the different telescopes; and thus, this allows observations of a celestial object that are made simultaneously by combining many radio telescopes. Instead of physically connecting the antennas, the data received at each antenna in the array is paired with timing information, usually from a local atomic clock, like a hydrogen maser, and then stored for later analysis on magnetic tape or hard disk. At that later time, the data is correlated with that from other antennas similarly recorded, to produce the resulting images. Using this method, it is possible to synthesize an antenna with the equivalent size of the Earth. The VLBI technique enables the distance between telescopes to be much greater than that possible with conventional interferometry, which requires antennas to be physically connected by various types of transmission lines. The large distances between the radio telescopes enable very high angular resolutions to be achieved. The synthesized beams with less than 1 milli-arc-second are possible at high frequencies.

There are two pre-eminent VLBI arrays operating today. One is the Very Long Baseline Array, where the radio telescopes are located across North America. Another is the European VLBI Network, where the radio telescopes are located, respectively, in Europe, China, South Africa and Puerto Rico, also referred to as Global VLBI. Since its inception, recording data onto hard media was the only way to bring the data recorded at each telescope together for later correlation. Today, the availability of worldwide, high-bandwidth networks makes it possible to do VLBI in real time. This technique is also referred to as e-VLBI, originally pioneered in Japan and more recently adopted in Australia and in Europe.

2.2.3 *Synthetic Aperture Radio Telescopes*

The two-element radio interferometers can only improve the resolution of radio sources at one-dimensional direction, and do not give two-dimensional visual image like optical telescopes. In 1952, Ryle at Cambridge University presented a radio imaging technique, called synthetic aperture radio telescope. Its basic principle is: any image can be decomposed into sine and cosine components of lots of luminance distributions; on the contrary, if the luminance distributions with sine and cosine shapes are known, the original images can be restored synthetically. The synthetic aperture radio telescope needs at least two radio antennas, similar to the two-element radio interferometer. Nevertheless, the basic method of the synthetic aperture is: one antenna is fixed as the center of circles and the circles are drawn at any distance as radius; another antenna is moved in turn to various locations on the different radial circles to perform the radio interferometry. A schematic diagram of the synthetic aperture radio telescope is illustrated in Fig. 2.5, where the actual operation processes are.

(1) Supposed that A and B, respectively, stand for two antennas, where A is a fixed antenna and B is a movable one.
(2) At any time, the two-element radio interferometer is composed of antennas A and B. The correlated amplitude and phase information can be obtained by correlation processing of the receiver, and thereby the sine and cosine components of the luminance distributions can be extracted from the information.
(3) Antenna A is fixed and antenna B moved in turn to various locations on the big circular plane to perform the radio interferometry. Considered the symmetry of circular plane measurement, generally it is only needed to measure half of the equivalent circular plane. And thus, the sine and cosine components of the luminance distributions of all directions and distances on the big circular plane are obtained completely.

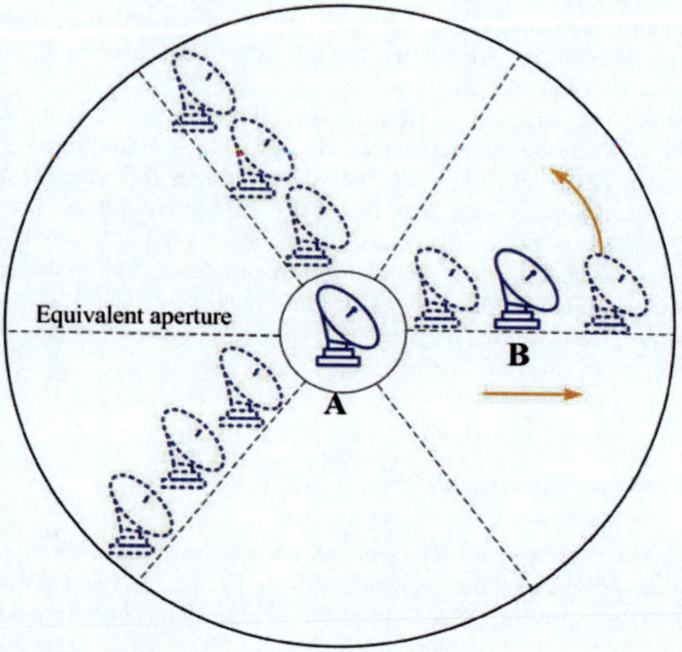

Fig. 2.5 Schematic diagram of synthetic aperture telescope

(4) After the sine and cosine data is processed by the Fourier transform, the image of the radio source in the sky area can eventually be gotten. The range of the sky area is determined by the Field of View (FOV) of the single antenna and the resolution is done by the diameter of the big circle.

Furthermore, the Earth's rotation effect is used to set up the radio antennas, which not only greatly reduce the frequency of movement of the antennas, but also increase the effective receiving area and improve the sensitivity of the synthetic aperture telescope.

The synthetic aperture telescope is a kind of high-sensitivity and high-resolution one to be applied to imaging and detecting the radio sources with constant intensities. From 1954 to 1971, a research team led by Ryle developed a series of synthetic aperture telescopes, and their sensitivities and resolutions were improved continually; the range of the sky survey was expanded from 1 to 15 billion light years, which almost reached the upper limit of the cosmic edge defined in the Big Bang theory and went back to the primary time of the universe; particularly, the team still obtained a fine image of the Cygnus's radio source, which is composed of two relatively distant extended radio sources and one compact galactic nucleus.

In practice, the movable antenna B can be replaced by multiple antennas, like Very Large Array (VLA), which is a centimeter-wavelength radio astronomy observatory located in central New Mexico on the Plains of San Agustin and a component of the

NRAO, USA. The VLA project was started in 1972 and formally built in 1980. And later, with the technology upgrading, the VLA has evolved into the Expanded Very Large Array (EVLA). In 2012, the array was officially renamed the "Karl G. Jansky Very Large Array".

The VLA comprises 27 independent antennas deployed in a Y-shaped array, each of which has a dish diameter of 25 m and weighs 209 metric tons. Each of the massive telescopes is mounted on double parallel railroad tracks, and thus the radius and deployment of the array can be transferred to adjust the balance between the angular resolution and the surface brightness sensitivity. Three arms of the Y-shaped array are 21 km long. Using the railroad tracks that follow each of these arms, the antennas can be physically relocated to a number of prepared positions, which allow aperture synthesis interferometry with up to 351 independent baselines. In essence, the array acts as a single antenna with a variable diameter. There are four commonly used configurations: the array A with a maximum antenna separation of 36 km, the array B with 10 km, the array C with 3.6 km and the array D with 1 km. The telescopes are switched among these configurations every 4 months or so. The available wavelength of the receiver at the VLA is from 0.7 to 400 cm, and the angular resolution between 0.2 and 0.04 arc-seconds. Using the VLA, with only 8 h of observation in the centimeter wave bands, a radio source image with a resolution of order of arc-seconds can generally be obtained.

2.3 Modern Large Radio Telescopes

In the electromagnetic wave spectrum, the range of frequency that makes up the radio spectrum is so wide that the types of antennas that are used as radio telescopes vary widely in design, size and configuration. According to the difference in the structures of radio antennas, the radio telescopes are usually divided into two major types, continuous and discontinuous aperture ones, where the former is a kind of classic radio telescopes with single parabolic-dish antennas, and the latter is a kind of combined antenna array based on the radio interferometry. According to the difference in mechanic equipment and driven mode, the continuous aperture radio telescopes are subdivided into fully steerable, fixed and combined modes. The fully steerable radio telescopes can be subdivided into the equatorial and horizontal, rotating in two coordinate directions. For the fixed radio telescopes, their antenna reflecting surfaces are always fixed at the ground sites, and the sky surveys are performed relying on the Earth's rotation. The combined radio telescopes can rotationally track along the declination direction, and sweep to observe along the right-ascension direction relying on the Earth's rotation. Moreover, according to the difference in the shapes of radio antennas, the radio telescopes are divided into parabolic, spherical and parabolic-cylindrical ones, as well as other antenna shapes, such as horn, spiral, traveling wave, dipoles, rotational-parabolic cross section and so on. In general, both the traceable parabolic antenna and the fixed spherical reflecting antenna are used most commonly.

At present, there are about 210 radio telescopes operating all over the world, among which there are 52 in Europe, 29 in North America, 18 in Asia, 14 in South America, 9 in Oceania, 8 in Africa, 5 in Pacific Ocean, 2 in Antarctica and 2 in Atlantic Ocean, and there is 1 in Indian Ocean. The names, affiliations or locations, operating wavelengths, diameters and typical characteristics of seven famous radio telescopes built before 1971 are listed in Table 2.1, where the Arecibo observatory represents the high-technical levels of the fixed radio telescopes.

The Arecibo observatory was designed by a physicist and astronomer at Cornell University, William Edwin Gordon (1918–2010), who is originally intended to use it to study the Earth's ionosphere. Therefore, a fixed parabolic was envisioned, pointing

Table 2.1 Seven famous radio telescopes built before 1971

Name	Affiliation or location	Wavelength (cm)	Diameter (m)	Built year	Notes
Arecibo Observatory	Cornell University / University of Central Florida, USA; Arecibo, Puerto Rico	3–100	305	1963	The largest diameter fixed spherical radio telescope before 2016
RATAN-600	Zelenchukskaya, Russia	1–50	576	1974	The circle of rectangular radio reflectors, the world's second largest-diameter individual radio telescope
Effelsberg 100-m Radio Telescope	Max Planck Institute for Radio Astronomy; Bad Münstereifel-Effelsberg near Bonn, Germany	0.35–73	100	1971	A fully-steerable radio telescope
Parkes Radio Telescope	Commonwealth Scientific and Industrial Research Organization; Parkes Observatory, New South Wales, Australia	0.7–75	64	1961	A fully-steerable radio telescope
Lovell Telescope	Jodrell Bank Observatory, Cheshire, England	5–200	76.2	1957	A fully-steerable radio telescope
Nancay Radio Telescope	Nançay, France	9, 18, 21	299×35	1965	A transit instrument of the unusual Kraus-type design, with two mirrors
Ooty Radio Telescope	National Centre for Radio Astrophysics, India	92	530×30	1970	A 530 m long and 30 m cylindrical parabolic antenna

in a fixed direction with a 150-m tower to hold equipment at the focal point. Later, to meet the application requirements in research areas of radio astronomy, atmospheric science and radar astronomy, the design was revised into a dish-shaped spherical reflector with a diameter of 305 m. Gordon selected a sinkhole with the karst topography located in the municipality of Arecibo, Puerto Rico, to put the very large dish. Areciboradio telescope was built in 1963, and its dish surface was made of 38,778 perforated aluminum panels, each about 1 m by 2 m, supported by a mesh of steel cables. Since the reflector is spherical, and not parabolic, the reflected signals will not focus on a focal point so that the method of adjusted mirror or linear feed source is used to collect energy. To aim the device, the receiver is moved to intercept signals reflected from different directions by the spherical dish surface of 270-m radius. The receiver is located on a 900-ton platform suspended 150-m above the dish by 18 cables running from three reinforced concrete tower, one 111-m tall and the other two 81-m tall, placing their tops at the same elevation. The platform has a rotating, bow-shaped track 93-m long, called azimuth arm, carrying the receiving antennas and secondary and tertiary reflectors. This allows the telescope to observe any region of the sky in a 40° cone of visibility at the local zenith. The observational strip formed in the sky with the Earth rotating is between $-1°$ and $+38°$ of the declination. The working wavelength range of the Arecibotelescope is from 3 to 100 cm, and its focal length is 132.6 m. From its construction in the 1960s until 2011, the observatory was managed by the University of Cornell. Under cooperative agreement with the U.S. National Science Foundation, it is currently being operated by the University of Central Florida, Yang Enterprises, and Metropolitan University (in Puerto Rico). For more than 50 years, the Areciboradio telescope had always been the world's largest single-aperture one. Until 2016, it was surpassed by the Five-hundred-meter Aperture Spherical Telescope (FAST) built in China.

Moreover, the Green Bank Telescope is the world's largest fully-steerable radio telescope, located in Green Bank, West Virginia, near the heart of the U. S. National Radio Quiet Zone, and completed in 2002. The telescope is 148 m tall and its working frequency range is from 290 MHz to 100 GHz; its surface is an elliptical cross section with 110 by 100 m, intercepted from the symmetric rotational paraboloid with a diameter of 208 m; it is unusual for radio telescopes that the primary reflector is an off-axis segment of a paraboloid, and its focal point and feed horn are just located at the side of the dish. The advantage of the off-axis design is that without any obstruction above the dish, the radiations from celestial objects can directly arrive at its surface to improve the effective collection areas, and cancel the reflection and diffraction effects possibly from the arms or holders of the telescope. The reflector's surface is together joined by 2004 active metal panels, each of which can adjust its position by an individual small motor. The surface deformation resulting from the discrepancies of gravity, wind and temperature can be determined in real time by the laser ranging system. And then, the computer sends control instructions to the motor to adjust the position of corresponding panel. In the telescope rotating, the parabolic shape can be kept by real-time adjusting the surface.

Beginning in the 1980s, a series of radio telescopes have successively been built in China, such as Miyun synthesis radio telescope with an array of 28 nine-meter antennas in Beijing Municipality, Delingha MM wave telescope with a diameter of 13.7 min Qinghai Province, Sheshan 25-m and 65-m radio telescopes in Shanghai Municipality, Ulumuqi 25-m radio telescope in Xinjiang Uygur Autonomous Region and Kunming 40-m radio telescope in Yunnan Province. Particularly, the world's largest fill-aperture radio telescope, FAST, located at the Dawodang depression in Pingtang County, Guizhou Province, southwest China, was completed in September 2016. The FAST has a fixed primary reflector put in a natural sinkhole in the karst topographical regions, focusing radio waves on a receiving antenna in a lightweight feed cabin suspended 140-m above it. The reflector is composed of 4450 perforated aluminum triangular panels, supported by a mesh of steel cables hanging from the rims. 2225 winches make the reflector an active surface, pulling on joints between the panels, deforming the flexible steel cable support into a parabolic antenna aligned with the desired sky direction. The feed cabin can be moved by a cable robot using winch servomechanism on six support towers. The receiving antennas are installed at the Stewart platform below the feed cabin, which provides fine position control and compensates for disturbances like wind. The working frequency range of the FAST is 70 MHz−3 GHz and its observable sky area is a 40° cone at the local zenith. The advantages of the FAST are large aperture, wide frequency band and high sensitivity. The FAST is mainly applied to surveying the large-scale neutral hydrogen extending to the edge of the universe, discovering the fainter radio pulsars, detecting the gravitational waves, searching the extra terrestrial intelligence and creating the pulsar timing array. On the one hand, the characteristics of high sensitivity and large sky coverage are very advantageous to the discovery of faint radio pulsars, millisecond pulsars, binary systems, planetary systems with pulsars, extra-galactic strong pulsars and non-globular cluster millisecond pulsars. On the other hand, the characteristics of non-integral high sensitivity are specially applied to studying polarization and single pulse of pulsars, accurately detecting interstellar medium and revealing radiation mechanism and genesis of pulsars.

2.4 Investigations into Radio Pulsars

2.4.1 Discovery of the First Binary Pulsar System

During several years since discovering the first pulsar, there had been more than 100 pulsars found successively. In 1974, Russell Alan Hulse (1950−), a research student from the University of Massachusetts Amherst, was working with his supervisor Joseph Hooton Taylor (1941−) on a large-scale survey for pulsars using the Arecibo 305-m antenna. They found out 40 new pulsars in 140 square degrees sky area. These pulsars had stable periodicity and their periodic change rates were mostly from 10^{-12} to 10^{-20} s/s. For the pulsars with so small periodic change rates, it is generally very difficult to find out their periodic changes during several days.

However, Hulse and Taylor noticed that one of these pulsars was always unpredictable during several days. After timing the radio pulses for some time, there was a systematic variation in the arrival-time of the pulses. Sometimes, the pulses were received a little sooner than expected; sometimes, later than expected. These variations changed in a smooth and repetitive manner, with a period of 7.75 h. Initially, they thought that the phenomenon possibly resulted from using an inappropriate data analyzing method. And thereby, Hulse shortened the sampling interval from 10 to 1 ms and made observations once again, but it was shown from the observational results that the periodic change rate of the pulsar was uncertain yet. By repeating observations and comparing researches, Hulse and Taylor eventually realized that the phenomenon could well be interpreted if the pulsar was in a binary orbit with another star. For the binary system, the pulsar is more and more away from the Earth in a half time of the orbital period, and thus its observed period will be longer and longer; during another half time of the orbital period, the pulsar is more and more close to the Earth, and thus its observed period will also be shorter and shorter. So, the periodic change rate of the pulsar in the binary system is far greater than that of other pulsars.

The radio pulsar in the binary system is named PSR B1913+16, and also called **Hulse–Taylor binary**, the first binary pulsar system to be discovered. Later, it was confirmed that the companion of the pulsar is a neutron star. Nevertheless, the pulses from the companion neutron star have not been detected, but it might only be the result of an unfavorable viewing angle. There is not material exchange between two sub-stars in the binary system, and the semi-major axis of the orbit of two sub-stars is 1.95×10^6 km, its period 7.75 h and its eccentricity 0.617. Consequently, this pulsar has extremely high orbital velocity, about 10^4 km/s, equivalent to one-tenth of the light velocity.

According to the general relativity, any object with mass will produce the gravitational waves as it is accelerating. The masses of objects on the Earth are so small that the gravitational waves resulted from them are extremely weak and thus cannot be detected by current technical means. For an ordinary star, although its mass is very huge, it is generally unable to be accelerated, and thus the gravitational wave radiated from it is also very weak. Based on the observed characteristics of the binary pulsar system, Hulse and Taylor speculated that this was possibly what physicists had been trying hard to find with extremely huge mass and extremely high velocity. It was shown from the investigation that the gravitational radiation from the binary system is so strong that the orbital period change is very obvious and gets to -2.6×10^{-12} s/s. To observe so small periodic change rate, it is required to highly accurately measure pulse arrival-time. Hulse and Taylor had continually observed and investigated PSR B1913+16 for more than 20 years by using the Areciboradio telescope. And finally, they got a result that there is only 0.4% difference between the actual observational values and the theoretic ones from the general relativity. In other words, they indirectly confirmed existence of the gravitational waves. In 1993, Hulse and Taylor won together the Nobel Prize in physics "for the discovery of a new type of pulsar, a discovery that has opened up new possibilities for the study of gravitation" while at this time they both were being employed at Princeton University.

2.4.2 Discovery of the First Millisecond Pulsar

In 1970s, there was radio source 4C21.53 in the Cambridge catalogue of radio sources, and it has basic properties of pulsars, such as strong polarization, reducing power spectrum and high compact, but its periodic pulse signals had never been detected. Since at that time it was not considered that the rotating periods of pulsars might get to millisecond order, most of radio telescopes were not sensitive to pulse signals with periods of less than 10 ms. In 1982, an American astrophysicist, Donald C. Backer (1943–2010), initiated a search in the region of 4C21.53, which would be sensitive to a wide range of pulse periods and dispersion measures, including very short periods. The initial search plan was to sample at a rate of 500 Hz, which would have been insufficiently fast to detect a pulsar spinning at 642 Hz. To simplify the search apparatus, Backer's research team sampled as quickly as possible, and time averaged the signal over a period of 0.4 ms, thus effectively sampling at 2500 Hz. As a result, they eventually determined that the radio source was a pulsar, PSR B1937+21, rotating every 1.558 milliseconds [1], a rate far beyond anything that astronomers studying pulsars had expected. The PSR B1937+21 is the first discovered millisecond pulsar, and its rotation along with other millisecond pulsars discovered later is very stable in their rotations. These millisecond pulsars are capable of keeping time as well as atomic clocks. In 1994, a Polish astronomer Aleksander Wolszczan (1946–) pointed out that the PSR B1937+21 has a dwarf planetary companion within a range of 2 AU, whose mass is about one-thousandth of the mass of Earth. Moreover, X-ray signals emitted from PSR B1937+21 were also detected by the BeppoSAX [2], an Italian–Dutch satellite for X-ray astronomy, successfully launched by an Atlas-Centaur rocket into a 4-degree inclination low Earth orbit on April 30, 1996 and ended its life falling into the Pacific Ocean on April 29, 2003. And thus, the PSR B1937+21 is also a X-ray binary system.

2.4.3 Acceleration Mechanism and Application of Millisecond Pulsars

Generally, it is shown from studies that all short periodic pulsars are young neutron stars with strong magnetic fields, and their periodic change rates are also larger. There is no longer thermonuclear reaction for an ordinary pulsar, and its radiations can be kept only by consuming the rotating energy. As a result, the pulsar's rotation speed will be slower and slower, and its pulse period gradually increases. In other words, with the pulsar's age increasing, its period will be longer and longer. However, the periods of millisecond pulsars are very short and stable, their characteristic ages even get to 1000 million years, higher 5 orders than other pulsars with short periodic periods. In the X-ray binary systems, the radiant energy of pulsars does not come from consuming the rotating energy, but from accreting matter from the companions. The accreted matter falls highly rapidly into neutron stars, and not only provides the radiant energy, but also accumulates the rotating energy. And thus, the rotations of neutron stars are be speeded up. In millisecond pulsars evolving, maybe they have ever undergone these rotating accelerations. After the companions evolve into supernovas, the neutron stars stop radiating X-rays due to being unable to capture matter from the companions, and then gradually evolve into the millisecond pulsars with radio radiations.

The rotating periods of millisecond pulsars are very stable and the periodic noises also smaller. And thus, the statuses and properties of the celestial objects near the millisecond pulsars might indirectly be investigated by observing the periodic disturbances of the pulsars. The millisecond pulsars become effective probes to investigate the characteristics of extra-solar celestial objects. In 1999, Wolszczan reported variations in the time of arrival of pulses from PSR B1937+21, and the companion having a mass almost equal to Cereus, located at 2.71 AU away from the pulsar [3]. More recent observations have not detected any regular periodic signal associated with the companion. Nevertheless, it is argued that the slight variations in pulse arrival-time are consistent with an asteroid belt having a total mass less than 0.05 earth-masses. Of course, to confirm the possible asteroid belt, the detection of periodicity in pulse timing variations associated with individual asteroids is still necessary [4]. At present, it has been discovered in the millisecond pulsar binary systems that the companions may be planets, optical stars, white dwarfs or neutron stars. The millisecond pulsars provide good space experimental conditions to detect gravitational waves and other celestial objects, which is also very advantageous to investigating the long-term evolutions of stars.

Furthermore, the periodic change rates of millisecond pulsars get to 10^{-19}–10^{-21} s/s, and their long-term stabilities have been even better than atomic clocks. And hence, the millisecond pulsars are ideal signal sources to study the pulsar timing models. The stable pulsar timing system is built as the datum of the atomic time and the astronomical timescale. A radio pulsar timing system was built at the Arecibo observatory in 1984. Two millisecond pulsars, PSR B1937+21 and PSR B1855+09, have been observed by using the Arecibo radio telescope for long term, and a large amount of data collected. As a result, the radio pulsar timing accuracy has reached the levels of 1 microsecond and even 0.2 microseconds.

In general, many factors to affect pulsar timing stability are unable to be separated effectively by using the individual pulsar timing, unless creating a Pulsar Timing Array (PTA), a set of millisecond pulsars which is analyzed to search for correlated signatures in the pulse arrival-time. That is to say, an ensemble pulsar time can be created by utilizing many millisecond pulsars with smaller periodic noises and from various directions. There are many applications for the PTAs. The most well known is to use an array of millisecond pulsars to detect and analyze gravitational waves. Such detection would result from a detailed investigation into the correlation between the arrival-times of pulses emitted by millisecond pulsars as a function of the pulsars' angular separations. Currently, there are three active pulsar timing array projects: the first is the Parkes Pulsar Timing Array (PPTA) at the Parkes radio telescope, which has been collecting data since 2005 and the second is the European Pulsar Timing Array (EPTA), using data from the Lovell Telescope, Westerbork Synthesis Radio

Telescope, Nancy Radio Telescope and Effelsberg Telescope, as well as the Sardinia Radio Telescope to be added upon completion; the third is the North American Nanohertz Observatory for Gravitational Waves (NANOGrav), using data from the Arecibo and Green Bank radio telescope. The three projects have begun collaborating under the title of the International Pulsar Timing Array (IPTA) project. The goal of the IPTA is to detect gravitational waves using an array of approximately 30 pulsars.

2.5 X-ray and Its Physical Characteristics

X-ray is a form of electromagnetic radiation, discovered by a German physicist, Wilhelm Röntgen (1845–1923). On November 8, 1895, Röntgen was investigating cathode rays from a Crookes tube, an early experimental electrical discharge tube invented around 1875, using a fluorescent screen painted with barium platinocyanide. The so-called **cathode rays** are the electron streams in which the electrons emitted from the cathode of low-voltage gas discharge tube are accelerated to form under the electronic fields. Although Röntgen convinced that the cathode rays would not cross through the wall of the vacuum discharge tube, the tube was still wrapped in black cardboard to avoid the interference of the visible light. After electrified, the cathode rays were obscured, while a faint green glow from the screen about 1-m away occurred. By repeating the experiments, Röntgen eventually concluded that it was an unknown type of radiation, and thus he referred to the radiation as "X-ray". Moreover, Röntgen found that the X-rays could also pass through books and papers on his desk, as well as penetrate the muscles to show the skeleton of the hand. He made a picture of his wife's hand on a photographic plate formed due to the X-rays. The photograph of his wife's hand is the first image of a human body part using the X-rays.

In fact, Röntgen was not the first to have observed the effects, but he was the first to systematically study them. He wrote an initial report "On a New Kind of Ray: A Preliminary Communication" and submitted it to Würzburg's Physical-Medical Society journal on December 28, 1895. This was the first paper written on X-rays. The discovery of X-rays not only has a significant impact on medical diagnosis, but also leads to subsequent major scientific discoveries. It was just inspired by Röntgen's discovery of X-rays that a French physicist Antoine Henri Becquerel (1852–1908) discovered the radioactivity of spontaneous materials in 1896. In 1901, Röntgen won the first Nobel Prize in Physics for his discovery of X-rays. In order to commemorate the great discovery of Röntgen, sometimes X-rays are also referred to as Röntgen rays.

X-rays are a kind of high-energy photons, and have usually a wavelength range from 0.01 to 10 nm, corresponding to frequencies in the range of 2.4×10^{16}–3.0×10^{19} Hz and energies in the range of 0.1–200 keV. Generally, the X-ray within energy range of 0.1–20 keV is called **soft X-ray**, and that within energy range of 20–200 keV is called **hard X-ray**. The X-ray has physical properties as follows:

(1) Penetrating Capability
The X-rays have the short wavelength, high-energy and no charged properties. When matter is radiated by the X-rays, only a small fraction of energy is absorbed by the matter while most of the energy is passed through the atomic gaps of the matter. And thus, the X-rays usually show so strong penetration capability that they can easily pass through a book with several thousand pages thick, plank or hard rubber with several centimeters and aluminum plate with less than 20 mm. The penetrating capabilities of the X-rays are related with their wavelengths and the densities of various matters. A plumbum plate only with 1.5 mm thickness can almost obstruct all X-rays to pass through. Moreover, X-rays can also not pass through the dense atmosphere.

(2) Ionizing Radiation
The X-ray photons carry enough energy to ionize atoms and disrupt molecular bonds. Under the irradiation of the X-rays, the electrons outside the atomic nuclei for matter will escape from the atomic orbits to bring up ionizing. By measuring the ionizing charges, the radiation dose of X-rays can be determined. This is a basic principle of detecting the X-rays. The ionizing radiations of X-rays can induce gas to conduct electricity, matter to happen chemical reaction and organism to bring up biological effect.

(3) Fluorescent Effects
The wavelengths of X-rays are shorter than those of visible lights and usually are invisible. However, if some chemical compounds, such as phosphorus, barium platinocyanide, cadmium zinc sulfide and calcium tungstate, are radiated by the X-rays, the fluorescents, like visible lights or ultraviolet rays, will occur in these materials. The strengths of the fluorescents are proportional to the radiation dose of X-rays. This is a basic principle of X-ray fluoroscope imaging.

(4) Thermal Effects
When a solid material has absorbed the energy of X-rays, most of the energy is transformed into the thermal energy resulting in the temperature rise. The amount of temperature rise is dependent on the energy of the X-ray photons.

(5) Interference, Diffraction and Reflection Effects
The X-rays have much shorter wavelengths than visible light, which makes it possible to probe material structures much smaller than can be seen using a normal microscope. The property is used in X-ray microscopy to acquire high-resolution images, and also in X-ray crystallography to determine the positions of atoms in crystals. In these applications, the interference, diffraction and reflection effects of X-rays can be shown sufficiently.

2.6 Mechanism of X-ray Detection

In essence, the interactions between X-ray photons and materials are an absorption process. High-energy photons interact with material atoms to lose all their energy, and with the electrons outside atomic nuclei to lose a part of the energy. In general, the X-ray photons have very strong penetration capability, and even the soft X-rays can also penetrate aluminum films. And thereby, usually the X-rays cannot be focused on by using normal optical system. When the X-ray photons interact with matter, the material atoms absorb the photons, and then the electrons outside the atomic nuclei escape from the atomic orbits to bring up ionizing. X-ray detectors are the devices used to measure the flux, spatial distribution, spectrum or other properties of the X-rays. By measuring the released energy in impact effects between the incident X-ray photons and the major materials of the detectors, the flux of the photons can be detected.

Generally, there are three typical ways in interactions between the high-energy photons and the major materials of the detectors: photoelectric effect, Compton scattering effect and electron-pair effect. The three interactions are naturally different from one another, and each of them dominates a corresponding energy range. Of course, each interaction is also closely related with the atomic number of the matter.

2.6.1 Photoelectric Effect

When the X-ray photons interact with matter and their energy is greater than the electrons' binding energy in material atomic inner shell, all energy of the photons will be transmitted to the electrons of the atomic inner shell so that the electrons have escaped from atomic bondage. This interaction between the X-ray photons and matter is usually known as **photoelectric effect**, and the electrons produced by the photoelectric effect are called **photoelectrons**, to distinguish from free electrons. The kinetic energy of photoelectrons is equal to the photons' energy minus the binding energy of the electrons in the atomic inner shell. The greater is the binding energy of the electrons in the atomic inner shell, the smaller the kinetic energy of the photoelectrons. Besides the photoelectrons, the photoelectric effect also produces an ionized atom, in which the holes in atomic inner shell will quickly be filled with the electrons transited from the outer layer to the inner layer, or with the captured free electrons. And thus, the characteristic X-rays are emitted. When the characteristic X-rays are absorbed by the matter, the photoelectrons with the same energy will be produced once again, which are also usually called **Auger electrons**. If all energies of both the photoelectrons and the Auger electrons were lost inside the matter, the detected energy loss will represent the energy of the incident photons.

According to the momentum conservation condition, the photoelectric effect can occur only between the photons and the bound-state electrons. By the binding force between electrons and atoms, the momentum of photons is transformed into the recoil momentum of the atom. The stronger is the binding force, the greater the probability of emitting the photoelectrons. A schematic diagram of the photoelectric effect is shown in Fig. 2.6, where the electrons in K layer have greater photoelectric absorption probability than those in L layer. The photoelectrons from K layer usually account for 80% of the total of photoelectrons. In general, the photoelectric cross

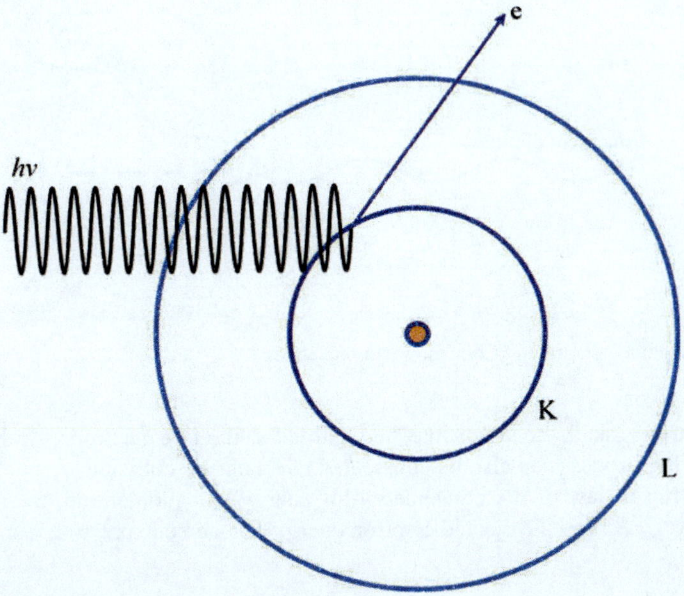

Fig. 2.6 Schematic diagram of photoelectric effect

section will rapidly decline with the increase in the energy of photons, and rapidly rise with the increase in material atomic number. And thus,

$$\sigma \propto Z^5 \cdot (h\nu)^{-\frac{7}{2}}, \tag{2.4}$$

where σ is the photoelectric cross section, Z is the material atomic number, h is the Planck's constant and ν is the frequency of the incident X-ray photons. The photoelectric effect is the dominant interaction mechanism in the soft X-ray band, or mostly occurs in the matter with high atomic number.

2.6.2 *Compton Scattering Effect*

When the incident X-ray photons are scattered by a static electron in matter, the photons transmit a part of their energy to the electron to make their kinetic energy and momentum change. The scattering of a photon by an electron is usually called **Compton scattering effect**. The incident photons are scattered by the electrons in absorption matter to lose a part of energy, and thus the incident direction of the photons is changed and deflected, which is illustrated in Fig. 2.7. Generally, the Compton scattering effect occurs only between the photons and the free electrons. When the photon's energy is far greater than the binding energy of the electrons, the

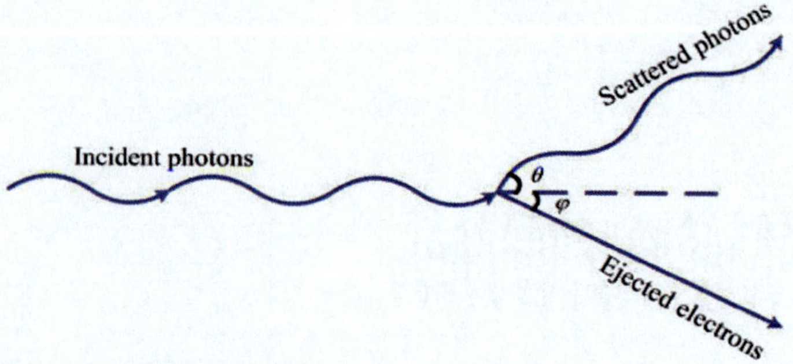

Fig. 2.7 Schematic diagram of Compton scattering effect

outer electrons inside the atoms may be regarded as the free electrons. And thereby, the scattering process can also be considered as an elastic collision.

According to law of the conversation of energy and momentum, the scattered photon energy E_p and the ejected electron energy E_e can be expressed, respectively, as

$$E_p = \frac{E_{p_0}}{1 + \alpha(1 - \cos\theta)}, \tag{2.5}$$

$$E_e = \frac{\alpha(1 - \cos\theta)E_{p_0}}{1 + \alpha(1 - \cos\theta)}, \tag{2.6}$$

where $\alpha = \frac{E_{p_0}}{m_0 c^2}$, E_{p_0} is the energy of the incident X-ray photons, θ is an angle between the incident photon direction and the scattering photon one, m_0 is the rest mass of electron and c is the velocity of light.

From formula (2.5) and (2.6), it is known that when $\theta = \pi$, the ejected electrons will get the greatest energy $\left(E_e^*\right)$. Supposed that the incident photons happen one Compton scattering, and all energies of the scattering photons are also lost in the matter, the continuous spectrum of the Compton scattering can be observed from 0 to E_e^*, and there is a greater cross section at the location corresponding to E_e^*. This cross section is also called **Compton edge**, a feature of the spectrograph that results from the Compton scattering in the scintillator or detector. If the material of X-ray detector is so thick that all energies of most photons are lost by multiple Compton scatterings, a full energy absorption peak will also be shown in energy spectrum. The cross section of the Compton scattering effect is far greater than that of the photoelectric effect, and thereby it has been an important research topic in data processing of X-ray astronomy and high-energy physics how to improve full energy peak detection efficiency or how to get the incident spectrum from the observed complex waves.

2.6.3 Electron–Positron Pair Effect

The so-called positron is the antiparticle of the electron. Generally, the electron–positron pair production is a phenomenon of nature where energy is directly converted to matter. The phenomenon of pair production can be viewed in two different ways: one way is as a particle and antiparticle, and another is as a particle and hole, in which the first way can be represented by formation of electron and positron, from a packet of electromagnetic energy traveling through matter.

When the mass of the X-ray photon incoming matter is greater than two times that of the rest electron, a pair of positive and negative electrons will be produced by the disappearance of the incident photon in a nucleon Coulomb field. This phenomenon is referred to as **electron–positron pair effect**. Typically, the threshold energy of the pair effect is 1.022 MeV. With the incident photon energy increasing, the probability generating the positive and negative electron pairs will gradually increase. According to law of the conversation of energy and momentum, the positive and negative electrons commonly share the energy of the incident photons, and certainly the recoil nuclei also share a part of the energy, but their momentums are very small. When a positron comes to rest, it interacts with another electron, resulting in the annihilation of the both particles and the complete conversion of their rest mass back to pure energy in the form of two oppositely directed 511 keV gamma-ray photons. That is to say, the positive and negative electrons are annihilated to produce two gamma-ray photons with the energy of 511 keV. Therefore, the electron–positron pair effects are mostly used to detect the photons from the higher energy spectrum bands, like hard X-rays and gamma rays. The probability of pair production, characterized by the cross section, is a very complicated function based on quantum mechanics. In general, the cross section increases approximately with the square of the atomic number and with the photon's energy, but this dependence is much more complex.

2.7 Classification of X-ray Detectors

In application of the XPNAV, the X-ray detectors are designed and made on the basis of the interaction ways between the X-ray photons and the detecting materials. Two interaction ways, the photoelectric effect and the Compton scattering effect, are mostly utilized in the detector design for the XPNAV. By measuring the secondary electrons produced in the interaction of the X-ray photons with the detecting materials, the incident photons can be detected indirectly.

2.7.1 *Typical X-ray Detectors*

X-ray detector system is usually composed of three parts: incident X-ray preprocessing, detector's body and readout electronics. According to the different properties of the materials applied to making the X-ray detector's bodies, the X-ray detectors can generally be divided into six basic types: gas proportional counters, Micro-Channel Plate (MCP) detectors, semiconductor detectors, Charge-Coupled Device (CCD) detectors, scintillation detectors and calorimeters.

2.7.1.1 Gas Proportional Counter

The X-ray detector developed by taking gas as the main material to absorb photons and then to collect the ionizing charges is called **gas detector**. When the X-rays go through gas, the gas will be ionized to produce positive ions and free electrons. If there is an electric field in the gas chamber, the ions and electrons will move in different directions and thus bring up a detectable current. A windowed cylindrical chamber is filled with an inert gas; a positively-charged thin anode wire is installed along the center line of the chamber and a high-voltage electric field is also added; when the incoming X-ray photons interact with the gas matter, the gas molecules are ionized to produce positive ions and free electrons; since an incoming photon creates a number of such ion pairs proportional to its energy, and the number of the incoming photons can be counted according to their energy loss. And thereby, this type of gas detector is usually called **gas proportional counter**. A schematic diagram of the gas proportional counter is shown in Fig. 2.8.

In Fig. 2.8, when the X-ray photons enter into the cylindrical chamber filled with xenon gas from the incoming window, the photons interact with the xenon gas and the gas is ionized to produce the photoelectrons or the ejected electrons. The number of the initial ion pairs that are produced is proportional to the energy loss of the incoming photons. With the initial electrons moving toward the anode under the high-voltage electric fields from hundreds to thousands of volts, the gas molecules

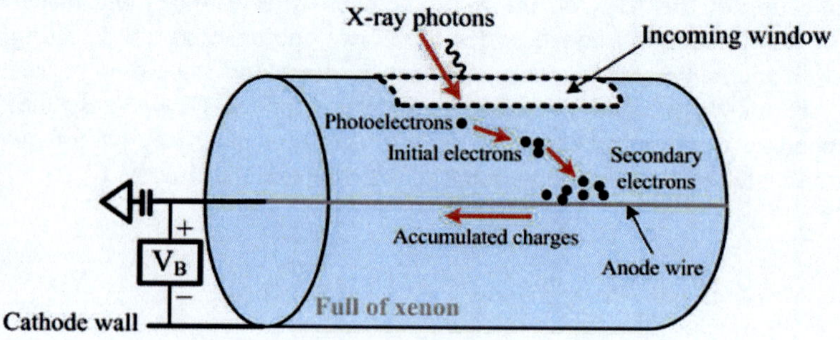

Fig. 2.8 Schematic diagram of gas proportional counter

will be ionized once again. The closer are the electrons to the anode, the stronger the electric fields and the greater the probability of ionization. The secondary ionizations will also produce more secondary electrons and positive electrons. The process of ionizing proliferation is called **electron avalanche effect**. The amplification of the electron avalanche effect is equal to the number of the charges collected in the anode, and then an electric voltage pulse signal is produced. Meanwhile, the amplitude of the pulse signal is proportional to the energy loss of the incoming photons. Generally, collimators may be added in front of the detector window to reject X-ray background photons.

The gas proportional counters are mainly applied to detecting X-rays with an energy range from 1 to 50 keV. Using this type of detector, the photon arrival-time accuracy may usually get to microsecond level, limited by the positive ion and anode–cathode spacing. However, since the gas easily loses or leaks in practical application, the detector's performance declines gradually, its lifetime shortens and its reliability is low relatively.

2.7.1.2 MCP Detectors

The MCP is a thin plate made from highly resistive material of typical thickness 2 mm with a regular array of micro-channels leading from one face to the opposite, densely distributed over the whole surface. It can be used to detect single-particle (electron, ion and neutron) and low-intensity impinging radiation (ultraviolet radiation and X-rays). The micro-channels are tiny tubes or slots, typically less than 25 μm in diameter, and spaced apart from each other by approximately 15 μm. The micro-channels are parallel to each other and often enter the plate at a small angle with its normal direction, about 8 degrees.

Generally, materials to excite photoelectrons are coated on the surface of the MCP and electrodes also installed on the upper and lower surfaces. There is a high-voltage electric field of about 2 kV between the upper and lower surface electrodes. By applying the strong electric field across the MCP, each individual micro-channel becomes a continuous-dynode electron multiplier. When an incident X-ray photon impacts on the surface of the MCP to produce one or two photoelectrons and enters into one of near micro-channels through a small orifice, the photoelectron is guaranteed to hit the wall of the channel due to the channel being at an angle to the plate. The impact starts a cascade of electrons that propagates through the channels, and thus a single impinging is turned into a cloud of electrons to amplify the original signal by several orders of magnitude. And finally, the electrons eject out on the opposite side of the plate where they are collected on an anode. Some anodes are designed to allow spatially resolved ion collection, producing an image of the photons incident on the plate. Although in many cases the collecting anode functions as the detecting element, the MCP itself can also be used as a detector. Obviously, the mechanism of the MCP detector is a photoelectric multiplier effect. A schematic diagram of the MCP detector is shown in Fig. 2.9. After the cascade of electrons, the micro-channel usually takes time to recover before it can detect another signal.

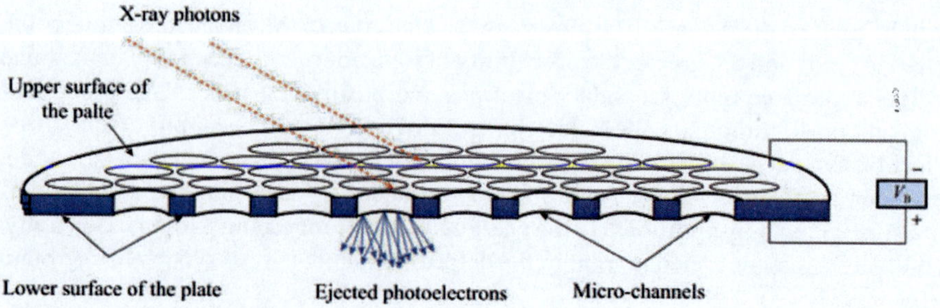

Fig. 2.9 Schematic diagram of MCP

This is the recharging and discharging time of the MCP, which can be decoupled from the high voltage applied to the plate and measured to directly produce a signal corresponding to a single photon.

The gain of a MCP detector is relied on the electric field strength and the ratio of the length relative to diameter of the micro-channels, in which the greater is the ratio, the higher the gain. A typical range of the ratio is 75:1–175:1. In order to obtain higher gain, most modern MCP detectors consist of two micro-channel plates with the angled channels rotated 180 degrees from each other, producing a chevron (v-like) shape. In a chevron MCP, the electrons that exit the first plate start the cascade in the next plate. The angle between the channels reduces ion feedback in the device, as well as producing significantly more gain at a given voltage compared to a straight channel MCP. The two MCPs can either be pressed together to preserve to spatial resolution or have a small gap between them to spread the charge across multiple channels which further increases the gain. By using the cascade of multiplier, the gain of more than two MCPs can usually get to 10^6–10^8. Since the micro-channels have better shielding effect on electromagnetic fields and can image without distortion, the MCP detectors have usually a better spatial resolution. It should be noted that the gain of the MCP is very noisy, meaning that two identical particles detected in succession will often produce wildly different signal magnitudes. The temporal jitter resulting from the peak height variation can be removed using a constant fraction discriminator. Employed in this way, the MCPs are capable of measuring photon arrival-time with very high resolution.

The MCP detectors can work at room temperature and detect the energy range from 1 to 5 keV. Moreover, the photon arrival-time accuracy using the MCP detectors can get to nanosecond level, and the imaging spatial resolution is also very high. However, the MCP detectors are generally complex to manufacture and also require very low pressure or vacuum to be effective. The detectors have no energy resolution and their detection efficiency is also very low. It is difficult for the MCP detectors to directly restrain background noise.

2.7.1.3 Semiconductor Detectors

In ionizing radiation detection physics, a **semiconductor detector** is a device that uses a semiconductor as main detection material to measure the effect of incident charged particles or photons based on p–n junction principle of the semiconductor. The detector types mainly include silicon detectors, Silicon Drift Detector (SDD), diamond detectors, germanium detectors, as well as cadmium telluride and cadmium zinc telluride detectors, in which a schematic diagram of the silicon detectors is illustrated in Fig. 2.10.

In general, the silicon detectors work by doping narrow strip of silicon to turn them into diodes, which are then reverse biased. As charged particles pass through the strips, they cause small ionization currents that can be detected and measured. A thin layer of pure gold is coasted on a piece of n-type mono-crystalline silicon. As atmospheric oxygen passes through the gold and seeps gradually into the n-type silicon, a thin layer of p-type silicon will be brought up and p–n junction is also produced on the opposite surface of the mono-crystalline silicon. In the interface of the p–n junction, positive and negative charge carriers will be recombined to grow a thin layer of high impedance without charge carriers, called **depleted layer**. And thus, an electric field can be produced through the junction from the n to p. The depleted layer is a sensitive region to detect X-ray photons. In order to increase the thickness of the depleted layer, a reverse bias voltage is put by the electrodes installed on two sides of the p–n junction. The higher is the reverse bias voltage, the thicker is the depleted layer and the higher the efficiency to detect X-ray photons. When a semiconductor is irradiated by X-ray photons, free electrons and holes will be produced. The number of electron–hole pairs is proportional to the energy of the radiation to the semiconductor. Under the influence of the biased electric field, electrons and holes, respectively, travel to the electrodes, where they result in a pulse that can be measured in an outer circuit. The holes travel in the opposite direction

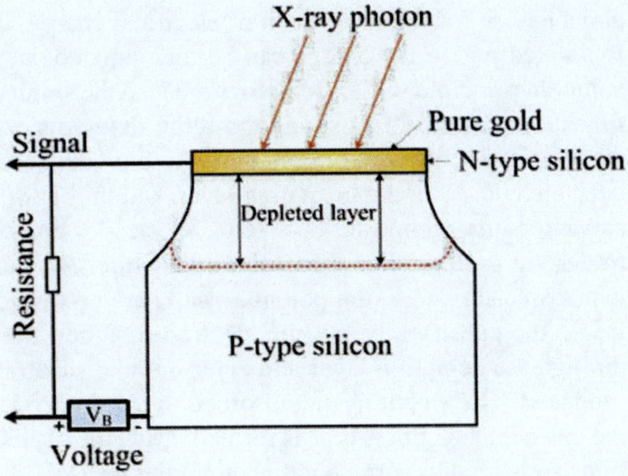

Fig. 2.10 Schematic diagram of semiconductor detector

and can also be measured. As the amount of energy required to create an electron–hole pair is known and independent of the energy of the incident photons, measuring the number of electron–hole pairs allows the intensity of the incident photons to be determined.

The SDD is fabricated from high-purity silicon with a large-area contact on the entrance side facing the incoming X-rays, and consists of a large volume of n-type silicon bulk depleted from both sides: a homogeneous, shallow p–n junction on the side where the incoming X-ray photons enter into the detector and a structure of circular p+ drift rings on the opposite side. By applying a negative voltage on the homogenous back side (X-rays incoming window) and an increasingly negative voltage on the drift rings, a potential field distribution is created inside the detector such that the electrons generated by the absorption of ionizing radiation drift toward a small-sized collecting anode located at the center of the device.

For the germanium detectors, as the germanium as main detecting material has very high electrical resistivity, the thickness of their depleted layers can generally get to several centimeters. Therefore, the detectors are also called high-purity germanium detectors. However, the high-purity germanium detectors require a low temperature to work normally, and to reduce the background noise of the detectors.

Obviously, the semiconductor detectors can directly detect X-ray photons by collecting the electron–hole pairs which result from ionizing radiation. An average energy to grow an electron–hole pair is 3.5 eV for the silicon, and 2.94 eV for the germanium. The detecting energy range is usually 2–100 keV for the semiconductor detectors, and their detection efficiencies and energy resolutions are higher. The photon arrival-time accuracy using the semiconductor detectors can often get to microsecond level.

2.7.1.4 CCD Detectors

The CCD is an integrated circuit containing an array of linked or coupled capacitors, and a device for the movement of electrical charge, usually from within the device to an area where the charge can be manipulated, such as conversion into a digital value. In principle, **CCD detectors** are also the semiconductor ones, and thus sometimes also called **CCD semiconductor detectors**. An insulating layer of SiO_2 is made on the substrate of doping p-type semiconductor. Grooves with about 0.1 μm deep are etched on the insulating layer, and thus a lot of metal oxide semiconductor capacitors are composed of a set of electrodes and the substrate. Each three electrodes are grouped into three-phase structure. As a corresponding positive voltage is put on each phase, the potential wells are produced by inducting depletion layer inside the substrate below the electrodes. When the incident X-ray photons pass through the depletion layer and enter into the substrate, electron–hole pairs will be produced. The electrons are absorbed to the electrodes under the electric fields and the inversion layer of n-type is formed inside the depletion region. And then, the electrons are stored in corresponding potential wells while the holes reversely enter into the ground to disappear. During whole exposure of the CCD, these photo-induced

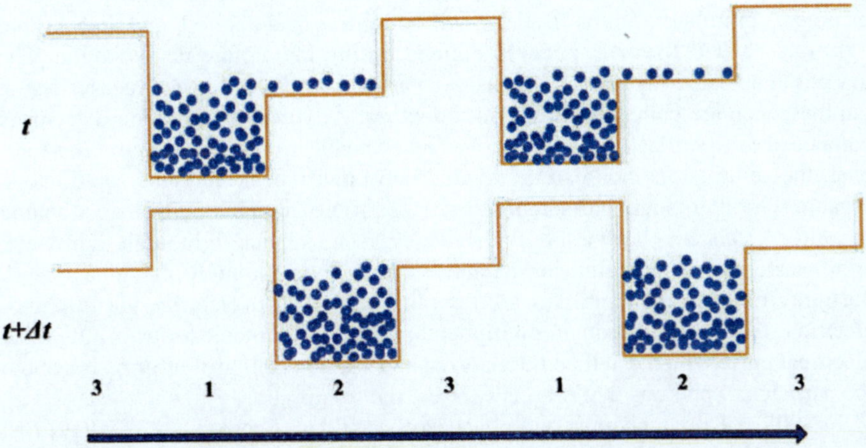

Fig. 2.11 Transfer of charge packet of CCD detectors

charges have always been accumulated in the potential wells, and their number is proportional to that of the incident photons. The CCD is covered with rows and rows of electrodes, forming the potential wells and two-dimensional photosensitive array. Ending the exposure, the charge images corresponding to the incident photon ones will be preserved in potential well array. The electrons (blue) are collected in potential wells (yellow) created by applying positive voltage at the gate electrodes (G), as shown in Fig. 2.11, where the positive voltage to the gate electrode in the correct sequence is used to transfer the electrons.

The charge packets (electrons) of all CCD pixels are recorded by using an amplifier in a time sequence, similar to shift registers in logical circuits. However, these output signals change with strength of the charge packets corresponding to the image location, and rather than 0 or 1. The work temperature of CCD semiconductor detectors is usually 170 K or so, the detecting energy range is 0.3–10 keV and the detecting efficiency gets to 90%. The CCD detectors have low self-noise, the imaging position resolutions using the detectors can get to 30 μm and the photon arrival-time accuracy to microsecond level. Using the pixel array configuration, the detectors provide a high-quality imaging capability. The array of pixels allows good two-dimensional positioning for the arriving photons. The timing accuracy is limited by the readout electronics and the deep depletion ways. The limitations of the CCD detectors are: a pixel is only read each time; deep depletion is required to clear the remaining energy; some devices may require backlit illumination.

2.7.1.5 Scintillation Detectors

A scintillator is a material similar to crystal that exhibits scintillation, the property of luminescence, when excited by ionizing radiation. As charged particles or photons income a scintillator, atoms inside it will be ionized and excited, and then visible light (also called **luminescence**) is emitted during the atomic de-excitation. The devices to detect X-ray photons or charged particles by collecting and measuring the luminescence are called **scintillation detectors**. A typical scintillation detector is composed of scintillator, photomultiplier tube and electronic readout circuit. Moreover, the scintillators can also be divided into three types: inorganic scintillators, organic scintillators and gas scintillators. Most of the inorganic scintillators belong to solid crystals, and have a lot of advantages, such as quick light, high light yield, appropriate wavelength, high density and low cost. The photomultiplier tubes absorb the light emitted by the scintillator and re-emit it in the form of electrons via the photoelectric effect. The subsequent multiplication of those photoelectrons results in an electrical pulse which can then be analyzed and yield meaningful information about the particles or photons that originally strike the scintillator.

In 1903, a British chemist and physicist Sir William Crookes (1832–1919) first built a scintillator in nuclear physics experiments. By a microscope in a darkened room, the naked eyes could see the luminescence in which alpha particles struck a ZnS screen. In 1944, when the photomultiplier tube to replace the naked eyes was used to measure the luminescence, the scintillators gained additional attention and thus the modern scintillation detector birthed. Currently, besides the photomultiplier tubes, the scintillators can still be coupled to other electronic light sensors like photodiode or silicon photomultiplier to build the scintillation detectors. Vacuum photodiodes are similar but do not amplify the signal while silicon photodiodes detect incoming photons by the excitation of charge carries directly in the silicon. Silicon photomultipliers consist of an array of photodiodes which are reverse biased with sufficient voltage to operate in avalanche mode, enabling each pixel of the array to be sensitive to single photons. The scintillation detectors have many advantages, like quick response time, flexible structure design, as well as high detection efficiency, sensitivity and energy resolution, but can only be used to detect the hard X-ray with the energy range of 20–200 keV.

2.7.1.6 Calorimeters

The calorimeters are composed of super-cooled solid matter. As the incoming X-ray photons enter into the absorbent solid material, the temperature pulse induced in the material is measured. The material must be kept near 0 K. The amount of temperature rise is proportional to the energy of the incoming photons, so the photons can be detected by measuring the temperature pulse. The types of devices mainly utilize the thermal effects of X-rays, in which most of the energy of the incoming X-ray photons absorbed by the material is transformed into such thermal energy so that its temperature increases. Typically, a calorimeter has three core components, namely, absorber, calorimetric resistance and heat conductor. The absorber is used to absorb the thermal energy resulting from the incoming X-ray photons; the calorimetric resistance is used to measure rising temperature, which is dependent on the energy of the incoming photons; the heat conductor is used to transfer out the amount of heat in order to keep the absorber cooling. When a calorimetric resistance of semiconductor is approximate to superconductivity and its temperature rises, its resistance value will quickly increase. Therefore, the energy of the incoming photons can be obtained by measuring the resistance value.

The calorimeters can detect a single photon. Due to the required cryogenic cooling, the detector power usage is high and must utilize a significant amount of supporting electronics and hardware. The absorber is typically very small, with an order of 1 mm^3, and thus the detection area is small. For these types of detectors, a multi-layer nested optical system is usually required to increase the collection area. The calorimeters have higher energy resolution, and their photon arrival-time accuracies are usually better than nanosecond level.

2.7.2 Positioning and Imaging Technologies of X-rays

It is well known that the larger is the effective detection area of a detector, the more the number of X-ray photons collected during a unit time and the higher its sensitivity also. However, there is not the capability positioning and imaging the X-ray photons for the detector's major body. According to the different incident X-ray preprocessing structures and imaging principles, the X-ray imaging detector systems can usually be classified into two categories: one is collimated aperture imaging detector, applied to soft X-ray imaging and another is modulated imaging detector, applied to hard X-ray imaging. For the collimated aperture imaging detectors, the concept of aperture has been extended. The aperture is a general term of devices, components or methods which made the X-ray detectors to have positioning and imaging capability. The collimators with a shape of grid or honeycomb, made by using a sheet of metal, are usually installed on the front end of the incident surface of the detectors. And thus, only X-rays within certain angle range in the collimators can enter into the detectors, forming the collimator's apertures, also called **field-of-view angles**. Furthermore, according to different aperture design functions, the aperture imaging detectors are subdivided into two basic types, collimated detector and focused detector. The former has simple design and can effectively control incoming background photons outside the aperture angle; the latter can not only control the incoming background photons, but also focus on incoming effective X-ray photons to increase detecting area and enhance the sensitivity and the space resolution.

The basic principle of the modulated imaging detectors is: the strength of X-ray source within sky coverage is modulated by designing the corresponding collimators and detectors while isotropic background counters within the field of view will not be modulated and are regarded as white noise; the modulated X-ray images can be got by the inversion calculation of the sky coverage. By using the modulated imaging technologies, it is easy to get wider field of view, accurately locate X-ray sources and simultaneously measure background and X-ray sources in the same direction.

Conventional optics-telescope designs require reflection or refraction in a manner that does not work well for X-rays. Visible light optical systems use either lenses or mirrors aligned for nearly normal incidence. That is, the light waves travel nearly perpendicular to the reflecting of refracting surface. The lenses for visible light are made of transparent materials with an index of refraction substantially different from 1. However, X-rays are different with visible light, and have many properties, like shorter wavelength, higher frequency, higher energy and stronger penetrability. As X-rays pass through all known transparent materials, their refractivity is almost equal to 1. Striking mirror surfaces nearly perpendicular, X-rays are either transmitted or absorbed, but not reflected. Therefore, it is very difficult to focus X-rays by using conventional optical focus techniques. Alternatively, considering the property of soft X-rays, they can be focused by using some special focus techniques, such as Wolter grazing incidence, reflection grating and micro-pore optics. Moreover, there are many modulated imaging techniques, such as coded-aperture mask, rotated slab and scanning observation. The Wolter grazing-incidence focus technique and the coded-aperture mask one will be introduced briefly in this section while other techniques can be seen further in the related reference [5].

In 1952, a German physicist Hans Wolter (1911–1978) designed an aplanatic system of grazing-incidence mirrors that satisfied the Abbe sine condition, also known as **Wolter telescope**. In geometric optics, the incident angle is an angle between the incident beam on a surface and the line perpendicular to the surface at the point of incidence. This line is usually called the normal. Sometimes, dealing with a beam that is nearly parallel to a surface, it is more useful to refer to the angle between the beam and the surface, rather than that between the beam and the surface's normal. That is to say, the right angle 90 degrees minus the angle of incidence. This small angle is called a glancing angle or grazing angle, and the incoming way at grazing angle is called **grazing incidence**. If soft X-ray photons glance through polishing metal surface at a smaller grazing angle (typical values between 10 arc-minutes and 2 degrees), most of the photons will be reflected to the focal points of detectors. Wolter showed that his telescopes could be made using a combination of a parabolic mirror with either a hyperboloid or ellipsoid secondary. According to different combination types and focus locations, Wolter telescopes were divided into three types: Wolter–I, Wolter–II and Wolter–III. Principle of Wolter's grazing-incidence optics systems is illustrated in Fig. 2.12. Wolter's key innovation was that by using two mirrors, it is possible to create a telescope with a usably wide field of view. In contrast, a grazing-incidence telescope with just one parabolic mirror could

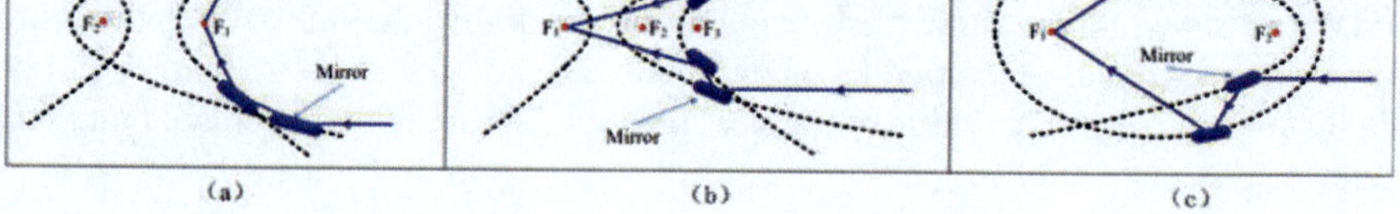

Fig. 2.12 Optical principles of typical Wolter telescopes, where **a** is Wolter–I, **b** is Wolter–II, and **c** is Wolter–III

focus X-rays, but only very close to the center of the field of view as it would suffer from extreme coma.

For X-ray astronomy observations, Wolter–I telescope is an annular sector mirror composed of parabolic and hyperbolic mirror combination, and its incident aperture is a thin ring; only X-ray photons that firstly come to the parabolic mirror and then the hyperbolic one can be focused on the detectors located at focal point. Hyperbolic mirrors with same focal length and different radius are nested by multiple layers to increase effective collection area of telescope. Furthermore, the sensitive detectors are installed on the focal point and the telescope can image for X-ray sources within a sky area of $1^\circ \times 1^\circ$. The space resolution of the telescope is usually relied on the mirrors' focusing capability and the detectors' positioning accuracy.

Coded-aperture masks are grids, gratings or other patterns of materials opaque to the various wavelengths of electromagnetic radiations, which are usually high-energy radiations such as X-rays and gamma rays. As mentioned earlier in the section, it is impractical that imaging is made at the X-ray and gamma-ray bands by lenses or mirrors. Modulation imaging by aperture is thereby often used instead. The pinhole camera is the most basic form of such a modulation imager, but its disadvantage is low throughput, as its small aperture allows through little radiation. Only a tiny fraction of the light passes through the pinhole, which causes the low Signal-to-Noise Ratio (SNR). To solve this problem, the mask can contain many holes, in one of several particular patterns, for example. Multiple masks, at varying distances from a detector, add flexibility to this tool. Specifically, the modulation collimator was used to identify the first cosmic X-ray source and thus to launch the new field of X-ray astronomy in 1965.

Obviously, the coded-aperture mask imaging technique originated from the pinhole camera. A mask with the pinholes is mounted ahead position-sensitive detector and incident X-rays pass through the pinholes. And thus, the X-rays within sky area will be cast upon the detector by the pinholes and an image or shadow of X-ray intensity is obtained. The imaging resolution depends on the pinhole size and the position resolution of the detector. The blocking and penetrating radiation of a coded-aperture mask can be expressed, respectively, as 1 and 0. As X-rays from a direction come to a coded-aperture mask, the shadow of the mask will be cast upon the position-sensitive detector. The shadow's position changes with the incoming direction. As the X-rays come from various directions, these shadows will be overlapped on the detector. If the autocorrelation value of the mask's modulation function is designed as 1, the intensity distribution of the X-ray sources within the sky area can be obtained by the correlation analysis. In other words, the properties of the original radiation sources can then be mathematically reconstructed from the shadows. There are three key problems for the coded-aperture mask: how to increase collecting area (or duty cycle), how to design the coded mask with autocorrelation and how to develop highly position-sensitive detector. The coded-aperture imaging technology is particularly used to detect hard X-rays and the higher imaging resolution can usually be achieved by the technology.

2.8 Development of X-ray Astronomy

2.8.1 Opening X-ray Astronomical Observations

During more than 50 years after X-rays discovered by Wilhelm Röntgen, X-rays were widely applied in industrial manufacture, medical diagnosis and scientific research. However, nobody knew anything about the X-rays from space. The X-rays can only be able to be detected in outer space as they cannot pass through the Earth's atmosphere. And thus, progress on cosmic X-ray research is very slow. American astronomers first detected the X-ray radiations from the Sun by sounding rockets until 1949. By utilizing three Geiger counters carried with the sounding rocket, an Italian-American astrophysicist Riccardo Giacconi (1931–2018) was leading a research team to detect lunar fluorescence X-rays which resulted from the Sun in 1962, but they unexpectedly found a strong X-ray source from Scorpius (Sco X-1), and cosmic X-ray diffusion background. And thus, the prelude for human beings to investigate X-ray astronomy was opened.

The cosmic X-ray radiations mainly come from four aspects: (1) the electron transitions of atoms in the rarefied gas whose temperature is more than an order of 10^5 K, like the coronae and accretion disks; (2) the cyclotron radiations of non-relativistic electrons in the extremely strong magnetic fields, like the accretion columns of neutron stars; (3) the synchrotron radiations of relativistic electrons in the extremely strong magnetic fields, like the Crab nebula; (4) the blackbody radiations in the dense gas whose temperature gets to an order of 10^6 K, like the neutron stars' surface and accretion disks.

For detecting the high-energy radiations from celestial bodies, there are unique advantage and important significance by using X-rays. On the one hand, there are the extensive ground-state neutron atoms in interstellar space, whose energies are more than 13.6 eV (ultraviolet frequency band); except Lyman series, these neutron atoms have continuous absorption effects, which gradually weaken with increase in frequency; the interstellar gas widely absorbs the ultraviolet rays whose energies are higher than 13.6 eV, but are very transparent for the X-rays whose energies are higher than 0.1 keV; thereby, X-rays can pass through interstellar space and bring information on the Milky Way and extra-galactic objects. On the other hand, for cosmic high-energy radiations, the number of photons gradually decreases with radiation energy increasing; therefore, a large amount of information from cosmic objects mainly concentrates on observable low-energy photons, which are the X-ray frequency bands in electromagnetic spectrum.

In earlier cosmic X-ray detecting activities, besides using the sounding rockets, hot air balloons are also utilized to carry large-scale detection equipment into the upper atmosphere in order to detect the X-rays. Although the flight altitude of the balloons is usually lower than that of the sounding rockets, the resident space time of the balloons is so long that can detect higher energy photons that pass through the upper atmosphere.

In the early 1960s, NASA launched the Orbiting Solar Observatory (OSO) Program, the name of a series of American space telescopes primarily intended to study the Sun, though they also included important non-solar experiments. There were eight satellites (OSO 1–8) for the program, launched successfully into the Low Earth Orbit (LEO) between 1962 and 1975. In March 1967, the OSO-3 was launched into nearly circular orbit with a mean altitude of 550 km, inclined at 33 degrees to the equatorial plane. The OSO-3 extensively measured the hard X-ray from solar flares and the cosmic diffuse X-ray background radiation from 10 to 40 keV, and particularly detected firstly an extra-solar X-ray source by an observatory satellite, Scorpius X-1. Thereafter, the OSO-5 launched in January 1969 measured the diffuse background X-ray radiation from 14 to 200 keV; the OSO-6 launched in August 1969 observed three instances of hard X-ray coincidences with gamma-ray bursts; the OSO-7 launched in September 1971 observed solar flares in the gamma-ray spectrum and collected data allowed for identification of Vela X-1 as a HMXB; the OSO-8 launched in June 1975 discovered an iron emission line in the X-ray spectrum of a galaxy cluster.

In addition, the first specific satellite for the purpose of X-ray astronomy, "Uhuru", was launched into an initial orbit of about 560 km apogee, 520 km perigee and 3 degrees inclination on December 12, 1970. The satellite was named in recognition of the hospitality of Kenya where it was launched by the Italian San Marco launch platform near Mombasa, and operated by the NASA. The payload of Uhuru consisted of two sets of proportional counters, each with about 840 cm^2 effective area. The two sets of counters were placed back to back and collimated to FOV $0.52° \times 0.52°$ and $5.2° \times 5.2°$, respectively, which were sensitive with more than 10% efficiency to X-ray photons within 2–20 keV range. Pulse-height analysis in eight energy channels was used to obtain information on the energy spectrum of the incident photons. The Uhuru performed the first comprehensive survey of the entire sky for X-ray sources, with a sensitivity of about 0.001 times the intensity of the Crab nebula. It is a scanning mission with a spin period of 12 min. The outstanding scientific advances achieved by the Uhuru are: to discover and study the pulsing accretion-powered binary X-ray sources, like Centaurus X-3, Vela X-1 and Hercules X-1; to identify Cygnus X-1, the first strong candidate for an astrophysical black hole; to observe many important extra-galactic sources. There were four successive versions of the Uhuru catalogue, in which 4U catalogue, the first comprehensive X-ray catalogue, contains 339 objects and covers the entire sky in 2–6 keV energy band.

Meanwhile, the Orbiting Astronomical Observatory (OAO) was another space observation program launched by NASA in the mid-1960s, and consisted of four satellites, OAO-1, OAO-2, OAO-B and OAO-3. The OAO-3 was a collaborative effort between the NASA and the British Science Research Council (currently known as the Science and Engineering Research Council), and carried an X-ray detector and an ultraviolet telescope. The OAO-3 was launched in 1972 and proved to be the most successful of the OAO program. To mark the 500th anniversary of the birth of Nicolaus Copernicus (1473–1543), the OAO-3 was named Copernicus after launching. The Copernicus operated until 1981 and returned high-resolution spectra of hundreds of stars along with extensive X-ray observations. Among the significant discoveries made by the Copernicus were those of several long-period pulsars, with the rotation periods of several minutes instead of the more typical seconds or less.

The Astronomical Netherlands Satellite (ANS) jointly funded by the Netherlands Institute for Space Research and the NASA, a space-based X-ray and ultraviolet telescope, was launched into the Earth orbit in August 1974. The ANS could measure X-ray photons in 2–30 keV energy band with a 60 cm^2 detector. The positions of the Galactic and extra-galactic X-ray sources were found by the ANS. The ANS also measured their spectra, looked at their variations over time, discovered X-ray bursts and detected X-rays from Capella, which is one of the brightest stars in the X-ray band with an X-ray luminosity about 10,000 times that of the Sun. Moreover, the UK and USA also jointly developed a space observatory dedicated to observing the sky in the X-ray band, a satellite called Ariel-5, launched from San Marco platform in the Indian Ocean in October 1974. The satellite has the all-sky monitor composed of two one-dimensional pinhole cameras, which can scan most of the sky area for every revolution. Its angular resolution is $10° \times 10°$, with an effective area of 3 cm^2, and an energy bandpass of 3–6 keV. The Aerial-5 detected the X-ray pulsars whose rotational periods get to several minutes.

In the late 1970s, NASA started to develop X-ray astronomy research program, the High Energy Astronomy Observatory (HEAO), which include a series of three large low-earth-orbiting satellites for X-ray and gamma-ray astronomy and cosmic-ray investigations. The three satellites were, respectively, denoted by HEAO-1, HEAO-2 (also known as the Einstein Observatory) and HEAO-3 after launching. The satellites were three-axis stabilized to arc-minute accuracy, with fixed solar panels. The HEAO-1 surveyed the sky in the X-ray portion of the electromagnetic spectrum from 0.2 keV to 10 MeV, providing nearly constant monitoring of X-ray sources near the ecliptic poles and more detailed studies of a number of objects by observations lasting 3–6 h. The HEAO-2 carried a single large grazing-incidence focusing X-ray telescope that provided unprecedented levels of sensitivity and arc-second angular resolution of point sources and extended objects. There were four focal plane instruments installed in the HEAO-2, including a high-resolution imaging camera in 0.15–3 keV energy range, an imaging proportional counter in 0.4–4 keV, a solid-state spectrometer in 0.5–4.5 keV and a Bragg focal plane crystal spectrometer. The HEAO-3 included three scientific instruments: one cryogenic high-resolution germanium gamma-ray spectrometer and two devoted to cosmic-ray observations. The normal operating mode of the HEAO-3 was a continuous celestial scan, spinning approximately once every 20 min about the satellite's Z-axis, which was nominally pointed to the Sun. Thereafter, some countries and international organizations successively developed and launched a series of X-ray astronomy satellites, such as Aerial-6, Hakucho, Tenma, European X-ray Observatory Satellite (EXOSAT), Ginga, Kvant/Mir, Granat, Röntgen Satellite (ROSAT) and Spartan, and the activities continued into the 1990s.

The cosmic X-ray detecting activities since 1970 are listed in Table 2.2, where famous X-ray astronomy satellites include ROSAT, Advanced Satellite for Cosmology and Astrophysics (ASCA), Rossi X-ray Timing Explorer (RXTE), Chandra X-ray Observatory and X-ray Multi-Mirror Mission-Newton (XMM-Newton).

Table 2.2 X-ray astronomy satellites launched since 1970

Name	Detection energy range (keV)	Country or international organization	In-orbital period
Uhuru	2–20	USA	1970–1973
HEAO-1	0.2–10000	USA	1977–1979
HEAO-2 (Einstein)	0.2–20	USA	1978–1981
HEAO-3	50–10000	USA	1979–1981
Hakucho (Astro-A)	0.1–100	Japan	1979–1985
Tenma (Astro-B)	0.1–60	Japan	1983–1985
EXOSAT	0.05–50	ESA	1983–1986
Ginga (Astro-C)	1–500	Japan	1987–1991
ROSAT	0.1–2, 0.042–0.21	Germany, USA & UK	1990–1999
ASCA (Astro-D)	0.7–10	Japan & USA	1993–2001
BeppoSAX	0.1–300	Italy & Netherlands	1996–2002
RXTE	2–250	USA	1995–2012
Chandra (AXAF)	0.1–10	USA	1999–
XMM-Newton	0.2–12	ESA	1999–
HETE-2	0.5–400	USA	2000–
Integral	3–10000	ESA	2002–
Swift	0.2–150	USA	2004–
Suzaku（Astro-E2）	0.3–600	Japan & USA	2005–

2.8.2 Röntgen Satellite

The ROSAT is a space-based X-ray telescope led by German Aerospace Center, with instruments jointly built by West Germany, the UK and the US. In Germany, X-ray is called Röntgenstrahlen for honor of Wilhelm Röntgen, and thus ROSAT is short for Röntgen Satellite. On June 1, 1990, the satellite was launched from Cape Canaveral into a nearly circular Earth orbit of 580 km altitude and 53 degrees inclination. ROSAT carried a German-built imaging X-ray telescope, with three focal plane instruments, including two German position-sensitive proportional counters and an US-supplied high-resolution imager. The X-ray mirror assembly was a grazing-incidence fourfold nested Wolter-I telescope with an 84-cm diameter aperture and 240-cm focal length. The angular resolution is less than 5 arc-seconds at half energy width. The X-ray telescope assembly is sensitive to X-rays in 0.1–2 keV energy range. In addition, an extreme ultraviolet telescope supplied by the UK is the wide field camera, co-aligned with the X-ray telescope and covered the energy band from 0.042 to 0.21 keV. The ROSAT was a three-axis stabilized satellite which could be used for pointing observations, for slewing between targets, and for performing scanning observations on great circles perpendicular to the plane of the ecliptic. The unique strengths of ROSAT are with high spatial resolution, low background and soft X-ray imaging for studying the structure of low surface features.

The ROSAT was damaged on September 20, 1998 when a reaction wheel in the satellite's attitude measuring and control system reached its maximum rotational speed, losing control of a slew and damaging the high-resolution imager by exposure of the Sun. Thereby it was difficult to control the satellite under the circumstances outside its initial design parameters. Finally, the ROSAT stopped working due to the failure of the attitude control system on February 12, 1999. After that, the satellite ran aimlessly in space and until October 23, 2011 re-entered the Earth's atmosphere to burn out, over the Bay of Bengal in the East of Indian Ocean. The ROSAT worked normally on the orbit for over 8 years, and obtained a large amount of observation results, including the X-ray all-sky survey catalogue of more than 150,000 objects, extreme ultraviolet all-sky survey catalogue with 479 objects, catalogue of serendipitous sources with approximately 100,000 from the pointing phase, detailed morphology of supernova remnants and galactic clusters, detection of shadowing of diffuse X-ray emission by molecular clouds, detection of pulsations from Geminga, detection of isolated neutron stars, discovery of X-ray emission from comets and observation of X-ray emission from the collision of Comet Shoemaker-Levy with Jupiter.

2.8.3 Advanced Satellite for Cosmology and Astrophysics

The Advanced Satellite for Cosmology and Astrophysics (ASCA), formerly named ASTRO-D, is the fourth cosmic X-ray astronomy satellite by Japan Aerospace Exploration Agency (JAXA), and the second for which the USA provided a part of the scientific payloads. On February 20, 1993, the satellite was launched from Kagoshima Space Center of Japan into an Earth orbit of 523 km perigee, 615 km apogee and 31° inclination. The ASCA operated successfully for over 7 years until its attitude control was lost on July 14, 2000 during a geomagnetic storm, after which no scientific observations were performed, and finally re-entered the Earth's atmosphere to burn out on March 2, 2001.

The ASCA carried four large-area X-ray telescopes, in which at the focal point of two is a gas imaging spectrometer in 0.7–10 keV energy band and at the focal point of the other two is a solid-state imaging spectrometer in 0.4–10 keV. The telescope belongs to Wolter type, with 1.2-m diameter, 3.5-m focal length, and collecting area of 1300 cm^2 at 1 keV and 600 cm^2 at 7 keV. The ASCA is the first X-ray astronomy mission to combine imaging capability with a broad passband, good spectral resolution and a large effective area. It also is the first X-ray astronomy satellite using the CCD. Therefore, the primary scientific objective of the ASCA is the X-ray spectroscopy of astrophysical plasma, especially the analysis of discrete features such as emission lines and absorption edges. The sensitivity of the instruments carried by the ASCA allowed for the first detailed, broadband spectra of distant quasars to be derived. Moreover, the suite of the instruments provided the best opportunity at the time for identifying the sources whose combined emission makes up the cosmic X-ray background.

2.8.4 Rossi X-ray Timing Explorer

The Rossi X-ray Timing Explorer (RXTE) is a satellite that observes the time variation of astronomical X-ray sources, named after Italian experimental physicist Bruno Benedetto Rossi (1905–1993). On December 30, 1995, the satellite was launched from Cape Canaveral into a nearly circular Earth orbit of 283-km altitude and 23-degree inclination, and hereafter operated by the NASA. There were three payloads installed on the satellite, an All-Sky Monitor (ASK), a Proportional Counter Array (PCA) and a High-Energy X-ray Timing Experiment (HEXTE), and their performance indicators are, respectively, listed in Table 2.3. The RXTE's mission is to observe X-rays from black holes, neutron tars, X-ray pulsars and X-ray bursts. The significantly scientific results from the observations of RXTE mainly included: to show an evidence for the existence of the frame-dragging effect predicted by the theory of general relativity; to locate a candidate intermediate-mass black hole, named M82 X-1; to prove that the diffuse background X-ray glow in our galaxy comes from innumerable, previously undetected white dwarfs and from other stars'

Table 2.3 Performance indicators of RXTE's payload

Indicator term	PCA	HEXTE	ASK
Energy range (keV)	2–60	15–250	2–12
Energy resolution	18%@6keV	15%@6keV	–
Time resolution	1μs	8ms	80% of the sky every 90 minutes
Spatial resolution	1°×1° (FWHM)	1°×1° (FWHM)	3′×15′
Sensitivity ($\times10^{-3}$Crab)	0.1	1000	30
Collecting area (cm^2)	6500	2×800	90
Detector	5 proportional counters	2 clusters of 4 NaI/CsI scintillation counters	Xenon proportional counter, position-sensitive

coronae; to infer the size of the smallest known black hole. Until January 3, 2012, the RXTE became invalid and ceased scientific observations.

2.8.5 Chandra X-ray Observatory

The Chandra X-ray Observatory (CXO) was named after an Indian-American astrophysicist Subrahmanyan Chandrasekhar (1910–1995), and previously known as the Advanced X-ray Astrophysics Facility (AXAF). On July 23, 1999, the CXO carried on Space Shuttle *Columbia* was successfully launched by the NASA into a highly elliptical orbit with the eccentricity of 0.744, the perigee of 14,308 km, the apogee of 134,528 km and the inclination of 76.7°. The CXO applied to detecting the soft X-rays from the universe is one of the four Great Observatories operated by the NASA. Other Great Observatories are, respectively, the Hubble Space Telescope (HST) applied to observing in the visible light and ultraviolet, the Compton Gamma Ray Observatory (CGRO) applied to observing the hard X-rays and gamma rays and the Spitzer Space Telescope (SST) applied to observing the infrared. Currently, except for the CGRO becoming invalid, others are ongoing to work normally in the orbits.

The CXO used a Wolter-type telescope with four pairs of nested mirrors, together with their support structure, called the high-resolution mirror assembly; the mirror substrate is 2-cm-thick glass, with the reflecting surface of a 33 nm iridium coating, and the diameters are, respectively, 65 cm, 87 cm, 99 cm and 123 cm. The thick substrate and particularly careful polishing allowed a very precise optical surface, which is responsible for CXO's unmatched resolution: between 80% and 95% of the incoming X-ray energy is focused into one arc-second circle. The telescope has 12 m focal length, 400 cm^2 collecting area and 0.5 arc-seconds resolution. There are two kinds of detectors on the focal plane of the telescope: an advanced CCD imaging spectrometer consisting of 10 CCD chips, and operating in 0.2–10 keV energy range; a high-resolution camera consisting of two micro-channel plate components and imaging over the range of 0.1–10 keV, whose time resolution is 16 microseconds. Moreover, there are also two transmission grating spectrometers, which swing into the optical path behind the mirrors to provide the CXO with high-resolution spectroscopy. The high-energy transmission grating spectrometer works over 0.4–10 keV and has a spectral resolution of 60–1000; the low energy transmission grating spectrometer has a range of 0.09–3 keV and a resolution of 40–2000.

The CXO is specially applied to observing fainter cosmic X-ray sources, and its scientific discoveries include: to give astronomers their first glimpse of the compact object at the center of supernova remnant Cassiopeia A, probably a neutron star or black hole; to show a never-before-seen ring around the central pulsar in the Crab Nebula; to observe the first X-ray emission from the super-massive black hole, Sagittarius A*, at the center of the Milky Way; to find much more cool gas than expected spiraling into the center of the Andromeda Galaxy; to discover a new type of black hole in galaxy M82; to take the earliest image in X-rays of the shock wave of a supernova, SN 1987A; to observe sound waves from violent activity around a super-massive black hole in the Perseus Cluster; to put limits on the cross section of the self-interaction of dark matter by observing the Bullet Cluster. The origin and evolution of the universe can deeply be understood by the mankind from the observations of black holes, supernovae and dark matters.

2.8.6 X-ray Multi-Mirror Mission

The X-ray Multi-Mirror Mission-Newton (XMM-Newton) was named after British physicist and astronomer Sir Isaac Newton, and also known as the high-throughput X-ray spectroscopy mission. On December 10, 1999, XMM-Newton was launched by the European Space Agency (ESA) from Guiana Space Centre into a highly elliptical orbit with the eccentricity of 0.816, the perigee of 5,663 km, the apogee of 112,878 km and the inclination of 67.1°.

The XMM-Newton is a space telescope with 10.8-m long, and with 16.16-m wide as its solar arrays are deployed. The satellite adopts the three-axis stabilized control, which allows it to aim at a target with an accuracy of 0.25–1 arc-seconds. This stabilization is achieved through the use of its attitude and orbit control system, which allow it to point at different celestial targets and can turn it at a maximum of 90 degrees per hour. The telescope assembly consists of three Wolter-I telescopes, in which each is also composed of 58 cylindrical nested mirrors. The diameter of the outermost mirror is 70 cm, and that of the innermost is 30.6 cm and the focal length is 7.5 m. The telescope assembly produces a total collecting area of 4425 cm^2 at 1.5 keV and 1740 cm^2 at 8 keV. Moreover, there are six instruments on the board of the XMM-Newton:

three European photon imaging cameras, two reflection grating spectrometers and one optical monitor. The photon imaging cameras consist of two MOS-CCD cameras and a single p–n CCD camera, with a total FOV of 30 arc-minutes and an energy sensitivity range between 0.15 and 15 keV. The reflection grating spectrometers are a secondary system on the spacecraft and are composed of two focal plane cameras and their associated reflection grating arrays. This system, operating in the energy range from 0.35 to 2.5 keV, is used to build X-ray spectral data, and it can determine the elements presented in the target, as well as the temperature, quantity and other characteristics of those elements. The optical monitor is a 30-cm optical-ultraviolet telescope designed to provide simultaneous observations alongside the spacecraft's X-ray instruments. The monitor is sensitive between 170 and 650 nm in the FOV of $17' \times 17'$, co-aligned with the center of the X-ray telescope's field of view.

In 2006, the XMM-Newton was used to discover the galaxy cluster XMMXCS 2215–1738, 10 billion light year away from Earth, whose mass is about 500 trillion solar masses. This cluster is surprisingly large and evolved for a cluster that existed when the universe was only 3 billion years old. In 2011, a flare that lasted 4 h at a peak intensity of 10,000 times the normal rate was observed by the XMM-Newton, from an observation of Super-giant Fast X-ray Transient IGR J18410-0535, where a blue super-giant star sheds a plume of matter that was partly ingested by a smaller companion neutron star with accompanying X-ray emissions. In 2013, the spin rate of a super-massive black hole at the core of galaxy NGC 1365 was firstly measured by the XMM-Newton along with the NuSTAR, a space-based X-ray telescope launched in 2012 that uses a conical approximation to a Wolter telescope to focus high-energy X-rays from astrophysical sources and operates in the energy band of 3–79 keV. At the same time, it verified the model that explains the distortion of X-rays emitted from a black hole. In 2014, the monochromatic signal around 3.5 keV was extracted from the spectrum of X-ray emissions observed by the XMM-Newton. This signal is coming from different galaxy clusters, and several scenarios of dark matter can justify such a line.

2.9 Observability of X-ray Pulsars

2.9.1 Early X-ray Detections

It is shown from studies that the visible and radio signals are emitted from the Crab Nebula by the synchrotron-radiation mechanism. If there are some higher energy charged particles in the Crab Nebula, it is also possible to emit high-energy X-rays by the synchrotron radiation. And thus, the X-ray radiations from the Crab Nebula should be similar to the visible light, and there is an extended radiation area. Meanwhile, there could be a compact center object in the Crab Nebula, from which X-rays are emitted.

The Crab Nebula was ever observed and investigated by an X-ray astronomical research team of the U.S. Naval Research Laboratory (NRL) in 1963. However, the spatial resolutions of X-ray detectors were so low at that time that it could not be decided by direct observations whether the radiating source was exactly a point or not a surface. And thus, a particular observation of celestial phenomena was sufficiently utilized by the research team. The Moon coincidently passed through the direction of the Crab Nebula in July 1964. Supposed that X-rays come from the extended Crab nebula, they will gradually be shaded by the Moon, and this duration will continue about 12 min; supposed that X-rays come from a central point source, all of them will be shaded instantaneously. It was shown from the observational experiment that the flux of X-rays was only changed slowly during the Moon passing through the Crab Nebula, and there were 90% of X-rays emitted from the extended radiation source for the Crab Nebula. Although it was possible from this conclusion that there was 10% of X-rays emitted from the central point source (in fact, only 4% confirmed later.), nobody focused on it so seriously that the opportunity to discover the neutron stars was lost as easily as that.

There was a neutron star with pulsating frequency of 33 Hz at the center of the Crab Nebula, discovered by radio astronomy observation in 1969. And thus, the point source of X-ray scintillation was searched in the Crab Nebula by using the method of X-ray scintillation detection, and separated from the extended radiation source. With launching the Uhuru in 1970s, two X-ray sources, Centaurus X-3 and Hercules X-1, were discovered successively. Both Centaurus X-3 and Hercules X-1 radiate periodic X-ray pulsating signals, and their periods are 4.84 s and 1.24 s, respectively. They belong to a new type of celestial objects, which have the observational features of periodic eclipses and the eclipsing durations are usually very short. It is shown reasonably that the two X-ray pulsars are located, respectively, in the binary systems with short-period orbits. In other words, the close binary systems are formed by the modes. The X-ray radiations are produced by accreting the materials from the companions in the close binary systems, and the spinning rates of pulsars are increased further to form the millisecond pulsars.

2.9.2 Flux Analysis of X-ray Pulsars

With launching and using a series of X-ray astronomy satellites, such as the Einstein Observatory, ROSAT and so on, a large amount of X-ray sources were discovered successively. Particularly, the millisecond pulsars with periodic X-ray radiations were also discovered. In general, many rotation-powered pulsars like the Crab pulsar are located at the centers of the supernova remnants whose outer layers are the extended radiation sources. In order to get the central point radiation sources, it is required to compare the flux intensities of X-ray pulsars with those of the nebulas around them. Moreover, there is also a large amount of diffuse X-rays in space, whose intensity distributions are not uniform in different energy ranges. Therefore, the observabilities of X-ray pulsars mainly depend on the ratio between the flux intensities of the pulsars

Table 2.4 Luminosities and flux intensities of the typical supernova remnants and their compact objects

Supernovae	Compact objects	Luminosity of supernova remnant $\lg(L_x)$ (erg/s)	Flux intensity of compact object F_x (erg/cm^2/s)	Luminosity of compact object $\lg(L_x)$ (erg/s)
Crab	PSR B0531+21	37.38	2.1×10^{-9}	35.98
SNR0540-69.3	PSR B0540–69	37.02	2.9×10^{-12}	36.38
VelaXYZ	PSR B0833–45	33.77	4.0×10^{-12}	32.91
MSH15-52	PSR B1509–58	35.27	6.5×10^{-14}	34.37
CTB80	PSR B1951+32	34.68	2.0×10^{-12}	33.80

and those of the supernova remnants, as well as the directions of the detected pulsars, the performances of the detectors and the integrated time of signals.

The statistical results of the luminosities and the flux intensities of typical supernova remnants and their compact objects are listed in Table 2.4, from which it is shown that the X-ray flux intensities radiated from the pulsars are far less than those from synchrotron radiations of the nebulae around them. The total of the radiations from the X-ray pulsars are very small in the supernova remnants, and the radiation ratios for the first four pulsars in Table 2.4 are 4%, 25%, 10% and 4%, respectively.

For example, there are about 780 X-ray photons each second received by using a detector with an effective collecting area of 0.5 m^2 for the Crab pulsar, and about 7800 photons from the Crab Nebula. Obviously, the background noise is 10 times more than the X-ray signals from the pulsar. Generally, it is impossible that the pulsating profiles of X-ray pulsars are directly obtained by using total of the photons from one pulsating period. Nevertheless, the background noises have no periodicity, and produce additional counts of the photons randomly and uniformly; the pulsating signals have periodicity, and their pulses will be accumulated to intensify. Therefore, the clearly distinguishable pulse profiles of X-ray pulsars can usually be obtained by period folding and synchronization averaging.

References

1. Backer DC, Kulkarni SR et al (1982) A millisecond pulsar. Nature 300(5893):315–318
2. Nicastro L, Cusumano G et al (2004) BeppoSAX observation of PSR B1937+21. Astron Astrophys 413(3):1065–1072
3. Arzoumanian Z, Van der Hooft F, van den Heuvel EPJ (1999) Pulsar timing, general relativity and the internal structure of neutron stars. Koninklijke Nederlandse Akademie van Wetenschappen, Amsterdam

4. Shannon RM, Cordes JM et al (2013) An asteroid belt interpretation for the timing variations of the millisecond pulsar B1937+21. Astrophys J 766(1):14
5. Wang SG, Zhou YY (1999) X-ray astrophysics. Science Press, Beijing (Chinese)

Chapter 3
A Brief History of Space Flight

Outer space (or simply **space**) is the expanse that exists beyond the Earth and between celestial objects, and it is not completely empty but a hard vacuum containing a low density of particles, predominant plasma of hydrogen and helium, as well as electromagnetic radiation, magnetic fields, neutrinos, dust and cosmic rays. There has not been a clear boundary between the Earth's atmosphere and space while the density of the atmosphere gradually decreases with altitude increasing. Conventionally, an altitude of 100 km above sea level, also called **Kármán line**, was used as the start of outer space in the Outer Space Treaty which entered into force in 1967. The Kármán line was calculated by Theodore von Kármán (1881–1963), a Hungarian-American mathematician, physicist and aerospace engineer because, at an altitude of about 100 km, an aero-craft would have to travel faster than orbital velocity to derive sufficient aerodynamic lift from the atmosphere to support itself. The altitude of 100 km is thereby used as a working definition for the boundary between aeronautics and astronautics. Outer space is the common resource of mankind, and the exploration and utilization of outer space is the unremitting pursuit goal of mankind.

Space flights (or **astronautic flights**) are various kinds of ballistic flight activities of the spacecrafts into or through outer space, including orbiting the Earth, interplanetary space flight, interstellar space flight, flying around celestial objects, flying over celestial objects, landing on celestial objects and returning from celestial objects. The objectives of space flights are to explore, develop and use the resources of outer space and celestial objects, and to benefit mankind on the Earth. Traveling in space and exploring the universe has always been the oldest, simplest and most beautiful dream of human beings. For thousands of years, a long and splendid human civilization has spread many wonderful myths and moving legends about space flights. At the beginning of the twentieth century, the pioneers dedicated to studying modern aerospace science had initially established a basic theoretical system of rockets and space flights, which had been increasingly widely disseminated. The societies to study space flights were successively built in many counties in the world, and they promoted to develop the modern astronautic sciences and technologies. The first artificial Earth satellite, Sputnik-1, was successfully launched by the Soviet Union in

P. Shuai, *Understanding Pulsars and Space Navigations*, Navigation: Science and Technology 11, https://doi.org/10.1007/978-981-16-1067-7_3

October 1957. It is a significant mark of mankind entering into the age of space flight. In the 1970s, Pioneer-10 and Pioneer-11, as well as Voyager-1 and Voyager-2, were successively launched by the United States. They are the first artificial spacecrafts to reach the edge of the solar system and have currently entered into interstellar space. From myth to fantasy, and from theory to practice, human beings have embarked on a long journey of space flight.

3.1 Concepts of Space Flights

In general, the flight activities of the spacecrafts into or through outer space are defined as the space flights, and similar concepts also include interstellar travel and aerospace. The **interstellar travel** is the term used for crewed or un-crewed travels between stars or planetary systems, namely including interplanetary spaceflight and interstellar spaceflight. In practice, the former is usually confined to travels between the planets inside the solar system, and the latter is done to travels between the stars outside the solar system. The interstellar spaceflight will be much more difficult than the interplanetary spaceflight. The distances between the planets in the solar system are less than 30 Astronomical Units (AU), whereas the distances between stars are at least hundreds of thousands of AU, and usually expressed at light-years. Because of the vastness of those distances, interstellar spaceflight would require a high percentage of the velocity of light, and extremely-long travel time, lasting from decades to millennia or longer, or a combination of both. The **aerospace** is also defined as the human efforts in science, engineering and business to fly in the Earth's atmosphere (aeronautics) and surrounding space (astronautics). Sometimes, aerospace especially refers to the general term for the activities of entering, exploring, developing and utilizing outer space and celestial bodies beyond the Earth.

Currently, astronautic activities are usually divided into three types: space technology, space science and space application. The **space technology** is a comprehensive engineering technology that provides technical means and guarantees conditions for various space activities; the **space science** encompasses all of the scientific disciplines to explore and study various phenomena and laws occurring in outer space by using the space technology; the **space application** refers to the utilization of the space technology and its development of space resources, in the national economy, national defense construction, scientific research, cultural education, as well as other fields. the **space resources** refer to various environments, energy and materials that can be exploited and utilized by human beings in outer space, including high-altitude space, high vacuum, ultra-low temperature, strong radiation and microgravity environment, solar energy sources, as well as rare material resources obtained from extraterrestrial celestial bodies. There is a mutual needing and supporting relationship among the space technology, space science and space application, and they are inseparable from each other to form together with a whole of space activities.

In order to carry out various space activities, a large-scale space engineering system needs to be built, which is usually composed of five major systems: transportation system, launch site system, spacecraft system, telemetry, tracking and control (TT&C) system, as well as application system [1]. Besides the five major systems, the manned space engineering system should also particularly be included other two systems, the astronaut system and the recovery and landing system. The spacecraft system is a core segment in the whole space engineering system and accomplishes specific space missions together by cooperating and coordinating together with other systems. According to whether there are crews involved in space missions, the spacecrafts can usually be classified into two categories: unmanned spacecraft and manned spacecraft, as shown in Fig. 3.1, where the former is also be divided into artificial earth satellites and space probes, and the latter into space stations, manned vehicles and space shuttles.

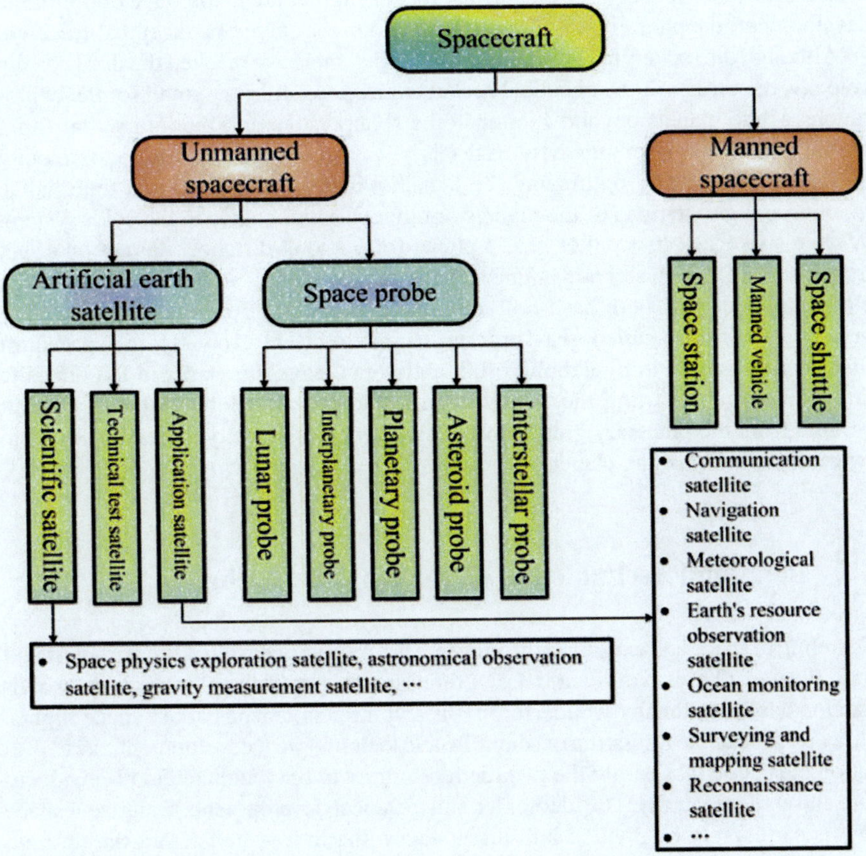

Fig. 3.1 Classification of the spacecrafts

The key of space flight is to make the spacecraft with enough flight speed to overcome and get rid of the constraints of the gravity of the Earth or Sun and fall back to the Earth's surface no longer. It is so-called **cosmic velocity**, a minimum initial velocity which the spacecrafts launched from the Earth's surface orbit the Earth, or escape away from the Earth to fly to interplanetary space, out of the Solar System or into interstellar space. More generally, the space flights require three characteristic velocities, the first cosmic velocity, the second cosmic velocity and the third cosmic velocity, which are respectively 7.9 km/s, 11.2 km/s and 16.7 km/s [2].

At such high speed, what types of engines are used as power sources? The rocket engines are the ideal power sources for space flights. According to different fuels and exhaust materials utilized by them, the rocket engines are generally divided into four types: chemical rocket engines, electric rocket engines, nuclear rocket engines and photon rocket engines. The chemical rocket engines are the most mature and widely used engines at present; the electric rocket engines are mainly applied to orbit keeping for some spacecrafts, the nuclear rocket engines are in the stage of prototype development; the photon rocket engines are still in the stage of conceptual research.

Although the rockets using the chemical propellants have reached the third cosmic velocity, the thrusts of the chemical rocket engines are still very small for the explorations of the planets beyond Jupiter in the solar system and the interstellar flight missions. The **planetary gravity assist** is a way to change the orbit and velocity of spacecraft flight by optimizing the launch windows of rockets and the relative motions and gravitations of the planets or other celestial bodies in the solar system. When a spacecraft firstly flies near a planet from a long distance, its motion effect is like that of the spacecraft bounced back by the planet, which is similar to an elastic collision, but both have not yet come into contact. And thus, the planetary gravity assist is also called **gravitational slingshot effect**. Actually, the spacecraft flies near the planet in hyperbolic orbit, and it can leave the planet in the opposite direction without starting the orbital engine. Moreover, as long as the spacecraft escapes from the planetary gravitation, it can achieve a velocity increment equal to twice the velocity of the planet.

3.2 Ideas and Technologies of Ancient Space Flights

Traveling in space and exploring the universe has always been the oldest, simplest and most wonderful dream of mankind. For thousands of years, the glorious human civilization has spread many wonderful myths and moving legends about space flights. However, due to the limitation of scientific and technological development, the exploration of space flight before the twentieth century was only addicted to blind adventure and endless fantasy. The ideologies and technical developments for ancient space flights can largely be divided into three stages: the first stage is from the thirtieth century B.C. to the sixth century A.D., the myths and legends about space flight as well as its idea sprouting; the second stage is from the seventh century to the sixteenth century, the invention of gunpowder and the development and application of the ancient rocket technology; the third stage is from the seventeenth century to the nineteenth century, the rise and spread of space science fictions.

3.2.1 Flying Myths, Legends and Fantasies

Our ancestors, in the long-long history of recognizing and transforming nature, had been longing for a pair of wings to soar like birds in the blue sky and travel in space, which gradually evolved into countless wonderful myths and moving legends. Among the many flying stories handed down from ancient China, "Chang'e Running to the Moon" is the most familiar and household for Chinese.

The original name of Chang'e was Heng'e, and later she was known as Chang'e to avoid the same name as Heng Liu (203 B.C.–157 B.C.), an emperor in the Western Han Dynasty of China. The taboo of name is a privilege of the imperial family in ancient China, an important cultural phenomenon. According to the long legend, Chang'e was Di Ku's daughter and Da Yi's wife in the Ancient Times of China. Di Ku was the third Sovereign during "Three Emperors and Five Sovereigns" (about from the thirtieth century B.C. to the twenty-first century B.C.). Chang'e is the Chinese goddess of the Moon, her beauty is extraordinary. Da Yi got an ever-young medicine from West Queen Mother, a goddess in charge of immortality, punishing evils and warning against disasters in the ancient Chinese mythology; Chang'e stole this medicine to eat and then flew to the Moon to become immortal; and thus, she lived in the Moon's palace forever. The earliest record of Chang'e's deeds is in *Gui Cang*, the *Book of Changes* in the Shang Dynasty (1600 B.C.–1046 B.C.), in which the hexagram Gui Mei reads: "Formerly, Heng'e (Chang'e) steals the immortal medicine from West Queen Mother, and then eats it for running to the Moon". In the Western Han Dynasty, the Huainan King An-Liu (179 B.C.–122 B.C.), et al., wrote a philosophical work entitled *Huai Nan Zi*, in which the allusion on Chang'e Running to the Moon was quoted: "Yi (Da Yi) got the immortal medicine from West Queen Mother, Chang'e stole it and ran to the Moon; and then, she felt regret, but was unable to redeem her fault". Chang'e's husband, Da Yi, is a divine archer, and the story of "Da Yi Shooting the Sun" is recorded in *the Classic of Mountains and Rivers*, an ancient book in the Pre-Qin Dynasty (before 221 B.C.). The story says: During the reign of Sovereign Yao, there were ten suns in the sky, burning plants and crops to scorch; in order to save common people, Da Yi shot down nine of the suns; from then on, everything grew in a suitable climate. In addition, other myths in the Ancient Times of China were also recorded in *the Classic of Mountains and Rivers,* such as "Kua Fu Pursuing the Sun", "The Goddess Patching the Sky", "Jing Wei Reclamation" and so on.

The legends in the Shang Dynasty are recorded in some ancient books, such as *Emperor's Deeds* and *Mystery Novels*. "Flying with the wind and traveling thousands of miles every day", and various fantastic pictures of flying by clouds and driving by fog, are depicted in these legends. During the Warring States Period, a great patriotic poet, Qu Yuan (340 B.C.–278 B.C.) portrayed in his monumental poem Li Sao that he yearned to ride the carriage made of jade and ivory, which was pulled by the flying dragons, and to travel in space. Meanwhile, he also portrayed in his another poem Yuan You that the wind God opened the way for him as he wandered in clouds. It fully demonstrates the poet's romance and imagination of traveling in space. By the Western Han Dynasty (202 B.C.–9 A.D.), with developments of social productive forces, the people began to believe that as long as equipped with wings, they could fly in the sky like birds. Therefore, the persons with two or four wings were engraved in the remaining stone carvings or bronze mirrors of the Western Han Dynasty.

Most of the flying myths and legends handed down in ancient Europe only tell the flights near the Earth, and do not yet imagine the scenes of flying to other celestial bodies outside the Earth, such as the Moon. In the mythology of ancient Greece and Roman, Apollo was a god in charge of the Sun, music, medicine and youth. According to mythology, Apollo drove the war carriage around the Earth every day and brought light and joy to the Earth. The science of Europe originated from ancient Greece. The academic discussions of the ancient Greeks were relatively free, and they could comment on any problems, such as the origin of heaven and earth, the structure of the universe and so on. And thus, the various academic schools were brought up in succession. It is the scientific spirit of ancient Greece: science is a purely speculative, theoretical and curious investigation [3].

Pythagoras (580 B.C.–500 B.C.), a mathematician and philosopher of ancient Greece, believed that the earth was spherical and that the Earth and other planets were moving together around a central fire. Another ancient Greek philosopher, Aristoteles (384 B.C.–322 B.C.) recognized from the view of aesthetics that the Earth was spherical, but he insisted that the Earth was a center of the universe and that both the Earth and other celestial bodies were composed of different matter, in which the Earth was composed of four elements, including water, gas, fire and soil, and besides the four elements, other celestial bodies still contained the fifth element, "ether". Moreover, an ancient Greek astronomer Aristarchus (315 B.C.–230 B.C.) approximately determined the ratios of the Sun and the Moon relative to the Earth. Meanwhile, he also believed: the Sun was very larger than the Earth; the Sun and other stars were all stationary; the Earth rotated around its rotation axis every day, and moved a circle each year around the Sun in a circular orbit; the Sun located in the center of the orbit. Unfortunately, the idea of "Heliocentric Theory" was not universally recognized and widely spread at that time.

Another ancient Greek astronomer Hipparchus (190 B.C.–120 B.C.) created a system of geocentric theory on the basis of a large number of astronomical observations, and used the deferent and epicycle to explain the apparent motions of celestial bodies. Later, the geocentric theory, further developed by Ptolemy (a Greco-Roman mathematician and astronomer, 90 A.D.–168 A.D.), gradually formed a complete theoretical system that ruled European astronomy for more than 1000 years.

Plutarch (46 A.D.–120 A.D.), an ancient Greek biographer, had ever pointed out that "the Earth should be at the center of the universe, but because the universe was infinite and unbounded, there was no such center", which coincides with Einstein's modern view of the universe. Furthermore, he also believed: the Moon was the second Earth, but less than the Earth; other aspects of the Moon were all similar to the Earth, and there were residents on the Moon. It was just under the guidance of this anti-traditional view that Lucian (120 A.D.–180 A.D.), a sophist and satirist of ancient Greece, wrote two fantasy fictions on traveling to the Moon: One was *The True Story*, in which a series of adventure flying stories of the voyage on the sea were told; the other was *The Man in the Cloud*, in which a philosopher named Monis in ancient Greece equipped with a pair of wings and flew to the Moon in order to prove that the Earth was round. Although Lucian himself admitted that all the stories in fiction were fabricated, many novelists got inspiration and fantasy from them.

The flying myths, legends and fantasies in the ancient times have been deeply seeded in people's hearts, which established an ideological foundation for space flight, and for those who are later full of imagination and creativity, are undoubtedly a power source to explore and discover space science and technology.

3.2.2 Development of Gunpowder and Rocket Technology

Gunpowder is one of the Four Great Inventions of ancient China, and a great contribution of the Chinese nation to the development of human civilization. Justly inspired by the myths and legends like "Chang'e Running to the Moon", "Da Yi Shooting the Sun" and "Kua Fu Pursuing the Sun", immortal magicians believed that they could make a kind of "immortal medicine" and thus developed alchemy. The **alchemy** is a kind of pharmaceutical method of heating and sublimating medicines, the traditional technique of making Chinese medicines in ancient times. The alchemy was founded in the Zhou Dynasty (1046 B.C.–256 B.C.) and flourished from the Qin Dynasty (221 B.C.–207 B.C.) to the Western Han Dynasty (202 B.C.–8 A.D.). Both emperors and nobles were so infatuated with a vision to become immortals forever young that some necromancers and Taoist-priests were driven to refine "elixir of life" at that time, and in doing so, the formulations of gunpowder were gradually invented. By the Sui Dynasty (581–618 A.D.), a form of gunpowder made from saltpeter, sulfur and charcoal dust had been developed very maturely. During the Tang Dynasty (618–907 A.D.), a black gunpowder formally came into being, whose proportion formula of "one saltpeter, two sulfurs and three charcoals" was adopted to greatly improve the efficiency of rapid combustion and explosion, and thus the military fire-arrows filled with the black gunpowder was developed. The gunpowder originated from the formulation of making the elixir and was ever used as a medicine. In *Compendium of Materia Medica* written by Shizhen Li (1518–1593 A.D.) during the Ming Dynasty, gunpowder had been used to treat psoriasis, insecticide, dampness and plague.

In ancient Chinese classics, the word "rocket" has wide meanings, which is different from the modern rocket, a space vehicle that carries its own propellant and uses the principle of jet propulsion to fly. The ancient rockets, as actual weapons applied to military wars, could be divided into three types in accordance with their different design principles and application effects and ranges: classical arson fire-arrows, gunpowder arson fire-arrows and jet fire-arrows. It was reported that the classical arson fire-arrows, which were the incendiary agents made from rosin, oil, sulfur and cured-hay, appeared before the Qin Dynasty. The classical arson fire-arrows, as fire-attack weapons in the ancient wars, were usually attached to arrow poles and launched with bows. In Brief History of Wei from *Annals of the Three Kingdoms* that was written by Shou Chen in the Western Jin Dynasty (265–316 A.D.), the historical facts were recorded: the armed forces led by Liang Zhuge from the Shu State attacked Chen Cang city of the Wei State; Zhao Hao, a general of guarding the city, used the fire-arrows to shoot toward the aerial ladders utilized by the army of the Shu State to climb the wall; as a result, the ladders were burning and all soldiers on them were burned to death. The fire-arrows recorded in this classical book are the classical arson ones. The gunpowder arson fire-arrows, early appeared in the Northern Song Dynasty (960–1127 A.D.), are the gunpowder cartridges made from a mixture of saltpeter, sulfur and charcoal, into which auxiliary agents like oil, dry paint and wax are also added. These fire-arrows were attached to the arrow poles and shoot to targets with the bows, and their arson and extension efficiency was improved more notably than the classical arson fire-arrows. According to the record of Military Annual of *History of the Song Dynasty*, Jisheng Feng, an official of the Ministry of War, presented the gunpowder arson fire-arrows to Kuangyin Zhao (927–976 A.D.), the founding emperor of the Song Dynasty and Zhao very liked the weapons and rewarded clothes and silk bundles as presents to Feng.

The jet fire-arrows, early appeared in the Southern Song Dynasty (1127–1279 A.D.), had been the rockets actually based on the principle of reaction force. The process of making and launching the jet fire-arrow is solid gunpowder containing much saltpeter is loaded into a cardboard paper tube; the upper part of the paper tube is sealed with a thin layer of mud, the inner part is designed as a cavity and the lower part has a small hole; a blasting fuse is inserted into the small hole; the paper tube is attached to the gun or the arrow pole to launch; when the gunpowder is burned through the blasting fuse, the flame or gas flow is ejected from the small hole to produce reaction force; as a result, the fire-arrow can autonomously be launched without any external mechanical force. The cavity of the fire-arrow's paper tube is similar to the combustion chamber of a modern rocket engine, the high-temperature and high-pressure gas flow generated as igniting gunpowder is the working medium and the small hole in the lower part of the paper tube is equivalent to the jet nozzle. The working medium generates a high-speed expanding gas flow through the special small hole, which can effectively transform the chemical energy into kinetic energy, and thus the thrust is produced to make the fire-rocket take off. Obviously, both the jet fire-arrows and the modern rockets belong to the same type of engine designed with the principle of reaction force. The jet fire-arrows have many advantages, such as easy manufacture, far-shooting range, good motility and strong destructivity.

After strengthening its regime, the Southern Song frequently warred with the Jin State (1115–1234 A.D.) and the Mongolian Empire (1206–1271 A.D.), and all of them used the fire-arrows as weapons in the wars. In November 1161, the army of Jin State invaded the Central Plains and suffered a heavy blow from the advanced rocket weapons "Thunderbolt Cannons" made in the Southern Song. The Thunderbolt Cannons were made from the paper tube, with which gunpowder, explosive and lime were filled; as igniting the blasting fuse, the gunpowder would burn to produce

Fig. 3.2 A military jet fire-arrow of the Southern Song Dynasty, China

the jet flame, and the Cannons were shot to enemy's position by the reaction thrusts; with the explosive exploding, the lime becomes into dust fog to attack the enemy. A military jet fire-arrow of the Southern Song Dynasty is demonstrated in Fig. 3.2. For years of warring, the fire-arrow weapons of the Southern Song Dynasty had played an important role, but its fire-arrow technology was also gradually mastered by the Jin State and the Mongolian Empire. The Mongols invaded the Southern Song Dynasty to the south, and at the same time, also entered into the vast areas of Western Asia and Eastern Europe on a large scale. So, China's gunpowder and fire-arrow technology were also introduced to European countries at the time.

The Ming Dynasty (1368–1644 A.D.) was the peak period of the development of the fire-arrow weapon technology in China, and thus the jet fire-arrow technology was advanced to a high level. There were three technical characteristics of the fire-arrows during the Ming Dynasty. Firstly, the types of fire-arrows were very various, including single fire-arrows, cluster fire-arrows, fire-arrow bombs and multiple stage fire-arrows. Secondly, the books on various fire-arrow technologies were written, in which their manufacture, application, assembling, charge proportioning and processing technique were described in detail. Thirdly, the multi-service units were equipped with the fire-arrows, including then navy, infantry and cavalry, and there were still professional fire-arrow forces and regulations on the use and management of the fire-arrows. The various types of fire-arrow weapons like the

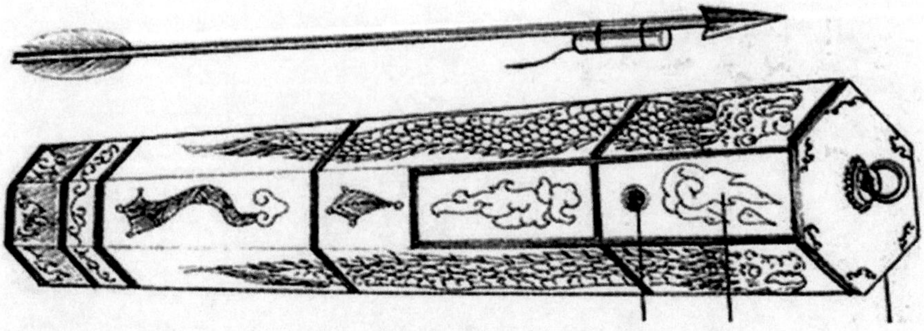

Fig. 3.3 Structure of fire-arrow "One-Swarm-Bee Arrow" recorded in *Wubei Zhi*

single fire-arrows, cluster fire-arrows, fire-arrow bombs and multiple stage fire-arrows were recorded in *Wubei Zhi*, a comprehensive military works written by Yuanyi Mao (1594–1644 A.D.), a military specialist in the Ming Dynasty.

The single fire-arrows are the types of rockets with feather-ling stability control, whose rocket barrels are 27-cm long and 4-cm in diameter, arrow poles are 2-m long and 2-cm in diameter, feather-lings are 23-cm long, as well as shot ranges can reach 500 steps (one step in the Ming Dynasty equal to about 1.5 m). The cluster fire-arrows are the types of rockets with special launch-barrel to shoot out multiple arrows at the same time, such as "Five-Tiger out-of-Hole Arrows (5 sets, range 500 steps), Seven-Star Arrows (7 sets), Nine-Star Arrows (9 sets), Crossbow-Meteor Arrows (10 sets), Fire-Dragon Arrows (20 sets), Long-Snake-Burst Arrows (30 sets, range 200 steps), One-Swarm-Bee Arrows (32 sets, range 300 steps, as shown in Fig. 3.3), Group-of-Leopard Running Arrows (40 sets, range 400 steps), Forty-Nine Sets Flying-Curtain Arrows (49 sets), Hundred-Tiger Running Arrows (40 set, range 400 steps)" and so on. The fire-arrow bombs are the types of rockets composed of the launch vehicle with a jet-propulsion device and the warhead, whose ranges are usually very far, such as "Flying-Fire Raven", "Shock Sky-Guns of Attacking Thieves in Flying Air" and so on. The multiple stage fire-arrows are mainly applied to sea warfare, such as "Fire-Dragon out of the Water" (a two-stage rocket, as shown in Fig. 3.4), and "Sand Barrel in Flying Air", which disturbed the enemy's mind and could be recycled to use again.

There were many kinds of fire-arrows in the Ming Dynasty, whose overall technical level was in the leading position all over the world at that time. Although the rocket technology in the Qing Dynasty (1636–1912 A.D.) developed to some extent, the general trend was gradually declining. After moving its capital to Beijing, the Qing Dynasty had been in a peaceful environment for a long time, passing through the prosperous age from Emperors Kang Xi to Qian Long (also called the High Qing era, 1681–1796), and thus the rocket weapons were seldom used and developed. At the beginning of the nineteenth century, the rocket technology of the Western countries eventually caught up with that of China.

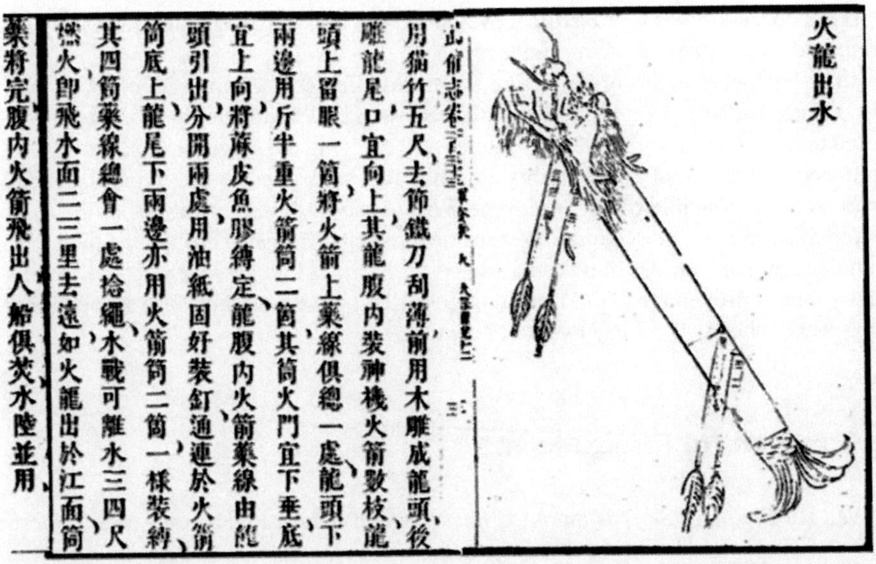

Fig. 3.4 Shape of fire-arrow "Fire-Dragon out of the Water" recorded in *Wubei Zhi*

In the thirteenth century, with the rise of the Mongolian Empire and its Western Expedition, the rocket technology of ancient China was introduced to Arabian Countries through India, and then to European countries. Italy was the first European country to develop and utilize rocket weapons. Later, many countries such as France, Germany, Poland, Britain and Russia all mastered the rocket weapon technology successively and made significant improvements in rocket's shape, size, guidance, material and gunpowder proportioning, which greatly enhanced operational effectiveness. Particularly, a British man William Congreve (1670–1829) made a great contribution to the development of rocket technology, making it rise to a new position and expanding its application. Congreve was born in a military family and had a strong interest in rocket weapons. From 1792 to 1799, the Indian army heavily attacked the British army by using rocket weapons in the region of Seringa-Patam, which made Congreve strongly touched and shocked. Since then, Congreve began to concentrate on the studies on rocket weapons, and successfully developed the famous Congreve rocket. From 1806 to 1807, the British army widely used the Congreve rockets in the battles of Sicily, Naples and Bologna against the French army and in the attacks on Copenhagen, Denmark. The Congreve rockets have a long-range, high hitting accuracy, and greatly enhanced warhead explosion efficiency, which has achieved remarkable combat results.

The performance of the Congreve rocket almost reaches the limit level of the jet rockets. With the rapid improvement of artillery range and accuracy, rocket weapons were gradually replaced in the mid-nineteenth century. Until 1890, the ignition system and launch way of rockets were improved, and a rocket weapon

"Aerial Torpedo" was successfully developed. And thus, the solid gunpowder rockets returned to the stage of history once again.

The jet rocket is the first reaction propulsion device used in practice. Although it is not an ideal space launch vehicle yet, its basic principle can completely be used to develop space vehicles. By intensively studying the solid rockets, astronautic pioneers realized that space flights are only able to depend on the type of rockets based on the principle of reaction force. With technological innovations of the solid gunpowder rockets, the liquid rockets came into being further. Through the pioneers' studies and tests, it was shown that space flights could be achieved by the rockets using liquid propellants. And thus, the close link between rockets and space flights was established in the early twentieth century.

3.2.3 Wan Hu Flying to Sky

"Wan Hu Flying to Sky" is an old Chinese legend, which told a story of a Chinese official in the middle Ming Dynasty, named Wan Hu or Wan Hoo, achieving his aspiration of space flight. On the day of the flight, Wan Hu sat in the chair where 47 fire-arrow rockets were attached to produce the thrust, and his two hands also took respectively one large kite to produce lift force. He commanded that 47 assistants, each armed with torches, simultaneously rushed forward to light the rockets. As a result, Wan Hu lost his life because of the rocket explosion. All versions of this legend have been spread to China by translating the English documents. However, up to now, the records about Wan Hu and his deeds have not yet been retrieved from the existing Chinese historical documents, which is also a reason why it is difficult to determine in Chinese whether his name is "万户" or "万虎" through the English transliterations. The foreign documents about "Wan Hu Flying to Sky" largely originated from the following versions:

(1) In October 1909, an article written by John Elfreth Watkins, which told a story similar to Wan Hu Flying, was published in the issue of *Scientific American*, but the protagonist of the story was Wang Tu instead of Wan Hu. The story said: "*Tradition asserts that the first to sacrifice himself to the problem of flying was Wang Tu, a Chinese mandarin of about 2000 years B.C. who, having had constructed a pair of large, parallel and horizontal kites, seated himself in a chair fixed between them while forty-seven attendants each with a candle ignited forty-seven rockets placed beneath the apparatus. But the rocket under the chair exploded, burning the mandarin and so angered the Emperor that he ordered a severe paddling for Wang*".

(2) A monograph written by Willy Otto Oskar Ley (1906–1969) who was a German-American science writer and space advocate, *Rockets–The Future of Travel beyond the Stratosphere*, was published in 1944. The name of "Wan Hu" first appeared in the monograph. In 1958, Willy Ley also said, in his another

book, that Wan Hu invents and tests a rocket aircraft in 1500 A.D., with quite spectacular self-sacrifice.

(3) In 1945, Herbert Spencer Zim (1909–1994), an American naturalist and author, published his monograph, *Rockets and Jets*. He told a story about Wan Hu Flying in this book: "*Early in the sixteenth century, Wan decided to take advantage of China's advanced rocket and fireworks technology to launch himself into outer space. He supposedly had a chair built with forty-seven rockets attached. On the day of lift-off, Wan, splendidly attired, climbed into his rocket chair and forty-seven servants lit the fuses and then hastily ran for cover. There was a huge explosion. When the smoke cleared, Wan and the chair were gone, and was said never to have been seen again*".

(4) In 1947, it was described in a poster of Civil Defense Association of America that: "*According to one ancient legend, a Chinese official named Wan-Hu attempted a flight to the Moon using a large wicker chair to which were fastened 47 large rockets. Forty-seven assistants, each armed with torches, rushed forward to light the fuses. In a moment, there was a tremendous roar accompanied by billowing clouds of smoke. When the smoke cleared, the flying chair and Wan-Hu were gone*".

(5) In 1956, two Soviet rocket scientists, Theodosiyev (В.И. Хеодосьев) and Sinalev (Г.Б. Синярев), published their book *Introduction to Rocket Technology*. In this book, they said that the Chinese were not only the inventor of the rocket but also the first dreamers to try to use the rocket to launch people into space.

Moreover, a British rocket specialist W. Maxwell said: "The story of Wan Hoo is an interesting event in the early history of the rocket". Up to now, only can rockets based on the principle of jet propulsion transport people into space. According to the principle, Wan Hu tried to make the activity of space flight early before more than 500 years, and thus became the world's first astronaut lifted explorationally by using rockets into outer space. Although a severe flight accident happened, Wan Hu was a flying hero by first using the thrust of rocket and made an important contribution to the development for mankind to explore the unknown field. So, it was decided in a conference of the International Astronomical Union (IAU) in the 1970s that a lunar impact crater located on the Moon's far side was named "the crater Wan-Hu" in the honor of Wan Hu.

However, it is puzzling that all documents depicting the Wan-Hu flying legend did not mark out where the quoted historical materials came from. Some scholars believe that the historical materials possibly originate from the manuscripts of Chinese, and thereby it can not be recorded where they come from. A lot of depictions of Wan Hu are usually paradoxical; and meanwhile, it is very difficult to search the historical documents about the Wan Hu flying published before 1909. Therefore, some scholars also believe that the stories of Wan Hu are very possibly fabricated through the Western scholars' fantasies about the Oriental civilization. What's more, the producers of the program "Rumor terminator" of the U.S. exploration channel once designed a launch experiment of simulating the Wan Hu flying seat. It was shown from the experimental result that enough thrust to lift the seat up can not be generated by using the bundled modern small rocket-propellers.

In the 1960s, Xuesen Qian (Hsue-shen Tsien, 1911–2009), the famous scientist and the founder of China's space industry, told the young students the Wan Hu flying story that his tutor Shizhuo Wang had ever told him. "In the nineteenth year (1483 A.D.) when emperor Chenghua was in power during the Ming dynasty, there was a rich man named Wan Hu, who had a strong interest in the gunpowder and rocket invented by the ancient Chinese. He imagined that he would use the huge thrust of the rockets to send himself into blue sky. After full preparation, the day of the flight finally came. Wan Hu seated a snake-shaped flying vehicle with forty-seven rockets attached, and his hands still took two large kites. Wan Hu commanded his attendants to ignite the first row rockets, and after sounding a tremendous roar with billowing clouds of smoke, the flying vehicle began to lift off the ground. And then, the second row rockets were automatically ignited, and the explosion suddenly happened. Finally, Wan Hu and his flying vehicle fell and broke in Wanjia Shan".

There are two facts to need to be explained further. One is about Shizhuo Wang (1905–1991), one of the pioneers of China's aviation industry, whoever studied in the United States in 1925, and received his master's degree in engineering from the Massachusetts Institute of Technology in 1928. After returning to China in 1933, he served as a professor at the aviation group of the Department of mechanical engineering, Tsinghua University. From 1934 to 1935, Shizhuo Wang with other two professors, Xinjue Qian and Zhu Wang, formed a tutor team to together serve as the training and tutoring for Xuesen Qian, then a preparatory student studying in the United States. Another is about Wanjia Shan, now located near Manshui River, Huoshan County, Anhui Province. Yuanzhang Zhu (1328–1398), the first emperor in the Ming dynasty, was born in Fengyang County, Anhui Province.

Although the historical facts of the Wan Hu flying legend need to be confirmed, and it is possible to become a mystery of history forever, the following historical facts are clear:

(1) The Ming Dynasty is the peak period of ancient Chinese rocket technology development. And thus, the flight using the thrust of the rockets naturally became people's most simple and universal thoughts at that time.
(2) In the late Qing Dynasty, especially in 1900 when the Eight-power Allied Forces invaded China, a large number of historical relics and ancient books were lost overseas. Perhaps, to a large extent, Wan Hu and his deeds were recorded in these historical materials.
(3) Although there are various versions of the legend described in the foreign documents, most of the detailed depictions of the story are accordant each other, such as the story happening in the middle Ming Dynasty (1480–1520), 1 chair, 47 rockets, 2 kites and the attendants for igniting. It is also clearly shown that all of the versions came from the same historical material.
(4) According to the time when Shizhuo Wang told Xuesen Qian the story, it could be speculated that the Wan Hu flying legend had widely been spread in the Western scholars before 1940s.

(5) The Wan Hu flying legend early came from the United States. It is unnecessary that American scholars fabricated an adventure story that an ancient Chinese attempted to fly into outer space by using rockets, and widely spread it.

3.2.4 Historical Context of Space Science Development

3.2.4.1 Science and Technology in Modern China

During more than 1000 years before the Ming Dynasty, the ancient Chinese science and technology represented by the Four Great Inventions, i.e., the papermaking, compass, gunpowder and movable-type printing technologies, were far ahead of the other countries in the world [4]. A large number of the scientific and technological works, from *The Nine Chapters on the Mathematical Art* (a classical mathematical book) revised by Cang Zhang and Shouchang Geng in the Western Han Dynasty, *Qimin Yaoshu* (a comprehensive agronomy book) written by Sixie Jia in the Northern Wei Dynasty, *The New Revision of Materia Medica* (a medical and pharmacological book) by Jing Su in the Tang Dynasty, *The Sketchbook of Dream Brook* (a technological encyclopedia) by Kuo Shen in the Song Dynasty, *Timing Service Calendar* (a new calendar) by Shoujing Guo and *Agricultural Book* (an agronomy and agricultural machinery book) by Zhen Wang in the Yuan Dynasty to *The Compendium of Materia Medica* (a comprehensive medical and pharmacological book) by Shizhen Li, *The Encyclopedia of Agricultural Policy* (an agricultural policy measures and technological book) by Guangqi Xu, *Xu Xiake's Travels* (a geographical investigation report) by Hongzu Xu, *Tiangong Kaiwu* (a comprehensive scientific and technological book on agriculture and handicraft production) by Yingxing Song and *Wubei Zhi* (a book on military science and technology) by Yuanyi Mao in the Ming Dynasty, had greatly promoted the development of politics, economy, military affairs and science and technology in ancient China.

During the 28 years since 1405, the fleets led by He Zheng in the Ming Dynasty had made seven voyages to the West, involving more than 100,000 people successively, and the furthest reaching the east coast of Africa. The unprecedented expedition of ocean-going voyage had made amazing achievements, which showed China's leading technology in the world's voyage industry at that time, as well as the peace, prosperity and confidence of the Ming Dynasty. However, in 1433, Zhanji Zhu, an emperor in the Ming Dynasty, suddenly ordered to stop Zheng's fleet from sailing again. From then on, China's power turned to the inside and then the world's oceans were handed over to the Western adventurers.

The science and technology in ancient China was produced and developed for the purpose of practicality, and the feudal autocracy of thousands of years had fettered people's thoughts. In the Ming Dynasty, The gate for Zheng's fleet to explore the unknown world was just opened but quickly closed. For the explorations of space flight, the ancient Chinese still stayed in the fantasy poetrics with supernatural power and the stories on immortals and ghosts. A famous poet in the Tang Dynasty, Bai

Li (701–762 A.D.), fantasied in his poetry *Gu Feng* that: as he was on the sea, he would travel across sky in the fairyland, but the birds and chariots that took him to the fairyland would not be found for some reason. Cheng'en Wu (1500–1583 A.D.), a famous novelist in the Ming Dynasty, wrote in the novel *Journey to the West*: Monkey King learned 72 changes, and could cover a hundred thousand miles by making one "somersault cloud"; and thus, he would fly to the heavenly palace at any time. Consequently, modern science did not come into being in China which had a splendid and brilliant oriental civilization, but in Europe after the Renaissance. Justly in this context, the scientific fictions on space flights were also quietly emerging in the West.

3.2.4.2 European Renaissance and Exploration Age

During the Middle Ages (476–1453 A.D.), there was not a strong political power to rule Europe, and thus the feudal separatism brought frequent wars, resulting in slow or stagnant development of science and technology and productivity in the whole of Europe. In Europe at that time, the classical civilization was almost smashed and could not be recovered, and the religious theology controlled people's spirit and restricted academic and speech freedom, but Europe began to rise in poverty and obscurity.

In 1271, Marco Polo (1254–1324), a young Italian merchant from Venice, followed his father and uncle to the East along the Land Silk Road. They passed through Mesopotamia, Iranian Plateau and Pamir Plateau, and finally arrived in Dadu (now Beijing) of the Yuan Dynasty in 1275. Marco Polo traveled in China for 17 years and served as an official of the Yuan Dynasty. He visited many regions of China at that time, such as the southwest and southeast coastal regions. In 1292, Marco Polo set out from Quanzhou city, Fujian Province, accompanied together with three envoys of the Yuan Dynasty to escort Princess Kuokuo Zhen through the Maritime Silk Road to Persia to get married. And finally, in 1295, he returned to his hometown Venice city, Italy. In 1298, there was a sea battle between Venice and Genoa, two Italian city-states at war over control of trade routes in the Mediterranean Sea. Marco Polo as a gentleman-commander of a galley in the Venetian navy joined in the warfare. Unfortunately, he was captured and put into a Genoese prison, where he met a fellow prisoner named Rustichello da Pisa, a sailor from Venice, who was an Italian romance writer. Marco told Rustichello the incredible stories that he had ever traveled through West Asia, Persia, India, China and other cities and countries between 1271 and 1295. Marco was released within a year due to his fame as an adventurer and then was little known of his life after his return to Venice. Rustichello recorded Marco's traveling stories in detail and later published the well-known travelogue, *The Travels of Marco Polo*, a book that described to Europeans the then wonderful culture and inner workings of the Eastern world, including the wealth and large territory of the Yuan Dynasty and its capital Dadu, giving a first comprehensive look into China, India, Japan and other Asian cities and countries. As a result, the book greatly inspired European's strong desire to explore the highly civilized and rich Eastern world, and finally broadened their horizons.

At the same time, the rising Ottoman Empire (1299–1922 A.D.) so frequently warred with the Byzantine Empire (also referred to as the Eastern Roman Empire, 295–1453 A.D.) that the people, which fled from the warfare, brought a large number of the ancient Greek and Roman cultural books and art treasures to the commercially developed cities of Italy. Some advanced intellectuals in the new bourgeoisie publicized the humanistic spirit through cultural creation by studying the art and culture of ancient Greece and Rome. So, a great revolution of emancipating the mind, science and art, well known as the **Renaissance**, which rose in the major cities of Italy and prevailed throughout Europe, opened the prelude of modern European history.

The fifteenth century, when several isolated and independent continents began to connect under the promotion of voyage explorations and trades, is an important watershed in human history. In 1434, when is just the second year after the emperor of the Ming Dynasty ordering to dismiss Zheng's fleet, greatly driven by Prince Henry (O Infante Dom Henrique Duque de Viseu, 1394–1460), who was with a pioneering spirit, the Portuguese fleet crossed through Cape Bojador on the west coast of Africa, which used to be called "death limit", and thus made the first step in the Age of Exploration. In 1492, Christopher Columbus (1451–1506), an Italian navigator and explorer, led his fleet across the then uncharted Atlantic Ocean with the help of the Spanish queen, and eventually arrived in the American continent after 70 days of hard sailing. In 1497, a Portuguese explorer Vasco da Gama (1460–1524) first opened a sea route to India by passing through the Cape of Good Hope at the southern end of Africa. In August 1519, with the support of the King of Spain, another Portuguese explorer Ferdinand Magellan (1480–1521) and his fleet set sail from the port of Seville in Spain, and entered into the Atlantic Ocean along the Guadalquivir river; they sailed west across the Atlantic toward South America, made landfall at Rio de Janeiro in December 1519, and from there, sailed south along the coast, searching for a route through or around the South American continent; they bypassed the Tierra del Fuego at the southern end of the continent, crossed through the Pacific Ocean and reach the Philippine Islands in April 1521; they passed through the Southern India Ocean and the Cape of Good Hope, and finally returned to Sanlúcar de Barrameda on the coast of Spain in September 1522. As a result, the fleet led by Magellan had achieved the great feat of the first circumnavigation of the Earth for mankind. The Magellan's fleet as setting out consisted of five ships and about 270 crew members, most of whom were Spanish and only around 40 Portuguese, and as returning to Spain, only one ship Victoria with ghostly 18 crews survived. Magellan himself died in a battle against the local troops of the Philippines in April 1521, and thereafter Juan Sebastián Elcano, a Spanish explorer aboard the ship, took the command of the carrack Victoria to finish the final voyage from the Moluccas in Indonesia to the Sanlúcar de Barrameda in Spain. The Magellan's fleet successfully accomplished the circumnavigation at a huge cost, which proves that the Earth is round and the oceans around the world are connected together with each other.

The Exploration Age not only brought huge wealth for Europeans, but also inspired their spirit of adventure and exploration, and they were accepting the baptism of innovative ideas. The dusty academic and scientific spirit of ancient Greece started to revive in Europe, which is of epoch-making significance in the scientific and technological history of human beings.

3.2.5 *Rise of Space Science Fictions*

Just under the epoch context of the Renaissance and the Exploration Age in Europe, some basic sciences like astronomy, physics and mathematics have been developed rapidly. In 1543, an astronomical book *On the Revolution of the Celestial Spheres* was written by Nicolaus Copernicus (1473–1543), a famous Polish astronomer, was officially published. The heliocentric theory proposed in this book overturned the status of the Ptolemy's geocentric system which had fettered people's thoughts for more than 1000 years, which had a revolutionary change and profound impact on modern astronomy. Thereafter, some famous astronomers and mathematicians, including Danish Tycho Brahe, German Johannes Kepler, Italian Galileo Galilei and British Isaac Newton, had made a lot of astronomical observation and studying work, and further established and perfected the heliocentric theory, which completely changed people's outlook on the universe. After the sixteenth century, through some new astronomical views and a series of major scientific discoveries, people generally realize: the Earth is no longer the center of the universe, but an ordinary planet in the solar system; both Mars and Venus are the planets similar to the Earth; both living and non-living bodies contain the same elements; the fundamental elements of matter are limited; both planets and stars have the same elements as the Earth; all living things are evolving; and so on. People are used to feeling and imagining the situation of planets, and thus some problems, for example, what other planets are like, whether there are human beings living on the planets and what the existing forms of extraterrestrial life are, always make people fascinate and fantasy. What fascinates people most is the Moon, the nearest neighbor of the Earth, and thereby, traveling to the Moon naturally became the first fantasy theme at that time.

In 1608, the German astronomer Johannes Kepler (1571–1630) wrote a novel *Somnium* ("*The Dream*") in Latin, but the novel was not officially published until 1634 by his son, Ludwing Kepler. Johannese Kepler attempted to use the new astronomical views and scientific knowledge at that time to indirectly lead out the scientific fantastic story. In Johannese Kepler's narrative, an Icelandic boy and his witch mother learn of an island named Levania (our Moon) from a daemon. Kepler presented the detailed imaginative descriptions through the daemon, such as the Moon and its phenomena, the phenomena between the Earth and the Moon, the planets' periods and the apparent motions of celestial bodies. In the meantime, he also believed that, in addition to some special features, the celestial bodies such as the Sun, Mercury, Venus, Mars, Jupiter and Saturn are as same as the Earth. Kepler's novel *The Dream* is considered the first serious scientific treatise on lunar astronomy. An English divine and Church of England bishop Francis Godwin (1562–1633) wrote a book, *The Man in the Moone*, which was first published posthumously under the pseudonym of Domingo Gonsales in 1638, and in which a story of Gonsales traveling to the Moon was described in detail. This pure fantasy and lack of scientific novel were very popular. It had been translated into many languages to publish, which had a great impact on arousing people's interest in space flight and writing space science fiction. In the same year, John Wilkins (1614–1672), a British clergyman, natural philosopher and author, published his first book, *The Discovery of a World in the Moon*, which was in fact appeared anonymously. In 1640, he also anonymously published a Discourse concerning a New Planets, and from this time the Discourse and the Discovery have been published together as a single work. Wilkins is the founder of the metric system and the first secretary of the British Royal Society. His work is addressed to general readers rather than experts in the subject, and its aim is to popularize the view of the universe due to Copernicus and Kepler. He believed that the Moon was not a shining disk but rather it was a world with a landscape like that of the Earth and that the Moon was a habitable planet and traveling to the Moon would be possible one day. Wilkins himself still insisted that by developing a flying car, people could fly into the air and even to the moon, and predicted that the technical principle of the flying car would soon be found by scientists. In 1657, *a Voyage to the Moon* written by Cyrano de Bergerac (1619–1655), a French satirist and dramatist, was published posthumously. It is a classic fantasy novel, in which with a very romantic and rigorous scientific attitude to explore a variety of possible flight ways, the arguments have a strong logic. Cyrano de Bergerac believed that the attraction of magnets could be used to achieve lift-off. If a person's feet were tied to a magnet and his hands threw another magnet into the air, the person would be dragged up under the attraction of the thrown magnet; when he rose to the height of the thrown magnet, he would grasp the magnet with his hands and continued to throw it up; and then, he was also dragged up to a higher height once again; like this, by circularly throwing the magnet, he could eventually fly to the Moon. Furthermore, another Cyrano de Bergerac's novel on the *Voyage to the Sun* was also published in 1662, in which the story of a prisoner who escapes from prison and is blown the Sun country by a whirlwind was told. It was shown once again that the author had a rich imagination and logical reasoning ability.

In 1703, a flying way of utilizing the idea of catapulting to the Moon was described in a novel written by David Russen, *Voyage to the Moon.* In 1705, Daniel Defoe told an ancient race's mastery of the lunar flight in his novel, *The Consolidator*, the way of which was to utilize the strong flame to drive the spring and wheel to drive the wings, as well as described various spaceships and legends of lunar flights. In 1751, *a Narrative of the Life and Astonishing Adventures of John Daniel* written by Ralph Morris was published by a pseudonym, in which a flight way with wings driven by a pressure pump was discussed. In 1752, Francois-Marie Arouet (pseudonym: Voltaire), a French Enlightenment writer, told a story in his short work *Micromégas*, that an inhabitant from the distant planet which circles the star Sirius and his companion from the Saturn visit to the Earth. In addition, there were many fantasies of space flight during the seventeenth and eighteenth centuries. For example, first flying out of the atmosphere, people could directly fly to the Moon by the lunar gravity; people imagined that the spacecraft would be driven by electricity. Limited to the level and condition of science and technology at that time, people realized that to achieve space flight was really difficult. And thus, the emerged space science-fictions ingeniously integrated science with non-science into a body, to imagine the space flight ways only with a supernatural force.

By the nineteenth century, with strengthening of the scientific discovery and technological level, the space science fictions emerged new styles. In 1827, George Tucker, a professor at the University of Virginia, proposed to invent a kind of anti-gravitation material in one of his science fictions, and thus the spaceship made from the material could fly to the Moon under the repulsion of the Earth. The spaceship was sealed and equipped with breathing oxygen and heating system. In 1863, Achille Eyraud, a French writer, published a science fiction about the voyage to Venus, in which according to the fact that gunshot bullets to generate recoil force, a new power device named reaction force motor was presented. It is just now the only feasible way to realize space flight. To some extent, Achille Eyraud's fiction is the first scientific prediction work on launch a vehicle engine, which contains profound scientific significance.

In 1865, a well-known science fiction *From the Earth to the Moon*, written by Jules Gabriel Verne (1828–1905) who was a French novelist, poet and playwright, was first published by the publisher Pierre-Jules Hetzel. It tells the story of the Baltimore Gun Club, a post-American Civil War society of weapons enthusiasts, and their attempts to build an enormous space gun and launch three men in a projectile with the goal of the Moon landing. When the launch was ready, the club received a telegram from an eccentric Frenchman, Michel Ardan, claiming that as an explorer, he had not been interested in any place on the Earth and hoped to travel to the Moon by the projectile. The club approved his request and redesigned the projectile into a spaceship with soundproof walls. On December 1, Ardan, Barbicane and captain Nicholl by the spaceship were launched to the Moon from Colombia by the Gun firing way. The operation of the spaceship was tracked by Cambridge Observatory utilizing a 20-inch telescope. It was observed that the spaceship did not land on the Moon, but orbited the Moon with an altitude of 2300 miles away from its surface, becoming a satellite of the Moon. It is a typical representative work of modern space science fictions. In order to write the novel well, Verne consulted and investigated many mathematical, physical and astronomical problems. The design parameters of the spaceship and the launch device were obtained through strict mathematical calculation. The novel was popular all over the world and widely welcomed by readers. Many people were worried about the safety of the three astronauts and asked Verne to write another sequel. Five years later, Verne wrote a follow-up, *Around the Moon*, about the situation that the three astronauts, Ardan, Barbicane and Nicholl, had to deal with while on their way to the Moon and back. Launching several auxiliary rockets to get rid of the orbit around the Moon, the spaceship returned to the Earth and splashed into the Pacific Ocean. When the capsule was picked up by the U.S. Navy, the three astronauts were playing poker as if nothing had happened. With rich imagination and rigorous scientific attitude, Verne correctly predicted the basic process of aerospace activities, including the rocket's launch site, the capsule of spaceship, the weightlessness, the rocket flight in orbital transfer segment, the brake rocket and the splash-down recovery on the sea.

As a curious fact, the novel finished in Verne's serialized form in 1869; exactly a hundred years later, man would land on the Moon.

A space fiction *The Brick Moon*, written by American writer Edward Everett Hale (1822–1909), was published serially in *The Atlantic Monthly* from 1869 to 1871. It tells a story about a young Boston man who imagined to erect a tremendous ring around the Earth's poles over the Greenwich Meridian Line, with a diameter large enough to be seen anywhere on the Earth. And thus, the local longitude could be determined by observing the ring, and the distance between the local site and the Greenwich be calculated. However, because of the extremely huge scale of the project, he wanted to replace it with an artificial satellite, the so-called "brick moon" built with bricks. The speculative science fiction pays more attention to scientific facts, especially in which the implementation plan of artificial satellite and its application were first presented. Kurd Lasswitz (1848–1910), a German author, scientist and philosopher, wrote an influential science fiction novel, *On Two Planets*, which was first published in two hardcover volumes in 1897. It tells that three earthlings went to the Arctic by balloon to explore, where they found a strange building. At this time, the balloons began to open and rise vertically, as if falling into a vortex, so they entered a reversed gravitational field. At the lower end of the gravitational field was the building, and at the upper end, about a radius of the Earth was another building. The three earthlings eventually knew that the two buildings were the bases established by the Martians as they first came to the Earth. The Martians rescued them from the reversed gravitational field and told that the buildings over the Arctic was an orbital capsule made from anti-gravity materials, which were located just right at the balance point of the gravity between Earth and Mars. Some space technological details, such as photoelectric sensor, photoelectric battery, orbital station, reaction engine and orbital transfer controller, are described in this novel, and its scientific value can be compared completely with Verne's novel.

From the seventeenth century to the nineteenth century is the golden age of writing space science fictions. On the basis of the remarkable imagination and prediction of space flight from the science fictions, the concepts of space flights have been established in people's mind and the confidence of achieving space flight strengthened further. Inspired by the space science fictions, many pioneers were decided to devote themselves to the greatest cause for human beings, the researches on space science and technology, so as to usher in the era of the great development of space flight in the twentieth century.

3.3 Four Founders of Modern Rocketry and Astronautics

The early scientific fantasy laid the ideological and technical foundation for the development of modern aerospace technology, and further, it would be realized to leap from quantitative change to qualitative change. The creative ability of scientists should not only have rigorous scientific methods but also the imagination of artists. Bold imagination and speculation often become the source of motivation for scientists to create. The Russian Konstantin Tsiolkovsky, the French Robert Albert Charles

Esnault-Pelterie, the American Robert Hutching Goddard and the German Hermann Julius Oberth are the prominent pioneers and founders of modern rocketry and astronautics in the early twentieth century [5]. Justly inspired by the space science fiction novels, they established the theoretical system of rocket and spaceflight and ushered an emerging discipline, astronautics.

3.3.1 Konstantin Eduardovich Tsiolkovsky

Konstantin Eduardovich Tsiolkovsky (Констант´ин Эдуáрдович Циолкóвский, 1857–1935) was the Imperial Russian and Soviet rocket scientist and the pioneer and founder of modern astronautics, known as the "father of astronautics". Konstantin Tsiolkovsky has created many world firsts in rocket theory and space flight, established a complete theoretical system of space science, published more than 500 academic papers and science fictions, and contributed his lifetime and energy to the development of human astronautic industry. In August 1911, Konstantin Tsiolkovsky wrote a famous saying in a letter to his friend: "the Earth is the cradle of mankind, but one cannot live in the cradle forever; in the pursuit of light and space will at first timidly penetrate beyond the limits of the atmosphere, and then will conquer all the space around the Sun". Konstantin's visionary ideas for the future of humanity in space were far ahead of his time.

On September 17, 1857, Konstantin Tsiolkovsky was born in the village of Izhevskoe in Ryazan Province, about 200 km in the south of Moscow, Russia. His father is Polish, a forester; his mother is Russian, an educated woman. He grew up in a middle-class family with 18 children, out of which he was the fifth child. At the age of 10, Tsiolkovsky caught scarlet fever and became hard of hearing. For this reason, he was suspended from the local school for boys and had to study at home by self-education. He had never completed any formal education. His father's library became his school, and his disability did his driving force. At the age of 16, his father sent him to Moscow to study. Although he was still unable to enter the school to study normally, there was better studying condition than that in the countryside, and he taught himself at Chertkovskaya Library, which held the Russian finest collection of books. He widely studied mathematics, analytical mechanics, astronomy, physics, chemistry, as well as classical literature. Particularly, he was strongly interested in the issues on space flight. Around this time, he also found the novels of French science fiction and adventure writer Jules Verne, *From the Earth to the Moon*. It is just the scientific fiction that inspired a generation of spaceflight pioneers. Nevertheless, Tsiolkovsky thought that the way of launching the projectile was not feasible, and an approach to realizing space flight would be based on the principle of action–reaction force.

After returning to his native place in 1876, Tsiolkovsky passed exams for his Teacher's Certificate and became a teacher to teach arithmetic and geometry in a school named Uezdnoe Uchilishe in the town of Borovsk in Kaluga Region, about 110 km in the south of Moscow, until 1892. During this period, while working on air balloon building, life in free space, aerodynamics and philosophy and building the centrifuge to investigate the influence of gravity, he married a local preacher's daughter, and they had seven children. Tenaciously teaching himself in a closed environment for a long time, he was used to independently think problems and question everything. Due to the lack of more and updated references and of understanding the situation of the latest scientific and technological development, he often acquired new knowledge through his creative experiments. In 1880, Tsiolkovsky presented a paper on the kinetic properties of gases to the Society of Physics and Chemistry in Saint Petersburg. Scientists of the Society were surprised when they reviewed the paper: his studying methods, mathematical derivation and calculation results were almost perfect, and he claimed that the problem was resolved after working on it for a long time. In fact, someone else had already published similar findings as early as 25 years ago. The scientists guessed that the young man might have few books for reference and the lack of contacting with the outside world, and would not know the research work done by predecessors. Therefore, a carefully worded reply was written by the Russian famous chemist Dmitri Mendeleev (Дм´итрий Ива́нович Менделе́ев, 1834–1907) to Tsiolkovsky, in which Mendeleev expressed his appreciation for Tsiolkovsky's research work and achievements, explained that the problem had been resolved and encouraged him to make further research results. On the basis of the experiments with gases, Tsiolkovsky wrote a manuscript titled "Free Space" in 1883, but it had not been published until 1956. The manuscript contained a sketch considered to one of his earliest depiction of a spaceship. A simple drowning shows what space-suited travelers in weightlessness look like; a cannon-like machine propels the spaceship through vacuum; a primitive gyroscope controls the orientation of the ship in space. He made the first attempt in his decades-long effort to expound the meaning of the cosmos for humanity and the effects that vacuum and weightlessness would have on future space travelers.

In 1892, as a result of promotion, Tsiolkovsky moved to the provincial capital of Kaluga, where he worked as a mathematics teacher at a high school until retiring in 1920. He would remain in Kaluga until his death in 1935, and it was there that he created the monumental body of work, making him become the most fruitful space scientist and a godfather of rocketry and space exploration. Tsiolkovsky started with the studies on aircraft and science fiction writing and published two science fictions, *On the Moon* and *Dream of the Earth and Sky* respectively in 1893 and 1895, in which his thoughts about space flight were developed and shown further. Since around 1896, Tsiolkovsky studied extensively the theory of jet propulsion. By systematically and intensively studying the theory of space flight, he derived in detail the rocket motion equation, the famous Tsiolkovsky Formula named after him, which provides the mathematical relationship between the changing mass of a rocket as it burns through fuel, the velocity of the exhaust gases, and the final speed of the rocket. It is considered a foundation of astronautics.

In 1898, Tsiolkovsky completed a manuscript of the classic space treatise, titled "The Exploration of Cosmic Space by Means of Reaction Devices", which was not published in the magazine *Scientific Review* until 1903. It is arguably the first academic treatise on the rocketry. He calculated that the speed required to orbit the

Earth is 8 km/s and that it could be achieved by means of a multi-stage rocket fueled by a mix of liquid oxygen and liquid hydrogen. Since then, Tsiolkovsky had successively published many academic papers on rocket theory and space flight in the Journal of Aviation Report from 1910 to 1914, thus building a completely theoretical system of Astronautics, in which his research findings or academic views have created many firsts in the history of space science and technology development.

(1) For the first time, it is clearly proposed that a liquid rocket is an ideal vehicle for space flight.
(2) For the first time, all kinds of liquid propellants are studied systematically and comprehensively, and it is proposed that liquid hydrogen and liquid oxygen are the best rocket propellants.
(3) For the first time, the equation of rocket motion in a vacuum is derived in detail, and the escape velocity of rocket is also calculated.
(4) For the first time, the concept of rocket's mass ratio is proposed, and its important significance is described.
(5) For the first time, the pumping way of a liquid rocket propellant is proposed.
(6) For the first time, the regenerative cooling method of the combustion chamber of rocket engine is proposed.
(7) For the first time, the aerodynamic heating of the rocket motion in the atmosphere is studied.
(8) For the first time, the sketch of a complete spaceship design is drawn.
(9) For the first time, it is proposed to use a gyroscope to realize the orientation control of spacecraft.
(10) For the first time, the effects of weightlessness on biological and human body are investigated, and the measures to reduce the adverse effects of weightlessness and overweight are provided.
(11) For the first time, the experiments on the effects of weightlessness and overweight on small animals are carried out.
(12) For the first time, the measures to improve the closed cabin environment by utilizing plants and to provide food for astronauts are proposed.
(13) For the first time, the idea of space station and space biosphere is put forward.
(14) For the first time, the idea of propelling the spacecraft by using solar pressure is put forward.
(15) For the first time, the idea of space colony is put forward.

Tsiolkovsky's early works were published in professional scientific journals. The readers were mainly scientists, and some people were skeptical. Yakov Isidorovich Perelman (Яков Исидорович Перельман, 1882–1942), a Russian and Soviet popular science writer, was interested in Tsiolkovsky's academic papers and adapted them into popular astronautic science fictions. In this way, the readership expanded rapidly to improve the popularity of Tsiolkovsky, but the ordinary readers did not think that Tsiolkovsky was a scientist. In the meantime, Tsiolkovsky also published a science fiction, "*Beyond the Earth*", which was widely welcomed by readers and won a high fame.

For many years, Tsiolkovsky's ideas were little known outside Russian, and the field lagged until German and other scientists independently made the same calculations decades later. After the Russian October Revolution, Tsiolkovsky's work was encouraged by the Soviet government. At that time, when the country was very difficult, it still actively supported to research the rockets, but it was yet insufficient to understand the importance of the astronautic theory early proposed by Tsiolkovsky. With the development of researching the rocket and spaceflight in Europe and the United States, many astronautic enthusiasts started to devote in the Soviet Union to studying rocket. In view of the rapid development of international rocket and spaceflight, the Soviet government started to pay more attention to the field. In 1924, the Soviet government organized experts to review Tsiolkovsky's articles, and published again his classic paper "The Exploration of Cosmic Space by Means of Reaction Devices" and popular science fictions. The special organizations to study rocket and spaceflight were set up in Moscow and Leningrad, and the associations of space flight were also created by the student groups of the universities. Tsiolkovsky was invited to join these organizations and provide guidance to study aerospace science.

For the final decade of his life, Tsiolkovsky was still working. In 1926, he published "Plan of Space Exploration", which consists of a bold 16-step program, from the very beginning of space conquest until the far future, including interstellar travel and the colonization of the solar system and beyond. Three years later, he proposed the thoughts about multistage rockets in his book, *The Space Rocket Trains*. His calculations proved that building a rocket with separate stages, each of which would be jettisoned as it finished consuming its propellants, would allow a payload to be accelerated further. In his later years, he wrote a comprehensive paper, "The Work of Space Rocket: 1903−1927", which was published 1 year after his death. He systematically summarized his work and achievements of studying rocket theory and space flight in this paper, and predicted the development steps in the future.

In the 1930s, with the upsurge of studying rocket and spaceflight all over the world, Tsiolkovsky's fame was also rising rapidly. In 1932, on his 75th birthday, his deeds and scientific achievements were published in the Soviet newspapers and magazines, and the Soviet Premier Joseph Stalin also sent him a birthday message. For a while, the old man became an outstanding hero of the Soviet Union. On September 19, 1935, Tsiolkovsky died at the age of 78 in Kaluga and was given a state funeral. He lived to see a younger generation of Soviet engineers and scientists begin to make his visionary concepts reality. It is gratifying to the great pioneer that his planned spaceflight practice was first achieved in his hometown the Soviet Union. The world's first artificial earth satellite was launched successfully on October 1, 1957, and since then human beings entered the age of space flight.

3.3.2 Robert Albert Charles Esnault-Pelterie

Robert Albert Charles Esnault-Pelterie (1881−1957) is a French aircraft designer and spaceflight theorist, and one of the founders of modern rocketry and astronautics.

He was granted more than 200 patents on inventions in such fields as aviation and aircraft-engine construction, and he made remarkable achievements in the theory of space flight. His lectures and academic papers on the theory of aerospace science can be compared with those of Tsiolkovsky, and are regarded as a symbol of the birth of modern astronautics.

On November 8, 1881, Robert Esnault-Pelterie was born in Paris to a textile industrialist. His father was a textile machinery manufacturer, so he had a strong interest in machinery since he was a little boy. At the age of 13, he designed and assembled a signal device for toy train; at the age of 17, he designed a physical and chemical experiment and started to study radiotelegraph; at the age of 21, when he graduated from the University of Paris with a degree in physics, he was granted the first patent for invention, "An Electronic Delay Device with High Sensitivity". Esnault-Pelterie had a wide range of research fields, including metallurgy, electronics, magnetism, hydraulics and thermodynamics, and especially in the fields of engineering technology, aviation and aerospace, he had made many pioneering contributions.

In 1906 and 1907, Esnault-Pelterie built the world's first monoplane and the first radial aircraft engine and invented the airplane control system based on the control stick. He worked on the theory of interplanetary navigation and calculated the most advantageous flight trajectories for spacecraft. In the beginning of the twentieth century, the isolated visionaries and thinkers sketched the sinews of the spaceflight concept, but the technical details of rocketry and space travel had precarious credibility. Consequently, the usual all-knowing intellectuals of the day dismissed them as ridiculous. Esnault-Pelterie was an accomplished engineer and also had the fame of an aviation pioneer, and his credibility helped him to gain acceptance by mainstream scientific audiences. In 1912, he gave lectures on the theory of Astronautics respectively in Saint Peterburg, Russia, and at the Physical Society of Paris, France. As a scientist, he emphasized the importance of addressing the physical foundations of spaceflight. On the basis of his lectures and presentations, he wrote and published a paper "Consideration of the Results of the Unlimited Lightening of Motors" in the *Journal of Physics*, French, in 1913, and at this time, he did not know Tsiolkovsky's 1903 paper. He pointed out in his paper: as there is no air between the stars, airplanes can not fly in the interstellar space; but the existing scientific knowledge tells us that there is a motor that does not need air to support its flight, which is the rocket. Esnault-Pelterie considered that the rocket works on the basis of the principle of momentum conservation, and its exhausting jet stream is similar to bullet but has continuity. Based on the law of conservation of momentum and energy, he derived the rocket equation and got the escape velocity of the rocket as 11.28 km/s. He considered the energetic properties of guncotton, hydrogen–oxygen mixture, and radium as propellants. Then he provided estimates of the required velocity increments and flight times for travel to the Moon, Venus and Mars and requirements to the propellants for such missions. Considering the interplanetary flight of humans, he proposed passive spacecraft temperature control. He divided it into three stages to launch the rocket from the ground to the Moon. In the first stage, the rocket is accelerated to reach the escape velocity to get rid of the gravity of the Earth, and the flight time of the stage is 24 min 9 s. In the second stage, the rocket engine turns off and continues to fly to the Moon at the initial accelerated velocity, and the flight time is 48 h 30 min. In the third stage, when the rocket flies to arrive at a predetermined point close to the Moon, the direction of the rocket's engine is reversed and restarted to reduce the flight velocity; the rocket velocity is exactly zero at landing on the Moon; the flight time is 3 min 46 s. Based on the scientific theory and calculation, Esnault-Pelterie's lecture reports and academic papers are scientific predictions rather than fantasies, which can be compared with the classic paper published by Tsiolkovsky in 1903. So, both laid together with the theoretical foundation of modern astronautics.

In 1927, Esnanlt-Pelterie delivered a lecture "Rocket Exploration of the Very High Atmosphere and the Future of Interplanetary Communications" in the annual general meeting of the French Astronomical Society, in which he emphasized the connection to his groundbreaking presentation 15 years earlier. His lecture aroused lively interest and the audience frequently interrupted it with applause. A half of a year later, the French Astronomical Society published the expanded version of his lecture in a special supplement, as a separate book, "*Rocket Exploration of the Very High Atmosphere and the Possibility of Interplanetary Travel*", in which the word "Astronautics" appeared for the first time as a scientific term and in quotation marks. In the meantime, together with a French industrialist André Louis-Hirsch, he established the first international prize in astronautics, *the REP-Hirsch International Astronautics Prize*. The Society formed a Committee of prominent scientists to award the prizes. In 1929, the first REP-Hirsch Award went to the German aerospace pioneer Hermann Oberth for publishing his book, "*Ways to Space Flight*". In the epilogue of his award-winning book, Oberth wrote: "*The French Astronomical Society recognized this book with the REP-Hirsch Award. In addition to great advantages that this award has brought me and will bring in the future, it had also a not-small moral effect. I did not sincerely believe that a German would be awarded such a prize in France, especially since good French, Russian, Italian, and English works were also submitted. One can see that science and education are capable of bridging national divisions. I think that I can not thank the French Astronomical Society better than to pledge here to also work for science and education and to judge others only their accomplishment*".

In 1930, on the basis of comprehensively and systematically summarizing more than 20 years' research results, Esnault-Pelterie wrote and published his culminating work, *Astronautics*, which covered virtually all that was then known of rocketry and space flight. This book involves celestial mechanics, astrophysics, physical chemistry, ballistics and biology, in which the problems of rocket engine, spaceship and space flight are discussed in detail, as well as the Theory of Relativity are also introduced into studying on rocket's motions for the first time, and thus is known as the encyclopedia of aerospace science. Four years later, he published its supplement edition, *Astronautics-Complement*, in which the principle of three-axis stabilized spacecraft by rapidly rotating gyroscope and the re-entry problem of spaceship into the atmosphere and its solutions are described in detail. Esnault-Pelterie presented the new science in a consistent and detailed way: discussing rocket motion in vacuum and air; considering gas flows in converging–diverging nozzles; applying thermodynamics to the combustion processes of various fuel-oxidizer combinations; pointing out the exceptional properties of atomic hydrogen as a propellant; suggesting to use reaction wheels for spacecraft's attitude control; as well as outlining possible rocket applications including studies on the aurora borealis and on the upper atmosphere, and missions to the Moon and to the planets. In addition to studying theoretical astronautics, he also initiated practical development of liquid rocket engines, so that he lost the ends of four fingers on his left hand in experiments with gasoline and tetranitro-methane as rocket propellants. Unfortunately, his researches have not been recognized and supported by French officials, while his rocket control idea has been successfully applied to the V-2 rockets developed by Germany later.

Esnault-Pelterie was one of founders of the French Astronautic Society and became a member of the French Academy of Natural Sciences in 1936, which shows that his hard work and achievements in the field of aerospace have been recognized by the scientific community. His research work was terminated in the Second World War, and he settled in Switzerland after the War. In 1946, he wrote a book *Dimensional Analysis*, at the end of which he proposed to set up branches for each scientific field in studies on the history of science, naturally including the histories of aeronautic and astronautic science. The purpose is to enable future generations to learn from the experiences and lessons of predecessors and better promote the development of science and technology. It can be seen that Esnault-Pelterie has really a profound thought that ordinary people cannot do. It is also a painful lesson that he learned when he saw that French rocket technology at that time was far behind Germany, the United States and the Soviet Union. After successfully launching the world's first artificial earth satellite in the Soviet Union, the Poland-Soviet astronautic scientist Ary Abramovich Sternfeld, who won the 1934 REP-Hirsch Award, wrote to André Louis-Hirsch, saying: "Robert Esnault-Pelterie's works have had a significant impact. The Soviet scientists used his theory of astronautics in their work".

Although Esnault-Pelterie had not actually developed a rocket, his rocket design idea was quite advanced. He had the honor of witnessing the successful launch of the world's first artificial earth satellite, turning his astronautic science theory into reality. On December 6, 1957, Robert Esnault-Pelterie passed away at the age of 76 in Geneva, Switzerland.

3.3.3 Robert Hutchings Goddard

Robert Hutchings Goddard (1882–1945)was an eminent American inventor and physicist, as well as one of the founders of modern rocketry and space flight. He developed and launched successfully the world's first liquid-fueled rocket, ushering in an era of space flight and innovation.

On 5 October 1882, Robert Goddard was born at Maple Hill in Worcester, Massachusetts, the United States. His parents were farmers, and that helped him acquire an interest in nature. He was the only surviving child of his parents, and the family moved to Boston soon after he was born. As a child, Goddard became interested in science and the parents encouraged him, gifting him a microscope, telescope and a subscription to the magazine of *Scientific American*. He grew up so weak and ill that he was often unable to go to school. Most of the time, he was self-taught and liked reading wonderful science fictions. Jules Verne's "*From the Earth to the Moon*" and Herbert George Wells's "*The War of the Worlds*" made him have a strong interest in space flight and unlimited imagination, and clear goals for it.

In 1898, Goddard together with his parents moved back to Maple Hill, Worcester. When his health slowly improved, Goddard attended South High Community School in Worcester in 1901. Graduating from High School in 1904, he then attended Worcester Polytechnic Institute to study physics. His fascination with space flight and means of attaining it continued into his college years. In an assigned theme, "Traveling in 1950", he was also intrigued with the notion of the fastest possible travel for living bodies on the Earth's surface and projected a plan for travel inside a steel vacuum tube in which cars were suspended and driven by the attraction and repulsion of electromagnets. Later, he was granted a patent on a vacuum-tube system of transport, with the thrust acceleration and deceleration as a chief principle. At the same time, he also studied the problem of gunpowder rocket and made the experiment of measuring rocket thrust and flight speed. In 1908, he graduated from college and earned a B.S. degree in physics.

Soon after remaining the college as an instructor, Goddard enrolled at Clark University in Worcester to study for a master's degree in physics, where his supervisor was the famous professor Arthur Gordon Webster, who was the founder of the American Physical Society. Webster was an open-mind scholar who encourages Goddard to develop his hobbies and strengths, and thus in addition to studying his thesis, he also considered studies on the related rockets. At this time, Goddard had realized the defect of low combustion efficiency of gunpowder rocket, and only liquid fuel to be able to provide the energy required for space flight. In 1910, Goddard earned an M.S. degree in physics by the thesis titled "Theory of Diffraction", and a year later received a Ph.D. degree by the dissertation "On the Conduction of Electricity at Contacts of Dissimilar Solids". He remained at Clark University for a year as an honorary fellow in physics, teaching courses and carrying out rocket experiments.

In 1912, Goddard accepted a research fellowship at Princeton University's Palmer Physical Laboratory, where he worked out the mathematics which would allow a rocket to travel upwards into space, proved that thrust and consequent propulsion could take place in a vacuum, and investigated the ratios of energy to thrust per weight of various fuels, including liquid hydrogen and liquid oxygen. Shortly thereafter, he had tuberculosis and became very ill. As a result, he had to resign from his position at Princeton University and went back to Worcester to treat his illness. He worked for an hour a day as his health improved. During this period, he filed two patents: one is on a multi-stage rocket with solid explosive material as fuel, and another on a rocket using solid fuel or liquid propellants as fuel. The two patents became instrumental in the history of developing rockets. In 1914, he accepted a part-time job at Clark University as a research fellow and instructor and was also able to continue his research on the rockets. He began to experiment and made significant headway over the next 3 years. He found that the combustion efficiency of gunpowder rocket was about 2% and the combustion speed only 300 m/s. By improving the combustion chamber and the nozzle's material and shape, and replacing the black powder with smokeless powder, the efficiency could reach 60% and the speed 1.8 km/s. In 1917, the Smithsonian Institution in Washington provided him a research funding of $5000 over 5 years, and the Clark University contributed $3500 to their physics laboratory as further sponsorship. On the basis of intensively studying and testing the combustion chamber of the rocket, he fully demonstrated that the gunpowder rockets are not only low in efficiency and jet speed but also extremely unstable in thrust. It was concluded eventually that only liquid rockets are the ideal vehicles for space flights.

When the United States announced its participation in the First World War, Goddard worked with the U.S. military to develop rocket weapons. He designed two kinds of rocket weapons: one is the weapon with a long range, similar to launching projectiles; the other is the weapon with a short range, known today as anti-tank rocket or bazooka. In 1918, he and his assistants carried out the test of launching the anti-tank rocket for the U.S. Army at a launching site in Maryland and achieved good test results. The Army showed interest, but the War came to an end, so the development did not go ahead. In 1919, Goddard wrote a research report entitled "A Method of Reaching Extreme Altitudes", which was published in the form of a treatise at the Smithsonian Institution. He described his mathematic theories of rocket flight as well as his experiments with solid-fueled rockets in the treatise, outlined the challenges of lifting mass high to the atmosphere, and explored how rockets could solve the problems of high-altitude studies. Such around 1750 copies were distributed worldwide that the publication of his treatise gained him national attention from newspapers in the United States, in which there were some objective reports, as well as doubts and sharp sarcasm about his scientific knowledge. Although his discussion of targeting the Moon was just a small part of the work as a whole and was only intended as a possibility rather than as intent, the newspapers overly sensationalized it, to the point of misrepresentation and ridicule.

Goddard continued to focus on the storage and transportation methods of liquid fuels and liquid oxidants and to look for possible fuels such as propane, ether and gasoline. Since 1921, he had devoted to develop liquid rocket engine and carry out a series of static tests but did not obtain the ideal results until 1925, a continuous 24-s test of the rocket engine at the Physical Laboratory of Clark University. On March 16, 1926, Goddard launched the world's first liquid-fueled rocket from his Aunt Effie's farm in Auburn, Massachusetts. Although the flight itself was modest, just reaching an altitude of around 12.5 m and covering a distance of 56 m, it represents the birthday of modern rocketry. From 1926 to 1941, Goddard carried out a total of 35 launching rocket tests and achieved a series of landmark test results, which are listed in Table 3.1. It is shown from Table 3.1 that 11 intensive launch-tests were carried out successively from 1935 to 1936 and a good flight test was achieved in 1937, with a flight altitude of 2700 m. Goddard had been suffered from technical problems such as the rocket stability control, the propellant transfer and the engine

Table 3.1 Launching rocket tests carried out by Goddard between 1926 and 1941

No.	Launching date	Rocket's Prototype	Flight Altitude (m)	Flight Time (s)	Notes
1	March 16, 1926	Goddard 1	12.5	2.5	The first liquid-fueled rocket
2	April 3, 1926	Goddard 1	15	4.2	
3	Dec. 26, 1928	Goddard 3	5		
4	July 17, 1929	Goddard 3	27	5.5	
5	Dec. 30, 1930	Goddard 4	610		
6	Sept. 29, 1931	Goddard 4	55	9.6	
7	Oct. 13, 1931	Goddard 4	520		
8	Oct. 27, 1931	Goddard 4	410		
9	April 19, 1932	Goddard 4	41	5	
10	Feb. 16, 1935	Series A	200		
11	March 8, 1935	Series A	300	12	
12	March 28, 1935	Series A	1460	20	
13	May 31, 1935	Series A	2300		
14	June 25, 1935	Series A	37	10	
15	July 12, 1935	Series A	2000	14	
16	Oct. 29, 1935	Series A	1200	12	
17	July 31, 1936	Series L-A	60	5	
18	Oct. 3, 1936	Series L-A	60	5	
19	Nov. 7, 1936	Series L-A	60		

(continued)

Table 3.1 (continued)

20	Dec.18, 1936	Series L-B	1		Turn to horizontal flight after launching
21	Feb. 1, 1937	Series L-B	570	20.5	
22	Feb. 27, 1937	Series L-B	460	20	
23	March 26, 1937	Series L-B	2700	22.3	Max flight altitude
24	April 22, 1937	Series L-B	2000	21.5	
25	May 19, 1937	Series L-B	990	29.5	
26	July 28, 1937	Series L-C	630	28	
27	Aug. 26, 1937	Series L-C	600		
28	Nov. 24, 1937	Series L-C	30		
29	March 6, 1938	Series L-C	160		
30	March 17, 1938	Series L-C	660	15	
31	April 20, 1938	Series L-C	1260	25.3	
32	May 26, 1938	Series L-C	40		
33	Aug. 9, 1938	Series L-C	1500		Optic altitude
			1000		Pressure altitude
34	Aug. 9, 1940	Series P-C	90		
35	May 8, 1941	Series P-C	80		

cooling. For this reason, a number of flight tests had done, but the problems had not been solved yet, even in the last flight in 1941.

At that time, the liquid rockets were being to a cutting-edge science and technology field, which is quite difficult. As a result, there were many deviations in Goddard's research methods. In addition, Goddard was too secretive about his studying work to disclose technical details and problems to the others, which led to many detours, while the key technologies affecting the rocket performance had not been resolved. In August 1936, Goddard was visited by Frank Malina, who was then studying rocketry at California Institute of Technology and whose mentor was the famous Hungarian-American mathematician and physicist Theodore von Kármán. Goddard declined to discuss any of his research, other than that which had already been published in the report, "Liquid-Propellant Rocket Development". This deeply troubled von Kármán, and later in the biography, *The Wind and Beyond: Theodore von Kármán, Pioneer in Aviation and Pathfinder in Space,* written by Theodore von Kármán and Lee Edson, he said: "*I believe Goddard became bitter in his later years because he had no real success with rockets, while Aero-jet General Corporation and other organizations were making an industry out of them. There is no direct line from Goddard to present-day rocketry. He is on a branch that died. He was an inventive man and had a good scientific foundation, but he was not a creator of science, and he took himself too seriously. If he had taken others into his confidence, I think he would have developed workable high-altitude rockets and his achievements would have been greater than they were. But not listening to, or communicating with, other qualified people hindered his accomplishments*". So, Goddard's achievements in the liquid rockets had little effect on the later development of rockets in the United States, and he was quickly surpassed by other rocket research teams.

After the incident of the Attack on Pearl Harbor in 1941, the United States declared its participation in the Second World War. Goddard was once transferred to the U.S. Department of Defense to participate in the development of rocket weapons, but after all, it was a short time, and his research results had not been applied to the War. Since the Army declined his offer to develop the rockets for them, Goddard temporarily gave up his preferred field to work on experimental aircraft for the U.S. Navy. When the War ended, Goddard was able to inspect captured German V-2 rockets, many components of which he recognized. Nevertheless, he would not design any more rockets of his own.

On August 10, 1945, Goddard passed away in Baltimore, Maryland, due to throat cancer at the age of 63. Goddard's wife, Esther Christine Kisk, gathered his papers and worked on securing patents after his death. Many of his original papers containing his seminal work on rockets can be seen in the Smithsonian Institution Archives. Many years later, his work was acknowledged by the U.S. government when $1,000,000 settlement was made for the use of his patents. The Goddard Memorial Library at Clark University was named in his honor. In 1959, NASA's Administrator announced the naming of Goddard Space Flight Center (GSFC) under construction near Greenbelt, Maryland, in commemoration of Robert Hutchings Goddard, the American pioneer in rocket research. There are also many streets and a park that bear his name, and the Blue Origin makers have named a reusable launch vehicle for him. Goddard's influence and impact continue to be felt throughout our current space exploration efforts, plus those in the future.

3.3.4 Hermann Julius Oberth

Hermann Julius Oberth (1894–1989) was an Austro-Hungarian-born German engineer and physicist, as well as one of the pioneers of modern rocketry and astronautics. He has made a great contribution to the theory of rocket motion and astronautics. His works on space flight have influenced the whole of Europe, attracting a large number of young people to devote themselves to studying rockets and astronautics.

On 25 June 1894, Hermann Oberth was born in Transylvanian town, Hermannstadt, Austria-Hungary (today Romania), and his father was a prosperous physician. Around the age of 11 years old, Oberth liked to read science fictions, especially "*From the Earth to the Moon*" and "*Around the Moon*" written by Jules Verne, re-reading them to the point of memorization. From this time, Oberth became fascinated within the spaceflight stories while he boldly raised the questions of flying ways. For example, launching the astronauts by the artillery in Verne's book, they would undoubtedly be smashed into pieces due to the action of huge gun-tube acceleration. At the age of 14, he constructed a model rocket as a middle school student, arriving independently at the concept of the multistage rocket, but he lacked then the resources to pursue his dream on any except a pencil-and-paper level.

At the age of 18, Oberth moved to Germany where he spent most of his adult life. In 1913, dutifully following his father's footsteps, Oberth enrolled at the University of Munich to major in medicine, but it was not his ambition, and thus he took elective courses like mathematics and astronomy in accordance with his own interest, exploring the possibility of space flight. Soon after the outbreak of the First World War, his education was interrupted by service in the Imperial German Army. He was assigned to an infantry battalion, and sent to the Eastern Front against Russia. Two years later, he was moved into a medical unit at a hospital in Sighisoara, Transylvania, in Austria-Hungary. In the meantime, he used his spare time to study spaceflight theory and weightlessness effect and designed a long-range rocket weapon. In 1917, he showed how far his studies had reached by firing his rocket with liquid propellant in a demonstration to the Prussian Minister of War, but it has not attracted official attention. As a medical student, he was more concentrated on space medicine, particularly the impact of weightlessness on human physiology and psychology. For this reason, he had conducted a series of experiments of weightlessness, intentionally drowning or injecting anesthetics when swimming to get the experience of weightlessness.

In 1919, Oberth once again came back to the University of Munich to major in physics rather than in medicine, and later went to the University of Göttingen to study scientific fundamental courses like mathematics, physics and astronomy. By 1922, he had preliminarily established a theoretical frame of rocket motion and space flight and submitted a doctoral dissertation on rocket science to the University of Heidelberg, but it was rejected as "utopian". Meanwhile, he learned from the newspaper that the U.S. rocket expert Robert Goddard had published the classic paper, "A Method of Reaching Extreme Altitudes", and wrote to Goddard for the full text. Concerned with the German interest in space flight, Goddard sent him a courtesy copy with some apprehension. Goddard's paper gave him great confidence, which the principle of his liquid rocket was not only similar to that of Goddard but also he had more in-depth and forward-looking in the studied on astronautic theory. The next year, Oberth was eventually awarded a Ph.D. degree in physics with the same rocketry dissertation that he had written before, by the University, Cluj-Napoca, Romania. Furthermore, he privately published his 92-page work as the somewhat controversial book, "*The Rocket* ", which is mainly composed of three parts as follows:

(1) The first part is the general problems of rocket motion, including the basic equations of rocketry, the escape velocity and the jet velocity of the engine.
(2) The second part is the detailed designs of high-altitude rockets, in which the rocket propellants with liquid fuels and the advantages of liquid oxygen and alcohol as propellants are focused on, as well as the feasibility of utilizing the rockets to studying high-altitude atmosphere is discussed.
(3) The third part is theoretically to describe the spaceships and to discuss the problems of flying to the Moon, Mars and Venus by the spaceships.

This book was very popular, especially among young people. The first edition soon sold out, and then the second edition was published once again at the end of 1923. Max Valier, a young freelance science writer and one of the founders of the German Spaceflight Society, had adapted the book into a popular work, "*The Advance into Space*", and published it many times, once again improving Oberth's popularity among the German and European publics. Under the influence of Oberth's works, Willy Ley, a 19-year-old student majored in biology, devoted to studying rocketry and later became the backbone of the German Spaceflight Society, writing many books on astronautic science and scientific history. Moreover, Oberth's works would also have a huge and life-changing impact on 10-year-old Wernher von Braun, later becoming a giant in both German and American rocket engineering. To support himself and his family, Oberth had taught physics and mathematics at the Stephan Ludwig Roth High School in Medias, Romania, from 1924 to 1938.

In 1928, a famous German director, Fritz Lang, was making a science fiction film called "*The Woman on the Moon*", in which there are scenes set in outer space, and invited Oberth as a scientific consultant. One of Oberth's main assignments was to build and launch a rocket as a publicity event just before the film's premiere. He also designed the model of the "Friede", the main rocket portrayed in the film. The film was of enormous value in popularizing the ideas of rocketry and space exploration. With Lang's strong support, Oberth then carried out rocket developments and tests, but they were not successful. In 1929, Oberth won the first REP-Hirsch Prize of the French Astronomical Society for his outstanding contribution to studying astronautic science. He used the 1000 Francs prize from the REP-Hirsch to publish his book, "*Ways to Space Flight*".

It was under the influence of Oberth's unremitting efforts and his astronautic monographs that the German military first realized the great military strategic significance of studying rockets and supported the relevant research work. After the outbreak of the Second World War, Oberth was invited to participate in rocket researches at the Technical College in Vienna, Austria and then was transferred to the Technical College in Dresden, Germany, to develop the fuel pumps of large rockets.

Both colleges were under the control of the German military, and Oberth was still a Romanian citizen at the time. Oberth joined the German nationality at the instigation of German authorities and was sent to the Peenemünde rocket research base led by Wernher von Braun in 1941, to work on Nazi German rocketry projects, including the V-2 rocket weapons. In 1945, after the Allied occupation of Peenemünde, Oberth was captured but soon released. Thereafter, Oberth had worked in education and was ever elected honorary chairman of the German Society of Space Research, writing scientific articles for magazines or serving as a scientific adviser. Oberth was allowed to leave Nuremberg, Germany, and to move to Switzerland in 1948, where he worked as an independent rocketry consultant and a writer. In the early 1950s, Oberth had also worked for the new Italian Navy. Returned to Germany in 1953, he published a book entitled "*Men in Space*", in which he described his ideas for space-based reflecting telescopes, space stations, electric-powered spaceships, and space suits. In 1955, at the proposal of von Braun, Oberth was invited to develop space rockets for NASA in Huntsville, Alabama, USA. Among other things, he was involved in writing the research report, "*The Development of Space Technology in the Next Ten Years*". Later, he also went to work for the Convair Corporation in the United States as a technical consultant on the Atlas rocket program, and then retired in 1962.

On December 29, 1989, Oberth died at the age of 95 in Nuremberg, Germany. Among the four great pioneers in the field of modern astronautics, only Oberth witnessed the brilliant era of space flight development, which the early astronautic theory has become reality. In 1957, the Soviet Union successfully launched the world's first artificial earth satellite, which is an important sign for humanity entering the era of space flight; in 1961, the Soviet Union successfully launched the first astronaut into space; in 1969, the United States successfully launched the first manned moon landing; in 1971, the Soviet Union successfully built the first space station; in 1982, the United States successfully launched a reusable space shuttle. Moreover, there was a series of deep-space exploration missions, including the probes that exploring the planets and interplanetary space and that were flying out of the solar system.

3.4 Early Astronautical Societies and Historical Merits

By the 1920s, the pioneers of modern astronautics had initially established the fundamental theory system of rocketry and space flight, and it had been spread widely [5]. In many countries, such as the Soviet Union, Germany, the United States and the United Kingdom, a generation of young people with ideals, interests and passions spontaneously organized and successively set up various societies of studying liquid rockets and space flights, which promoted the development of modern space science and technology.

3.4.1 Soviet Societies on Rocketry and Astronautics

Justly under the effects of Tsiolkovsky's works on aerospace science, the Soviet Union was the first country to set up the societies on rocketry and astronautics and became the country of the most that absorbed and trained the young reserve talents in the field of space flight. Fridrikh Arturovitch Tsander (Фридрих Артурович Цандер, 1887–1933), a gifted rocketry expert in the Russian Empire and the Soviet Union, was the early follower of Tsiolkovsky. He had systematically studied the astronautic theories, including the energy required to reach the certain planet, the work is done to launch a certain mass from the Earth to a far distance, the oxygen needed for a person to fly in outer space and the metal materials used as rocket fuels, and actively publicized and popularized the knowledge of space flight. Tsander saw himself taking the theoretical groundwork laid at the turn of the century by Tsiolkovsky and extending it to the practical reality of taking man into space. By studying and publicizing space science, Tsander increasingly realized that space flight, a long-term mission and an arduous work, would be achieved only by relying on collective strength and wisdom.

In January 1924, Tsander gave a lecture entitled "Flight to Other Planets" to the amateur astronomical association in Moscow, when he suggested setting up the research organization on interplanetary travel, which received wide responses from the audiences. Shortly thereafter, the research organization was set up officially and named Society for Studies of Interplanetary Travel, which was a branch of the Military Science Department at Zhukovsky Air Force Academy. The Society became a platform for scientists and space explorers to meet for discussions and organize public educational events. The Society's aims are to bring together researchers from all over the Soviet Union devoted to space flight, investigate the progress on rocketry and astronautics in Europe and the United States, sort out and publish the status of space flight studying, promote the researches on rocket's military applications, and encourage individual research works. It is the first society of studying space flight in the Soviet Union. At first, the Society had only 25 members and soon expanded to almost 200, including the astronautic pioneer Tsiolkovsky and the popular science writer Perelman. Tsander had attempted to lead the Society to develop and test liquid rocket engine, but due to the lack of research funding support, the Society's main activities could only be theoretical studies, academic lectures and exhibitions on astronautic science, and finally, the Society was disbanded in 1927. Under limited conditions, Tsander independently developed the OR-1 reaction engine, which used compressed air and gasoline and was able to produce a thrust of 0.145 kgf. In 1931, a glider pilot who was keen on rocketry and space flight, and a 24-year-old graduate from the Moscow Bauman Higher Technical School, Sergei Pavlovich Korolev (Сергéй Пáвлович Королёв, 1907–1966), visited Tsander, hoping to jointly develop the liquid-propellant rockets.

At the beginning of the 1930s, there were a number of amateur rocket research groups and lone researchers in the Soviet Union. In order to unite the rocket enthusiasts, the Soviet government sanctioned to create a professional rocketry organization, named Group for the Study of Reactive Motion (GIRD in its Russian acronym), with branches in Moscow and Leningrad, and later with those in Baku, Tiflis, Arkhangelsk, Novocherkassk, Bryansk and other cities. With the support of Mikhail Nikolayevich Tukhachevsky (Михаил Николаевич Тухачевский, 1893–1937), who was then the vice-chairman of the Revolutionary Military Commission of the Soviet Union and was made a marshal of the Soviet Union in 1935, the Moscow branch of GIRD by led Tsander and Korolev obtained the contract research funds. The Moscow branch was organized into four brigades and ten projects to study rocket engines and winged and un-winged missiles.

The first brigade, led by Tsander, was responsible for the development of liquid rocket engines, and worked on the three projects: project 01 was the continuation of the OR-1 experimental engine, investigating the energetic fuels of including powdered metals mixed with gasoline and the combustion chamber cooled regeneratively by air entering at the nozzle end; project 02 was to develop the OR-2 engine for Korolev's RP-1 rocket-powered glider; the project 10 was to develop the engine of burning liquid-oxygen and gasoline for the GIRD-X rocket. The second brigade, led by M.K. Tikhonravov, was mainly responsible for the development of the rockets and worked on the four projects: project 03 was the development of a pump-fed engine called RDA-1, with the liquid-oxygen and gasoline as fuels; project 05 was actually the third rocket designed by the second brigade, using the ORM-50 nitric acid and kerosene engine developed by V.P. Glushko; project 07 was the first rocket designed by the second brigade, using the OR-2 engine developed by Tsander, and its purpose was to test a liquid-oxygen and kerosene propulsion system and its unusual design included fuel tanks in the oversized fins; project 09 produced the first rocket to actually be flown by GIRD. The third brigade, led by Y.A. Pobedonostsev, was responsible for the development of the ramjet engines and the solid-fuel rockets respectively in projects 04 and 08, one of which was the development of the supersonic wind tunnel with a flow of Mach 3.2. The fourth brigade was led by Korolev, whose mission was to develop PR-1 manned rocket-powered glider and the unmanned cruise missile for project 06.

On March 28, 1933, after completing the design and first bench test of the GIRD-X rocket, Tsander died unexpectedly from typhoid fever. In August of this year, the launch test of the project-09 rocket, the first semi-liquid rocket in the Soviet Union, which was designed under the leadership of Tikhonravov, got successful. Its flight altitude reached 400 m, and the maximum thrust of the project-09 engine was 53 kgf, with a duration of 18 s. In November of this year, the GIRD-X rocket was launched, and its flight altitude reached 80 mbefore its engine failure. It was the true liquid-fuel rocket to fly in the Soviet Union, although not particularly successful. The engine of the GIRD-X rocket burned liquid-oxygen and gasoline, and it could be cooled regeneratively. Korolev was the over-all chief of GRID, as well as the fourth brigade's leader and the chairman of its technical council. It was an early demonstration of his impressive skill as a project manager. Later, he became the chief head of the Soviet space program.

In addition, the Gas Dynamics Laboratory (GDL) was the most influential organization on studying rocketry and space flight in the early Soviet Union. Its predecessor was N. I. Tikhomirov's Laboratory which was created by the Soviet government in Moscow to develop the idea of Tikhomirov, a chemical engineer, for using slow-burning solid explosive substances for self-propelled mines for air and water. In 1925, Tikhomirov's Laboratory was moved to Leningrad. In 1928, it was officially renamed Leningrad Gas Dynamics Laboratory, and a military engineer I.T. Kleimenov served as the director of the Laboratory. In 1929, the second division of the Laboratory was established to initiate a new research direction, i.e., electrothermal and solid-fuel, liquid-fuel rocket engines. Valentin Petrovich Glushko (Валент´ин Петрóвич Глушкó, 1908–1989), who was then an only 21-year-old engineer and just graduated from Leningrad University, served as the director of the second division. The GDL was built with the strong support of Tukhachevsky, with powder missile research, rocket-assisted takeoff, liquid-fuel rocket and ORM engine series as the major directions. Consequently, it was directly managed by the Technical Headquarters of the Soviet Red Army, and its research work was strictly kept secret. This marked the beginning of the official liquid rocket research activities in the Soviet Union. For the 5 years since establishing the second division, Glushko had completed the design, development and test of 30 types of the engines, and solved a series of technical problems, such as propellant materials, pump-feeding, fuel-supply unit, combustion chamber cooling, nozzle's shape and engine performance measuring. The ORM-34 engine designed by Glushko, which uses nitric acid and kerosene as propellants, could produce a thrust of 300 kgf. Furthermore, he also developed the world's first electro-thermal rocket engine.

In September 1933, the Jet Propulsion Research Institute (RNII) was officially created in Moscow by Decree Number 113 of the Revolutionary Military Soviet, signed by Tukhachevsky as a result of a merger among the GDL, the Moscow branch of GIRD and the Leningrad branch of GIRD. The GDL officially became a Leningrad branch of RNII. Kleimenov was assigned to lead the RNII, and for a short time, Korolev worked as deputy chief of the RNII. It was an important sign of the normalization of Soviet rocket research. At the beginning of 1936, Korolev developed a concept for a manned glider, which would be powered by both liquid-fuel and solid-fuel rocket engines. The glider was expected to achieve a speed of 300 m/s at an altitude of 3 km and to climb up to a maximum altitude of 25 km and would be piloted by two people wearing pressure suits. A heavy transport plane would tow the rocket from takeoff and up to the altitude of 8–10 km. Considering the implementation of the project, Korolev proposed to install Glushko's ORM-65 engine on the SK-9 glider, which Korolev had completed the development a year earlier. The resulting configuration was designated RP-318-1 glider. At the end of 1936, the jurisdiction over RNII was transferred from Narkomat of Heavy Industry to a spin-off ministry named Narkomat of Defense Industry, and the Institute was renamed Scientific-Research Institute 3 (NII-3). The same year on ballistic missiles, which employ liquid-fuel propulsion, was consolidated in a single department led by Korolev.

In June 1937, Marshal Tukhachevsky was tried with seven other top Red Army commanders in conjunction with the Stalinist purges, on charges of conspiracy with Germany. All eight were convicted and executed. With the "Anti Soviet Movement" expanding, the Soviet intelligentsia with any apparent or assumed links to the old revolutionary Bolshevik elites became one of the primary targets of the purges. Because Tukhachevsky was an original patron of the NII-3, the arrests within the Institute followed. The Institute's director and his deputy were executed while leading engineers Glushko and Korolev were given long prison terms.

Since Korolev was arrested in prison, the NII-3 management ordered the mothballing of the RP-318-1 glider, and the project remained stalled until December 1938. In the wake of the purges at the NII-3, a competing team of engineers took over the RP-318-1 project. Glushko's ORM-65 engine was replaced with the RDA-1-150 engine developed by L.S. Dushkin. In 1940, the first powered flight of the RP-318-1 glider was conducted and its flight speed reached 140 km/h. As a result of the experiments, the Soviet Defense Committee issued a decree authorizing the development of a rocket-powered aircraft. The BI-1 experimental fighter plane, develop V.F. Bolkhovitinov's team and equipped with liquid-fuel rocket engines, conducted its first flight in 1942.

In the late 1930s, the NII-3 also designed and developed a variety of rocket engines. Tikhonravov developed an air research rocket for the Soviet Association of Aeronautic Technology, and its flight altitude reached 3 km. Its engine was the 12-K engine developed by Dushkin, with liquid oxygen and alcohol as propellant, and with a thrust of 300 kgf and the duration of 60 s. Glushko took charge of the development of ORM series rocket engines, in which the thrust of ORM-64 engine reached 680 kgf, and the variable-thrust engine ORM-65 was with a thrust range of 50−150 kgf. The director of the seventh design office at the NII-3, L. K. Korneev, took charge of the design of R01−R10 series rockets, of which R05 was a cosmic ray exploration rocket designed for the Institute of Geophysics, the Soviet Academy of Sciences, and R-10 was a two-stage liquid rocket, with four solid boosters tied around the first stage and the designed takeoff thrust of 1000 kgf.

After the outbreak of the Second World War, the technicians of the NII-3 were demobilized or persecuted, and the rocket research work was almost stagnant. At this time, Stalin soon recognized the importance of aeronautical engineers in preparing for the impending war with Hitler. To sufficiently exploit the jailed talents, a system of prison design bureaus, known as sharashkas, was set up soon after. Glushko was rose to head of a sharashka in the city of Kazan (officially called TsKB-4), which was developing rocket installations to improve maneuverability and takeoff performance of conventional aircraft. Despite the austere conditions, Glushko developed new engine series RD-1, RD-2 and RD-3, with nitric acid and kerosene as propellants and with pump-feeding ways, and their thrusts reached 300 kgf, 600 kgf and 900 kgf respectively. Around the early 1941s, Korolev, Glushko's former colleague from the NII-3, was also transferred to TsKB-4 as his deputy. In 1943, it was successful in a flight test that the RD-1 engine assisted the aircraft takeoff.

In general, during the 1920s and 1930s, many societies studying rocketry and astronautics had been established in the Soviet Union. Besides the more influential societies such as the Group for the studies of Interplanetary Travels, Group for the Studies of Rockets, Gas Dynamic Laboratory, Group for the Studies of Reactive Motions and Jet Propulsion Research Institute, other ones such as the Ukrainian Space Research Society and the Interplanetary Research Department at the Soviet Inventor Association, had also carried out the extensive publicities, popularities, studies and experiments on aerospace science. Except for the well-known Katyusha rocket, a kind of rocket weapon using solid propellants, the liquid rockets developed by the astronautic societies were not used in the Second World War. Nevertheless, the astronautic societies cultivated a large group of specialists with rich theoretical research and practical experience, laying a consolidated foundation for the successful launch of the first artificial earth satellite and making indelible contributions to the development of human space efforts.

3.4.2 German Society for Space Flight

In the 1920s, Hermann Oberth's book "*The Rocket into Interplanetary Space*" became so widely popular in German-speaking Central Europe, inspiring a group of young German people to be fascinated and love for astronomy, rocketry and astronautics, among which Johannes Winkler (1897–1947) was a big fan of Oberth and his works. During the First World War, Winkler joined the German army and was severely wounded at the Battle of Lake Narocz in 1916, leading to a lengthy hospitalization. After seven months in the hospital, he attended the Danzig Technical College, graduating with a degree in mechanical engineering in 1918. While working as an engineer for Junkers Aircraft Company in the city of Breslau, Germany (today Wroclaw, Poland), Winkler became seriously fascinated with Oberth's calculations proving the feasibility of manned space travel using liquid propellant rockets.

On June 1927, the Junkers Aircraft engineer Winkler together with other two astronautic enthusiasts, a paleontology student Willy Ley and a construction engineer Max Valier, founded the Society for Space Flight (Verein fuer Raumschiffahrt–VfR) in Breslau, Germany. In addition to the three founders, among its first members were also Hermann Oberth, Walter Hohmann and Rudolph Nebel, who are now legendary names in the history of early space flight. Winkler was elected as the first president of the Society, and the Society decided to create an academic journal entitled "*The Rocket*". The Society's aims are: to carry out theoretical and experimental researches on rocket and space flight; to verify and apply the theories presented by Oberth in his book; to develop a large-size spaceship that can carry people to outer space or other planets. In order to disseminate the idea of space flight to the public, the Society decided to edit and publish an intellectual book, "*The Possibility of Interplanetary Travel*", by Willy Ley as its chief editor. To devote to studying liquid rockets, Winkler quitted his job at the Junkers Aircraft Company. With the Society expanding rapidly, its membership had reached 900 people by 1929, in which not only were there the famous experts, besides Oberth, such as the French astronautic pioneer Robert Esnault-Pelterie, the Austrian rocket theorist Franz von

Hoefft, the Austrian rocket pioneer Guido von Pirquet and the Soviet astronautic scientist Nikolai Alekseevich Rynin but also many young people such as the 17-year-old Berlin Technical University student Wernher von Braun, who later became a core leader in the development teams of the V-2 rocket in Germany and the Saturn-5 huge-rocket in the United States. However, the establishment of the VfR had not been supported and funded by the German government yet. Due to the lack of research funding, it had to face survival and developmental problems. As a result, the organizational structure and research work of the VfR were relatively loose, and actually, only several core members had always studied liquid rockets.

In order to raise research funding, Valier persuaded a German automaker to provide funding for rocket car development. Consequently, he had developed a series of solid-fuel rocket vehicles with a speed of over 112 km/h and was highly praised by manufacturers and major newspapers. However, some members from the VfR criticized Valier: it has been proved that only liquid rockets are a unique feasible vehicle to realize space flight; Valier should not spend time and energy in solid-propellant rocket research, which is also a kind of misguidance for the public. Later, with the support of a manager of the German Industrial Gas Application Society, Valier and his two assistants, Walter Riedel and Arthur Rudolph, started to develop liquid rocket vehicles with liquid-oxygen and gasoline as propellants. In March 1930, they developed a non-cooled liquid rocket engine with the thrust of 8 kgf, which was not an ideal performance specification. To make the engine thrust reach 100 kgf, they improved the designs of the nozzle, combustion chamber and propellant proportion of the engine. On 17th May 1930, they conducted a series of tests for the improved engine, using kerosene mixed with water in conduction with liquid oxygen; during testing the combustion chamber pressurization, the engine exploded, and Valier was broken through his pulmonary artery by a piece of shrapnel, resulting in blood loss and death. At this time, he was just the age of 35. It was the first fatal accident in aerospace technology experiments, which the Society deeply regretted. Since then, the relevant experiments had been particularly cautious and the professional safety assessment links were also added.

Two years after founding the VfR, Winkler resigned from the post of the VfR's president, and Oberth was elected as the second president. Later, Winkler conducted liquid rocket researches on his own with funding from a number of small development contracts. Winkler developed a "three legged" rocket HW-1, with a strange shape and a stable tail-control device, which had three storage tanks with liquid oxygen and methane propellants, 60-cm high and 5-kg weight. In February 1931, Winkler conducted the first flight test of the HW-1 rocket at Dessau, Germany. Although its flight altitude only reached 3 mdue to a malfunction, it was the first liquid-propellant rocket flown in Europe. In March of the same year, the HW-1 rocket was conducted the second flight test, which flew to an altitude of 500 mover Dessau. At that time, the members of the VfR did not know the liquid rocket launch experiments made as early as 1926 by Goddard in the United States while thought that it was the first to set a world record for liquid rocket flight, which extremely excited them. Shortly thereafter, Winkler started to develop the second large liquid rocket, HW-2, with 2-m high and two large propellant tanks loaded with liquid nitrogen and methane.

The rocket's combustion chamber was connected in parallel with a slender nozzle, with the designed flight altitude of 5000 m. In October 1932, the launch test of HW-2 rocket was carried out from East Prussia, on the coast of the Baltic Sea. However, due to the leakage of propellant, the rocket exploded after igniting and rising only 3 m, which hit Winkler hard. Since then, Winkler returned to his old job at the Junkers Aircraft Company in silence and broke away from the VfR informally. During the Second World War, he worked for the government at an Aeronautical Research Institute. Although he designed a number of Jet-assisted takeoff units and sounding rockets, none were ever put into production. On 27 December 1947, Winkler died in Braunschweig-Querum, Germany.

Furthermore, under the guidance of Hermann Oberth, the research team, composed of some key members of the VfR, such as Willy Ley, Wernher von Braun, Klaus Riedel and Rudolph Nebel, had been devoted to the research of small liquid rockets. In the autumn of 1929, Oberth conducted a static firing of his first liquid-fueled rocket engine named Kegeldüse, which was built by Klaus Riedel and von Braun in workshop space provided by the Reich Institution of Chemical Technology. The engine used liquid oxygen and gasoline as fuels, and despite lacking a cooling system, it could run briefly. In July 1930, Nebel arranged a demonstration on the static firing tests of the Kegeldüse for the director of the National Chemical Institute in an attempt to secure financial support. In the demonstration, the engine worked for 90 s, generated a thrust of 7 kgf, and consumed the liquid oxygen of 6 kg and the gasoline of 1 kg. Nevertheless, all rocket flight tests based on the engine ended in failure until 1931. For this reason, the rocket engine was redesigned, in which high-pressure nitrogen was injected and a better fuel valve was installed; the liquid-oxygen tank and the gasoline tank were connected in series, four stabilizers were installed at the tail of the rocket body, and parachutes and timing-opening devices were installed; the top engine was connected with four conduits, and a stabilizer bar was installed at the center of the rocket to guide. The research team also named the new rocket "Propeller" and developed four types of rocket series. At the end of 1931, the flight test of the Propeller rocket reached an altitude of 1650 m, and the liquid rocket experiment achieved great success, winning the reputation for the VfR, highly praised by the American Interplanetary Society.

With the gradual growth of the Nazi Party in Germany, most of the VfR's members lost their enthusiasm for rocket research and began to look for their own ways. Willy Ley went to Britain alone, while Wernher von Braun, Rudolph Nebel, Klaus Riedel and Walter Riedel were recruited to join the Research and Development Department of the German Army Artillery Bureau. By 1933, the German Society for Space Flight basically died out. During just 5 years, the Society had done a lot of pioneering works, holding 23 rocketry and astronautic exhibitions and carrying out 270 rocket engine ignition tests and 87 launch tests, which lay a technical foundation for the successful development of the German V-2 rockets.

3.4.3 American Space Research Organizations

In the 1930s and 1940s, the space research organizations in the United States mainly including the Interplanetary Society in New York, the Jet Propulsion Laboratory of Caltech, Cleveland Rocket Society and Yale University Rocket Club, among which the Interplanetary Society (later renamed Rocket Society) and the Jet Propulsion Laboratory played an important role in promoting the U.S. rocketry and astronautics development.

3.4.3.1 The American Rocket Society

In the early twentieth century, the explosion of space fantasy was a double-edged sword for spaceflight advocates, which inspired young people to believe in the possibility of space travel but convinced many adults that the idea was absurd, especially in the United States. In April 1930, the American Interplanetary Society (AIS) was founded in the city of New York by 12 science fiction writers or spaceflight advocates, including David Lasser (1902–1996) who was an editor at the monthly fiction magazine *Science Wonder Stories*, George Edward Pendray (1901–1987) who was an editor at the newspaper *New York Herald-Tribune*, and Murray Fletcher Pratt (1897–1956) who was a science fiction and fantasy writer. Lasser was elected as the first president of the AIS, and Pendray as the first secretary-general and Lasser's deputy. Among the 12 members of the AIS, there was one woman member, Leatrice May Gregory, who was Pendray's first wife. The purpose of founding the AIS was to promote the American interests and experiments toward interplanetary expedition and travel. However, the fact that the magazine editors or fantasy writers predominated was unique to the United States, which reflected that with the exception of Robert Goddard, the genre's flourishing and the dearth of serious space theoreticians. As soon as the AIS was founded, Lasser sent an invitation letter to Goddard, asking him to make an academic lecture for the membership and join the AIS and to provide help and guidance for them to carry out the studies on rocket's technology, but Goddard replied that he refused. The AIS was lack of both members with a professional background in science and the expert's help, and consequently, it could only carry out some publicity and popularization of knowledge related to space flight. Meanwhile, they also created a *Journal of the American Interplanetary Society*, which mainly published the articles on the interplanetary travels and the latest development in international rocket researches.

In 1931, both Pendray and his first wife visited the German Society for Space Flight and met Willy Ley in Berlin, as well as looked around the rocket test conducted by the French astronautic pioneer Robert Esnault-Pelterie. They were excited by the achievements of European liquid rocket researches and experiments. Pendray began to recognize that the AIS should not only stay in scientific fantasy and endless discussion but also started to develop rockets. By popularizing and publicizing the knowledge on rocketry and astronautics for more than one year, the AIS had a certain popularity, and with the new members joining, the membership's profession formation had also been changed significantly. At the time, there had been some scientists and engineers joining the AIS, such as young engineers Alfred Africano, Hugh Franklin Pierce, John Shesta and Lovell Lawrence. In addition, Goddard also agreed to join the AIS, but he few took part in the activities of the AIS.

In 1932, Pendray replaced Lasser as the second president of the AIS, and both he and Pierce took charge of the design of the first rocket, AIS-1. They attempted to copy the small rocket of the VfR, including the structure shape, layout and propellant. The AIS-1′s motor was expected to produce 27 kgf thrust for 20—30 s. In November of the same year, they were preparing for the test of the AIS-1 rocket at an empty farm field near Stockton, New Jersey. Unfortunately, the rocket slipped while being mounted, and fell to the ground; it was twisted out of alignment, ending its short career. After a half year, the AIS-2 rocket, built of salvaged parts from the AIS-1, was allowed to launch from the Marine Park on Staten Island. The rocket roared 76 m into the air, where the oxygen tank burst, and then fell into the Lower New York Bay.

In April 1934, in order to highlight the substantial research on liquid rocket technology, the AIS was renamed the American Rocket Society (ARS), when most of the original science fiction crews had left, to be replaced by a group of scientists and engineers. Meanwhile, on the basis of improving the design of the AIS-2 rocket, both Pendray and Shesta had led two teams respectively to develop ARS-3 and ARS-4 rockets, of which the ARS-4 rocket had made significant improvements, whose length had been increased to 2.1 m, and a single combustion chamber with four canted nozzles designed. In September of the same year, ARS-4 rocket was launched from Staten Island, reached 130-m altitude and crashed into the New York Bay 407-m downrange, which achieved a good flight test result. Afterward, the ARS decided to stop the development of following-on series rockets and gave a priority to the development of high-thrust engine and test-bed, of which Shesta was responsible for the test-bed and his assistant was a student named James Wyld who just graduated from Princeton University. Moreover, other members of the ARS also carried out extensive researches on rocket's motion theory, structural design, cooling and application. In 1936, the ARS and its member Alfred Africano were awarded the REP-Hirsch Prize by the French Astronomical Society in recognition of their pioneering tests with liquid-fueled rockets. In 1938, the 23-year-old rocket expert James Wyld designed a new type of liquid rocket with a gyro stabilizing device, a parachute and a control system. Particularly, its engine was a regeneratively cooled motor, using a new type of propellant, which could produce a thrust of 41 kgf.

With the outbreak of the Second World War, the U.S. government increasingly focused studies on rocket technology. The ARS was mostly an East Coast Group, while new rocket centers were emerging in the West. At first, it had never been considered to create an independent professional rocket society on the West Coast, but this idea was finally abandoned due to some rocket enthusiasts from the West Coast joining the ARS. The ARS's periodical publication, *Astronautics*, played an important role in information exchange among rocket enthusiasts. Several young members, including Wyld, Pierce, Shesta and Lawrence, had contributed many refinements to the motors and designs. Based on the work that they had done with the ARS, they formed a company, Reaction Motors, Inc., in the end of 1941, which was the first American company founded to produce liquid-propellant rockets. They had formed the company to do business with the U.S. government, which would not contract with individuals. For example, the company undertook the contract projects from the U.S. Navy, which developed high thrust liquid rockets for heavy aircraft assisted takeoff, and finally completed the development of the motor with a thrust of 2700 kgf.

By 1947, the ARS passed new by-laws that included official grades of membership, added four regional sections, and established a subcommittee for technical specialties including reaction-motor development, fuels and combustion, instrumentations and communications, as well as aerodynamics and space problems. As a technical research and development organization, the ARS has grown increasingly, whose members are involved in aviation, aerospace and related scientific fields, providing technology and talent reserves for the United States to enter the age of space flight.

It is worth mentioning that just 2 years later after founding the AIS, another important science organization, the Institute of Aeronautical Sciences (IAS), was set up at the Yale Club in New York, to promote the application of science in the development of aircraft. As same as both the AIS and following-up ARS, the Institute soon became the main American scientific and engineering society in the field of aeronautics. On 1 February 1963, considering the rapidly growing importance of research and development in the emerging rocket-related areas beyond the traditional realm of aeronautics, the ARS was merged with the IAS to form the American Institute of Aeronautics and Astronautics (AIAA). At this time, the ARS had more than 20,000 members, a spectacular transformation of an original group of a dozen spaceflight dreamers in New York. At present, the AIAA is the world's largest professional society of aerospace scientists and engineers.

3.4.3.2 The Jet Propulsion Laboratory

Frank Joseph Malina (1912–1981), an American rocket specialist almost forgotten, had ever made a significant contribution to the birth of the Jet Propulsion Laboratory (JPL) in the 1930s and 1940s. He and his unlikely research team laid the foundation for the U.S. space program. In 1934, after getting a bachelor's degree in mechanical engineering from Texas A&M University, Malina received a fellowship from the California Institute of Technology (Caltech), where he had the good fortune to begin working for the renowned aerodynamicist Theodore von Kármán (1881–1963) who was leading one of the world's foremost centers of aeronautical instruction and research, the Guggenheim Aeronautical Laboratory, Caltech (GALCIT). At that time, as the limit of propeller propulsion for high-speed flight became obvious, von Kármán and others eagerly sought alternatives. Malina was conducting experiments with model propellers in the wind tunnel for his theses and turned more and more to the possibilities of rocket propulsion while he analyzed the characteristics of propellers.

Coincidently, in March 1935 at one of the weekly GALCIT seminars, William Bollay, then a graduate assistant under von Kármán, reviewed the possibilities of a rocket-powered aircraft based upon a paper published by Eugen Sänger. The following October, Bollay gave a lecture on the subject before the Institute of the Aeronautical Science in Los Angeles. Local newspapers reported on Bollay's lecture, which resulted in attracting to GALCIT two rocket enthusiasts, John W. Parsons and Edward S. Forman. Parsons was a self-trained chemist who, although he lacked the discipline of formal higher education, had an uninhibited and fruitful imagination. Forman, a skilled mechanic, had been working for some time with Parsons on powder rockets. They were going to develop a liquid-propellant rocket motor but found that they lacked adequate technical and financial resources for the task. They found Bollay at Caltech for help while Bollay directly sent them to Malina, who was considering the problem on rocket [6].

In February 1936, Malina proposed a detailed program to his mentor von Kármán that he with Parsons and Forman would design a sounding rocket carrying scientific instruments into the upper atmosphere to an altitude of about 40 km, with a plan for a combination of theoretical studies and experiments. Despite the fact that both Parsons and Forman had no Caltech affiliation, von Kármán agreed to support the trio, and only to provide advice and use of the facilities but no actual funding. The trio was the first members of the rocket research team at the GALCIT, and soon after some graduate students were attracted to join the team, including Apollo M.O. Smith who would go on to become chief aerodynamics engineer at Douglas Aircraft Company, and Xuesen Qian who led Chinese aerospace industry from the late of 1950s to 1970s and was known as the father of Chinese rocketry and spaceflight. The team rapidly came up with a number of inventions and innovations on the design of rocket nozzles, composite solid propellants and rocket design like the regenerative cooling of the motors and rocket-stabilized control system. Qian worked with Malina mainly on the thermodynamic characteristics of the motors. Although Malina as leader of the team imposed repeatedly scientific rigor, there were extraordinary risks in the experiments. In a memorable experiment, a small rocket motor that they were testing misfired, releasing a corrosive cloud of nitrogen dioxide–alcohol mixture that permeated most of the GALCIT and left behind a thin layer of rust on much of the permanent equipment of the laboratory, and then there was a serious explosion of gas apparatus, but fortunately, none was hurt. Thereafter, they were known at Caltech as the "Suicide Squad". Since then, they were told to move the apparatus outside the campus at once, and eventually, a series of tests on rocket engines were conducted at Arroyo Seco, a dry river basin on the western edge of Pasadena. Their theoretical calculations, including fuel consumption, thrust, as well as temperatures and pressures inside the motors, could be backed up from the crucial data provided by the tests. By January 1937, the team had successfully developed and tested a rocket motor with a duration-running of 40 s. Over the next year and a half, the team continued their experiments as time and funding allowed, and Malina with Smith published a landmark paper on the flight performance of a constant-thrust sounding rocket in the *Journal of Aeronautical Science*. Although the experiments were advanced steadily, their research results drew little outside interests.

In December 1938, under von Kármán's behest, Malina presented a report on "Jet Propulsion" to the National Academy of Sciences Committee on Army Air Corps Research in Washington, D.C., where he intentionally avoided the word "Rocket" which still had a poor reputation in scientific circles at that time. The report impressed the U.S. Army Air Corps's commanding general, Henry A. Arnold. As a result, Arnold gave Malina's team $1000 to develop rockets for Jet-Assisted Takeoff (JATO) in January 1939, and half of a year later, gave them an additional $10,000. Arnold hoped that the JATO rockets, mounted on an airplane's wings, might help a heavily laden aircraft takeoff from the short runways at islands on the Pacific Ocean [7]. With the support of the project funding, the team made every effort to develop solid and liquid propellant units for the JATO.

For developing the solid-propellant rocket motor, the problem that the solid propellants fuel during a short time needed to be solved. Starting from the thermodynamic equations, Malina and others had proved theoretically that it is feasible to use rocket motor to assist heavy aircraft takeoff, and Parsons was responsible for the specific development of the motor. By examining the theoretical stability of burning a solid fuel under pressure, Malina and von Kármán obtained a conclusion, now known as the von Kármán–Malina theory of constant-thrust long-duration engines, which showed that the process could be made stable if the pressure inside the chamber remained constant. For developing the liquid-propellant rocket motor, the problems on the combustion efficiency and stability of the propellants needed to be studied intensively. In 1940, Malina received a Ph.D. degree in aeronautics by his dissertation entitled "Characteristics of the Rocket Motor and Flight Analyses of the Sounding Rocket".

In August 1941, the team's gunpowder-based propellant was performing well enough to warrant flight tests on an actual aircraft. They developed a solid rocket motor with 4.7-kg weight that could produce a thrust of 12.7 kgf within 12 s. At March Field, in California, their JATO rockets cut the takeoff distance of an Ercope, a fighter-sized civilian aircraft, by half. The U.S. Navy officials watching the assisted takeoff test were interested in the rocket motors, and required the development of the solid rockets with a thrust of 90 kgf, a working time of more than 20 s and a long storage time. Especially, the JATO rockets stored for more than a few days would explode upon ignition. For the better part of a year, Parsons and Malina searched in vain for a solution. Until June 1942, Parsons got an inspiration from a construction crew mixing molten asphalt, which he could cast asphalt with an oxidizer like potassium perchlorate to create a solid propellant. The combustible asphalt would act as both fuel and binder. Parsons's idea, now known as composite propellant technology, has been widely applied in torpedo, deep-water bomb, missiles and launch vehicles. The rocket motors filled with the composite propellants are able to produce a thrust of 1000 kgf and a working time of more than 30 s. Meanwhile, the team had successfully developed the liquid rocket engine with a thrust of 454 kgf and a working time of 25 s. Two liquid rocket engines were mounted to the lower part of the A-20 Attacker to carry out the flight firing test, and no explosion occurred in the 44 tests. In the same year, Malina together with von Kármán, Parsons, Forman and their colleague Martin Sommerfield founded Aero-jet General Corporation to massively produce and commercialize their rockets for the U.S. Army and Navy. The founders pooled an initial investment of $250 each and the company was incorporated by Andrew G. Haley, a prominent rocket scientist and lawyer.

In 1943, according to the British intelligence reports that the Germans were constructing an extraordinarily large rocket at Peenemünde on the Baltic Sea coast, the U.S. Army required von Kármán to develop a kind of long-range guide missile similar to the Germans and soon provided the project research funding with $3 million for the first year of operation. It was "The Research and Development Program for Long-range Jet Propulsion Missile Weapons" issued by the U.S. Army Ordnance. The research contents of the program include rocket propellants and materials, missile remote control units and missile aerodynamics, etc. and its technical specification requirements are: the weight of high explosive and propellant is 454 kg; the range of missile is 120—160 km; the impacting point deviation is not more than 2%; the flight speed exceeds the detection capability of enemy fighter. As the war effort ramped up, the production requirements for rockets quickly outgrew the context of Caltech. Consequently, the rocket research team at the GALCIT was officially named the Jet Propulsion Laboratory in November 1944. At the beginning of the JPL, von Kármán was named the JPL's director, and Malina was its chief engineer, and two months later, as spending much of his time in Washington, von Kármán resigned the JPL's post, and Malina, who had been overseeing daily operations anyway, was soon named acting director.

For the research program from the U.S. Army, the JPL was going to firstly build a smaller, lighter version of the V-2 called for a small solid-fueled rocket, and then to develop a larger liquid-fueled guided missile. The former, dubbed the Private, stood 2-m tall, which used asphalt as a fuel with potassium perchlorate as an oxidizer. The latter, called the Corporal, was 11-m high, which burned hydrazine with red fuming nitric acid as the oxidizer, a combination the JPL team had developed and patented. Since the innovative fuel remained liquid at room temperature, it did not need a cooling system, and it was hypergolic, meaning it ignited spontaneously when its constituents were brought into contact, so no ignition system was required.

Considering the differences in the shape and structure of the engine and missile, the program initially includes four types, Private A, Private F, Corporal E and Corporal F. The first two types of missiles use the solid rocket motors, and each type was bundled four motors to reach total thrust of 756 kgf; the missiles fly without remote control system and their designed rang are required up to 18 km. The last two types use the liquid rocket motors, and each has a total weight of 5000 kg; the thrust of the liquid engine is required up to 9100 kgf and its working time is 60 s; the missile's range is required up to 120 km. Obviously, the last two types had a large technical span, which needed to take a long time to complete the mission. For this reason, before building the guided version of the Corporal, Malina suggested that the development of an unguided rocket should be carried out as a transition, and the rocket was dubbed the WAC, standing for either "Woman's Army Corporal" or "Without Attitude Control", whose length is 4.9 m, diameter 20.5 cm, total weight 302 kg and empty weight 136 kg. In September 1945, when is 2 weeks after the official end of the Second World War, the WAC was firstly launched at the White Sands Missile Range in New

Mexico, and in the following October, the subsequent launch reached an altitude of 70 km, nearly to the boundary of the upper atmosphere. Later, a more advanced version of the Corporal would be deployed in Europe as the American first nuclear missile. Two months after the WAC launch test, Malina's team returned to the White Sands once again to test the Private missile. In all, they launched 24 missiles without a single failure. Although it was never used as a weapon, the Private was the direct predecessor of all solid-fueled missiles that came after it.

For Malina, the success of the Private and WAC Corporal was bittersweet. He finally developed a high-altitude sounding rocket, but it was already clearly that the same machine was also a stepping-stone to a nuclear-armed ballistic missile. In 1947, due to growing tired of the predominance of military missile work, Malina eventually left the JPL for Europe and worked for the United Nation Educational, Scientific and Cultural Organization (UNESCO) in Paris. Of course, maybe there is something else to say about his departure. Malina resigned from his post at UNESCO to become a studio artist in 1953. Malina never returned to research although in 1960 he helped found the International Academy of Astronautics (IAA) with von Kármán and others, to foster international cooperation in space exploration.

In December 1958, the JPL was transferred from the Army jurisdiction to NASA's control shortly after the establishment of the new organization. However, NASA's contractual arrangement with Caltech for the performance of research and development at the JPL began in 1962. Currently, the JPL becomes a unique national research facility that carries out robotic space and Earth science missions, located at Pasadena, 32 km northeast of Los Angeles, California, USA.

The JPL is a relatively late-founding rocket research institution, which has completed a series of high-performance tasks of the solid rocket engine, liquid rocket engine and sounding rocket in a relatively short time. However, its achievements far exceed those of other research organizations and the rocket pioneer Goddard. What is the secret of the JPL's success? In addition to the military funding support, stable scientific research environment and first-class R&D team, it is more important for the approaches to solve the problems, which by using theoretical analysis and research, the nature of the problem is explored, and finally, the physical realization method to solve the problem is found out.

3.4.4 British Interplanetary Society

In the early 1930s, Philip Ellaby Cleator (1908–1994), a young office worker in the Liverpool area, England, had been interested in space flight since he was a teenager. He liked reading various space science fiction novels, collecting and sorting out other countries' news about rocket technologies and space flight researches published in newspapers and magazines. He later recorded that his interest in space was kindled in the middle of the 1920s by a film about radium and its proposed use as a nuclear fuel, in which the closing scene was of a rocket-like vehicle heading for the empyrean, trailing a radioactive glow astern. He was later to relate: "I have no idea how other members of the audience reacted to this spectacular vision of the future, but it certainly made a lasting impression upon me". As learned about the establishment of the American Interplanetary Society and its related activities from the magazine *Science Wonder Stories*, Cleator immediately contacted its Secretariat to inquire about the status of its establishment and membership. Meanwhile, a prediction from a researcher of the U.S. Academy of Naval Science, that humanity would be able to make a rocket flying to the Sun, was reported in a local newspaper *Liverpool Echo*, which greatly inspired Cleator.

In January 1933, Cleator published an article entitled "The Possibilities of Interplanetary Travel" in *Chambers Journal*, which was subsequently picked up by the Editor of the *Liverpool Echo* and eventually by the national press in the form of the Daily Express, leading to a front-page feature. At the same time, Cleator also advertised in the *Liverpool Echo* for anyone interested in the interplanetary idea to get in touch with him. As a result, some local responders and enthusiasts, who dreamed of using rocket propulsion to fly to the Moon and the planets, frequently visited Cleator to discuss how to found a learned group about interplanetary travel. Eventually, on Friday 13 October 1933, Cleator with Leslie J. Johnson, H. C. Binns and other four responders together held the founding conference of the British Interplanetary Society (BIS) at Binns's office on the second floor of No. 81 Dale Street, Liverpool. Cleator was elected as the first president and Johnson as the first executive secretary in the founding conference. Like Cleator, many of the first members were young, intelligent and avid science fiction fans. Cleator worked hard to promote the group's development and soon created the *Journal of the British Interplanetary Society*. In 1934, Cleator made a special trip to Germany to visit Willy Ley. At this time, the German Interplanetary Society had been nearly defunct, and Ley introduced to him the developments and experiments of German liquid rockets, as well as the relevant researches carried out by other countries. The Society grew from those early foundations and some of the earlier members included people like Arthur C. Clarke, Les Shepherd, Archibald M. Low, Eric Burgess, Ralph Smith, Harry Ross, Kenneth W. Gatland, Val Cleaver and Patrick Moore. At Cleator's invitation, the German Willy Ley, the Soviet Y. I. Perelman and N. A. Rynin, the French R. Esnault-Pelterie, the Austrian G. von Pirquet and the American George Ed. Pendray were also enrolled as foreign members. For anyone who knows their space history, all of these people have had huge impacts on the development of space technology or in helping to communicate and advocate for the exploration of space. However, due to the limitation of the Explosive Act 1875, a law amended in the UK in 1875, forbidding individuals to manufacture, keep, sell, carry and import gunpowder, nitro-glycerine and other explosive substance, the Society could only conduct purely theoretical researches while did not carry out actual rocket experimentation, though of high level as many scientists and engineers had become its members. The teenage member Burgess was going to launch a solid-fuel rocket from his hometown in Manchester while was promptly charged.

For the BIS, 1936 was proved to be a landmark year. In the same year, Cleator published his book entitled *Rockets through Space*, which introduced the public to basic spaceflight idea and attracted some public notice; a film, *Things to Come*, also inspired a flurry of interest in spaceflight; the member Archibald M. Low established a branch of the BIS in London, where soon became the location for the Society's headquarter; the Explosive Act 1875 was repealed by the British government; a Technical Committee of the BIS, led by John H. Edwards, was organized whose tasks were studies on spacecraft's design and equipment, including two core members Ross and Smith [8].

For the Technical Committee, the first of these tasks was the Lunar Spaceship Project, in which both Ross and Smith made significant contributions to looking at the feasibility of landing men on the Moon. Comparing to more than 100 combinations of liquid and solid propellant rockets, it was considered that the escape velocity could be achieved by using the multi-stage solid rocket in parallel. They had designed a tremendous rocket for landing on the Moon, with a height of 32 m, a diameter of 6 m and a total weight of 1000 metric tons (including propellants of 900 metric tons), of which a total of 2490 solid-fueled motors are connected in a six-stage parallel structure, installing a group of small liquid-fueled motors between each stage shell and the rocket body to precisely control flight attitude. Ross was responsible for a conceptual design of the Lunar Spaceship [9], whose head is equipped with skin to block meteors and solar radiation. In the cabin of the Lunar Spaceship, there are three elastic seats and the devices processing air, water and food. The spaceship would carry a crew of three safely to the Moon, permit them to land for a stay of 14 days, provide for a safe return to the Earth with a final payload of half a ton. In order to facilitate the safety of astronauts on the surface of the Moon, the Lunar Spacesuit was specially designed by Ross. The 68-kg Lunar Spacesuit, equivalent to 11 kg on the Moon, could be worn for up to 12 h, within the temperature range from plus 120 degrees to minus 150 degrees Celsius, suiting to the night and day on the lunar surface. The Spacesuit design was four-ply, made up of a thin exterior skin of closely woven cloth, and had 1 cm layer of cellular heat-resisting material and a 1–2 mm main airtight sheath of fabric-backed natural or synthetic rubber. It also had an interior lining of non-hygroscopic material, mainly for comfort, to manage contact between the rubber and skin and absorption of the water vapor. During the flight, the astronauts could maintain the observation of a specific sky area all the time by using a device "Orientation Scope", to determine the flight direction. The navigation parameters were obtained by using the altimeter, gyroscope, accelerometer and stable control device to realize the autonomous flight of the Spaceship. In order to make the best use of the centrifugal force of the Earth's rotation, and avoid or reduce casualties and property losses as the explosion accident occurs possibly, it was assumed that the launch site was selected near Lake Titicaca in the Andes on the border of Bolivia and Peru, at 16-deg south latitude and 3800-m altitude. It is an extremely detailed design of the spaceship landing on the Moon. The object of the exercise was to demonstrate that, within the capabilities of propellants that could theoretically be specified at that time, such a mission was not merely technologically possible but be economically viable.

During the Second World War, the BIS suspended its activities, and its members could only independently carry out researches within their capabilities. Cleator was upset that he could not persuade the military to support the development of rocket weapons; based on the theoretical researches and calculations, Clarke thought that one of the main applications of space technology would be to realize long-distance intercontinental communication, and firstly proposed the concept of geostationary orbit and its communication satellite; Gatland and others studied the multi-stage rocket technology and initiated the establishment of a wartime rocket research group, and finally, their researches showed that a small artificial satellite could be put into orbit by using a three-stage rocket.

In June 1945, the BIS was re-founded by Low, and L. J. Carter drafted a formal constitution to give the Society a solid legal basis. In the year, Clarke published the famous prediction paper in the magazine of *Wireless World*, entitled "Extra-terrestrial Relays—Can Rocket Stations Give Wide-world Radio Coverage". A few months later, Burgess noted that unmanned repeater satellites would be just as effective and they eventually became a reality in the 1960s. In 1951, at the 2nd International Astronautical Congress (IAC) in London, Gatland and others presented the paper "Minimum Satellite Vehicles", by which it was inspired that the U.S. Naval Research Institute developed a three-stage launch vehicle that was used to launch in 1958 the American first satellite Explorer-1 with only 13.97-kg weight. Furthermore, the paper also affected France and Britain, whose first satellite was a small satellite, weighing 41.7 kg and 66 kg respectively.

The British Interplanetary Society is the oldest significant profession-spaceflight organization still existing under its original name. Currently, the Society continues to promote the exploration and use of space for the benefit of humanity, by connecting people to create, educate and inspire, and advance knowledge in all aspects of astronautics, and is one of the most influential academic research organizations in the field of astronautics. Its professional journal, *Journal of the British Interplanetary Society*, also becomes the famous academic journal in the field.

3.4.5 Other Countries' Groups and Individuals

Before the Second World War, in addition to the above-mentioned astronautic research societies with certain scale and influence established by the Soviet Union, Germany, the United States and the United Kingdom, other countries also set up some small-scale astronautic organizations, such as the French Astronautic Society, the Dutch Astronautic Society, the Austrian Rocket Research Group, the Argentine Astronautic Research Center, the Japanese Rocket Society, and so on.

In addition, there were individuals in Europe, which independently investigated rocket technology and its application. A German engineer and pilot, Reinhold Tiling (1893—1933), had successively developed 12 kinds of large-size solid-fueled rockets with gunpowder as propellant. In 1931, the flight tests of his rockets were conducted, one of which the flight altitude reached 2 km at a maximum speed of 1127 km/h. He rejected to cooperate with the German Society for Space Flight and had ever tried to apply his rockets to mail post, but the delivery distance of several thousand meters was of little practical significance. In 1933, he started to develop a long-range rocket that could fly over the English Channel. As he and his two assistants was compressing the gunpowder charges for the rocket motor, unfortunately, an explosion occurred at the workshop, and the trio was killed.

An Austrian rocket pioneer, Friedrich Schmiedl (1902–1994), had also attempted to use solid-fueled rockets to deliver mails between two villages separated by mountains or between two cities. He had developed and tested a series of recoverable solid rockets, V1-V6, from 1928 to1930, and refused to participate in the development of any military rockets. In 1931, he successfully conducted an official rocket-mail service between Austrian towns and then began with what he called regular mail service using his new type rocket G1. Schmiedl had ever dreamed of using enormous rockets to transport long-range mails between two cities or even two continents. However, the Austrian Post Office obtained a law prohibiting further rocket-mail activities in 1934, and a year later, a general government ban on private rocket development was promulgated. With the breakout of the Second World War, he destroyed his rocket work, not wanting it to be used for military purposes. Similarly, he refused an offer of the postwar to go to the United States and conduct further rocket research.

Another Austrian-German aerospace engineer, Eugen Albert Sänger (1905–1964), was born in Pressnitz, Austria-Hungary, on September 22, 1905, and was best known for his contribution to lifting body and ramjet technologies. In 1933, he began pioneering tests of rocket engines at Vienna and published his original thesis "*Rocket Flight Engineering*", which covered in great mathematical detail the design of rocket planes, space stations, and even interplanetary travel using ion engines that could attain relativistic velocities. It is a book far leaping ahead of other works on astronautics at that time. For example, the rocket-plane design was a horizontal takeoff-and-landing and hypersonic glider in this book; assuming a liquid propellant rocket engine with an exhaust velocity of 3700 m/s and a mass ratio of 7:1, he calculated that the rocket-plane would have a range of 4000–6000 km while gliding at Mach 13 in 40–60 km altitude. In 1934, Sänger developed a liquid-propellant rocket engine with regenerative cooling, which researched a thrust of 30 kgf and demonstrated exhaust velocities of 3000 m/s at 50-atm chamber pressure. In 1936, Sänger was invited to work for the Herman Göring Aviation Research Institute, Germany, and studied a manned intercontinental hypersonic bomber with its own rocket engines that would climb to the fringe of space and then skip along the upper atmosphere, not actually entering orbit. This remarkable design was called "Silver-Bird", which would rely on its fuselage creating lift as a lifting body to fly along a sub-orbital path. He also designed the rocket engines applied to this space-plane, which would be required to generate a thrust of 10,200 kgf. By 1942, Sänger had tested the rocket engine at a chamber pressure of 100 atmospheres, powered by high-energy propellants with exotic additives, and components of the engine with 10^5 kgf thrust. Since Sänger estimated that it would take twenty years of development to produce an operational system of the Silver-Bird, the Hitler government suspended the program until 1944, but too late to be of any interest as the Nazi Germany entered its death throes. Meanwhile, Sänger also did important work on ramjet engine technology during the War, studying the projects like interceptor SK P14. It was shown from the results of ground and flight tests that the ramjet engines promised a more efficient method of propulsion than the rocket ones at the high speeds and altitude that Sänger hoped to attain. After the War, Sänger had worked for the French armament ministry, directed the Institute of Jet Propulsion Physics, Stuttgart and taught at the West Berlin Technological University. On 10 February 1964, Sänger died of a heart attack in Berlin while lecturing to his students.

3.5 Entering the Age of Space Flight

The early spaceflight research societies and individuals played a significant role in promoting the development of rocketry and astronautics, and since then humanity began to enter the age of space flight. From the development and application of liquid rockets and V-2 missiles to the launch of the first artificial earth satellite, the manned earth orbit flights and the manned landing on the Moon, humanity were opening step by step the great gate of space flight age.

3.5.1 Liquid Rockets and V-2 Missiles

On June 28, 1919, the victorious countries, led by Britain, France and Russia, signed the Treaty of Versailles at the Palace of Versailles in Paris, aiming at punishing and reducing the military equipment of the defeated countries, which marked the official end of the First World War. Although the armistice, signed on 11 November 1918, ended the actual fighting, it took six months of the Allied negotiations at the Paris Peace Conference to conclude the peace treaty. By the Treaty of Versailles, as the main defeated country in the War, Germany was restricted to develop military equipment such as combat aircrafts, tanks, artilleries and machine guns, especially the Army equipment. Under such an international context, the German army had to develop a new weapon system, liquid rocket weapon, which leaps over the conventional army equipment.

Karl H. E. Becker, head of the ballistics and munitions section of the Weapons Testing Division of the Reich Ministry of Defense, was a key early supporter of German development of rockets as weapons. In 1929, Colonel Becker (later became an Artillery General) got the permission of the Ministry of Defense, leading the research on the possibility to jet propulsion rockets and their applications to the delivery of bombs, in which he appointed Lieutenant Walter R. Dornberger to take charge of the specific work. In December 1930, the Ministry of the Army held a special conference on the development of rocket weapons, indicating that German officials began to support the liquid-fueled rocket research program. At the end of 1932, Dornberger selected some standouts from the VfR, including Wernher von Braun, Walter Riedel, Arthur Rudolph and Heinrich Grünow, to formally establish a liquid rocket research group. The 21-year-old von Braun was responsible for the design and development of the first small liquid rocket with a thrust of 270 kgf, which was named Assembly number 1 (A-1). Although von Braun and others had participated in the development of the "Propeller" rocket of the VfR, and had certain technical basis and experimental experience, due to the lack of systematic rocket theory research and detailed test data analysis, it was not smooth to develop the A-1 rocket and all of many static firing tests ended with the engine explosions. Eventually, von Braun abandoned to further develop the A-1 small rocket, and directly design the A-2 rocket with a engine thrust of 1500 kgf, the exhaust velocity of 3000 m/s and the duration of 45 s. Moreover, some problems of the engines, such as the ignition delay and too-high mass center, were also deeply studied and solved. In December 1934, the launch tests of two A-2 rockets got successful, in which the flight altitudes reached 2400 m.

The test success of the A-2 rockets gave the Dornberger's group great confidence and strong support from the military authorities. Consequently, Dornberger and others started to build a completely closed and streamlined large-scale rocket research base, Peenemünde Army Research Center. In April 1936, the Aviation Ministry paid 750,000 marks to the town of Wolgast for the whole Northern peninsula of the Baltic island at Usedom, to founding the Peenemünde Army Research Center as one of five military proving grounds under the German Army Weapons Office. By 1938, the Center was built and put into use, with a total cost of 300 million marks. Meanwhile, with the base facilities improving, the number of people was growing, and the total reached 17,000 people in the peak period. During the construction of the Center, von Braun had also been taking charge of developing the A-3 rocket, in which the magnetic servo system and gyroscopes were used to control the exhaust-gas cabin and achieve flight stability. The detector and parachute were installed at the A-3 rocket's head to sample test data. For all that, the A-3 rocket tests were not smooth, and all three test rockets had different faults, eventually leading to failures. On the basis of the A-3 rocket, the design of stability control system was further improved to develop the A-5 rocket. In 1939, the launch test of the first A-5 rocket equipped with the full guidance system and the parachute was successful. Dornberger took the opportunity to request from the German Army Staff Headquarters for the program of developing the A-4 rocket weapons with real combat capability, and thus the Center was granted the highest priority in personnel, material and financial supports. However, after inspecting the Center, Adolf Hitler did not show a special interest in the practical application prospects for the liquid rockets. At this time, as the Second World War was already started, Hitler still believed in the strong power of the German Air Force, and urgently needed enormous amounts to accelerate the development of military aircrafts and other military equipment and facilities. Shortly thereafter, Hitler ordered to cancel the priority of the A-4 rocket weapon research program. In this way, the development of the A-4 rocket weapon had been pushing forward difficultly and slowly from the end of 1939 to 1943.

According to Dornberger's idea, the new generation of rocket weapons must surpass the famous "Paris Gun" to show its superiority, which would be favored by the superior leaders. The Paris Gun was any of several long-range cannon produced by the German arms manufacturer during the late of the First World War, and was so-called because it was specially built to shell Paris at a range of approximately 121 km with a hit accuracy of about 5%, which was never attained before. Dornberger required that the new generation of rocket weapons should carry 1000 kg of explosives and have a range of 250 km, which became the overall technical specifications for the design of the A-4 rocket weapons. By further calculating and analyzing, it was shown that the thrust of the A-4 rocket engine was required to reach 25,000 kgf, nearly 17 times higher than the 1500 kgf thrust of the A-5 rocket, with the duration of 65 s, and the accuracy of the radio telemetry and track system was required to far exceed the technological level at that time. It needed systematically to study and solve a series of the problems on basic theories and engineering techniques, mainly including three aspects: firstly, to study aerodynamics, involving aerodynamic stability, supersonic aerodynamics and aerodynamic heating; secondly, to study the guidance and control system technology; thirdly, to develop the huge thrust rocket engines. Certainly, under Dornberger's strong leadership, the work was carried out with limited funds.

Von Braun was specifically responsible for the overall design of the A-4 rocket project. In the entire process of studying, designing, testing and manufacturing the A-4 rocket, there were many universities and research institutes in Germany undertaking the related subsystem developments and experimental tasks. In October 1942, the launch test of the A-4 rocket was successful, with an altitude of 85 km, a range of 190 km and a hit accuracy of 4 km. Apparently, the performance specifications of the A-4 rocket was better than those of the Paris Gun. It is a milestone in the development history of liquid rocket and aerospace technology. Both the superior officials from the German military and the arms dealers came to watch the launch test of the A-4 rocket.

With the German army's failures step by step in the battlefields, some superior military officials began to place high hopes on the Army Force's A-4 rocket weapons and the Air Force's V-1 cruise missiles. In the summer of 1943, Hitler assigned a senior military mission to the Peenemünde Army Research Center to investigate the launching demonstrations of A-4 rocket weapons and V-1 cruise missiles, of which the range of A-4 rocket reached 260 km. After the military mission reported its findings, Hitler summoned Dornberger and von Braun in person and said to Dornberger: "I thank you. Why don't I believe your work will succeed? I should apologize to two men in my life, and only two men. Field Marshal Walter von Brauchitsch, he has repeatedly reminded me of the importance of your research work. Another is you, because I have never believed your work will succeed". The German authorities regarded the V-4 rockets as the strong weapons to recover the defeat and specifically renamed it retribution 2, namely V-2 missile, in which "V" is the initial letter in German word "Vergeltungswaffen", and promoted Dornberger to Major General and awarded Professor von Braun the title. At the same time, the German Army began to build a V-2 missile launch site near Hague, Netherlands, and develop mobile missile launch vehicles.

From 8 September 1944 when Nazi Germany launched the first V-2 missile to London, England, to 27 March 1945, a total of 3745 V-2 missiles was launched, 1115 of which hit the British mainland, killing 2724 people and injuring 6476 people, and causing huge damage to buildings. It fully demonstrated the strong power of the V-2 missiles but did not save the defeat of the German army. The V-2 missiles are the new types of weapons in a hurry, whose many technical problems have not been solved completely, and their hit accuracy and reliability are relatively low, in which only 44% of the missile fall point errors are less than 10 km. Some missiles exploded in the rising flight sections, or in the air thousands of meters away from the targets, or even after drilling into the ground. Due to the delay in the delivery and use of telemetry units, the problems had not been thoroughly identified. Even so, in general, the German liquid rocket technology was far ahead of other countries in the world at that time and had entered the stage of weaponization application, whose contribution is worthy of affirmation, purely technologically, to the development of modern rockets.

In the late of the Second World War, Dornberger, von Braun and others also worked out a top-secret "Group Program", which aimed to design manned spaceship and realize manned spaceflight. For this reason, on the basis of improving A-4 rockets, they designed series of rockets from A-6 to A-11 to carry out the special tests on rocket technology and manned flight. Von Braun once envisaged that the multi-stage manned liquid rocket technology could be an application to the development of long-range commercial aircrafts and bombers, as well as to the construction of specific space stations; the large-scale solar reflectors could be installed in space to gather the Sun's light and heat to the selected areas on the surface of the Earth for uses; humanity could be sent to the planets in the solar system for travels, and of course, the first step would go to the Moon. For all that, in wartime, it can only be an assumption.

In April 1945, Dornberger, von Braun and others carried a large number of test equipment, V-2 missiles and technical documentation, and led a group of German rocket specialists to surrender to the U.S. Army. These technical specialists were interrogated by the U.S. Army after surrendering. As Dornberger was a senior German officer, he was transferred to London for interrogation by the British War Crimes Investigation Unit in connection with the use of slave labor in the production of V-2 rockets, subsequently detained for 2 years at Bridgend in South Wales, and then released to the United States. According to the secret program "Operation Paperclip" formulated by the U.S. Department of Defense, hundreds of German rocket scientists and engineers, including von Braun and his V-2 rocket team, were allowed to the United States work for the Army, contributing to the development of American aerospace industry.

In June 1945, the Soviet Red Army advanced from the East Line to occupy the rocket base at Peenemünde, but the highly confidential base had been deserted, and only some ordinary technicians and V-2 missile parts were seized. In addition, Britain and France also obtained some rocket experts, equipment and technical information from Germany. The German liquid rocket technology has been transferred to many countries, laying a foundation for the development of space technology in the world.

3.5.2 Launching the First Artificial Earth Satellite

After the end of the Second World War, the international situation and pattern changed greatly. The United States implemented the "Cold War" policy of "containing" the Soviet Union. The world was divided into two camps, the capitalist camp led by the United States and the United Kingdom, and the socialist camp led by the USSR. On July 16, 1945, the United States successfully tested and exploded the first atomic bomb in the world, and 20 days later dropped two atomic bombs respectively in two Japanese cities, Nagasaki and Hiroshima, forcing Japan to surrender quickly and unconditionally. This aroused the high vigilance of the general secretary of the Communist Party of the USSR and premier of the USSR, Joseph Stalin (1878–1953), who immediately ordered to speed up the development of the Soviet atomic bomb. At that time, although the Soviet military strength had made great progress, it was unable to compete with the United States yet. Especially in the long-range strategic bombers, the USSR was difficult to catch up with the United States in a short period of time, and it was difficult to constitute a deterrent force to bomb the U.S. mainland. So, Stalin could only hope to develop the long-range rocket weapons, the intercontinental ballistic missiles.

In July 1944, S. P. Korolev was paroled from the Kazan sharashka, and in September 1945, went to Germany together with other Soviet specialists, such as V. P. Glushko and N. A. Pilyugin, to review and receive the German rocket technical documentation and equipment. Moreover, they also captured some German specialists on the V-2 missiles, such as the guidance and control expert H. Groettrup, the rocket engine expert W. Baum and the manufacturing expert O. Putze. Under the supervision of the Soviet soldiers and technicians, the German technicians led by Groettrup assembled a V-2 missile once again. In August 1946, while still in Germany, Korolev was appointed chief of a department in the newly created the Scientific Research Institute NII-88 in Podlipki, northeast of Moscow, which was responsible for the development and industrial production of missile technology based on the German hardware. In October 1947, the V-2 missile assembled by the German technicians was successfully launched in Stalingrad with a range of 270 km. For this way, the German rocket experts had smoothly accomplished their missions, and the following missile development works would all be done by the Soviet specialists themselves. At first, the Soviet specialists started to copy a short-range V-2 missile, named R-1 missile. On the basis of improving the designs of the cooling system, nozzle, combustion chamber and propellant of the R-1 missile engine, the R-2 missile had been successfully developed with a range of 590 km. Shortly thereafter, the medium-range ballistic missile SS-3 which was the improved R-2 missile, on which the RD-101 engine developed by Glushko was mounted, was successfully developed with a range of 1800 km.

In the meantime, after nearly 4 years of hard developing and catching up, the Soviet Union eventually successfully tested and exploded its first atomic bomb on August 27, 1949. Subsequently, on October 1 of the same year, the People's Republic of China was born under the leadership of the Communist Party of China. The two great events provided the chips for the Soviet Union in the confrontation against the West, and enough shocked the United States. On 1 November 1952, the United States detonated the world's first hydrogen bomb at a large atoll called Eniwetok of the Marshall Islands in the South Pacific Ocean and accelerated to develop the large-size liquid rockets. Under the context of the Cold War between the East and the West, the United States and the Soviet governments attached great importance to the development of new weapons, as well as large-size and long-range vehicles, which has created conditions for speeding up the development of launch vehicles and artificial satellites. In the postwar, from the perspective of scientific research, there were scientists in many countries, studying the feasibility of launching man-made satellites. In September 1951, at the 2nd IAC in London, many scholars put forward the initiative of launching artificial earth satellites and manned space stations, and Gatland from the BIS presented the proposal that the man-made satellites could be launched by using three-stage rockets. At the same time, the International Radio Science Association (IURS) and the International Union of Geodesy and Geophysics (IUGG) had adopted a resolution to launch an artificial earth satellite during the International Geophysical Year 1957—1958. This resolution had been positively responded to by the United States, the Soviet Union and other countries.

After Stalin passed away on March 5, 1953, Nikita Khrushchev (1894—1971), who then served as the first secretary of the Communist Party of the Soviet Union, came to power over the USSR and set out to reform and reverse many aspects of Stalin's wasteful doctrine on the military. Despite all this, Khrushchev would continue and even accelerate the rocket development program, as the only hope for the USSR to overcome the U.S. air supremacy. For this reason, he strongly supported Korolev's idea on the development of intercontinental ballistic missiles.

While developing an intercontinental ballistic missile, Korolev modified the German Groettrup's concept that started out as a cluster of engines, and conceived of a lengthened core sustainer stage, well known as the R-7 missile, which would allow all engines to be ignited on the launch pad, eliminating the problems of air start in Groettrup's design. The R-7 faced enormous development problems that Glushko's RD-105/RD-106 engines had seemingly insoluble combustion chamber instability. Since the weight of the thermonuclear warhead to be carried was increased from 3 metric tons to 5.5 metric tons, the entire vehicle had to be scaled up proportionately. It meant that Glushko's troublesome single-chamber engines had to be replaced by newly designing RD-107/RD-108 engines, with a cluster of four nozzles sharing common fuel pumps and with liquid oxygen and kerosene as propellants. It could assure stable combustion but greatly increased the complexity of the booster, with a total of 20 main motors and sixteen vernier motors firing at lift-off. Under Korolev's enormous drive and personal attention to detail, all obstacles were overcome ultimately. The R-7 adopted the two-stage rocket design, of which the first stage consists of four strap-on boosters using the RD-107 engines and the second stage is a long core rocket using the RD-108 engine. The R-7 has a 31.07m length with a warhead, and 280 metric tons lift-off weight. On August 21, 1957, The R-7 rocket was launched from the Baikonur launch site and perfectly flew nearly 8800 km. Despite the warhead disintegrated at an altitude of around 10 km over the Kamchatka

Peninsula, it is almost a successful launch test. Subsequently, the TASS news agency officially announced "…the creation of a long-range multistage ballistic missile in the Soviet Union". It meant that the Soviet missiles could be launched to any corner of the Earth and attained the strategic goal of launching an intercontinental missile ahead of the United States, which also lain a solid technical foundation for the Soviet Union to take the lead in entering the age of space flight.

As early as March 1950, M. K. Tikhonravov made a presentation entitled "Rocket Packs and Prospects of Their Development" at a scientific conference held by his research institute, advocated a concept of multistage rockets and reported the results of calculating the mass of vehicles for various flight ranges. Particularly, in this presentation, he considered a launch vehicle consisting of three R-3 rockets, which would offer practically an unlimited flight range and the capability to reach orbit, possibly with the satellite heavy enough to carry a pilot. He finally concluded outlined various applications of the satellites and the possibility of manned space flight, which paved the way to serious considerations into the problems of the artificial satellites. With Tikhonravov's gift of vision, his research into long-range multistage missiles allowed that the concept of orbiting satellites matures into a real space program. Firstly, he conceptualized a possible plan for space exploration, which would start with simple satellites designed primarily for science and for testing rockets carrying them. Secondly, the manned rockets would be launched to test their possibility, and in turn, these missions would pave the way to a small experimental satellite with one or two people onboard, capable of long-duration flights. Thirdly, a large-scale "Station Satellite", which could be applied to scientific research and served as a launchpad for deep space missions, would ultimately be launched, to be maintained via regular transports from the Earth. However, there were many people who held objections to Tikhonravov's idea and space program, including the superior leaders from the Soviet military and the specialists from the academic circles of the USSR. They considered that the creation of a satellite is a useless idea, not even worthy of discussion, let alone its actual development and that the development of satellite and space exploration should be a matter decades later due to the limitation of the technological level.

In order to support Tikhonravov to develop the space program, Korolev lobbied around and sincerely got the suggestions from the specialists of the Soviet Academy of Sciences, including its chairman A. N. Nesmeyanov, the academician M. V. Keldysh and the academician L. I. Sedov. With scientific establishment on his side, Korolev was ready to bring the satellite into politics. In May 1954, Korolev submitted a formal proposal entitled "On the Artificial Satellite of the Earth" to Dmitriy F. Ustinov, who was then the Minister of Defense Industry of the USSR. The proposal was accompanied by the research results and a review of the American work in the field compiled by Tikhonravov. Clearly, with Ustinov's help, the proposal on the development of problems of space flight was eventually approved three months later by Soviet of Ministers USSR, including V. Ryabikov at the Soviet of Minister, G. Pashkov at Sate Planning Ministry, and others. At the beginning of 1955, the Soviet Academy of Science created a special commission chaired by Keldysh, with

Korolev and Tikhonravov as deputies, which was tasked to oversee the satellite development program. Meanwhile, the messages for the Americans to develop the satellites were also exciting the USSR to speed up its space program. On July 29, 1955, President Dwight D. Eisenhower formally announced the intention of the United States to launch a satellite during International Geophysical Year, and the *National Geographic* then confidently assured Americans that "History would record that date as the day the White House announced that the United States would build the first true space vehicle". Subsequently, on August 3 of the same year, at the sixth Congress of the International Astronautical Federation (IAF), Academician Sedov, not to be outdone, also said that the USSR intended to launch a satellite. Not doubt, Sedov's statement was sanctioned in Moscow in response to the U.S. challenges.

The project on the Soviet first artificial earth satellite consists of four major tasks as follows:

(1) The first task is to develop a special launch vehicle, which was designed under the Korolov's leadership and named "Sputnik" (the pronunciation in Russian, Спутник) rocket. The Sputnik rocket, with a number of 8K71PS, was modified and accomplished on the basis of the R-7 missile and was composed of a core stage, four strap-on boosters and a fairing, with a total length of 29.169 m, lift-off weight of 267.3 metric tons, lift-off thrust of 398,000 kgf and thrust-weight ratio of 1.49.

(2) The second task is to build a rocket launch site. The Sputnik rocket launch site is located in the region of Tyuratam, the Republic of Kazakhstan, with a geographic location of 63.4° E and 45.9°N, and with 2100 km away from Moscow, 150-km east of the Aral Sea, 288-km southwest of Baikonur town, and an average altitude of 90 m, belonging to semi-desert grassland, saline–alkali beach and hilly area with few people. In the middle of the 1950s, to meet the requirement of testing the R-7 missile with a range of 8800 km, Korolev set out selecting a new launch site and insisted on choosing the location as far south as possible, which could not only increase the range of the missiles but also fully use the assistance from the earth's rotation. By comparing with other several sites, the Tyuratam was finally selected as the launched site, whose construction began shortly thereafter. On June 2, 1955, the Chief of Staff of the Soviet Ministry of Defense approved the organizational structure of a new test range near Tyuratam, and the date became an official "birthday" of the launch site. To maintain its secrecy, the launch site was named Baikonur Cosmodrome rather than Tyuratam Cosmodrome. The first launch pad was built in March 1957 and subsequently applied to launching the R-7 missile. The launch site is 120-km long from the east to the west and 80-km wide from the south to the north, which is composed of the launch area, support area and monitoring stations.

(3) The third task is to develop the satellite body and its onboard scientific devices. The satellite is called PS-1 or Sputnik-1, and its design goals are to ensure absolute success of the launch and make the world feel its existence. The satellite

was designed as an aluminum-alloy-sealed ball-shaped container with a diameter of 0.58 m and a total weight of 83.62 kg, filled with dry nitrogen with a pressure of 0.12 MPa. Its main body was made out of two hemispheres 2 mm thick, connected with 36 bolts. Four spring whip antennas were installed symmetrically around the satellite, one pair of which is 2.4 m long and the other 3.9 m long. The special spring mechanism was designed to deploy antennas to the angle of 35 deg toward the major axis of the container, immediately after the spacecraft had separated from the rocket. Two radio transmitters working alternately are equipped with the satellite, with working frequencies of 20.005 MHz and 40.002 MHz respectively. The scientific objectives of the satellite are to determine the density of the upper atmosphere as its orbit decays, detect the radio wave distribution throughout the ionosphere, and verify theoretical calculations and technical solutions associated with delivering the satellite into its orbit.

(4) The fourth task is to build the ground control network. In order to measure and control the R-7 missile equipped with radio control, an extensive ground control network with two groups of measurement sites had been deployed before 1957, spreading over a vast distance of more than 6000 km across the USSR. The measurement sites were known as IP in its Russian acronym. A group of tracking stations responsible for the flight control of the R-7 missile, known as Taiga region, was stretched downrange from the launch site in Tyuratam along with the flight path, including nine measurement sites with the number form IP-1 to IP-9. Another group of stations, known as Kama region, were responsible for tracking of the warhead reentry over the target range in the Kamchatka Peninsula, including six measurement sites with the number from IP-12 to IP-17. The command and control station in the network, IP-1, was located at Site 18 of the NIIP-5 test range in Tyuratam, only about 1.5 km behind the launchpad for the R-7 missile. In May 1957, on the basis of the R-7 control network, the Soviet Ministry of Defense officially ordered to form a "Center on management and coordination of work of complex of means of measurements, communications and unified time service", and 13 detached Scientific Measurement Stations (NIP), to control and communicate with Sputnik-1. The entire network was designated as Command and Measurement Complex (KIK), including 13 NIP stations: NIP-1 at Tyuratam (IP-1), NIP-2 at Makat near the town of Guriev, NIP-3 at Sary-Shagan on the western coast of Lake Balkhash, NIP-4 at Yeniseisk in Western Siberia, NIP-5 at Iskhup, NIP-6 at Yelizovo on the Kamchatka Peninsula, NIP-7 at Klyuchi (Barnaul), NIP-8 at Bolshevo on the northeast of Moscow, NIP-9 at Krasnoe Selo near St. Petersburg, NIP-10 at Simferopol in the Crimean Peninsula, NIP-11 at Cape Shartushalu near Tbilisi in Georgia, NIP-12 at Novosibirsk (Kolpashevo) in Western Siberia and NIP-13 at Ulan-Ude on the east of Lake Baikal.

On October 4, 1957, the Soviet Union successfully launched the world's first artificial satellite of the Earth, Sputnik-1 (Простейший Спутник-1). The satellite entered into an elliptical orbit with a perigee altitude of 215 km, an apogee altitude of

939 km and an inclination angle of 65.1 deg, with an orbital period of 96.2 min. The satellite had been in orbit for 92 days, flying around the Earth in nearly 1440 circles, and on January 4, 1958, it was reentered into the atmosphere and finally burned out. In the scientific researches, the satellite was used for the on-orbit temperature and pressure tests, atmospheric density measurement, ionosphere detection, as well as high-altitude atmospheric resistance research. On the political side, it was shown that the Soviet Union won the first round of the Space Race. The successful launch of Sputinik-1 marks that humanity has entered into the age of space flight, which is of epoch-making significance to the development of space science and technology.

Since then, America, France, Japan, China and Britain have used their own rockets to launch their first artificial earth satellites respectively. On January 31, 1958, the United States successfully launched its first satellite, Explorer-1, using the Jupiter C rocket. The Explorer-1 has a slender cylindrical shape, with a length of 2.03 m, a diameter of 0.159 m and a weight of 13.97 kg. On November 26, 1965, France successfully launched its first satellite, Asterix-1, using the Diamant-A rocket. The structure of the Asterix-1 is the spindle shape of the upper and lower cones and the middle cylinder, with a weight of 41.7 kg. On February 11, 1970, Japan successfully launched its first satellite, Ohsumi, using the Lambda 4S rocket. The Ohsumi consists of the fourth-stage engine, spherical engine and conical measuring instrument cabin, with a total length of 1 m and a weight of 23.8 kg. On April 24, 1970, China successfully launched its first satellite, DFH-1, using the Long March-1 rocket. The DFH-1 is a spherical polyhedron shape, with a diameter of 1 m and a weight of 173 kg. On October 28, 1971, the United Kingdom successfully launched its first satellite, Prospero, with a weight of 66 kg, using the Black Arrow rocket. After that, the UK gave up its own rocket development program and focused on developing space science and application satellites. Today, all kinds of satellites are widely applied to the fields of meteorology, communication, navigation, remote sensing, space science and technology test, which promote the development of human civilization and the progress of science and technology.

3.5.3 Manned Earth Orbit Flights

The manned Earth orbit flight, a real space travel, has been the dream of humanity for thousands of years, and it is also the cause for the modern astronautics pioneers have strived for their lifetime. In the late 1950s, humanity has entered the age of space flight, launching a series of artificial earth satellites and interplanetary probes. However, there are still many fundamental problems to be solved, such as space biomedicine and space high-energy particle radiation, which are very risky for human life safety. For the manned space flight, there is a great technological span, at least involving three key technologies: one is the technology of high thrust launch vehicle; the other is the technology of spacecraft environment control and life support; the third is the technology of reentry and safe landing. In the 1960s, under the international environment of the Cold War between the East and the West and the Space

Race between the USSR and the US, the manned space flight programs were carried out apace. The USSR implemented the Vostok Program and the Voskhod Program, while the US did the Project Mercury and Gemini Program. And then, a series of manned and cargo orbital vehicles, such as manned spacecrafts, cargo spaceships, space stations and space shuttles, was developed and launched successively in the USSR (later Russia), America, China and other countries.

3.5.3.1 The Soviet Vostok Program and Voskhod Program

In November 1956, the Soviet Union drew up a development plan for manned space flight, including a manned landing on the moon. In fact, the lift-off thrust of the R-7 missile and Sputnik rocket, designed under Korolev's leadership, has a more margin, providing technical reserves and test experience for the design of high thrust launch vehicle. In November 1957, the second artificial satellite, Sputnik-2 with a small dog Laika onboard, launched by the USSR, to conduct technical tests on the tightness of the cabin, life safety assurance and the impact of weightlessness on animals. With the increasing number of satellites launched and scientific achievements made by the United States, considering the perspective of political interests and diplomacy, the Soviet government must continue to maintain its leading position in the field of space technology.

In 1958, the Soviet Union officially started to conduct the manned space flight plan, Vostok Program (Восток). The Vostok rocket, also called 8K72K rocket, is a three-stage launch vehicle, which is a slightly improved design version based on the rocket carrying the Soviet first unmanned probe to the Moon in January 1959. The Vostok rocket consists of five components, including the first stage with four strap-on boosters, the second stage (core stage rocket), the third stage rocket, the spacecraft adapter and the payload fairing, with a total length of 38.63 m, a lift-off weight of 287 metric tons, a lift-off thrust of 408,000 kgf and a thrust-weight ratio of 1.42. The rocket's structural changes were limited primarily to the stage originally developed for sending its payload to the Moon. For the Vostok missions, the key discrepancy was the replacement of the RD-105 engine on the third stage of the rocket with the RD-109 engine. Moreover, a larger cylindrical adapter was fitted onto the third stage to accommodate a manned spacecraft and its protective payload shroud. The new fairing had a big circular opening to enable Vostok pilots to eject from the cabin in case of an emergency on the launch pad or in the early phase of the launch. The third stage rocket with a length of 7.53 m would enter into orbit with the spacecraft to fly together.

The Vostok spacecraft consists of two main components: a descent module with the pilot cabin and an instrument module equipped with a braking engine. The descent module, with a spherical shape, is equipped with life safety equipment, food, ejection seat, television camera, optical directional device, observation device and transponder, etc., which can ensure the survival of astronauts for ten days. There are the windows used to observe and eject on the bulkhead of the descent module, and there are two antennas that applied remote control and communication outside the module. The instrument module, with a cylindrical shape, is equipped with 18 spherical high-pressure nitrogen and oxygen cylinders, retro-thrust engines, internal propellant tanks and communication antennas, etc. As a purely experimental vehicle, the Vostok spacecraft had to safely carry an astronaut into orbit and return the astronaut to the Earth. The main flight control function was limited to orienting the spacecraft tail first for the firing of the braking engine prior to reentry into the atmosphere. The braking engine, whose firing would be commanded by a programming timer, had to slow down the spacecraft by about 140 m/s to ensure the safe return of the vehicle. Since the successful completion of the reentry maneuver meant life or death for the astronaut and the braking engine could not be backed up by any other hardware, the Vostok spacecraft would be launched into an orbit with 180 km by 235 km. At that altitude, the density of the upper atmosphere would be enough to slow down and send the spacecraft back to the Earth roughly 5 days after launch, with a margin of error of 2.5 days. It is the reason why the spacecraft must carry enough food, water, air and power onboard to support a 10-day mission.

From May to December 1960, the Soviet Union had successively launched five unmanned prototypes of the Vostok spacecrafts, officially named from Korabl-Sputnik-1 to Korabl-Sputnik-5, to practically test the reliability of the spacecraft's design performance, life safety, and recovery and landing. However, except for the successful test of Korabl-Sputnik-3, the tests of the other four spacecrafts were ended in various failures. For this reason, by improving the spacecraft's designs, two unmanned flight tests were added. Two spacecrafts Korabl-Sputnik-6 and Korabl-Sputnik-7 were launched respectively on March 9 and 25, 1961, carrying with a dog each. Eventually, the two flight tests were very perfect and both dogs safely returned to the ground.

Although the success rate of the spacecrafts certified to carry a human was so dismal, at 9:07 Moscow time on April 12, 1961, the astronaut Yuri Alekseyevich Gagarin (΄Юрий Алексе́евич Гага́рин, 1934–1968)by the Vostok-1 spacecraft was launched from site 1 of the Tyuratam (Baikonur launch site) on schedule. Gagarin flew one circle around the Earth for one hour 46 min, and at 10:53 Moscow time, safely landed at a position of approximately 25 km southwest of Engels town in the Ternovsky District near Saratov, the USSR. As a result, Gagarin achieved the dream of human space flight for thousands of years for the first time. It was also another significant leap in space technical development history for humanity. Apparently, it was shown that the Soviet Union beat the United States again in the Space Race on the manned Earth orbit flight.

Since then, from August 1961 to June 1963, five Vostok series spacecrafts were sequentially launched to carry out a series of scientific researches and tests, including impacts of long-time in-orbit flight on the astronauts, spacecrafts on-orbit formation flight, space medicine and earth observation. Particularly, the world's first female astronaut, Valentina Vladimirovna Tereshkova (Валент΄ина Влад΄имировна Терешко́ва, 1937–) by the Vostok-6 spacecraft, was launch on June 16, 1963, and flew in orbit for three days. Meanwhile, Tereshkova conducted the tests of formation flight and space medicine together with the astronaut Valery Bykovsky who was on the Vostok-5 launched before.

After the success of the first manned space flight of the Soviet Union, Korolev began to work out a new space technology development plan, Vostok-ZH Program and Soyuz Complex Program, to further carry out the technical tests of spacecraft orbit maneuver, rendezvous and docking, space walk and on-orbit propellant filling, and finally achieve the manned landing on the Moon. Nevertheless, considering the complexity of technology, the long-term nature of the task and the political pressure, the Soviet government abandoned the two programs and replaced them with a short-term transitional mission, Voskhod Program (Восхóд). The goals of the Voskhod Program are to take the lead in carrying more than two astronauts on orbit before the United States and to carry out space walk and continuously on-orbit flight for 2 weeks.

On October 12, 1964, the Voskhod-1 spacecraft carrying with the three-member crew, including an engineer Konstantin Feoktistov, an astronaut Vladimir Komarov and a medicine doctor Boris Yegorov, was launched into orbit, and safely returned to the ground after flying in the orbit for 24 h, which first achieved the multi-member crew flying in the orbit. On March 18, 1965, the Voskhod-2 carrying with two astronauts Pavel Belayev and Aleksei Leonov was launched into orbit. After the spacecraft was put into its orbit for 1.5 h, Leonov started to make the first spacewalk for human beings and the duration of extravehicular walk reached 12 min. On the structural shape, the Voskhod-2 apparently different from the Voskhod-1 is to add a spacewalk exit channel. The Soviet Union had ever intended to launch the Voskhod-3 in the spring of 1966, to test the in-orbit flight for 20 days. However, the mission was eventually canceled because the United States had achieved the test goal before the launch time planned by the USSR. Unfortunately, on January 14, 1966, Korolev died from a botched hemorrhoid operation in Moscow at the age of 59, which is a huge loss for the following development of the Soviet aerospace industry.

3.5.3.2 The American Project Mercury and Gemini Program

In the middle and late 1950s, studies on aeronautics and space science in the United States were conducted under the auspices of some military agencies, mainly including the Department of Defense's Advanced Research Project Agency (DARPA), the Naval Research Laboratory (NRL), the Army Ballistic Missile Agency (ABMA) and the National Advisory Committee on Aeronautics (NACA). In the wake of Sputnik-1 in October 1957, President Dwight D. Eisenhower responded to the Soviet challenge and to public concern and excitement by reorganizing the American space effort. A key step was to create a new government agency to conduct civilian space exploration. In April 1958, President Eisenhower sent a draft law to the Congress that called for a civilian National Aeronautics and Space Agency based on the existing NACA to oversee the U.S. space program. In July of the same year, he signed into law legislation creating the National Aeronautics and Space Administration (NASA), and one month later, appointed T. Keith Glennan as NASA's first administrator. When NASA formally started operations on October 1, 1958, the agency and its administrator had broad authority to select goals and implement programs of space exploration. As a new enterprise, space exploration confronted an array of difficult scientific and technical problems, such as developing rockets that could carry machines and humans into space, learning about the space environment and taking the first steps toward human exploration. Under Glennan's leadership, the NASA made advances in all of these areas, among which the greatest public excitement came with his formation of the first human exploration effort, Project Mercury. At the same time, with the first selection of Mercury astronauts in 1959, public fascination in space reached a fever pitch. The scientific objectives of Project Mercury were to place a manned spacecraft in orbital flight around the Earth, investigate man's performance capabilities and his ability to function in the environment of space and recover the man and the spacecraft safely.

In order to speed up the Project Mercury, "Space Task Group" was soon formed at the NASA's Langley Research Center. The Group was headed by Robert R. Gilruth, chief of the Langley Research Center and Manager of the Project Mercury, and consisted of 33 members, including Maxime A. Faget, Paul E. Purser, Charles J. Donlan, Alfred J. Eggers, James A. Chamberlin, Raymond L. Zavasky and George M. Low, etc. On the selection of launch vehicles, the Space Task Group decided to choose four types of rockets: one is the nicknamed Little Joe rocket specially designed by Faget and Purser, a four-cluster, solid-fuel booster, which is used for the quick escape-off and life-saving in case of an emergency at launching the spacecrafts; the other is the Redstone rocket, which is used for the flight test of the spacecraft ascending and descending along the trajectory or orbit; the third is the Jupiter A rocket, which is used for the tests of reentry and recovery of the spacecraft's prototypes; the fourth is the Atlas rocket, which is used for launching manned spacecraft into orbit.

The design, construction and testing of the Mercury Spacecraft were undertaken by McDonnell Aircraft Corporation. The Mercury Spacecraft was designed as a cone-shaped capsule with a cylinder mounted on the top, which has a total length of 8.5 m (including the escape tower of 5.8 m), a diameter of 1.9 m at the arc-shaped bottom and a weight of around 1.6 metric tons. A small solid rocket with three nozzles was mounted on the top of the escape tower, which could generate a thrust of 24,000 kgf in one second. Consequently, in case of emergency, the Spacecraft would rapidly be separated from the rocket to ensure the life safety of astronauts. The flight tests of Project Mercury were implemented in three stages as follows.

The first stage was to verify the technology and design from August 1959 to April 1961. In order to verify the performance of the escape and life-saving system of the Mercury Spacecraft, the launch tests of eight Little Joe (LJ) rockets and one Big Joe (BJ) rocket had been carried out successively. However, except for two successful tests of LJ-2 and LJ-1B, there were various technical failures without exception for the other seven tests, and the expected test goals had not been achieved. For all that, there had been a deep understanding of the impact of weightlessness and overload on animals, the spacecraft recovering, the spacecraft's splash-down on the sea, as well as the design of escape and life-saving system. Moreover, the flight test data also provided a reference basis for the further optimal design. In the meantime, the combinations with the Mercury vehicles and the Redstone rockets

(MR), the unmanned vehicles BA-1 (Beach Abort with the Redstone), MR-1, MR-1A, MR-2 and MR-BD were launched sequentially, of which the flight tests of the BA-1, MR-1A and MR-BD were fully successful, verifying the suborbital flight performance of the Redstone rocket, the technology of spacecraft heat protection, and the technology of splash-down and recovery on the sea. By combining the Mercury crafts with the Atlas rockets (MA), three unmanned crafts MA-1, MA-2 and MA-3 were launched sequentially, of which only the MA-2 test was successful, verifying the entire technical flow and operational process: rocket-firing launch, engine shutdown, escape tower separation, craft separation, retro-thrust engine firing, craft reentry, parachute opening, and finally splash-down on the sea. Until the first space flight of Gagarin in April 1961, for the Project Mercury, there were still many technologies and designs that need to be verified by the flight tests.

The second stage was to test the manned suborbital flights from May to August 1961. On May 5, 1961, the first American astronaut Alan B. Shepard by the MR-3 spacecraft Freedom-7 was launched from the Cape Canaveral launch site, Florida, by using the Redstone rocket. The spacecraft was normally separated from the rocket, and its maximum altitude only reached 186 km. And finally, the spacecraft reentered into the atmosphere along its trajectory and recovered safely. The duration of the entire flight was 15 min and 28 s. This is a typical suborbital manned flight test. On July 21, the MR-4 spacecraft carrying an astronaut Virgil I. Grissom was launched successfully, whose flight process was basically as same as the MR-3. For the United States with a backward position in the Space Race, although the publics with dissatisfaction were slightly calmed down by the two suborbital manned flights, the Redstone rocket did not have the capability of launching manned spacecraft into orbit yet.

The third stage was to carry out the manned orbit flight from September 1961 to June 1963. In order to fully test the rocket's performance, NASA decided to conduct again three unmanned flight missions. On September 13, 1961, the MA-4 craft was launched by using the Atlas rocket, which although was partially successful, the rocket performance, orbital flight capability and control system had been verified fully. Subsequently, the MS-1 craft combining the Mercury spacecraft and the Scout rocket (a series of solid-fueled rocket designed at the Langley Research Center) was launched on 1 November 1961, but immediately after lift-off, the rocket developed erratic motions, after 28 s the booster began tearing apart, at 43 s after launch the safety officer of the launch site sent a self-destruct order, and thus the mission was ended in failure. Finally, on November 29, the MA-5 craft carrying a chimpanzee was launched by using once again the Atlas rocket, and at 3 h and 22 min after launch, the capsule was spotted by a P5M search airplane at an altitude of 15 km. At this time, the ships Stormes and Compton were contacted and they moved into position from a distance of 48 km away. The "Stormes" arrived in the area 1 h and 15 min after landing and hauled the capsule aboard via its lanyard, cracking the onboard window. The chimpanzee fared well in the orbit, who withstood a peak of gravitation acceleration of 6.8 G during booster-engine acceleration and 7.6 G with the rush of the sustainer engine. Consequently, the flight test of the MA-5 was successful. The two perfect demonstrations of the Atlas rockets strengthened NASA's confidence in launching a manned spacecraft for orbital flight. On February 20, 1962, the astronaut John H. Glenn by the MA-6 spacecraft Friendship-7 was successfully launched into orbit with an altitude of 260 km, and after three circles around the Earth, the Friendship-7 returned to the ground safely, lasting for 4 h and 55 min and 23 s. As of the time, the United States had completed a real sense of manned space flight. Since then, three missions of the manned space flights, MA-7, MA-8 and MA-9, were conducted sequentially, of which the longest flight duration reached 34 h and 19 min and 49 s with 22 orbits around the Earth.

Although the advance on Project Mercury in the United States is delayed, the key techniques of manned space flight have been verified systematically and thoroughly, the various types of launch vehicles tested and improved, the achievements on space medicine made, and the abundant experience of in-orbit flight accumulated, all of which created conditions for implementing the subsequent large-scale space flight program. At the beginning of Project Mercury, the NASA and the Space Task Group initiated a systematic argument of the future development plan of space technology in the United States. At that time, the manned flight around the Moon and the manned landing on the Moon had been focused on commonly for the members of the argument team. For this reason, based on the implementation of Project Mercury, a series of key technologies including long-time orbit flight, space-walking and rendezvous and docking was still required to break through. By adding an instrumental cabin, the Space Task Group proposed an improved design for the Mercury Spacecraft, known as MK-II plan.

During the early years of the Soviet and American race into space, their competitions were measured by headlines-making "firsts", such as the first satellite, the first robotic spacecraft to the Moon, the first man in space, and so on. To the dismay of the United States, the USSR had first achieved each of these feats. For this reason, on May 25, 1961, President John F. Kennedy (1917–1963) announced before a special joint session of Congress the dramatic and ambitious goal of sending an American safely to the Moon before the end of the decade, after which the NASA decided a much larger organization to perform the function of the Space Task Group. As a result, the Group was transformed into the Manned Spacecraft Center (now the Lyndon B. Johnson Space Center), located in Houston, Texas. On the basis of the MK-II plan, the Manned Spacecraft Center put forward a more complete spaceflight plan, Gemini Program, whose main goals were: to test an astronaut's ability to fly long-duration missions (up to two weeks in space), to understand how spacecraft could rendezvous and dock in orbit around the Earth and the Moon, to perfect re-entry and landing methods, and to further understand the effects of longer space flights on astronauts. On November 22, 1963, President Kennedy was shot and killed in the City of Dallas, Texas and subsequently, Vice President Lyndon B. Johnson succeeded him. One week later, President Johnson renamed both the Launch Operation Center and the Cape Canaveral Auxiliary Air Force Station to the John F. Kennedy Space Center with Executive Order 11129.

McDonnell Aircraft Corp. as a manufacturer continued to undertake the missions of developing the Gemini Spacecrafts. The Gemini Spacecraft, similar to the Mercury Spacecraft in shape, was an enlarged and redesigned spacecraft for two astronauts, with a conical configuration, a total length of 5.7 m, a maximum diameter of 3.05 m and lift-off weight of 3.0–3.9 metric tons. In the Gemini Spacecraft, the escape tower was canceled while replaced with the ejection life-saving seat design. The Spacecraft had two major units, the reentry module with the crew cabin and heat shield, as well as the adapter composed of two sections, in which the equipment section carried fuel, oxygen and power supplies, and the retro-grade section carried retro-rockets that slowed the spacecraft to make it fall out of orbit. By using small rockets on the adapter, the astronauts could not only change their orientations in space but also their orbital paths. Before the Spacecraft reentered into the atmosphere, the adapter was discarded. The nose as the rendezvous and recovery component came off when the main parachute was deployed, and finally, the crew cabin splashed down horizontally, with the two hatches on the top. The Gemini rocket was derived from the U.S. Air Force Titan-2 intercontinental Ballistic Missile, generally known as Titan-2 GLV (Gemini Launch Vehicle), with a height of 33 m, a lift-off thrust of 193,500 kgf and a launch capability of 3.62 metric tons (for an orbital altitude of 384 km).

From April 1964 to November 1966, a total of 12 GT series spacecrafts from the complexes of the Gemini spacecrafts with the Titan rockets were developed and launched sequentially under the leadership of NASA, among which the GT-1 and GT-2 were unmanned test spacecrafts and each of the spacecrafts from GT-3 to GT-12 took two astronauts to carry out the technological tests like space walking, long-term orbit flight and rendezvous and docking. On June 3, 1965, two astronauts James McDivitt and Edward H. White by the GT-4 spacecraft were launched into orbit, of which White carried out the space walking test for 23 min. As the target vehicle Agena failed to launch into orbit in time, the GT-7 and GT-6 conducted the rendezvous and docking tests in space on December 15, 1965, and a joint body formed by the two spacecrafts flew together for 5 h and 19 min.

On July 18, 1966, the GT-10 spacecraft with the astronauts John Yong and Michael Collins was launched from the Kennedy Space Center and respectively followed by the tracking and rendezvous and docking tests with the upper stage target vehicles, Agena-10 and Agena-8, which were launched into the orbits by the Atlas rockets in advance. The Agena-10 was docked with the GT-10 in an orbit with an altitude of 298 km, and the joint body formed by both had flown along the orbit for 4 h before both were separated. Subsequently, the astronauts manipulated the GT-10 again to track and approach the Agena-8 spacing 37 km apart. At 3 m away from each other, Collins walked out of the GT-10's crew capsule to arrive at the Agena-8 by using the mechanism system and achieved the sampling and returning test task. The GT-10's astronauts returned to the ground safely on July 21 and successfully completed all test tasks including the space rendezvous and docking and the orbital maneuver. At this time, the United States had surpassed the Soviet Union in the field of manned space flight technology, clearing the key technical obstacles for the manned landing on the Moon.

On September 12 and November 11, 1966, two manned spacecrafts GT-11 and GT-12 were launched respectively to further test and verify the technologies of the space rendezvous and docking and the extravehicular space walking, and to carry out the experimental researches on gravity gradient, artificial gravity and long-term space biomedicine. The astronauts in the two spacecrafts returned to the ground safely on September 15 and November 15 respectively which finally marked the successful completion of the Gemini Program.

3.5.3.3 Developments of the Manned and Cargo Orbital Vehicles

After successful completion of the Vostok and Voskhod Program, the Soviet Union continued to improve the development of the Soyuz (Союз) manned spacecraft, further develop the Progress (Прогресс) cargo spacecraft, and finally build the Space Stations, Salyut (Салют) and Mir (Мир).

The Soyuz, which is currently the longest serving manned spacecraft in the world, was originally designed by Sergei Korolev Design Bureau for the Soviet effort to explore the Moon at the beginning of 1960s. The Soyuz spacecraft consists of a nearly sphere-shaped habitation module, a bell-shaped reentry capsule and a cylinder-shaped instrument compartment, with a total length of 7.6 m, a maximum diameter of 2.7 m and a lift-off weight of about 6.5 metric tons. It is able to take three astronauts, with long-term autonomous flight, orbital maneuver and rendezvous and docking capabilities. There are six types of the Soyuz spacecraft series developed successively, including Original Soyuz, Soyuz T, Soyuz TMA, Soyuz TMA-M and Soyuz MS, all of which are launched from the Baikonur Cosmodrome by the Soyuz rockets. On April 23, 1967, in the competitive atmosphere of the Moon Race, the first Original type Soyuz-1 with an astronaut Vladimir Komarov was hastily launched for an ambitious mission, calling for orbital rendezvous and docking with the second Soyuz scheduled to follow it, but the problems onboard Soyuz-1 forced to cancel the second mission. More unfortunately, the Soyuz-1 flight itself ended in a disaster during landing. The problems with the parachute deployment caused the Soyuz-1 reentry capsule to hit the ground with such crashing speed, that the astronaut Komarov was killed on the spot. In the wake of the catastrophe, all manned flight missions were canceled until October 1968, and a series of unmanned missions followed, during which two pairs of vehicles docked in orbit automatically. On October 8, 2010, the Soyuz TMA-01 M spacecraft with three astronauts was launched and successfully docked with the MIM-1 Poisk module on the International Space Station (ISS). On July 7, 2016, the newly modified type Soyuz MS-01 spacecraft was launched for the first time, and 2 days later, successfully docked at the ISS, delivering three members of the ISS crew, a Russian astronaut Anatoly Ivanishin, an American astronaut Kate Rubins and a Japanese astronaut Takuya Onishi, who would serve as the long-duration expedition on the outpost during their 128-day flight. The Soyuz spacecrafts have heavily been used for the ISS program, and have served as only means for Americans to make crewed space flights in the world since the retirement of the U.S. Space Shuttle in 2011.

In the early of developing the space stations, the Soviet designers had realized that long-duration missions in the Earth orbit would demand a constant supply of consumable materials from the Earth. For this purpose, at first, they attempted to employ the Soyuz spacecrafts, but the problems were not solved. As a result, by modifying the

Soyuz spacecraft, the materials and supplies for the station crews would be placed into a pressured section of the cargo craft which loosely resembles the habitable module. The docking port equipped with a hatch would allow the station crew to load and unload the spacecraft in a pressurized environment. The reentry capsule or descent module of the Soyuz was replaced with an un-pressurized propellant and refueling compartment, to safely store the poisonous propellants outside the pressurized compartments and avoid the pollution on the station's environment from any leaks of gaseous supplies. As in the Soyuz, the propulsion and service systems were installed in the tail section of the craft. Thereby, the cargo craft was composed of Cargo section, Tanker section, and Instrument and Propulsion section. Unlike the Soyuz, the cargo craft was inseparable during its entire flight and upon the conclusion of its supply mission to the station would be directed into the atmosphere to burn up. On January 20, 1978, the first cargo craft was successfully launched for the supplies of the Salyut-6 space station, and immediately named Progress. Up to now, the Progress cargo spacecrafts have been designed and developed six types by the Soviet Union and later Russia, including Original Progress, Progress M, Progress M1, Progress M2, Progress M-M and Progress MS, of which Progress M2 series have not been launched as the launch vehicle Zenit could not be produced on schedule in independent Ukraine after the disintegration of the USSR. On July 17, 2016, the Progress MS-03 spacecraft was launched, heading to the ISS, and two days later, successfully delivered more than 2.4 metric tons of various supplies for the long-duration expedition on the station. Furthermore, a new type of cargo ship called TGK-PG is developing in Russia and will take full advantage of the payload capacity of the Soyuz-2-1b rocket, delivering around 8.2 metric tons to the low Earth orbit, nearly one metric ton more than the mass of the Progress cargo craft series launched on previous versions of the Soyuz rockets.

As financial and technical problems held the USSR behind the United States in the Moon Race, the leaders of the Soviet space industry had to look for a new direction in the manned space program, which was to launch and build the Salyut space station. On April 19, 1971, the world's first crewed space station Salyut-1 was successfully launched into orbit, and 3 days later, the Soyuz-10 spacecraft with three astronauts was launched, docking to the Salyut-1, but as failing to open the hatch in the station, the first crew was not able to enter it. After the crew onboard the Soyuz-10 safely returned to the ground, the Soyuz-11 with three astronauts was launched on June 6 and the astronauts successfully entered and stayed in the Salyut-1 from June 7 to June 29, but unfortunately, they were killed due to depressurization during landing. Finally, On October 11, the Salyut-1 started to de-orbit and fall into the Earth's atmosphere to burn out. Thereafter, there were six space station series Salyut-2, Salyut-3, Salyut-4, Salyut-5, Salyut-6 and Salyut-7, built sequentially from April 1973 to April 1982, of which the Salyut-2/-3/-5 were used for military purposes and also called Almaz space stations. For the entire Salyut Program, over a period of 15 years from 1971 to 1986, a number of civil scientific researches including the long-term living in space and a variety of astronomical, biological and Earth-resources experiments, as well as the secretive military tasks, were successively carried out in seven Salyut space stations.

Since the core module was launched by the USSR into the Earth orbit on February 20, 1986, over the next decade the world's first modular space station Mir had been building by continually sending the additional modules aloft to assemble to the core unit. The Mir's core module resembled its simpler predecessors in the Salyut series, with the form of a stepped cylinder, a length of about 13 m and a maximum diameter of 4.2 m and had six docking ports that accommodated not only a succession of crewed spacecrafts and cargo ferries but also permanently attached expansion modules equipped for scientific research. During 14 years in orbit, the Mir space station had been docked with 39 vehicles, including 1 Soyuz T-15, 29 Soyuz TM series, 1 Discovery Shuttle, 7 Atlantis Shuttles and 1 Endeavour Shuttle, and conducted one approaching and maneuvering demonstration with the Discovery Shuttle. The Mir hosted more than 100 people from 12 countries, of which included 57 American astronauts during 1995–1998 as a part of the Mir-Space Shuttle cooperative endeavor, and served as a microgravity research laboratory in which the crews conducted experiments in biology, human biology, physics, astronomy, meteorology and spacecraft systems, with a goal of developing technologies required for permanent occupation of space. On March 23, 2001, the abandoned Mir space station made a controlled reentry, with the surviving pieces falling into the Pacific Ocean.

In the 1970s, the United States had built its first space station Skylab, and began to conceive a reusable spacecraft, Space Shuttle. In May 1973, the Skylab was launched into orbit, s which three successive three-man crews lived respectively for 28, 56 and 84 days in the orbit. The astronauts aboard the Skylab conducted 270 experiments in biomedical and life science, solar astronomy, Earth observations and materials processing, among which the most important experiments were investigations on the astronauts' physiological responses to long-duration space flight. In July 1979, the Skylab re-entered the atmosphere due to its orbit decaying and burned up. Finally, its remnants scattered over the Indian Ocean and a settled area of Western Australia, but fortunately nobody was hurt.

The Space Shuttle, whose official program name was Space Transportation System (STS), was taken from a 1969 plan for a system of reusable spacecraft of which it was the only item funded for development by NASA. It consists of three major components: a winged orbiter that carries both crew and cargo; an external tank containing liquid hydrogen and liquid oxygen for the orbiter's three main rocket engines; a pair of large, solid-propellant, trap-on booster rockets. At the lift-off, the entire system weighed 2000 metric tons and stood 56-m high. On April 12, 1981, the first orbiter, the Space Shuttle Columbia, was launched from the Kennedy Space Center to carry out orbital flight tests, leading to operational flights beginning in 1982. Besides the Columbia, Challenger, Discovery and Atlantis together formed a fleet of four orbiters, and were sequentially put into service. On January 28, 1986, the Challenger with seven crews exploded shortly after lift-off, killing all aboard including six astronauts and a schoolteacher who had been selected to fly via NASA's "Teacher in Space" program. In 1992, a replacement orbiter for the destroyed Challenger, Endeavour, flew its first mission. The operational missions of the five complete Space Shuttles included launching numerous satellites, interplanetary probes, and the Hubble Space Telescope (HST); conducting science experiments in orbit and a series of shuttle missions to the orbiting Russian Mir; participating in construction and servicing of the ISS. On February 1, 2003, the Columbia broke up catastrophically over north-central Texas at an altitude of about 60 km as it was return from an orbital mission, resulting in all seven crew members killed. The shuttle fleet was immediately grounded until July 2005 the flights resumed with the Discovery's launch. On July 8, 2011, the Atlantis was launched to conduct the last flight mission, which is also the 135th flight of the American Space Shuttle fleet. Since then, the NASA announced that subsequent crewed missions would use the Russian Soyuz spacecrafts as well as the spacecrafts built by the American companies.

The ISS program is a joint project between five participating space agencies from Russia, America, Europe, Japan and Canada. On 20 November 1998, as the initial module Zarya, developed by Russia to provide attitude control and solar power arrays, was successfully launched into orbit by a Proton rocket, the nascent station began to be under assembly and construction for the following decade. A month later, the Space Shuttle astronauts ferried up and attached the first American-built module Unity, a connecting node with multiple docking systems. In the middle of 2000, the Russian-built module Zvezda (a habitat and control centre) was added, and on November 2 of that year, the ISS received its first resident crew, comprising two Russian cosmonauts and an American astronaut. The ISS has been continuously occupied since then. In the wake of the catastrophic explosion of the Columbia in 2003 and the consequent temporary grounding of the space shuttles, the activity aboard the ISS was reduced virtually to a caretaker level, until 2005 other modules built by Europe, Japan and Canada began to be gradually mounted on the ISS. In May 2009, the ISS became fully operational and began hosting a six-person crew. Just after completion of the ISS, the shuttle fleet was retired from service. Thereafter, the supplies for the ISS are delivered by the Russian Progress spacecraft, the European Jules Verne Automated Transfer Vehicle (ATV) and Japanese H-2 Transfer Vehicle, as well as two kinds of cargo vehicles which were made in the American commercial companies, SpaceX's Dragon and the Orbital Science Corporation's Cygnus, while until the American companies develop crewed spacecraft, the astronauts to and from the ISS can only be ferried by the Soyuz spacecrafts. The space agencies that are partners in the ISS have not definitively decided when the program will end, but in 2014 President Barack H. Obama administration indicated that the program would receive the U.S. support until "at least 2024".

In 1992, the Chinese government decided to initiate a manned space engineering project and implement a three-step development plan: the first step is to launch a manned spacecraft, build a preliminary supporting experimental manned space-flight engineering system and carry out space application experiments; the second step is to break through the rendezvous and docking technology between a manned spacecraft and a space vehicle, and built a space laboratory with a certain scale and short-term caretakers to solve the problem of its space applications; the third step is to build a space station, to solve the problem of large-scale and long-term attended space applications. On November 20, 1999, the Chinese first unmanned spacecraft, Shenzhou-1, was launched from the Jiuquan Satellite Launch Center of China for the orbital flight test, and then three flight tests were conducted sequentially by launching the unmanned spacecrafts Shenzhou-2, Shenzhou-3 and Shenzhou-4. On October 15, 2003, the Chinese first manned spacecraft, Shenzhou-5 with the astronaut Liwei Yang, was successfully launched from the Jiuquan Satellite Launch Center by using Long March-2F rocket, flew 14 circles around the Earth, and safely return the ground, with the flight duration of 21 h and 23 min. It is a symbol that China is the third country with the capability of carrying out manned earth orbit flight independently after the USSR (later Russia) and the US. At the time, the first step development plan was completed successfully. From 2005 to 2017, there were six spacecrafts Shenzhou series, two space laboratories Tiangong series and a cargo spaceship Tianzhou-1 launched sequentially, Among which, the Shenzhou-6 carried two astronauts to conduct the flight test with 76 circles around the Earth, lasting 4 days, 2 h and 33 min; three astronauts onboard the Shenzhou-7 carried out the tests of extravehicular activity, a small satellite-accompanying flight and other technologies; the Shenzhou-8 unmanned spacecraft carried out autonomously the rendezvous and docking test with the Tiangong-1, which is an important milestone for the China's manned space flight project; both the Shenzhou-9 and Shenzhou-10 that carried with three-astronaut crew each (including one female astronaut each) were docked with the Tiangong-1 to conduct many space experiments, which marked the successful completion of the second step development plan; in October 2016, two astronauts taken by the Shenzhou-11 were launched from the Jiuquan Satellite Launch Center to enter the Tiangong-2 launched in September, and lived in space for 33 days to complete a series of scientific and technical test tasks; in April 2017, the Tianzhou-1 was launched from the Wenchang Satellite Launch Center, Hainan Province, China, by using the Long March-7 rocket, and docked with the Tiangong-2 to first carry out the in-orbit filling propellant test. So far as 2020, China is implementing the third step development plan of manned space flight.

3.5.4 Manned Landing on the Moon

The manned landing on the Moon is the greatest space flight activity of humanity in the twentieth century, which shows the great power of human conquering nature. Ironically, it is also the product of the Space Race between the United States and the Soviet Union. In the wake of the first satellite launched by the USSR, the US quickly set up the NASA to concentrate the national advantageous resources and try to seize the strategic commanding point in the field of manned space flight. However, as the Soviet cosmonaut Gagarin successfully completed his first space mission, the American dream was frustrated once again. As a result, the U.S. government had to make new choice. On May 25, 1961, President Kennedy made an address to Congress, and officially proposed an ambitious goal: "landing a man on the Moon by the end of this decade and returning him safely to the Earth". Shortly thereafter, the NASA initiated the well-known as Apollo Program, whose objectives are: to establish the technology to meet other national interests in space; to achieve preeminence in space for the United States; to carry out a program of scientific exploration of the Moon; to develop human capability to work in the lunar environment.

With the help of companies like Boeing, North American Rockwell, McDonnell Douglas, International Business Machines (IBM) and Grumman, the NASA designed and built the Apollo spacecraft, which could transport a small crew to the Moon and back again. The Apollo spacecraft is composed of three sections, a Command Module (CM), a Service Module (SM) and a Lunar Module (LM), of which the compartment of combining the CM with the SM is also called CSM.

The CM is the cockpit where the astronauts live and works in space flight, the control center of the spacecraft and the only section of the spacecraft designed to return to the Earth intactly, with a cone-shaped structure, a height of 3.9 m, a diameter of 3.9 m at the bottom, and a lift-off weight of 5.6 metric tons. Inside the CM, the habitable room held a three-astronaut crew is about 5.9 cubic meters while the rest is dedicated to control panels and displays. The CM is divided into the front cabin, the crew cabin and the rear cabin. The front cabin is equipped with docking detection devices, parachute recovery devices and attitude control engines; the crew cabin is a closed pressurized capsule equipped with necessities and life-saving equipment for astronauts to live 14 days; the rear cabin is equipped with 10 attitude control engines, various instruments and fuel tanks, as well as some systems like guidance and navigation control, telecommunication, stabilization and control, reaction control, environmental control and Earth landing. At launch, the three astronauts would sit on a couch that they could later fold and stow away when they entered in space.

The SM adopts light metal honeycomb structure, with a cylindrical shape, a height of 7.5 m, a diameter of 3.9 m, and a lift-off weight of 23.2 metric tons. Its front end is connected with the CM, and the rear end is a main engine nozzle. The SM is equipped with the main engine, attitude control engine group, fuel cells, propellant tanks, pressurization and electrical devices. The main engine is used for orbit transferring and maneuvering; the attitude control engine group is composed of 16 small thrust motors, used for the separation between the spacecraft and the three-stage rocket, the docking between the LM and the CSM, and the separation between the CM and the SM; the fuel cells provide the power for most of the crew's needs during the mission, but both the SM and the CM also carried batteries to supplement power.

The LM includes two sections, ascent stage and descent stage, with a combined-shape of two prisms, a height of 7.0 m, a width of 4.3 m, and a weight of 15 metric tons. As the main body of the LM, the ascent stage consists of a two-astronaut crew compartment, system display and control devices, S-band and radar antennas, reaction control system, engines and propellant tanks. The descent stage is a support platform of the LM, and consists of landing deceleration engines, landing gear with four legs, propellant tanks and instrument cabins. When the spacecraft enters into the parking orbit around the Moon, the LM is separated from the CSM, the deceleration engines are started to make it land softly on the surface of the Moon. When the astronauts complete the scientific investigation tasks on the Moon's surface and return to the ascent stage, the descent stage is used as a launch pad for the ascent engine firing, and left on the original landing position. The ascent stage leaves the Moon's surface for the parking orbit to rendezvous and dock with the CMS.

The Saturn family of launch vehicles was applied to the Apollo Program. As early as 1957, the U.S. Army Ballistic Missile Agency (ABMA) in the city of Huntsville, Alabama, under the technical direction of Wernher von Braun, was already working on a new large launch vehicle, Saturn rocket. In July 1960, the Department of Defense transferred the ABMA and its large booster project to the NASA. This formed the basis for the George C. Marshall Space Flight Center, where von Braun served as the first director until his transfer to the NASA Headquarters in 1970. The Saturn family of launch vehicles was mainly composed of the Saturn I, Saturn IB and Saturn V. The Saturn I was used for two kinds of missions, known as Block-1 and Block-2, of which several of the Block-2 missions orbited Apollo boilerplate payload. The Saturn I for the Block-2 was a two-stage rocket, with a height of 57.3 m and a lift-off thrust of 681,960 kgf. Subsequently, the design of improving the Saturn I evolved into the Saturn IB for the flight tests of Apollo's command and service module and lunar module. The Saturn IB was also a two-stage rocket, with a height of 68.63 m and a lift-off thrust of 743,730 kgf. Finally, a type of Heavy Lift Vehicle, the Saturn V was successfully developed at the NASA's Marshall Space Flight Center in 1967, to send human to the Moon. All of the Saturn V rockets were almost the three-stage ones except one two-stage rocket used in a flight test. The first stage of the Saturn V, also called S-IC, had five F-1 engines with liquid oxygen and kerosene as propellants, all of which could produce the lift-off thrust of 3,458,810 kgf to push the entire vehicle assembly more than 58 km vertically at a speed of 2.752 km/s; the second stage, called S-II, had five J-2 engines that could together produce a thrust of 510,637 kgf, and after jettisoning the S-I, the rocket reached an altitude of 162.5 km at a velocity of 6.932 km/s; the third stage, called S-IVB, had a single J-2 engine that could produce 120,128 kgf, eventually to put the Apollo spacecraft into the Low Earth Orbit (LEO). Once into the orbit, the third stage engine powered down temporarily. When the spacecraft achieved the proper alignment after a few rotations around the Earth, the engine would reignite. The resulting thrust guided the Apollo spacecraft into the trajectory flying to the Moon. After the second ignition, the S-IVB was separated from the spacecraft. The entire Saturn V rocket including the escape tower stood 110.6 m tall, with a lift-off weight of 2945.95 metric tons and a thrust-weight ratio of 1.18, whose launch capacity was 118 metric tons for the LEO and 47 metric tons for the orbit around the Moon. There were a total of 32 launches of the Saturn rockets between 1961 and 1975, including 10 Saturn I rockets, 9 Saturn IB rockets, 12 three-stage Saturn V rockets and 1 two-stage Saturn V rocket. Although some flights experienced significant problems, no Saturn rocket failed catastrophically in flight. The Saturn rockets were used in support of the Apollo lunar missions, the launch of the Skylab space station and ferrying crews to and from the Skylab, and to launch the American half of the Apollo-Soyuz Test Project.

The arguments on the manned landing on the Moon lasted for nearly three years at the NASA, and at first there were four schemes put forward in technical feasibility. The first scheme was the rendezvous and docking in the lunar orbit. The LM was separated from the CSM in the lunar orbit, and then descended and landed softly on the surface of the Moon; after completing the tasks of investigating the Moon, the ascent stage of the LM blasted off the descent stage that was abandoned on the lunar surface, to dock with the CSM, and then return to the Earth. The second scheme was the rendezvous and docking in the Earth orbit, which each cabin of the spacecraft was sent respectively into the Earth orbit by using small rockets for docking and assembling and then the assembled spacecraft would fly to the Moon. The third scheme was the assembling on the surface of the Moon. Firstly, a spaceship carrying the propulsion system was launched to the Moon; secondly, a manned spacecraft was sent to the Moon, and assembled with the propulsion system on the lunar surface; finally, the assembled spacecraft was launched from the Moon to return to the Earth. The fourth scheme was the direct landing on the Moon. The spacecraft was directly launched to the Moon by using a new type of giant launch vehicle; after landing on the Moon and completing the exploration tasks, the spacecraft took off by firing the rocket again, and finally return to the Earth.

By further arguing the technical and economical feasibility, the NASA finally approved the first scheme, the lunar orbit rendezvous and docking, proposed by an aerospace engineer John C. Houbolt (1919–2014) and his team at the Langley Research Center. A specific operation process of the scheme is as follows:

(1) Using the Saturn V launch vehicle, the Apollo spacecraft is sent to the Earth orbit, and then the S-IVB rocket engine shut off.
(2) When the S-IVB engine is fired again, the Apollo spacecraft is sent to the Earth-Moon transfer orbit.
(3) The LM is separated from the CSM by an ejection using the adaptation controller, and the CSM is guided to rotate 180 degrees.
(4) The S-IVB rocket's shell is jettisoned to expose the LM, and then the LM is docked with the CSM to form a combination body for together flight to the Moon.
(5) As the CSM's engine is fired to brake, the Apollo spacecraft enters into the orbit around the Moon, with an orbital altitude of 110 km.
(6) The CSM is separated from the LM, and continued to fly in the orbit around the Moon. Meanwhile, the landing gear of the LM began to deploy, and the descent engine is ignited to enter into the decent orbit. The LM will fall with a powered-mode toward the lunar surface.
(7) When the LM descends to a position of 15 km away from the Moon's surface, the descent engine is fired again. In this way, the astronauts can manually operate the LM to avoid obstacles and vertically land at the safe site on the lunar surface.
(8) The astronauts walk out of the LM to carry out the investigation of the lunar surface and the collection of the lunar samples.
(9) At the end of the lunar investigation, the astronauts return to the ascent stage of the LM. At this time, the descent stage is used as the launch pad, on which the ascent stage engine is ignited for the takeoff. As a result, the descent stage is abandoned on the surface of the Moon.
(10) The ascent stage is rendezvoused and docked with the CSM in the lunar orbit.

(11) When the astronauts return to the CSM with the collected lunar samples and investigated results, the ascent stage is separated from the CSM, and jettisoned and gotten rid of the orbit. Finally, the ascent stage falls to the lunar surface.
(12) As the engines of the SM are fired, the CSM is put into the Moon-Earth transfer orbit, and begins to back toward the Earth.
(13) Before the reentry into the Earth's atmosphere, the SM is separated from the CM and jettisoned, and the CM is rotated 180 degrees to reenter the atmosphere with its cone-shaped bottom forward the Earth's direction.
(14) Under the action of the atmospheric resistance, the CM is decelerated to fall down, and in the period, the ionization blackout effect will be generated due to aerodynamic heating.
(15) The parachutes are deployed and the CM is slowly splashed onto the Pacific Ocean. Shortly thereafter, the astronauts will be rescued by the military helicopters and escorted to the standby aircraft carrier.

The Apollo flight tests were carried out in four phases as follows.

In the first phase (1961.10−1965.7), a total of 10 Saturn I rockets were launched, of which the first four were suborbital flight tests and the last six were orbital flight tests. For the suborbital flight tests, the Saturn I rockets were the two-stage ones with an S-I first stage, a dummy upper stage as the second stage, and a cone-shaped head as same as the Jupiter rocket. The test purposes were to investigate the performance of H-1 engines, multi-stage parallel technology and control system. For the orbital flight tests, the Saturn I rockets included the up-rated S-I rockets as the first stages and the S-IV rocket as the second stages, and equipped with the Apollo spacecraft's simulators as the payloads. The test purposes were to investigate the performance of the second stage rocket engine RL-10, the coupling capability between the rocket and the spacecraft and the aerodynamic characteristics. Meanwhile, fully using the two-stage Saturn I rockets, several space science exploration satellites were put in the orbits, such as the Pegasus satellites that were designed to detect micrometeoroid hits.

In the second phase (1964.5−1966.1), four single-stage Little Joe II rockets were launched, carrying the Apollo spacecraft's simulator and the escape tower. For the first two flight tests, the altitude of the rocket flight reached 5 km, whose purpose was to investigate the performance features of the escape tower. For the third flight test, the rocket exploded and ended in failure. For the fourth flight test, the rocket flight altitude reached 22 km, whose purpose was to investigate the capability of rapidly separating the escape tower from the rocket.

In the third phase (1966.2−1968.4), using four Saturn IB and two Saturn V rockets, six unmanned spacecrafts, including three Apollo-Saturn combination (AS) series and three Apollo series, were respectively launched to carry out the flight tests. On February 26, 1966, the AS-201 spacecraft was launched by using the Saturn IB rocket from the Kennedy Space Center, which was a suborbital flight test, with the maximum altitude of 488 km and the flight duration of 37 min. The mission objectives were to achieve structural integrity and compatibility of launch vehicle, to launch payload, and to demonstrate various technical performances, such as coupling and separation of rocket, spacecraft and other units, propulsion systems, guidance and control systems, electrical systems, start-up and shutdown of the SM's engines, heat shield of the CM, and reentry and recovery of the CM. The operations of support facilities were demonstrated completely. There were several malfunctions, mostly minor, but three were serious, including the fault of service propulsion system, the fault of electrical power system and the loss of data during the reentry. Thereafter, the AS-202 and AS-203 were sequentially launched with the Saturn IB rockets, of which the former conducted an orbital flight test and the latter did a suborbital flight test. With the two flights, the combination tests between the rockets and the spacecrafts had been accomplished. On January 27, 1967, three astronauts by the AS-204 (or Apollo-1) spacecraft were making a preflight test on the launch pad at the Kennedy Space Center. On the schedule, they were going to carry out the first crewed flight on February 21. Unfortunately, three astronauts lost their lives as a fire swept through the CM. For this reason, the exhaustive investigation of the fire and extensive reworking of the CM postponed crewed launches until the NASA officials cleared them for flight. The Saturn IB schedules were suspended for nearly a year, and the rocket that finally bore the designation AS-204 carried the LM as the payload instead of the CM. As a result, on November 9, 1967, the unmanned Apollo-4 spacecraft was first launched into a predetermined earth-orbit by the Saturn V rocket, which fully verified the performance of the rocket, the coupling characteristics of the rocket with the spacecraft, the spacecraft's main propulsion system, the attitude control system, the mission support system, and the reentry and recovery of the CM. Meanwhile, the Apollo-4 was also conducted a simulation test of Earth-Moon transfer in the parking-orbit, and pushed to an orbital altitude of 18,079 km by its main engines. On January 22, 1968, the Apollo-5 was put into the earth orbit by the Saturn IB rocket, to verify the operation of the LM and its ascent and descent propulsion systems. At the altitude of 222 km, the second stage of the rocket separated from the LM, and the descent test of the LM was carried out. At the altitude of 961 km, the ascent test of the LM was carried out. On April 4, 1968, the Apollo-6 was launched into the orbit by using the Saturn V rocket to conduct all scheduled integrated flight tests. The Apollo-6 was pushed to an orbital altitude of 22,209 km by its main engine, but during the test, from the telemetry feedback there was a failure that the S-IVB engine shut down in advance.

In the fourth phase (1968.5–1969.5), four series of the Apollo spacecrafts carrying with the astronauts were launched sequentially, to conduct the crewed flight tests. On October 11, 1968, the Apollo-7 with three astronauts Walter M. Schirra, Donn F. Eisele and R. Walter Cunningham was launched into orbit by the Saturn IB rocket, to investigate the adaptability for the astronauts to fly for a long time in the spacecraft, as well as in-orbital maneuver of the CSM, attitude adjustment, and rendezvous and docking function. After flying for 11 days, the Apollo-7 returned to the ground safely on October 22. Subsequently, three crewed spacecrafts Apollo-8, Apollo-9 and Apollo-10 were successively launched to carry out a series of flight tests, such as astronaut's adaptability, spacecraft performance, circumlunar flight, long-distance communication, separation and docking between the LM and CSM, and environmental detection near the Moon. Among them, three astronauts John W.

Young, Thomas P. Stafford and Eugene A. Cernan by the Apollo-10 were launched on May 18, 1969, to carry out the flight test mission of encompassing all aspects of an actual crewed lunar landing, except for the landing. As the Apollo-10 was accelerated to the second cosmic velocity by the S-IVB stage of the Saturn V rocket, and it entered the Earth-Moon transfer orbit, the CSM was separated from the S-IVB rocket with an ejection mode, and turned 180 degrees to dock to the LM which was exposed by jettisoning the S-IVB shell. The recombined Apollo-10 turned again and flew to the Moon along the Earth-Moon transfer orbit, with a flight distance of 399,200 km. When the Apollo-10 entered into the lunar parking-orbit by the brake engine ignition, Young stayed in the CSM and continued to fly around the Moon while by using the descent engine, Stafford and Cernan in the LM was slowly descending toward the lunar surface. At an altitude of 15.2 km away from the lunar surface, they jettisoned the descent stage, and then returned to the lunar parking-orbit by using the ascent engines to dock to the CSM. As Stafford and Cernan returned to the CM, the ascent stage was separated from the CSM, and finally three astronauts returned to the Earth safely.

In order to ensure the successful implementation of the manned lunar landing program, the NASA had also carried out four auxiliary projects. Firstly, the Ranger Program was a series of unmanned space missions by NASA between 1961 and 1965, launched a total of nine vehicles, whose objective was to take the close-up images of the surface of the Moon and argue the feasibility of the crewed-craft landing on the Moon. Secondly, the Gemini Program was the NASA's second human spaceflight project, which was a bridge between the Project Mercury and the Apollo Program. A total of ten spaceships with two-astronaut crew each were launched successively between 1965 and 1966, to carry out a series of space experiments, such as biomedical research, maneuvering flight, rendezvous and docking, and extravehicular activity. Thirdly, the Lunar Orbiter Program was a series of five unmanned lunar orbiter missions launched by the NASA from 1966 to 1967. Intended to help select Apollo landing sites by mapping the lunar surface, they provided the first photographs from lunar orbit and photographed both the Earth and Moon. Fourthly, the Surveyor Program was a NASA project from 1964 to 1968, sent seven robotic vehicles to the lunar surface, whose primary goal was to demonstrate the feasibility of soft landing on the Moon. The capability of the spacecrafts to make midcourse corrections was demonstrated, and the Landers carried instruments to help evaluate the suitability of their landing sites for manned Apollo landings. Moreover, several Surveyor spacecrafts had robotic shovels designed to test lunar soil mechanics, and conducted to make the in-situ detection on the lunar surface and to investigate the physical and chemical properties of the lunar soil.

The Apollo flight tests and its auxiliary project missions were all successful, and the entire process of landing on the Moon was comprehensively practiced, so the NASA decided to carry out the mission of manned landing on the Moon, whose specific tasks are to achieve the astronauts landing on the surface of the Moon and safe-returning to the Earth, and to complete the investigation and sampling of the lunar surface. On July 16, 1969, the Apollo-11 spacecraft was put into orbit by the Saturn V rocket at the launch pad 39Aof the Kennedy Sapce Center. Three astronauts

Neil A. Armstrong (1930–2012), Edwin E. Aldrin (1930–) and Michael Collins (1930–) took the Apollo-11 to carry out the lunar landing mission. At about 109 h and 24 min after the launch, Armstrong first stepped onto the surface of the Moon, and announced by radio: "That's one small step for (a) man, one giant leap for mankind". About 19 min later, Aldrin followed him. During their moon walk of more than two hours, Armstrong and Aldrin inserted the U.S. national flag, and set up a device to measure the composition of the solar wind reaching the Moon, a device to receive laser beams from astronomical observatories on the Earth to determine the exact distance between the two bodies apart from one another, and a passive seismometer to measure moonquakes and meteor impacts long after the astronauts had returned home. They also took about 23 kg of rock and soil samples, took many photographs, and maintained constant communication with the mission control center in Houston, Texas. After Armstrong and Aldrin spent 21 h and 36 min on the lunar surface at a site named Tranquility Base and at the ascent stage of the LM, The ascent engines ignited and took off while the descent stage was left on the lunar surface. The ascent stage rendezvoused and docked with the CSM in the lunar orbit, and the trip back to the Earth began soon afterward. On July 24, the CM of the Apollo-11 splashed down in the Pacific Ocean about 1400 km west of Hawaii, and the three astronauts returned to the Earth safely. The Apollo-11 successfully completed the mission to the Moon, and finally turned human's dream into reality. It is also a symbol and peak for humanity to enter the age of space flight in a real sense.

Thereafter, from November 1969 to December 1972, six series of the spaceships from Apollo-12 through Apollo-17 were launched sequentially by the NASA. Among them, on April 11, 1970, the Apollo-13 was launched into the Earth orbit. However, the oxygen tanks in the SM exploded on the way to the Moon, and consequently the control center instructed the astronauts to give up the mission to the Moon. The three astronauts by the Apollo-13 relied on the power, water, air and food provided by the LM to successfully bypass the Moon and finally return to the Earth safely. The other five spacecrafts successfully landed on the surface of the Moon, and the astronauts carried out relevant scientific experiments and equipment placements on the surface. On July 26, 1971, the Apollo-15 with three astronauts was launched successfully, to explore the Hadley-Apennine region, set up and activate lunar surface scientific experiments, make engineering evaluations of new Apollo equipment, and conduct lunar orbital experiments and photographic tasks. The astronauts of the Apollo-15 first used the Lunar Roving Vehicle on the Moon's surface, making its range of activity reached 35 km, and widely carried out sample collections and related scientific experiments. The Apollo-15 set several new records for crewed spaceflight, such as the heaviest payload and the longest time in the lunar orbit, the maximum radial distance traveled on the lunar surface, and the longest crewed lunar mission. On August 7, 1971, the CM of the Apollo-15 carrying three-astronaut crew returned to the Earth safely. The Apollo-17 was launched on December 7, 1972, and 12 days later, returned to the Earth safely, which was the last mission of the Apollo Program. Its crew consisted of Commander Eugene A. Cernan, Lunar Module Pilot Harrison H. Schmitt, and Command Module Pilot Ronald E. Evans, of which Schmitt was the first scientist-astronaut to land on the Moon. In the mission, the sixth automated research station was set up; the Lunar Roving Vehicle traversed a total of 30.5 km; the lunar surface-stay time was 75 h; the astronauts gathered 110.4 kg of material.

The Apollo Program is the largest-scale astronautic project so far, which took 11 years and cost $25.4 billion in 1973 dollars. More than 20,000 enterprises, more than 80 scientific research institutions and more than 200 universities participated in the project, with a total number of more than 300,000 people. The successful implementation and achievements of the Apollo Program have a significant and far-reaching impact on politics, economy, science and technology in the United States and even the world. As the greatest space project and feat in the twentieth century, the Apollo Program has made its mark in history.

3.6 Flying Out of the Solar System

In January 1959, the world's first vehicle to explore the Moon, Luna-1, was successfully launched by the Soviet Union, whose primary objectives were: to measure temperature and pressure inside the vehicle and the magnetic fields of the Earth and Moon; to study the gas components of interplanetary matter and corpuscular radiation from the Sun, the meteoric particles in space, the distribution of heavy nuclei in primary cosmic radiation, and other properties of cosmic rays. However, due to an incorrectly-timed upper stage burned during its launch, Luna-1 only passed through within 5995 km away from the Moon's surface, and went into an orbit around the Sun, between the orbits of the Earth and Mars, becoming the first spacecraft to leave geocentric orbit. Despite not becoming a satellite around the Moon, Luna-1 opened the prelude of human space flight into the solar system. So far, more than 200 deep-space probes have been launched successively, to explore the Moon, planets in the solar system, comets and asteroids, and even land on the celestial bodies and sample back to the Earth. Nevertheless, for the vast universe, the interplanetary space flight in the solar system is insignificant. Flying out of the solar system, to sail into the interstellar space, explore the origin, evolution and current situation of the universe, and even look for extraterrestrial civilization, is the ultimate goal pursued by human beings.

In the 1970s, Pioneer-10, Pioneer-11, Voyager-1 and Voyager-2 launched successfully by the United States, were the first man-made vehicles to fly to the edge of the solar system, roaming in the vast universe, carrying a lot of information about the Earth's development history and human civilization, and looking forward to meeting with extraterrestrial civilization one day. Although such opportunity is very slim, humanity on the Earth still place great hopes on it. In addition, the New Horizons probe was launched by the NASA in 2006, to investigate the outer space of the solar system in detail.

3.6.1 Boundaries of the Solar System

To judge whether a man-made vehicle flies out the solar system or enters the interstellar space, it involves the determination of the boundary of the solar system. Where is the boundary of the solar system, and what criteria should be adopted to determine the boundary of the solar system? In the past twenty years, there has been a popular debate about whether Voyager-1 enters the interstellar space or flies out of the solar system. Currently, in the field of astronomy and astrophysics, there are two main criteria to determine the boundary of the solar system: one is the gravitational boundary, and another is the physical boundary.

3.6.1.1 Gravitational Boundary

The **gravitational boundary** refers to a spherical boundary based on the gravitational action range of the Sun; within the spherical boundary, the gravitational attraction of the Sun rather than that of the rest of the Galactic System controls the orbit motions of celestial bodies, such as planets and comets. In general, the gravitational action range of a celestial body is relative to that of another celestial source with stronger gravitation field. The Sun is rotating around the center of the Milky Way system, and the center is the stronger gravitational source compared to the Sun. From numerical analysis results, it is shown that the gravitational range of the Sun relative to the Milky Way is approximately 6.0×10^4 AU, which is equal to one fourth of the distance (about 4.22 light years) between the Sun and the nearest star, Alpha Centauri.

According to the view of dynamics, under the action of solar gravitation, the celestial bodies in the solar system move around the Sun. As a result, the orbits of the outermost planets in the solar system had been ever used as the boundaries. With the discovery of Uranus, Neptune and Pluto, the solar system boundary was expanded respectively from 19.2 AU to 30.1 AU and 39.5 AU. Of course, the Pluto has been downgraded to a mere dwarf planet at the IAU General Assembly in 2006. In 1943, an Irish astronomer, Kenneth E. Edgeworth (1880–1972), put forward the hypothesis that the material distribution of the nebula beyond the Neptune was so sparse that a large amount of the celestial bodies with smaller mass could be brought up. Later, a Dutch-American astronomer, Gerard P. Kuiper (1905–1973), further developed Edgeworth's hypothesis, and believed that there was a zone of ice-filled objects running on the edge of the solar system beyond the Neptune's orbit, where the remnants of the original solar nebula gathered, and also the birthplace of asteroids and short-period comets. This is the famous **Edgeworth–Kuiper Belt**, simply referred to as **Kuiper Belt**, which is a flat ring of icy small bodies that revolve around the Sun beyond the orbit of the Neptune, with 30–50 AU away from the Sun. Kuiper considered that comets should also be included within the solar system's boundary to be determined. In this way, the boundary of the solar system was extended by 10 AU. In the search for the solar system boundaries, the Pluto had been the only known object in the Kuiper Belt until another object, 1992QB1, was discovered outside the

Neptune in 1992, when the existence of the Kuiper Belt was first confirmed by the observation. Up to now, about 2000 small celestial bodies have been discovered in the Kuiper Belt, and their surfaces are covered with icy-ice made up of nitrogen, methane, ammonia and water.

In 1951, the Dutch astronomer, Jan H. Oort (1900–1992), put forward a conjecture that in the region about $5.0 \times 10^4 - 1.5 \times 10^5$ AU away from the Sun, there might be a huge cloud composed of long-period comets, including 100 billion comets. Later, the region was officially called **Oort cloud** by astronomical community. The comets with the orbital periods of millions of years move around the Sun in very flat elliptical orbits. As a result, they move very fast near the perihelion and stay for a short time; they move very slowly near the aphelion (the Oort cloud) and stay for a long time. The long-period comets spend most of their time near the aphelion, so although there are many comets in the Oort cloud, few are observed near the Sun. Certainly, for such distant objects, since they are greatly affected by the gravitation from other stars, it can not be determine whether they will return to the solar system again. The Oort cloud is the critical region of the gravitational interaction between the Sun and other stars, and also the gravitational boundary of the solar system. Although the Oort cloud has not been observed directly, the conjecture is still accepted by most of astronomical specialists.

3.6.1.2 Physical Boundary

The **physical boundary** of the solar system is the interface between the solar wind and the interstellar medium. In general, the Sun is divided into six layers from the center out, subsequently solar interior core, radiative zone, convective zone, photosphere, chromo-sphere and corona. Most of the Sun's mass is concentrated on the solar interior core, which occupies the innermost quarter or so of the Sun's radius, and more than 99% of the Sun's energy comes from this zone. The corona is located in the outermost layer of the Sun, belonging to the outer atmosphere of the Sun. Under the action of millions of degrees high temperature in the corona, hydrogen and helium atoms are ionized into high-speed charged particle streams such as positively charged protons, helium nuclei and negatively charged free electrons, which constantly break free of the Sun's gravity and shoot at the speed of 200–800 km/s to the outer space of the Sun, thus forming so-called **solar wind**. The **interstellar medium** refers to the general term of matter and radiation field between galaxies and stars, which is composed of the interstellar gas and dust. The interstellar gas mainly includes atomic, molecular, electronic and ionic states of gaseous hydrogen and helium. The interstellar dust mainly includes solid particles with a diameter of about 0.1 μm, such as ice, graphite and silicate, which are dispersed in the interstellar gas, and its mass accounts for about 10% of the interstellar gas.

In general, the vast, bubble-like region of space "blown" by the solar wind is called **heliosphere**, as shown in Fig. 3.5. Outside the heliosphere, there is the tremendous interstellar medium that the solar wind can no longer shake, where the solar wind (also known as **solar plasma**) gives way to the interstellar plasma permeating our

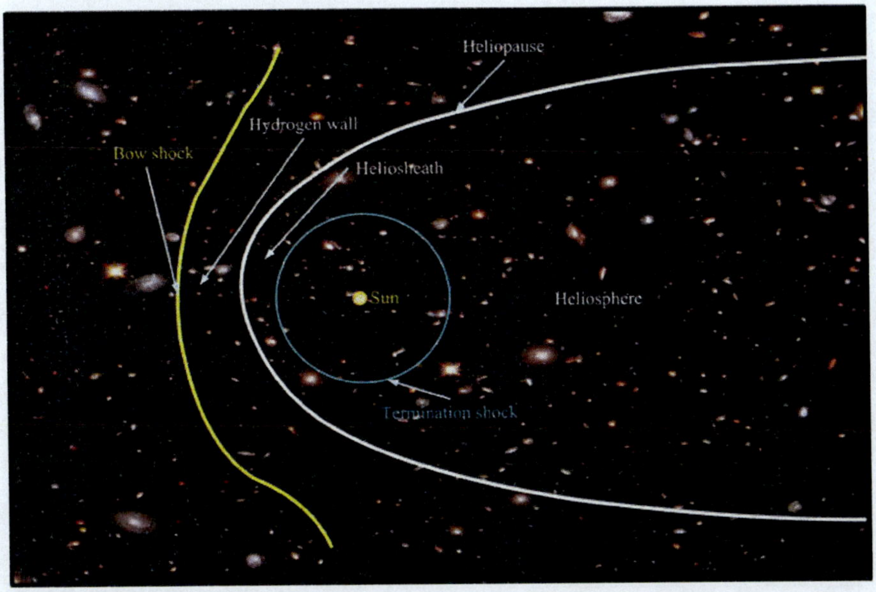

Fig. 3.5 Sketch map of the heliosphere's structure

galaxy. Here is called **heliopause**, which is the boundary where the solar wind stops as encountering the interstellar medium. The heliopause is usually considered as the outer physical boundary of the solar system, with a distance of about 120 AU away from the Sun. Apparently, the physical boundary of the solar system defined by the heliopause is much smaller than the gravitational boundary defined by the Oort cloud. Within the heliopause, the zone where the charged particles are decelerated from the supersonic speed to the subsonic speed by the interstellar medium is called **termination shock**, which is also the position where the particles of the solar wind begin to contact with the interstellar medium. The zone between the termination shock and the heliopause is called **heliosheath**. Outside the heliopause, **bow shock** will be brought up along the direction of the solar wind, due to the interaction between the interplanetary medium and the heliopause. There is a hydrogen wall between the bow shock and the heliopause. The so-called **hydrogen wall** refers to the area composed of hot hydrogen, which is formed under the interstellar medium acted on the edge of the heliopause.

In fact, the physical boundary defined by the heliopause is not unchangeable, but change with the change of the plasma density in the space. The so-called **plasma** is a kind of material form mainly composed of free electrons and charged ions. The matter from the sun is blown to the deep space in the form of solar wind, which produces the heliosphere represented by the plasmas. The plasmas from the Sun and galaxies carry information about the directions of their respective magnetic lines while the direction of the solar magnetic field is different from that of the interstellar magnetic field. Once the solar wind and its plasmas no longer play a leading role, instead of the interstellar medium and its plasmas, which means that the solar wind has reached interstellar space. Therefore, based on the definition of the physical boundary of the solar system, it can be determined by the number of charged particles and the change of the direction of the magnetic field in the environment where the spacecrafts are located, to evaluate whether they enter the interstellar space or fly out of the solar system. In other words, the basic criteria for judging the man-made probes flying out of the solar system are as follows:

(1) Whether does the number of charged particles from the Sun around the probes drop abruptly?
(2) Whether are the charged particles from galaxies and stars increasing rapidly?
(3) Whether does the direction of magnetic field change suddenly?

3.6.2 *Pioneer Probes*

3.6.2.1 Pioneer Program

The Pioneer program is one of a series of unmanned exploration programs for the planets and interstellar space in the solar system implemented by the United States in the 1950s–1970s. There were 19 probes planed and launched successively in the Pioneer program, of which only 10 successfully completed the scheduled exploration missions, mainly exploring the Moon and the Earth-Moon space, as well as the planets of the solar system, such as Venus, Jupiter and Saturn, and its interstellar space. The Pioneer program was implemented in the two phases as follows.

In the first phase (1958–1960), the exploration missions, including the Earth-Moon space, the space near the Moon and the Earth-Venus space, were carried out, whose goals are to demonstrate and verify the feasibility of the probes to reach the escape velocity of the Earth, and to conduct the lunar exploration research. During the phase, nine series of Pioneer probes, including Pioneer-0, 1, 2, 3, 4, P-1, P-3, 5 and P-30, were launched successively, of which except for the successful launch of Pioneer-4 and Pioneer-5, the other probes ended in failures because the rocket engines failed to reach the escape velocity of the Earth. On March 3, 1959, the cone-shaped Pioneer-4 probe which is made of glass fiber material, with the cone height of 0.58 m, the bottom diameter of 0.25 m and the total weight of 6.1 kg, was launched by Juno II rocket. The Pioneer-4 is the American first lunar probe to successfully break away from the Earth's gravity. It flew over the Moon and successfully entered into the heliocentric orbit, with a perihelion altitude of 0.98 AU, an aphelion of 1.13 AU, an orbital eccentricity of 0.072, an orbital period of 398 days, and an orbital inclination of 1.5 degrees. On March 11, 1960, the Pioneer-5 probe with the ball-shaped main structure was launched by the Thor-Ables IV rocket. The diameter of the Pioneer-5 is 0.66 m, the solar sails 1.4 m long and the weight 43 kg. The pioneer-5 entered into the heliocentric orbit with a perihelion altitude of 0.71 AU, an aphelion altitude of 0.99 AU, an orbital eccentricity of 0.1689, an orbital period of 312 days and an orbital inclination of 3.35 degrees. The pioneer-5 had been operated normally in the orbit for 106 days, to detect the distribution of magnetic field in the space between the Earth and Venus.

In the second phase (1965–1978), a series of the exploration missions of planetary and interplanetary space was carried out, whose goals are to explore and photograph the inner planets in the solar system, to fly over the Jupiter and Saturn, and finally to fly out of the solar system. During the phase, ten series of Pioneer probes, including Pioneer-6, 7, 8, 9, E, 10, 11, H, 12 and 13, were planned and launched sequentially, of which the Pioneer-E failed to launch; the Pioneer-H was the probe planned but not launched, whose structure and mission were as same as the Pioneer-10 and Pioneer-11 and which is now stored at the National Air and Space Museum in Washington, D.C.; the other eight probes had been successfully launched and completed the scheduled exploration tasks. The Pioneer-6, 7, 8 and 9 were a series of the space probes with similar missions and functions, which are used to explore the solar-wind ions and electrons, the cosmic rays and the interplanetary magnetic field. Each of the four probes had a cylinder-shaped structure, with the weight of 138–147 kg and with the spin-stabilized at 60 revolutions per minute. The Pioneer-6 and Pioneer-9 were located in the orbit of 0.8 AU away from the Sun, and both orbital periods are slightly shorter than that of the Earth, while the Pioneer-7 and Pioneer-8 were located in the orbit of 1.1 AU away from the Sun, and both orbital periods are slightly longer than that of the Earth. The Pioneer-10 and Pioneer-11 have been the most eye-catching exploration missions in the entire program. Both probes first explored the outer planets of the solar system, Jupiter and Saturn, and their interplanetary space, and then continued to fly out of the solar system. Both Pioneer-10 and Pioneer-11 are the first emissaries sent by humanity to visit the extraterrestrial planets, and carry a gold-anodized aluminum plaque each, on which there are the sketches of a man and woman and information such as the location of the Earth, looked forward to being intercepted and understood by the extraterrestrial civilization one day. The Pioneer Venus exploration project was composed of two probe missions, the Pioneer-12 and Pioneer-13. The Pioneer-12 was also known as the Pioneer Venus orbiter, with a total weight of 517 kg, and located in the orbit 142 km by 253 km away from the Venus's surface. The objectives of the Pioneer-12 were to measure the Venus's magnetic field and ionosphere and its space environment, and obtain the image of the Venus's surface by radar data. The Pioneer-13 was also known as the Pioneer Venus Multi-probes, with a total weight of 290 kg, and carried with four scientific instruments for exploring the atmosphere of Venus, including the Large probe, North probe, Night probe and Day probe. The Pioneer-13 descended to the atmosphere of Venus to carry out the exploration in detail, until finally hitting the surface of Venus. For all that, there had still been the measurement data sent back to the Earth by an undamaged detector after hitting.

3.6.2.2 Pioneer-10 and Pioneer-11

On March 2, 1972, the Pioneer-10 probe (also known as Pioneer-F) was successfully launched by the NASA. It was the first man-made vehicle to fly over the asteroid belt between Mars and Jupiter, and further explore the Jupiter. Its main tasks were to study the asteroid belt and the surrounding environment of the Jupiter, as well as the solar wind, cosmic rays, and heliosphere's farthest position. The probe had a hexahedral ring structure, with a total length of 2.9 m and a weight of 258 kg, equipped with a high-gain antenna with a diameter of 2.74 m, and carried more than ten kinds of scientific exploration instruments. The specific flight paths and key events of the Pioneer-10 were as follows:

(1) At 1:49:00 UTC on March 3, 1972, using an Atlas-Centaur rocket, the Pioneer-10 was successfully launched from the Launch Complex 36A at the Cape Canaveral, Florida. At 17 min after launching, the Pioneer-10 reached a velocity of 14.3 km/s; 11 h later, it flew by the Moon; 77 days later, it reached near the Mars.
(2) On July 15, 1972, the Pioneer-10 entered into the asteroid belt between the Mars and the Jupiter.
(3) On December 4, 1973, the Pioneer-10 arrived at the orbital position of 132,252 km away from the Jupiter. When the Pioneer-10 flew over the Jupiter, many pictures were taken and sent back to the Earth. By analyzing and processing the data sent back by the Pioneer-10, the radiation belt of the Jupiter was discovered, and it was clearly demonstrated that the Jupiter is a gaseous planet without solid surface and the density of gaseous matter increases with the increase of depth. The mass of the Jupiter is about 1/1000 of the Sun, but it is 2.5 times the sum of mass of the other seven major planets in the solar system. As a result, under the huge acceleration of Jupiter's gravity, the Pioneer-10 flew straight to the outside of the solar system.
(4) On June 13, 1983, the Pioneer-10 flew across through the Neptune's orbit, becoming the first man-made vehicle to fly out of eight planetary ranges.
(5) On May 24, 1989, the Pioneer-10 flew across through the Pluto's orbit and to interstellar space.
(6) On March 31, 1997, the Pioneer-10 officially ended its exploration mission in outer space of the solar system. With the Ground Deep Space Network, the Pioneer-10 was tracked by the NASA, to train flight operators how to capture and track extremely-weak radio signals from deep space.
(7) On February 17, 1998, the Pioneer-10 was 69.419 AU away from the Sun, the same distance as the Voyager-1 launched in 1977. Prior of that, the Pioneer-10 had been the farthest man-made spacecraft away from the Earth.
(8) On January 23, 2003, when the Pioneer-10 is 1.22×10^{10} km (about 80 AU) away from the Earth, its weak signal was received for the last time. With the increase of flight distance, the weak signal sent by the Pioneer-10 is not enough to be detected. Perhaps, the radioisotope thermoelectric generator onboard had decayed, so that the Pioneer-10 might not have enough power to send additional transmissions to the Earth.
(9) On September 9, 2012, using the orbital propagation method, the Pioneer-10 was predicted to be 106.676 AU away from the Sun.

(10) On January 1, 2016, the Pioneer-10 was predicted to be 114.07 AU from the Earth, traveling at 12.04 km/s relative to the Sun and outward at about 2.54 AU per year.
(11) On March 29, 2020, the Pioneer-10 was predicted to be 125.96 AU from the Earth.

At present, the Pioneer-10 is advancing at the velocity of 2.6 AU per year to the star Aldebaran in the constellation of Taurus, 68 light years away from the Earth. Supposed that the speed of the Aldebaran relative to the solar system is zero, the Pioneer-10 will arrive in the Taurus after about two million years.

The Pioneer-11 (also known as Pioneer-G) was the second ultra-deep space probe successfully launched after the Pioneer-10 to explore the Jupiter and the outer solar system. The Pioneer-11 used the Jupiter's powerful gravity to change its orbit and fly to the Saturn; and then, it flew by the Saturn, and out of the solar system along its escape orbit. The structure and weight of the Pioneer-11 is as same as those of the Pioneer-10, and its specific flight paths and key events were as follows:

(1) At 2:11:00 UTC on April 6, 1973, the Pioneer-11 was successfully launched from the Launch Complex 36B at Cape Canaveral, Florida.
(2) On December 4, 1974, the Pioneer-11 arrived at the distance of 34,000 km away from the Jupiter. The Great Red Spot of the Jupiter and the clouds over its polar region were photographed and the mass of its moon Callisto was measured. With the Jupiter's powerful gravity assist, the probe changed its orbit and got the acceleration, flying to the Saturn.
(3) On September 1, 1979, the Pioneer-11 passed through the Saturn's outer rings and arrived within 21,000 km of the planet, where two new moons and a new "F" ring were discovered. The Pioneer-11 also discovered and charted the magnetosphere and magnetic field, and mapped the general structure of the Saturn's interior. The heat radiation from the Saturn's interior was measured by the onboard instruments, and its planet-sized moon Titan was also found, which was too cold to support life. The Pioneer-11 flew through the Saturn's rings to test whether the dim rings caused damage to the devices onboard. In the meantime, the Voyager-1 and Voyager-2 have crossed the Jupiter and headed for the Saturn. As a pioneer, the Pioneer-11 provided the basis for the actual flight path selection of the Voyager-1 and Voyager-2.
(4) In February 1985, the batter energy storage of the Pioneer-11 decreased, which made it difficult to maintain the normal operation of the instruments and devices at the same time.
(5) On September 30, 1995, the radioisotope thermoelectric generator of the Pioneer-11 ran out of energy, which was not enough to support the exploration mission and stopped sending telemetry information. At this time, the Pioneer-11 was located in the distance of 44.7 AU away from the Sun, and the inclination angle between the spacecraft's orbital plane and the Sun's equator plane was 17.4 degrees.
(6) On March 6, 2011, the Pioneer-11 was predicted to be 82.805 AU away from the Sun.

(7) On July 19, 2015, the Pioneer-11 was 91.672 AU away from the Sun, 90.716 AU from the Earth, and was flying at a velocity of 11.376 km/s.
(8) On March 29, 2020, the Pioneer-11 was 102.82 AU away from the Earth.

At present, the Pioneer-11 is moving at 2.4 AU per year to the direction of Scutum constellation, and flying in the opposite direction of the Pioneer-10. In other words, the Pioneer-10 is flying away from the center of the Milky Way Galaxy, while the Pioneer-11 is flying towards the center of the Galaxy.

There were 11 scientific instruments mounted on the Pioneer-10 and Pioneer-11 respectively, including imaging Photopolarimeter, Helium Vector Magnetometer, Infrared Adiometer, Quadrispherical Plasma Analyzer, Ultraviolet Photometer, Charged Particle Instrument, Cosmic Ray Telescope, Geiger Tube Telescope, Sisyphus Asteroid-Meteoroid Detector, Meteoroid Detector and Trapped Radiation Detector. These scientific instruments were mainly used to detect and analyze the magnetic field of planets and interplanetary space, the cosmic rays, the solar wind, the neutral hydrogen at the heliosphere's edge and interstellar space, the distribution, size, mass, flux and speed of interstellar dust particles, the aurora and electromagnetic waves of the outer solar system planets, the atmosphere of outer planet's moons, as well as the surface morphology of the Jupiter, Saturn and their moons. As the first man-made spacecrafts to fly out of the solar system, the Pioneer-10 and Pioneer-11 carried a gold-anodized aluminum plaque of drawing the spin transition of a hydrogen atom, the location of the Sun and the Earth in the Galaxy, the flight path of the spacecraft, and the naked figures of a man and woman. If the spacecraft would be intercepted by the extraterrestrial intelligent life, the plaque could be used as the card which conveys the greetings and related information of mankind on the Earth.

3.6.3 Pioneer Plaque and Pulsar Positioning

The Pioneer-10 and Pioneer-11 were manufactured by the Systems Development Division of Thompson Ramo Wooldridge (TRW) Inc. in Redono Beach, California. In November 1971, when the Pioneer-10 was being tested finally prior to shipment to the Kennedy Space Center, a group of science correspondents from the U.S. national press were invited to the TRW systems, Inc. to see the Pioneer-10 under the harsh lights from an artificial Sun of the space simulator, and attend its briefing. Looking at the Pioneer-10 through the portholes of the simulator, Eric Burgess (1920–2005), who had ever been the early membership of the British Interplanetary Society and was then the correspondents at the *Christian Science Monitor*, was imagining the scene of the spacecraft to pass through the solar system and become the mankind's first emissary to another star [10]. He thought that the spacecraft should carry a special message, to tell its finder a million or even a billion years later hence that the planet Earth had evolved an intelligent species, who could think beyond its own times and beyond its own solar system. He immediately mentioned his idea to Richard Hoagland who was a freelance writer and Don Bane who was then the correspondent from the

Los Angeles Herald-Examiner, and both agreed enthusiastically. As a result, Burgess and Hoagland approached Carl E. Sagan (1934–1996), Director of the Laboratory of Planetary Studies at Cornell University, who was then visiting the JPL in the city of Pasadena, California, in connection with Mariner-9. Carl Sagan, together with Frank D. Drake (1930–) who was Director of the National Astronomy and Ionosphere Center at Cornell University, had ever designed a type of message that might be used to communicate with an alien intelligence, and a short while earlier, also been invited to attend a conference in the Crimea devoted to the problem of communicating with extraterrestrial intelligences. Sagan was enthusiastic about the idea of a message carried with the Pioneers, and invited Drake to take part in designing the plaque with the mankind's message patterns. Sagan's second wife, L. Salzman Sagan who was an American artist and writer, was in charge of the pattern drawing and artistic processing. The design pattern was presented to the NASA and soon approved, and etched into a 6061-T6 gold-anodized aluminum plate, with a size of 15.2 cm by 22.9 cm, a thickness of 0.127 cm and a weight of 0.12 kg. Its mean depth of engraving is 0.0381 cm to ensure that the pattern will still be legible after millions of years of space flight. The gold-anodized aluminum plate took only three weeks from design to completion, and was attached to the spacecraft's antenna support struts in a position to help shield from erosion by interstellar dust. The detailed design pattern of the Pioneer plaque is shown in Fig. 3.6, which mainly includes the following four aspects.

(1) **Hyperfine Transition of the Neutral Hydrogen**

The **neutral hydrogen** is a normal hydrogen atom composed of one proton and one electron orbiting the proton, while the proton and the electron also make the spin motion respectively. Generally, the corresponding energies of the electrons moving around the nuclei in different orbits are discontinuous values, which are called the **energy levels** of atoms. When an electron is in the orbit closest to the nucleus, the atomic state is the most stable, which is called **ground state** of the atom. When an electron is in the high-energy level of other orbits and has the tendency of returning to the ground state, the atomic state is unstable, which is called **excited state** of the atom. The spin motion of the hydrogen atom in the ground state forms two **hyperfine structures** of the atomic energy levels. One is that when the spin directions of the electron with the proton are same, the hydrogen atom is in **hyperfine high energy state**. The other is that when the spin directions of the electron with the proton are opposite, the hydrogen atom is in **hyperfine low energy state**. In the quantum mechanics, the process that an atom jumps from one energy level state to another is usually called **transition**. Transiting from the hyperfine high energy state to the hyperfine low energy state, the neutral atomic hydrogen will release the energy in the form of stable electromagnetic radiation, with a frequency of 1420.40575177 MHz and a wavelength of 21.10611405413 cm, which is the so-called **21-centimeter line**, also known as **hydrogen line**.

The neutral hydrogen is the most abundant element in the universe. The proton and electron usually have their axes aligned and spin in the same direction, giving the

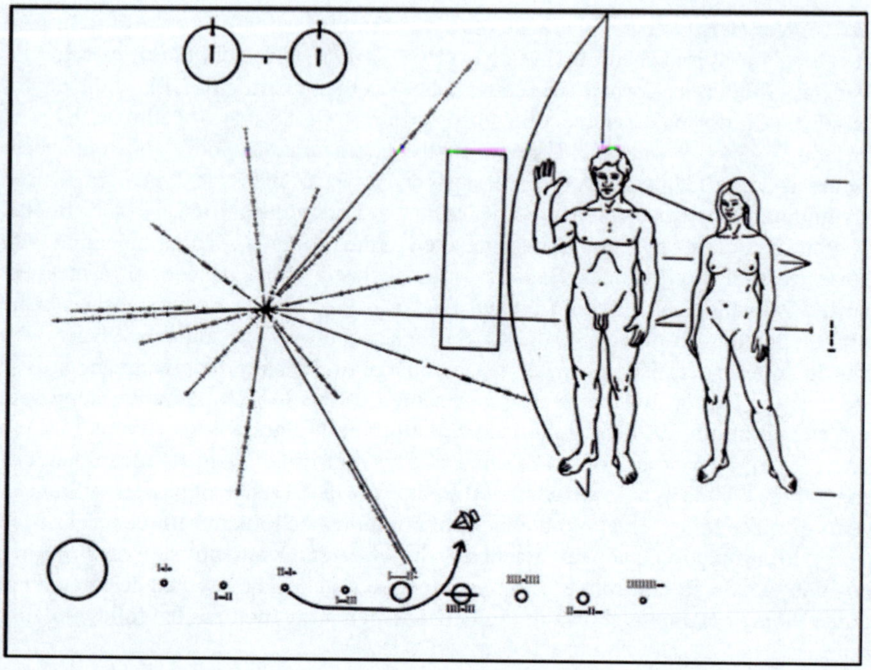

Fig. 3.6 Pattern etched on the gold-anodized aluminum plate of Pioneers. (*Credit* NASA Science, Solar System Exploration, https://solarsystem.nasa.gov/resources/706/pioneer-plaque/)

hydrogen atom a high energy state. However, in the vast gaseous clouds that pepper the galaxies, sometimes hydrogen atoms can knock together causing their protons and electrons to suddenly spin out in opposite directions, flipping the direction of their respective axes. With the proton and electron now spinning in different directions with their axes unaligned, the hydrogen atom acquires a low energy state, and its excess energy is radiated at the 21-cm line, which falls within the microwave region of the electromagnetic spectrum. The 21-cm line can penetrate the large clouds of interstellar cosmic dust that are opaque to visible light, and be observed frequently in radio astronomy. As a result, the oldest stars in the universe can be explored by the radiation of neutral hydrogen at 21-cm wavelength.

In Fig. 3.6, the two circles drawn at the top left represent the hyperfine low-energy state and the hyperfine high-energy state of the hydrogen atom respectively. The connection line between both indicates a hyperfine energy-level transition, and the short vertical line below the connection line indicates a binary digit "1". It means that the hyperfine energy-level transition is taken as a basic unit. In other words, this spin-flip transition of a hydrogen atom from the electron state spin-up to the electron state spin-down specify a unit of length (21 cm) and a unit of time (1420 MHz or 0.704 ns), which can be used as the measurements for the other symbols in the plate.

(2) Positioning and Timing by Using Fourteen Pulsars

At the middle of the left side in Fig. 3.6, a radial pattern with fifteen lines emanating from the same origin is shown in a polar coordinate system. There are fourteen radial lines with a string of binary numbers, representing the positions of fourteen pulsars relative to the Sun. With the hydrogen spin-flip transition frequency as the unit, the binary numbers at the end of each radial line stand for the period of corresponding pulsar, with 10 significant figures in the decimal notation. There is a tick mark near the end of each line, and the length from the origin to the tick mark stands for the relative distance of the pulsar, while that from the tick mark to the line's end gives the Z coordinate perpendicular to the Galactic plane. Because the earth-pulsar distance is not accurately known, the radial lines are shown with breaks. According to the actually-observed pulsars, it could be determined whether the Z coordinates are positive or negative. The fifteenth line extends to the far right, behind the human figures. From the tick mark near the end of the fifteenth line, it may be shown that its Z coordinate is zero, so the line indicates the Sun's relative distance to the center of the Galactic System. In an astronomical interpretation, the two most-likely origins would be the home star of the launch civilization and the center of the Galactic System. The reconstruction of pulsar periods will indicate that the origin of polar coordinates is not the Galactic center.

The relevant parameters of the fourteen selected pulsars, including the PSR number, binary period, decimal period, second's period, azimuth and distance, are listed in Table 3.2, where the azimuth is accounted in the clockwise direction and the line from the Sun to the Galactic center regarded as the azimuth of zero; the distance between the Sun and the Galactic center is with a unit of 100. The fourteen pulsars were selected to be distributed as evenly as possible in Galactic longitude, with the shortest period, the greatest longevity and the highest luminosity, to well receive and use their radiation signals in the far future. For any position and time solution, the fourteen pulsars would be able to provide enough redundancy, but also to allow for the good possibility that pulsar's emissions are highly beamed and that not all pulsars are visible at all view angles. If the gold-anodized aluminum plate is intercepted by the extraterrestrial civilization with the same or more advanced technical levels as mankind on the Earth, which have also kept the historical observation records of the pulsars, the launch epoch of the Pioneer can be derived from the properties of the pulsars' periods changing over time. Even if the glitches of pulsars occur and are not recoded, the launch epoch may be calculated with an error range of 100—1000 years. With triangulation, only three pulsars need to be observed to determine the earth's position of launching the spacecraft. Fortuitously, two of the pulsars are very near the Earth. If either is correctly identified, it can be used to place the position of our solar system in the Galactic System to approximately 20 parsecs, thereby specifying our location to approximately 1 in 1000 stars [11].

It should be pointed out that the listed period of one of the pulsars in Table 3.2 is misleading. At the time when the plaque was designed, pulsar "B1240–64" (now know as J1243–6423) had a known period of 0.388 s, only with the three significant digits, in the then astronomical community. Unfortunately, to be useful on the Pioneer

Table 3.2 Parameters of fourteen selected pulsars

No.	PSR Number (B1950)	Binary period	Decimal period (1/1420MHz)	Period at epoch 1970 or 1971 (s)	Azimuth from Galactic center (deg)	Distance (100, to Galactic center)
1	B1727−47	1000110001111100100011011101010	1178486506	0.8296830003	17	27
2	B1451−68	10110010011000101011101101111	374101871	0.2633767643	49	2
3	B1240−64	100000110110010110001001111000	551117432	0.3880000002	58	56
4	B0833−45	111100011011011001010100111	126726823	0.0892187481	95	15
5	B0950+08	10101011011001101100101000011	359455043	0.2530650432	129	1
6	B0823+26	101100111011010101011110001011	753751947	0.5306595992	162	2
7	B0531+21	10110011100000101010000010	47057538	0.0331296448	174	18
8	B0525+21	100111101000110101000100111000100	5320116676	3.7454907997	177	11
9	B0329+54	111100011111100011111000010110	1014906390	0.7145186427	215	7
10	B2217+47	101101100101101001000010110001	764842161	0.5384673780	263	10
11	B2016+28	101111001111001110011000001101	792520205	0.5579533903	292	3
12	B1933+16	11110010111110001110100011110	509549854	0.3587354200	308	40
13	B1929+10	10011001011010111010010111000	321746104	0.2265170382	315	1
14	B1642−03	100000110100101010001110101100	550675372	0.3876887792	344	4

Plaque, the binary listing for this pulsar had to be far more precise, to beyond 10 significant digits. In other words, Sagan and Drake should have estimated the pulsar's period to a more precise degree, and in the subsequent decades of years, it has been shown that their estimates are more than slightly wrong. Aliens may not recognize PSR J1243–6423 as the pulsar described on the Plaque map. Fortunately, the Aliens that could find and decode the map probably have a pretty fair knowledge of pulsars throughout the Galaxy, and they have the other thirteen pulsars to help them along.

(3) Trajectory of Pioneer Spacecrafts

In order to determine the specific position of the Earth more accurately, the relative position of the planets in the solar system and the trajectory of the Pioneer spacecraft flying out of the Earth are shown at the bottom in Fig. 3.6, where the large circle stands for the Sun, and the other small circles stand for the major planets of the solar system, in order of Mercury, Venus, Earth, Mars, Jupiter, Saturn, Uranus, Neptune and Pluto.

The relative distances of planets are indicated with the binary numbers near them. It is noted that the serifs on the binary digits "1" are presented to stress that the units are different from those of the pulsar's distances and periods, while the binary numbers represent the semi-major axes of the planetary orbits with one tenth of the semi-major axis of the Mercury's orbit as a unit, namely about 0.0387 AU. The distances of the planets relative to the Sun are listed in Table 3.3, respectively represented by the binary and decimal numbers. The Saturn's rings drawn sketchily could highlight the feature of the solar system, so that the solar system is easily distinguished from thousands of near stars. The Jupiter, Uranus and Neptune were then unknown, their ring systems are not as easily visible and apparent in the diagram as the Saturn. The Pluto had been considered to be a planet until 2006 the IAU reclassified it as a dwarf planet. A small picture of the spacecraft is shown in the diagram, standing for the Pioneer-10, which flies from the Earth to Mars, over Jupiter, and out of the solar system. Its antenna is shown pointing approximately back to the Earth. Both the Pioneer-10 and Pioneer-11 carried the identical plaque. However, after launching, the Pioneer-11 was redirected toward the Saturn and from there it left the solar system, with the trajectory different from that of the Pioneer-10. In this regard, the Pioneer-11 plaque is somewhat inaccurate. Flying by the Saturn, the Pioneer-11 would also greatly influence its future direction and destination as compared to the Pioneer-10, but this fact is not depicted in the plaques.

(4) Spacecraft's Silhouette and Naked Sketch

On the right side in Fig. 3.6, the human bodies of a man and woman in standing posture are sketched, and the silhouette of the Pioneer spacecraft is shown behind them. Between the brackets on the far right, the height of the woman is indicated in the binary number "1000", namely the decimal number "8". In unit of wavelength of the hyperfine transition of hydrogen, it means that 8 by 21 cm is equal to 168 cm. The right hand of the man is raised as a sign of good will. Although this gesture may not be understood, it offers a way to show the opposable thumb and

Table 3.3 Distances of the planets relative to Sun

Planets	Binary encoded	Decimal number
Mercury	1010	10
Venus	10011	19
Earth	11010	26
Mars	100111	39
Jupiter	10000110	134
Saturn	11110111	247
Uranus	111101111	495
Neptune	1100001100	780
Pluto	1111111100	1020

how the limbs can be moved. There is mainly a message to the extraterrestrials that the average height of human women on the Earth is 168 cm or so, while men are generally higher than women. Moreover, the spacecraft's silhouette is drawn in the same scale so that the size of the human bodies can be deduced by measuring the actually-intercepted spacecraft. The measurements of the woman's height and the actual spacecraft dimension are also a cross-check for the basic length unit with the hydrogen's hyperfine transition.

3.6.4 *Voyager Probes*

3.6.4.1 Programs from Mariner to Voyager

Between 1962 and 1973, an interplanetary exploration project, known as the **Mariner Program**, was conducted by the NASA in conjunction with the JPL, to investigate the inner planets of the solar system, including Mars, Venus and Mercury. There were ten series of Mariner probes sequentially launched in the program, from the Mariner-1 to Mariner-10, of which except the Mariner-1, 3 and 8, the other seven probes were launched and their exploration missions completed successfully. The Mariner-2 was launched in August 1962 and passed as close as 34,773 km to the Venus in December. The Mariner-4 was designed to conduct close-up scientific observation of the Mars and to transmit these observations back to the Earth, as well as to measure the fields and particles in interplanetary space in the vicinity of the Mars and to provide experience and knowledge of the engineering capabilities of interplanetary flights for long duration. Launched in November 1964, the Mariner-4 performed the first successful flyby of the Mars, returning the first close-up pictures of the Martian surface. It captured the first images of another planet ever returned from deep space, and their depiction of a cratered, seemingly dead world largely changed the scientific community's view of life on the Mars. The Mariner-5 launched in June 1967 was to measure the interplanetary and Venus' magnetic fields, charged particles, plasma and radio refractivity, as well as the ultraviolet radiations of the Venus' atmosphere, and carried out the explorations in the orbit of 4000 km from the Venus. Sequentially launched in February and March 1969, both Mariner-6 and Mariner-7 were the first dual missions to the Mars as a part of the NASA's wider Mariner program, to study the surface and atmosphere of the Mars during close flybys, to demonstrate and develop technologies required for future Mars missions, and to establish the basis for future investigations, particularly those relevant to the search for extraterrestrial life. The twin spacecrafts flew over the Martian equator and its South Pole region, analyzed its atmosphere and surface with remote sensors, and revealed signs of erosion on the Martian surface. The Mariner-9, launched in May 1971, was a robotic space probe that contributed greatly to the investigation of the Mars, becoming the first spacecraft successfully to orbit another planet. Launched in November 1973, the Mariner-10 was the first spacecraft to perform flybys to multiple planets. It flew by the Venus with the acceleration of solar gravity while entered into the orbit around the Mercury with the deceleration of Venus's gravity. The Mariner program had achieved a number of firsts, such as the first planetary flyby, the first planetary orbiter, and the first gravity assist maneuver. The structures of Mariner spacecrafts were developed by improving those of the Ranger probes. All Mariners had a hexagon-shaped base with a diameter of 1.04 m and a thickness of 0.36 m, and on the back of the base, there was a dish-shaped TT&C antenna with a diameter of 0.48 m. A pyramid-shaped metal frame was mounted on the front, and the solar sails with a size of 5.05 m by 0.76 m were deployed on the both sides or around.

In 1965, an aeronautics graduate student named Gary A. Flandro (1934–), who was then working part-time at the JPL, noted that an alignment of the Jupiter, Saturn, Uranus and Neptune which would occur in the late 1970s would enable a single spacecraft to visit all of the outer planets by using gravity assists. The particular alignment occurs once every 176 years. As a result, he proposed a concept of the Grand Tour, and one year later, the JPL was promoting the project, noting it would allow a complete survey of the outer planets in less time and for less money than sending individual probes to each planet. In 1969, the NASA created the Outer Planets Working Group, and officially initiated the **Grand Tour Program** that would have sent two groups of robotic probes to all the giant planets of the outer solar system. It called for four spacecrafts, two of which would visit the Jupiter, Saturn and Pluto, while the other two would visit the Jupiter, Uranus and Neptune. By 1971, the enormous cost of the project, around 1 billion dollars, led to its budget cut and cancelled.

After developing the Mariner-10, originally the Mariner-11 and Mariner-12 were designed and built by the JPL to explore two planets of the outer solar system, the Jupiter and Saturn. It was called the Mariner Jupiter-Saturn project in early 1972, with an estimated cost of 360 million dollars, and with the intention that the twin probes would last long enough to complete the original Grand Tour of the four giant planets, but be advertised as mission to only the Jupiter and Saturn to reduce estimated total project costs. As the project progressed, the name was later changed to **Voyager Program** because the probe designs began to differ greatly from previous Mariner missions. The Voyager Program was composed of twin probes, the Voyager-1 and Voyager-2, corresponding to the Mariner-11 and Mariner-12.

The design idea of the Mariner probe series was widely used in later space exploration and space flight missions. Besides the Voyager-1 and Voager-2, the Mariner-9 was expanded into the Viking-1 and Viking-2 probes to explore the Mars. Moreover, on the basis of the Mariner probes, the Magellan probe was designed to explore the Venus, and the Galileo probe was designed to explore the Jupiter. The second generation of Mariner probe was eventually evolved into the Cassini-Huygens spacecraft orbiting the Saturn. The New Horizons probe, launched in 2006 to explore the Pluto, mainly came from the simplified design of the Pioneer-10 and Pioneer-11, but some designs also came from the Mariner probe series, such as the three-axis stabilizer and the dish-shaped TT&C antenna.

3.6.4.2 Voyager-1 and Voyager-2

Both the Voyager-1 and Voyager-2 were sister probes concurrently developed by the NASA, to explore the outer planets of the solar system and achieve interstellar space flight. With the gravity assist from the giant planets, the probes could change their flight trajectory to explore multiple planets like the Jupiter, Saturn, Uranus, Neptune and Pluto, and their moons. The Voyager inherited the structural design from the Mariner probe, including the polygon base and the high-gain dish-shaped TT&C antenna with a diameter of 3.7 m. and had 721.9-kg lift-off weight and 420-W power consumption. The Voyager carried eleven scientific instruments onboard, including the imaging science system, radio science system, infrared interferometer spectrometer, ultraviolet spectrometer, tri-axial fluxgate magnetometer, plasma spectrometer, low energy charged particle instrument, cosmic ray system, planetary radio astronomy investigation, photopolarimeter system, and plasma wave system.

In 1977, originally it was intended to first launch the Voyager-1 and then the Voyager-2. However, because a malfunction was checked temporarily before launching, the Voyager-1 was delayed to be launched on September 5, while the Voyager-2 was launched on August 20 on the original schedule. For all that, the

Voyager-1 was greatly accelerated by the gravity assist, eventually catching up with the Voyager-2 before flying over the Jupiter and Saturn. The Voyager-1 flew by the Jupiter, measuring the Jupiter's magnetic field, magnetosphere, atmosphere and inner structure, discovering its aurora, rings and five new moons, and investigating in detail its Galilean moons. With the gravity assist trajectory at the Jupiter, the Voyager-1 successfully visited the Saturn, detecting complex structures of its rings, investigating its atmosphere and its giant moon Titan, and discovering new dynamic phenomenon. 22 moons of the Saturn together constituted a complex system of motion, which showed that the giant planet has its own power, and there was a sea of liquid hydrogen on its surface. The Voyager-2 flew over the Jupiter, Saturn, Uranus and Neptune for exploration. It measured the Uranus' atmosphere and magnetic field, correcting its rotation period and inner structural model, and discovering its 10 moons and 11 rings. It measured the Neptune's magnetic field, magnetosphere and inner structure, discovering the Great Dark Spot on its surface, verifying its five rings and six new moons, and investigating its moon Triton. After completing the explorations of the outer planets, both the Voyager-1 and Voyager-2 then left the solar system respectively in different directions, and sailed into interstellar space.

Considering the special alignment among Jupiter, Saturn, Uranus and Neptune that occurs once every 176 years, the launch window was chosen in 1977 for the Voyagers. In this way, the spacecrafts could be accelerated by using the gravity assist from the giant planets, and only need a small amount fuel for adjusting orbits, so they would visit four giant planets of the outer solar system. In order to take advantage of this opportunity, the specially-designed Voyager probes took only 12 years to complete the missions that four outer planets were visited successively. Otherwise, it needs to take 30 years to achieve such explorations.

The specific flight paths and key events of the Voyager-1 were as follows:

(1) At 12:56:00 UTC on September 5, 1977, by using the Titan IIIE rocket, the Voyager-1 was launched from Launch Complex 41 at the Cape Canaveral, Florida. On December 10, it arrived in the asteroid belt between the Mars and Jupiter. On December 19, it surpassed the Voyager-2 which was launched two weeks earlier.

(2) On September 8, 1978, the Voyager-1 flew out of the asteroid belt.

(3) On January 6, 1979, the Voyager-1 flew by the Jupiter for the exploration. On March 5, it reached about 349,000 km away from the Jupiter's center, which was the closest approach to the Jupiter. In the Meantime, it also photographed the volcanic eruption of the Jupiter's moon Io. The eruption plume of the volcano Loki rose 160 km over the limb of the Io.

(4) On August 22, 1980, the Voyager-1 flew by the Saturn for the exploration. On November 12, it reached the closest approach to the Saturn, 124,000 km away from the Saturn's cloud-tops. On December 14, it started the extended exploration missions.

(5) On February 14, 1990, the Voyager-1 took the first ever family portrait of the solar system as seen from outside, including the Venus, Earth, Jupiter, Saturn, Uranus and Neptune. Soon afterwards, to conserve power and computer

resources for other equipment, the onboard cameras were shut down and their software also removed from the spacecraft.

(6) On February 17, 1998, the Voyager-1 caught up with the Pioneer-10 at a velocity of 17 km/s, and reached the position about 69.419 AU (1.04×10^{10} km) away from the Sun.

(7) On December 17, 2004, the Voyager-1 passed through the termination shock, reaching the position of 94 AU away from the Earth, and entered into the heliosheath.

(8) On February 2, 2007, the onboard plasma wave system stopped operating. On April 11, the heater of the plasma wave system no longer also worked.

(9) On January 16, 2008, the radio astronomy experiments ever conducted by the Voyager-1 were stopped.

(10) On August 25, 2012, the Voyager-1 passed through the heliopause, reaching 121 AU away from the Sun, and sailed into interstellar space.

(11) On August 2, 2013, the Voyager-1arrived in the interstellar space of 125 AU away from the Sun.

(12) On December 6, 2014, the Voyager-1 reached 130.62 AU away from the Earth.

(13) On May 18, 2016, the Voyager-1 reached 135 AU away from the Sun.

(14) On April 9, 2019, the Voyager-1 was predicted to be 144.82 AU away from the Earth.

(15) On April 9, 2020, the Voyager-1 was predicted to be 148.39 AU (2.22×10^{10} km) away from the Earth, which is the most distant man-made vehicle away from the Sun and Earth.

At present, the Voyager-1 is flying towards the Oort cloud at a velocity of 3.6 AU per year, and will arrive in there about 300 years later. And then, it will take about 30,000 years to pass through the Oort cloud. Although it is not heading towards any particular star, about 40,000 years later, it will pass within 1.6 light years of star Gliese 445, 17.6 light years away from the Earth. The Gliese 445 is currently located in the constellation of Camelopardalis, and is generally moving towards the solar system at a velocity of 119 km/s. Perhaps eternally, the Voyager-1 is destined to wander the Milky Way. The Voyager-1 is powered by radioisotope thermoelectric generators, which create power by converting heat released by the decay of radioactive material. The Voyager-1 will run out of power around 2025, at which point the onboard scientific instruments will stop working, no longer transmitting data back to the Earth [12].

Compared to the Voyager-1, the Voyager-2 expanded the exploration missions for the planets of the outer solar system, and not only flew sequentially by the Jupiter, Saturn, Uranus and Neptune, but also deeply investigated the Kuiper Belt in the edge of the solar system, the heliosphere and the interstellar space. Its specific flight paths and key events were as follows:

(1) At 14:29:00 UTC on August 20, 1977, by using Titan IIIE rocket, the Voyager-2 was launched from Launch Complex 41 at Cape Canaveral, Florida. On December 20, it entered into the asteroid belt between the Mars and Jupiter.

(2) On October 21, 1978, the Voyager-2 flew out of the asteroid belt.

(3) On April 25, 1979, the Voyager-2 flew by the Jupiter for exploration. On July 9, it came within 570,000 km of the Jupiter's cloud tops.

(4) On June 5, 1981, the Voyager-2 flew by the Saturn for exploration. The closest approach to the Saturn occurred on August 26, and the temperature and density of its atmosphere was measured.

(5) On November 4, 1985, the Voyager-2 flew by the Uranus for exploration.

(6) On January 24, 1986, the Voyager-2 came within 81,500 km of the Uranus' cloud tops, investigating its rings, atmosphere and moons in detail.

(7) On June 5, 1989, the Voyager-2 flew by the Neptune for exploration. On August 25, the Voyager-2 reached the closest approach to the Neptune, and performed a close flyby of its larger moon Triton, so as to gather as much information on the Neptune and its Triton as possible, regardless of the spacecraft's departure angle from this planet. On October 2, it began to carry out the interstellar exploration mission.

(8) On August 20, 1997, the Voyager-2 had continuously flown and normally operated for 20 years.

(9) On November 3, 1998, the Voyager-2 terminated the scan platform and ultraviolet observations.

(10) On November 29, 2006, the onboard computer was unable to correctly decode the telemetry command from the Earth, with a random error. As a result, the electrical heaters of the spacecraft's magnetometer were turned on by mistake. The heaters remained turned on until December 4, and during the time, there was a resulting high temperature above 130 °C, causing the damage of the onboard magnetometer.

(11) On August 30, 2007, the Voyager-2 passed through the termination shock and then entered into the heliosheath. On September 6, the data tape recorder onboard of the Voyager-2 stopped working.

(12) On February 22, 2008, the Voyager-2 terminated the radio astronomy experiment operations.

(13) On April 22, 2010, the Voyager-2 encountered scientific data format problems. On May 23, it resumed sending science data from deep space after the flipped bit of the onboard computer was fixed.

(14) On November 7, 2011, the Voyager-2 switched to the backup boosters to conserve power.

(15) On July 25, 2012, the Voyager-2 was traveling at a velocity of 15.4 km/s relative to the Sun, and arrived at 99.13 AU from the Sun, at −55.29° declination and 19.888-h right ascension, and also at −34.0° ecliptic latitude, placing it in the constellation of Telescopium as observed from the Earth.

(16) On November 7, 2012, the Voyager-2 arrived at a position of 100 AU away from the Sun, becoming the third man-made vehicle to reach this distance. At this time, the Voyager-1 was 122 AU away from the Sun, and it was predicted that the Pioneer-10 and Pioneer-11 were respectively at the positions of 107 AU and 83 AU. While the twin Pioneers have ceased communications, both of the Voyager spacecrafts were well performing and still communicating with the Earth.

(17) On December 11, 2014, the Voyager-2 reached 106.65 AU away from the Sun.

(18) On November 5, 2018, the Plasma Science Experiment onboard of the Voyager-2 observed a steep decline in the speed of the solar wind particles. Since then, the plasma instrument has not observed the solar wind in the environment around the Voyager-2, which shows that the Voyager-2 has left the heliosphere.

(19) On April 9, 2020, the Voyager-2 is predicted to be 123.55 AU (1.848×10^{10} km) away from the Earth.

The Voyager-2 is not headed towards any particular star, although about 40,000 years later it should pass within 1.7 light years of star Ross 248, which is 10.3 light years away from the Earth and is located in the constellation of Andromeda. If undisturbed for 296,000 years, the Voyager-2 should pass by star Sirius at a distance of 4.3 light years. The Sirius, also known as Alpha Canis Majoris, is the brightest star in the night sky, with the apparent visual magnitude of −1.46, and a binary star in the constellation of Canis Major, located in about 8.6 light years away from the Earth. The Voyager-2 is expected to keep transmitting weak radio message until least 2025, over 48 years after it was launched.

It is noted that there are different flight directions for the Pioneers and the Voyagers. The Pioneer-10 is flying away from the center of the Galaxy, almost 180 degrees from the Pioneer-11, Voyager-1 and Voyager-2. The Voyager-2 is flying below the Galactic plane, while the other spacecrafts are flying above the Galactic plane.

3.6.4.3 Voyager Golden Record

The gold-anodized aluminum plaque carried on the Pioneer-10 and Pioneer-11 is similar to a greeting card or a telegram, which transmits the time and location of launching the probes and the information of human beings on the planet Earth to the extraterrestrial intelligence life. Based on the design idea of the Pioneer plaque, the NASA had specially set up a committee chaired by Carl Sagan to design a gold-plated copper record, known as "Sounds of Earth", which was mounted on one side of the decagonal base of the Voyager probes [13]. The golden record contains abundant information of human civilization on the Earth, with the mission of communication between human beings and extraterrestrial civilization. The record's cover was made of the material aluminum, and an ultra-pure sample of the isotope uranium-238 was electroplated upon it. It is well known that uranium-238 is the most abundant radioisotope found in nature, with a half-life of 4.468 billion years. Therefore, it is possible that the extraterrestrials that intercept the golden record will be able to use the ratio of remaining uranium to daughter elements to determine its age.

The patterns drawn on the back of the golden record are mainly divided into four parts. Firstly, the circular pattern drawn in the upper left corner represents the phonograph record and the stylus carried with it. Secondly, the line pattern drawn in the middle of the left hand is a side view of the phonograph record and stylus, with the binary numbers giving the time to play one side of the record, about one hour. Thirdly, the lower-left and lower-right corners show the same patterns as the Pioneer plaque, respectively indicating the location of the solar system with respect to fourteen pulsars and the center of the Galaxy, and the hyperfine transition of neutral hydrogen atom. Fourthly, a series of patterns drawn in the upper right-hand portion mainly show how pictures are to be constructed from the recorded signals.

There are 116 images and a variety of natural sounds collected and assembled in the golden record, of which the images are encoded in analogue form and composed of 512 vertical lines while the sounds is presented as audio and designed to be played at about 16.667 revolutions per minute. The 116 images include many photographs and diagrams both in black and white and color. The first images are of scientific interest, showing mathematical and physical quantities, the solar system and its planets, Deoxyribonucleic Acid (DNA), and human anatomy and reproduction. Care was taken to include not only pictures of humanity, but also some of animals, insects, plants and landscapes. The images on humanity depict a broad range of cultures, including food, architecture, and humans in portraits as well as going about their day-to-day lives. Many pictures are annotated with one or more indications of scales of time, size, or mass. Some images contain indications of chemical composition. All measures used on the pictures are defined in the first few images using physical references that are likely to be consistent anywhere in the universe.

The natural sounds made by surf, wind, thunder and animals were collected in the golden record. In addition, the musical selections are from different cultures and eras, and there are the spoken greetings in 55 ancient and modern languages, including a English greeting from the then U.N. Secretary General Kurt J. Waldheim, the printed message from the then U.S. President Jimmy E. Carter, and other human sounds like footsteps and laughter, etc. The musical selection is also varied, both the Eastern and Western classics, as well as various performances of indigenous music from around the world, including works and performances by the featuring artists such as Bach, Mozart, Beethoven, Stravinsky, Guan Pinghu, Blind Willie Johnson, Chuck Berry, Kesarbai Kerkar and Valya Balkanska.

For the vast universe, the chance that the Voyager probes will be intercepted by the extraterrestrial civilization is extremely slim. Even if the Voyagers are discovered by the extraterrestrials, it will be in the far future. The golden records onboard of the Voyagers are more like a kind of time capsule, trying to tell the story of human beings on the Earth to the extraterrestrials, as President Carter said: "*This is a present from a small, distant world, a token of our sounds, our science, our images, our music, our thoughts and our feelings. We are attempting to survive our time so we may live into yours*".

3.6.5 Debates on Voyager-1

Since it was shown by analyzing the measurement data that the speed of the solar wind had slowed down to subsonic speed, scientists at the Applied Physics Laboratory of Johns Hopkins University considered that the Voyager-1 entered the termination shock region in February 2003. However, some other scientists expressed doubt because the solar-wind detector onboard of the Voyager-1 ceased functioning in 1990, which meant that the termination shock detection would have to be inferred from the data from the other instruments onboard [14, 15]. On May 25, 2005, in a scientific session at the American Geophysical Union meeting in New Orleans, Edward C. Stone, as a project scientist for the Voyager program, presented evidence that the Voyager-1 crossed the termination shock on December 15, 2004, at a distance of 94 AU away from the Sun. At the same month, a NASA press release said that the consensus was that the Voyager-1 was then in the heliosheath.

On December 13, 2010, the Voyager-1 was at the distance of about 116 AU away from the Sun, when it was confirmed that the Voyager-1 had passed the reach of the radial outward flow of the solar wind as measured by the Low Energy Charged Particle device. It is suspected that the solar wind at this distance turns sideways due to the interstellar wind pushing against the heliosphere. In fact, the solar wind detection onboard of the Voyager-1 had been consistently at zero since June 2010, which also provided a conclusive evidence for the event.

In March 2011, to measure the sideways motion of the solar wind at the location of the Voyager-1 in space, the spacecraft's orientation and attitude was changed by the command from the ground control center. Based on a test roll done in February, the spacecraft's ability to maneuver and reorient itself had been verified. The spacecraft's course was not changed, while the spacecraft rotated 70 degrees counterclockwise with respect to the Earth to detect the solar wind. Since taking the family portrait photograph of the planets in 1991, it was the first time that the spacecraft had done any major maneuvering. After the first roll, the Voyager-1 had no problem in reorienting itself with Alpha Centauri, its guide star, and resumed sending transmissions back to the Earth. It was expected that the Voyager-1 entered into the interstellar space at any time. At this time, although the Voyager-2 was still detecting outward flow of the solar wind, it could be estimated that the Voyager-2 would experience the same environment as the Voyager-1 in the following months or years. On May 21, 2011, the Voyager-1 was at 12.44° declination, 17.163-h right ascension and 34.9° ecliptic latitude, located in the constellation of Ophiuchus as observed from the Earth. On December 1, 2011, a veteran scientist of the Voyager Program at the University of Arizona in Tucson, Bill R. Sandel reported that the Voyager-1 had detected the first Lyman-alpha radiation originating from the Milky Way. The Lyman-alpha radiation from other galaxies had been detected previously, but the radiation from the Milky Way was not detectable due to the interference from the Sun. On December 5, 2011, the NASA announced that the Voyager-1 had entered a new region referred to as "cosmic purgatory". Within the stagnation region whose inner edge is at the distance of about 113 AU away from the Sun, the charged particles streaming from the Sun slow down and turn inward, and the magnetic field of the solar system is doubled in strength as interstellar space appears to be applying pressure. The energetic particles originating from the solar system decline by nearly half while the detection of high-energy electrons from outside increases 100 folds.

Since May 2012, the Voyager-1 had detected that the energetic particle accounts with an energy spectrum of above 70 MeV were increasing dramatically. In general, it is considered that the energetic particles are generated by the cosmic rays from the distant supernovae outside the solar system. In June 2012, the NASA announced that the Voyager-1 was detecting changes in the environment that were suspected to correlate with arrival at the heliopause. Based on the detection data from the Voyager-1, it had been reported that the charged particles from interstellar space increased significantly. The charged particles from the Sun can not be detected normally as they are deflected by the solar wind within the heliosphere. On August 25, 2012, particularly the Voyager-1 detected that the high-energy particles from interstellar space increased steeply while the low-energy particles from the solar wind decreased suddenly. In September 2012, the Voyager-1 was traveling at a velocity of 17.0 km/s relative to the Sun, with the distance of 121 AU away from the Sun. A space scientist at Johns Hopkins University and principal investigator for the Low-Energy Charged Particle instruments onboard of the Voyager spacecrafts, Edmond C. Roelof declared: "Most scientists involved with Voyager-1 would agree that these two criteria have been sufficiently satisfied". However, the last criterion that the magnetic field direction was significantly changed from the Sun to the interstellar space had not been observed. From the actual measurement, the change of magnetic field direction was only two degrees. In other words, the Voyager-1 was traveling at the physical boundary of the solar system and would enter interstellar space at any time. On December 3, 2012, Ed Stone said: "Voyager has discovered that a new region of the Heliosphere that we had not realized was there. We are still inside, apparently. But the magnetic field now is connected to the outside. So it is like a highway letting particles in and out". In this region, the magnetic field was 10 times more intense than the Voyager-1 encountered the termination shock. It was expected to be the last barrier before the Voyager-1 exited the solar system completely and entered interstellar space.

Until September 2013, it was being debating whether the new region where the Voyager-1 was traveling was interstellar space or an unknown region of the solar system. Finally, the NASA officially confirmed that the Voyager-1 had reached the interstellar space in August 25, 2012. The final conclusion from the NASA made most of space scientists had abandoned the assumption that the direction of magnetic field must be changed when passing through the heliopause, and improved the heliopause's model. A key evidence to persuade the scientists that the Voyager-1 had crossed through the heliopause was an indirect measurement of an 80 folds increase in electron density. It was originated from a solar outburst that had occurred in March 2012, triggering the frequency of plasma oscillations in April 2013, and thus the electron density could be measured indirectly, which was generally expected to be two orders of magnitude higher outside the heliopause than within. Actually, additional data was also provided with the weaker sets of oscillations measured in October and November 2012. As early as 1980, the plasma spectrometer onboard of the Voyager-1 had ceased working, so the indirect measurement was required to support a correct decision. In the meantime, the NASA released audio renditions of these plasma waves, which represented the first sounds to be captured in interstellar space.

It is noted that the Voyager-1 having left the solar system with the heliosphere are two different expressions. The solar system is usually defined as the vastly larger region of space populated by bodies that orbit the Sun. As of April 2020, the Voyager-1 is traveling the distance of 148 AU away from the Earth, which is a distance of slightly more than one sixth to the aphelion of Sedna. The Sedna is also known as 90377 Sedna, a small celestial body discovered in 2003 in the outer reaches of the solar system, with an aphelion of 937 AU and a perihelion of 76 AU. Apparently, the spacecraft has not yet entered the Oort cloud with a distance of above 50,000 AU away from the Sun, which is regarded by astronomers as the outermost zone of the solar system. So, for the gravitational boundary of the solar system defined in astronomy, the Voyager-1 is still within the range populated by the Sun's gravity, but it has crossed through the physical boundary of the solar system and entered interstellar space.

3.6.6 New Horizons Probe

In July 2002, the NASA initiated an Integrated Exploration Strategy to tackle specific solar system exploration goals identified as top priorities by the planetary science community, known as **New Frontiers Program**, whose strategy is to explore the solar system with the medium-class spacecraft missions that conduct high-science-return investigations, and to deeply understand the solar system. The Program builds on the innovative approaches used in the NASA's Discovery Program, but provides a mechanism for identifying and selecting strategic missions that can not be accomplished within the cost and time constraints of the Discovery Program. At present, the New Frontiers Program mainly includes four missions. The first mission is the New Horizons probe launched in January 2006, to obtain a close look at the Pluto and its moons and further to enter into the Kuiper Belt, which help scientists answer basic questions about the surface properties, geology, interior makeup and atmosphere of the ancient, icy mini-worlds in that vast region. The second mission is the Juno probe launched in August 2011, to explore the Jupiter in detail, such as the gas giant's gravitational and magnetic fields, vast magnetosphere, intense aurora and the swirling clouds that form the Jupiter's colorful, trademark atmosphere, which help scientists understand the history of the solar system and provide new insight into how planetary systems form and develop in the Galaxy and beyond. The third mission is the Origins, Spectral Interpretation, Resource Identification, Security-Regolith Explorer (OSIRIS-REx) launched in September 2016, to retrieve a sample of an asteroid and carry it to the Earth for further study, which will help scientists understand the planetary formation and the origin of life, and aid in understanding asteroids that can impact the Earth. The OSIRIS-REx is currently orbiting the asteroid, 101955 Bennu, and will spend two years mapping it before collecting a pristine sample and returning to the Earth in 2023. The fourth mission is the Dragonfly probe slated for a 2026 launch and a 2034 arrival, to visit the Saturn's richly organic moon, Titan. The Dragonfly is an eight-bladed rotorcraft that will sample and examine dozens of sites around the icy Titan in the search for the building blocks of life.

The New Horizons probe is the fifth spacecraft to achieve the escape velocity needed to leave the solar system after the Pioneer-10, Pioneer-11, Voyager-2 and Voyager-1. As early as June 2001, the NASA announced to complete the proposal of the New Horizons design, and one year later, the New Horizons probe was listed in the New Frontiers Program as the first mission. The New Horizons probe is comparable in size and general shape to a grand piano, having been compared to a piano glued to a cocktail bar-sized satellite dish. Its body forms a triangle, with a thickness of about 0.76 m and a lift-off weight of 478 kg, and the main structural column is formed with a 7075 aluminum alloy tube, between the launch vehicle adapter ring at the rear and the high-gain dish antenna affixed to the front flat side. The high-gain antenna has a Cassegrain layout, composite construction, and a 2.1-m diameter, and in addition to it, there are a medium-gain dish antenna and two low-gain antennas. The prime-focus, medium-gain antenna, with a 0.3-m aperture and 10-deg half-power beam-width, is mounted to the back of the high-gain antenna's secondary reflector. The forward low-gain antenna is stacked atop the feed of the medium-gain antenna. The aft low-gain antenna is mounted within the launch adapter at the rear of the spacecraft. This antenna was used only for early mission phases near the Earth, just after launch and for emergencies if the spacecraft had lost attitude control. The spacecraft carried seven types of scientific instruments, including an ultraviolet imaging spectrometer, long range reconnaissance imager, solar wind at-Pluto analyzer, Pluto energetic particle spectrometer science investigation, radio science experiment, Ralph telescope, and Venetia Burney student dust counter. In addition to the scientific instruments, there are some cultural artifacts traveling with the spacecraft, such as an ounce of the ashes of Clyde W. Tombaugh who first discovered the Pluto in 1930, a Florida-state quarter coin whose design commemorates human exploration, a compact disc on which a collection of 434,738 names were stored, a piece of Scaled Composites Space-Ship-One, a flag of the US, and other mementos. The New Horizons probe first explores the Pluto and its largest moon, Charon. And then, it continues to fly to the outer solar system and to reach the Kuiper Belt composed of comets and cosmic debris, and will spend 5–10 years investigating small objects in the Kuiper Belt. Finally, provided it survives that far out, the spacecraft is likely to follow the Voyager probes in exploring the outer heliosphere and mapping the heliosheath and heliopause. It is expected that it will reach the heliopause in 2047. The specific flight paths and key events of the New Horizons are as follows.

(1) At 19:00:00 UTC on January 19, 2006, by using the Atlas V 551 rocket, the New Horizons probe was launched from Space Launch Complex 41 at Cape Canaveral, Florida, and directly into a solar escape trajectory at a velocity of 16.26 km/s. It flew by the Moon only 9 h after launching.

(2) On April 7, 2006, the spacecraft passed the orbit of the Mars, moving at roughly 21 km/s away from the Sun at a distance of 2.43×10^8 km. On September 4, the spacecraft flew by the Jupiter, taking its first photographs of the Jupiter from a distance of 2.91×10^8 km.
(3) On February 28, 2007, the spacecraft received a Jupiter gravity assist with a closest approach, with a distance of 2.3×10^8 km from the planet. The flyby increased its speed by 4 km/s, accelerating it 23 km/s relative to the Sun and shortening its voyage to the Pluto by three years. On June 28, the spacecraft started the first hibernation mode cycle.
(4) On June 8, 2008, the spacecraft crossed the orbit of the Saturn. On December 16, the spacecraft began the second hibernation mode cycle.
(5) On August 27, 2009, the spacecraft began the third hibernation mode cycle.
(6) On March 18, 2011, the spacecraft passed through the orbit of the Uranus.
(7) On July 1, 2013, the spacecraft photographed the images, from which the Pluto and its moon Charon were resolved as separate objects for the first time.
(8) On July 14, 2014, the sixth trajectory correction maneuver was performed after launching, to enable the spacecraft to reach the Pluto. Between July 19 and 24, the spacecraft snapped 12 images of the Charon revolving around the Pluto, covering almost one full rotation. On August 24, the spacecraft crossed through the orbit of the Neptune. On August 29, the spacecraft began the fourth hibernation mode cycle after a ten weeks test. On December 6, the spacecraft was waked up from its final Pluto-approach hibernation to begin regular operations.
(9) On February 12, 2015, the spacecraft was more than 2.03×10^8 km away from the Pluto when it began taking the photos, which showed the Pluto and Charon. On April 2, the spacecraft was at a distance of 1.23×10^8 km (0.82 AU) away from the Pluto, and at a distance of 32.06 AU away from the Sun, traveling at a velocity of 14.58 km/s or about 3.0 AU per year relative to the Sun. Radio signals take 4.5 h to travel to the spacecraft from the Earth. On July 14, the spacecraft passed within 10,000 km of the Pluto at a relative velocity of 13.78 km/s, to conduct the approaching scientific exploration.
(10) On January 1, 2019, the spacecraft made a flyby of the most distant solar system object ever explored, the Kuiper Belt object 2014 MU69, also known as Ultima Thule.
(11) On April 10, 2020, the spacecraft was at a distance of 46.97 AU from the Earth, and was exploring the Kuiper Belt in detail.

As of April 2020, the New Horizons probe is in the constellation of Sagittarius, at a position with the declination of −20.2875 degrees and the right ascension of 19.2767 h, flying towards the Sagittarius direction at a speed of roughly 3.0 AU per year. Although the New Horizons probe had ever the fastest launch speed to date, it will never catch up with the Voyager probes. The Voyager-1 gained a great gravity assist during flying by the Saturn and its moon Titan. When the New Horizons probe travels a distance of 100 AU away from the Sun, its speed will be reduced to

13 km/s, 4 km/s slower than that of the Voyager-1 at the same distance. The Pioneer-10, Pioneer-11, Voyager-1, Voyager-2 and New Horizons probe, as the messengers sent to interstellar space by mankind on the Earth, are flying out of the solar system. At present, the communication of the Earth with the Pioneer-10 and Pioneer-11 has ceased, while the other three probes are still able to communicate with the Earth.

References

1. Xu FX (2004) Introduction to satellite engineering. China Astronautic Publishing House, Beijing (Chinese)
2. Qian XS (2008) Introduction to interstellar flight. China Astronautic Publishing House, Beijing (Chinese)
3. Zeng T (2010) Revelations of science. Learning Press, Beijing (Chinese)
4. Needham JTM (2000) History of chinese science and technology (volume of mechanic engineering). Science Press, Beijing (Chinese)
5. Gu SF, Shi CL (2000) The history of the world space science development. Henan Science and Technology Press, Zhengzhou (Chinese)
6. Malina FJ (1968) The rocket pioneers. Eng Sci 2(5):8–13
7. Johnson JL (2014) America's forgotten rocketeer. IEEE Spectr 51(8):50–53
8. Watson N (2013) Reaching for the stars: nigel watson celebrates 80 years of the british interplanetary society. History Today 63(1):236–239
9. Ross HE (1939) The BIS space-ship. J Br Interplanet Soc 5(1):4–9
10. Fimmel RO, Van Allen J, Burgess E (1980) Pioneer: first to jupiter, saturn, and beyond. NASA SP-446, Scientific and Technical Information Office, NASA, Washington, D. C, pp 247–249
11. Sagan CE, Salzman L, Drake FD (1972) A message from earth. Science 175:881–884
12. Ludwig R, Taylor J (2002) Voyager telecommunications. The manual of deep space communications and navigation systems, Pasadena: California Institute of Technology, JPL
13. Sagan CE, Drake FD et al (1978) Murmurs of earth: the voyager interstellar record. Random House, New York
14. Fisk LA (2003) Planetary science: over the edge. Nature 426(6962):21–22
15. Krimigis SM, Decker RB et al (2003) Voyager 1 exited the solar wind at a distance of ~85au from the Sun. Nature 426(6962):45–48

Chapter 4
Navigations from Ground to Space

In human history, the fifteenth century A.D. is a significant watershed that human activities began to move from the mainland to the ocean. Driven by the maritime exploration and trade, people living on the originally isolated continents are increasingly frequent to and from each other. In the early twentieth century, humanity entered the age of air flight. In the late 1950s, humanity entered the age of space flight. Under such a context of times, the problem of guiding movable objects how to move safely and economically was increasingly prominent, which is on the navigations of the movable objects. The so-called **movable objects** refer to the moving man-made objects such as vehicles, ships, aircrafts, missiles and spacecrafts, as well as the activities that extend from human beings to other organisms. The so-called **navigation** refers to the process of monitoring and controlling the movable object from starting position to the destination in timely, safe and efficient ways according to the scheduled plans and regulations. With the development of science and technology, the navigation has gradually become a subject that specially studies the theories, methods and technical devices for the voyage, resulting in various means of navigation technologies and implementation approaches which are widely applied to all kinds of the movable objects from ground to space, including land navigation, marine navigation, aeronautic navigation and space navigation.

4.1 Connotations Related to Navigation

Although there are two professional terms in Chinese, "导航" and "航行", originating from the same English word "navigation", the two terms have very different connotations and denotations in Chinese. For the Chinese character "航", its original meaning come from the driving of "boat" on the water surface. With the promotion of modern disyllabic words, the character "航" is evolved into the term "导航", which is widely used in the displacement of the movable objects on sea, in air or in space relative to a certain site or the Earth, and even in the orbital motion of the spacecraft

P. Shuai, *Understanding Pulsars and Space Navigations*, Navigation: Science and Technology 11, https://doi.org/10.1007/978-981-16-1067-7_4

relative to a certain celestial body in the universe [1]. For the vehicles moving on land, it is often spoken to travel by cars or drive for journeys, and thus in Chinese, the character "行" is used more commonly than the character "航". The terms navigation, aeronautics and astronautics (or aerospace) are used in the different professional and technical fields, corresponding to sea voyage, air flight and space flight, respectively. Therefore, defined from the general technical field, the **Navigation** ("导航") refers to the process of monitoring and controlling the movable object from starting position to the destination in timely, safe and efficient ways according to the scheduled plans and regulations, and the **Voyage** refers to the process or activity of moving and changing the position of movable objects, which means a long journey involving travel by sea or in space and has close to the connotation of the term "航行" in Chinese.

Based on the general definition of navigation, an entire navigation process should include four tasks: to plan the path, to determine the current position, to monitor the track deviation and to correct the deviation. For the early sea-voyage and air-flight activities, it is extremely difficult to determine the position of the movable object, which is usually completed by a professional navigator. By analyzing the real-time position information of the movable object, the navigator can identify whether the movable object deviate from the scheduled planning route, the degree of deviation from the route and the current environment, to determine its states; then, the corresponding measures to correct the route deviation are taken to make the movable object return to the scheduled route; finally, the movable object can be guided to the destination safely, efficiently and just in time, according to the plan. In the Chinese comprehensive dictionary, *Cihai*, published in Chinese by Shanghai Dictionaries Publishing House in 1999, the definition of navigation is: "A method and technology to guide an aircraft or ship from one site to another along a certain route". With the application of modern advanced technology like radio navigation, the degree of automation has been improved increasingly, and the navigator's work has been replaced by various navigation systems or taken into account by drivers or pilots. To complete the navigation task, a key link is that the movable object's position can be determined just in time. As a result, the methods and techniques to determine the positions of the movable objects are customarily referred to as **navigation**, which is a special definition for navigation.

There are also the other two terms, "guidance" and "control", closely associated with the navigation. The so-called **guidance** refers to monitoring and guiding the movable object to the planned route or flying toward the targets based on certain law, and the so-called **control** refers to the process and method of adjusting the deviation between the current position of movable object and the planned route to make it return to the route. From the general definition, the navigation process should contain the guidance and control [2]. From the special definition, the navigation is the fundamental of the guidance and control, and the guidance and control are the purpose of the navigation, containing and correlating with each other. So, the navigation, guidance and control systems are often designed in an integrated way in practical engineering applications. In addition, "positioning" is also another term related to the navigation. In geodesy, the **positioning** refers to the technology and method of precisely determining the coordinates of a site under a given terrestrial reference frame. Real-time, dynamic position determination is the basic requirement of navigation. The positioning is the basis and special case of the navigation, and generally results from static location determination or post-processing, without real-time requirement, but has the requirement for high measurement accuracy, getting to the level of centimeters or even millimeters.

Generally, in order to successfully complete a scheduled navigation task, besides the position information, it is necessary to determine velocity, time, heading and attitude of the movable object. All these information are usually called **navigation parameters**. The entire devices or facilities that are used to determine the navigation parameters of the movable objects and guide them to voyage according to the planned routes are called **navigation system**. The navigation system can only provide a variety of navigation parameters, but not directly participate in controlling the movable objects for voyages, which is an open-loop system. In a sense, the navigation system is an information processing system that the data measured by the navigation equipment is processed into various navigation parameters to guide the voyage. The information provided by the navigation system is converted into the control instructions for the movable object, which are sent to the system's actuators to automatically control the movable object for voyage, and thus a complete close-loop system is constituted.

4.2 From Ancient South-Pointing Chariot to Modern Navigation System

During the Western Jin Dynasty (266–316 A.D.), a Chinese scholar named Bao Cui wrote a book, *Notes on Ancient Times and Today*, for interpreting all kinds of things in ancient times and at that time. In this book, he had recorded a war scene in ancient Chinabetween Yellow Emperor (Huangdi, about 2717 B.C.–2599 B.C.), a leader of the Huaxia tribal alliance, and Chiyou, a leader of Jiuli tribal alliance: "As driving the south-pointing chariots, Huangdi was fighting with Chiyou in the campaign of Zhuolu. Suddenly, Chiyou's army encountered a dense fog, and all of his solders lost orientation, while Huangdi used the south-pointing chariots to indicate the orientation for his army. As a result, Huangdi captured Chiyou and his solders, and finally became the Emperor". The **south-pointing chariot** was an ancient Chinese two-wheeled vehicle that carried a movable pointer to keep indicating the south by using a gear driving system, no matter how the chariot turned. The pointer usually took the form of a doll or figure with an outstretched arm. In about 2700 B.C., Huangdi used the south-pointing chariot as a navigation device to beat Chiyou, which is the first application of navigation technology in the ancient warfare recorded.

During the Warring States Period (475–221 B.C.), magnetite was found in the area of Cishan. Today, it is located in Cishan Town, Wu'an City, Hebei Province, China. In the practice of recognizing the magnetism of objects, a device to indicate the north–south direction, Sinan, was invented in ancient China [3]. The Sinan is a scoop-shaped device made of natural magnetite ore, which is placed on a smooth plate engraved with directions, and thus the orientation can be identified by the south-pointing property of magnet. On the basis of understanding the property that the magnets can attract the iron objects, the magnetic directivity was further known, and then the Sinan brought up. By repeating studies and experiments, a practical Compass was finally invented. Apparently, the Sinan is the most original compass used by people in ancient China. In terms of physical structure, the Compass as one of the Four Great Inventions of ancient China is composed of the Sinan, compass plate and magnetic needle. During the Northern Song Dynasty (960–1127 A.D.), it was recorded in a work, *Records in Pingzhou,* written by Yu Zhu who was a geographer that: "the master driving a boat is familiar with the geography, and is able to observe the stars at night, the Sun in day, and the Compass in cloudy day". It can be shown that the navigation techniques such as the terrain, astronomy and Compass were widely used by ship sailing near the coast of China at that time. As an all-weather navigation way, the Compass can make up for the lack of the astronomical navigation and the terrain navigation, of which the former is required to observe the Sun, Moon and stars, and the latter is required to observe topographic features on the ground. The Compass ushered humanity into a new era in the history of sea voyage. In the fifteenth century, the human history entered the Age of Exploration, and human activities began to expand from isolated continents to vast ocean voyage, thus promoting the rapid development of navigational system technology.

With the rapid development of modern ocean voyage and trade, the lighthouses as navigation marks were built at the prominent places of islands, coasts, headlands, ports and rivers. The lighthouse is generally composed of base, tower body and illuminator, with remarkable shapes and color features. To safely guide the voyage of ships, the terrain navigation is used, in which the ships' position can be determined on the sea chart by observing the topographic features and the lighthouses. The sea charts are necessary basic data for the voyage of modern ships, and mainly include such elements as the shore shapes, islands, reefs, water depths, navigation marks and radio navigation stations. With the development of modern navigation technologies, such as radar transponder, navigation satellite system and Automatic Identification System (AIS), the function of lighthouse navigation is weakening day by day. For all that, the lighthouses, carrying thousands of years of history and culture, have become the cultural landmarks protected by all countries in the world.

As a kind of fixed navigation mark, the lighthouse can directly be seen in daytime, and its lights with different colors observed at night, so as to guide the ships to bypass dangerous shoals and reefs, and safely enter the destination ports. In the coastal countries where the sea-voyage industry had been developed, various types of traditional and modern lighthouses can be seen everywhere. In ancient times, the lighthouse's light was generated by burning charcoal or coal, and later, by burning olive oil, whale oil and fossil oil, but the distances of their light transmissions were very limited. At the end of the eighteenth century, there appeared successively the lighthouses with reflectors and concave lenses. The kerosene lamp flame was placed at the focal point of a concave lens. The flame reflected on the concave lens to form a light column, whose transmission distance was increased greatly. The modern lighthouses use incandescent electric lamp as the light source, the center of the luminescent body is located at the focal point of spotlight lens. The light source radiates the spherical light through the spotlight lens to become light beam, whose range can reach 30 nautical miles and whose intensity equates to the light from burning hundreds of millions of candles simultaneously. The navigation lighthouse usually uses white light, but also uses the flash with different colors, and particularly the light that warns the dangerous directions near the port, mostly uses red, green or blue, so that the ships that are about to enter the port can clearly decide whether the navigation is safe or not.

According to ancient Greek poetry, the beacons with navigation function had been built along the Mediterranean coast in about 660 B.C. These are the earliest "Lighthouses" of human civilization. In 280 B.C., an Egyptian Royal wedding ceremony ship hit a reef and sank as it was sailing into the harbor of Alexandria. For this reason, Egyptian King Ptolemy II Philadelphus (283–246 B.C.) ordered to build a navigation lighthouse at the eastern end of the island called Pharos near the harbor of Alexandria. It is the famous Pharos Lighthouse, also known as the Lighthouse of Alexandria, which has been listed as one of the Seven Wonders built in ancient times along the eastern Mediterranean coast. The Pharos Lighthouse was designed and built by the ancient Greek architect Sostratus of Cnidus, and made of white marble with three layers. The lower, middle and upper layers were quadrangular, octagonal and cylindrical shape, respectively. The statue of Poseidon, a God of the sea, was mounted on the top of the Lighthouse. A large fire basin was installed below the statue at the top of the Lighthouse. Behind the fire basin, there was a polished and bright granite slab with reflective function. At night, the fire light can be reflected to far sea surface to guide ship's voyage direction. The Pharos Lighthouse had been estimated to be 137 m in overall height, and had been a symbol of the great reputation of the Egyptian monarch. It has stood for more than 1500 years at the harbor of Alexandria, and was completely destroyed by two major earthquakes in 1303 and 1323. In May 2015, the Egyptian government was planning to rebuild the Pharos Lighthouse. Up to now, "Pharos" in many European languages still means "lighthouse", from which the English word "Pharology" originates, which means the scientific studies on lighthouses and signal lights, and their construction and illumination.

Afterward, the ancient Romans built a series of lighthouses and created the earliest lighthouse network system. Arabs, Indians and Chinese also gradually mastered the construction technology of lighthouse. At present, the oldest and still in use lighthouse in the world is the Tower of Hercules built at the entrance of La Coruna Harbor in northwestern Spain, by the ancient Romans in the late first century A.D. The Spanish government has repaired and rebuilt it many times. The Tower of Hercules was built on 57-m-high rock, with a height of 55 m and a light range of 23 nautical miles. In 2009, the Tower of Hercules, as only architecture of lighthouses, was listed in the World Cultural Heritage List. Moreover, located at the mouth of the Gironde Estuary in southwestern France, the Cordouan Lighthouse is the oldest existing lighthouse in France, the tenth tallest traditional lighthouse in the world, with a height of 68 m, which has the design style of Renaissance and the reputation of "Versailles of the Sea" and "King of Lighthouses". In addition, there are many famous lighthouses, such as the Maiden's Tower located at the southern entrance of Bosphorus strait 200 m from the coast of Uskudar in Istanbul, Turkey; the Lighthouse at the harbor of Floroe on the west coast of Norway; the Louisburg Lighthouse at Cape Breton Island in the province of Nova Scotia, Canada; the Lighthouse in the city of Karlskrona along the Baltic Sea of Sweden; the Ontonagon Lighthouse in Michigan of the United States; and so on. These lighthouses carry the history and culture of the development of marine civilization in the Western world [4].

The pagodas and watchtowers built along the coast in ancient China have always been regarded as "man-made beacons" to guide the voyages of ships on the sea. The Light Tower of Huaisheng Temple in the city of Guangzhou, Guangdong, China, built in the reign of Zhenguan in Tang Dynasty (627–649 A.D.), stands on the Bank of the Pearl River. At night, the lanterns hung to the light tower are lighted up to guide ships into the city of Guangzhou. The Chongwu Lighthouse located in Hui'an County, Fujian Province, which is the first light tower with navigation function in China, was built in 1387. To safely guide ships into and out of the Yangtze River, a Firework Tower made of mounds was built in 1412, on the beach at its estuary. In 1868, with the establishment of Shipping-Tax Unit by the Qing Dynasty customs, a series of lighthouses began to be built from the Yangtze River Estuary to the North and South in the coast of China. By 1927, there were 59 coastal lighthouses under the jurisdiction of the Chinese customs, and another 6 offshore lighthouse boats.

In the 1980s, by using the advanced light-emitting devices, a number of lighthouses were newly built or rebuilt at important coastal locations in China, forming a contemporary "lighthouse chain". In 1997, 100 lighthouses of the world's historical relics were named by the International Association of Lighthouse Authorities (IALA), of which there are five lighthouses of China: Qingpu Tower in the City of Shanghai, Twin Towers at Jiangxin Island in Wenzhou City of Zhejiang Province, Huaniao Mountain Lighthouse in Shengsi County of Zhejiang Province, Laotie Mountain Lighthouse in Dalian City of Liaoning Province and Lingao Lighthouse in Lingao County of Hainan Province.

With the development of modern science and technology, the navigation has gradually become a specialized subject of studying the theories, methods and devices for voyages, widely used in all kinds of movable objects, such as vehicle, ship, aircraft, rocket, missile and spacecraft. According to the different information interaction and dependence between the navigation systems and the movable objects, the modern navigation systems can be divided into two categories: autonomous navigation system and non-autonomous navigation system. The **autonomous navigation system** refers to the system that the movable object does not interact any two-way information with the surrounding environment, completely relies on the equipment it carry to obtain the navigation information, and automatically accomplish the navigation task. On the contrary, the navigation system without the basic characteristics of autonomous navigation is called **non-autonomous navigation system**. According to the different development sequence and principle of navigation system, the modern navigation systems are also divided into the celestial navigation system, inertial navigation system, radio navigation system and navigation satellite system, and will be described sequentially in the following of this chapter. In addition, the pulsar navigation system, which uses the X-rays radiated by pulsars as natural beacons to provide long-time and high-precision autonomous navigation information for all kinds of spacecrafts, will be discussed in detail in Chap. 5 of this book, and thus the pulsars are known as the lighthouses for space flights. The other navigation technologies, such as the terrain matching navigation, the navigation based on the Earth's magnetic field, the gravity matching navigation and the navigation based on the biological polarized skylight [5, 6], are still in the phase of theoretical exploration, and thereby they will not be involved in this book.

4.3 Celestial Navigation System

4.3.1 Basic Principle

The **Celestial Navigation System (CNS)** is a system in which the movable object used its own celestial observation equipment to measure the elevation angles and azimuth angles of celestial bodies, such as the Sun, Moon, planets and stars, determine its position and attitude, and guide itself to safely voyage according to the scheduled plans and requirements. Its basic principle is: in local geographical coordinate system, both **heading angle** and **elevation angle** are measured by the star-observed devices, of which the former is the angle, on the local horizontal plane, between the direction of celestial body and the motion direction of movable object, and the latter is the angle between the direction of celestial body and the local horizon; the real heading of the movable object can be obtained by the azimuth angle of the celestial body looked up from the star almanacs at the corresponding epoch, subtracting the heading angle; the complementary angle of the elevation angle is called **zenith distance**, and the point that the celestial body is projected on the surface of the Earth is called **star point**; thereby, a circle with the star point as the circular center, and the spherical distance corresponding to the zenith distance on the Earth's surface as the radius can be drawn, and called **circle of equal altitude**; two circles of equal altitude are obtained by simultaneously observing two celestial bodies, and the intersections of the two circles form the real and false positions of the movable object; finally, with the prior information of the movable object's position or with the third equal-altitude circle, the false position can be eliminated, so that the longitude and latitude of the movable object can be solved, as shown in Fig. 4.1. In general, the CNS consists of the star-observed device, time generator, star-chart database and navigation computer. The star-observed devices usually include star tracker, celestial compass and sextant. The star tracker (also known as star sensor) is used to search, identify and track the celestial bodies in the background sky, and to measure the angle of the star tracker's aiming line direction relative to reference coordinate system. The celestial compass can indicate the heading direction of the movable object by measuring the direction

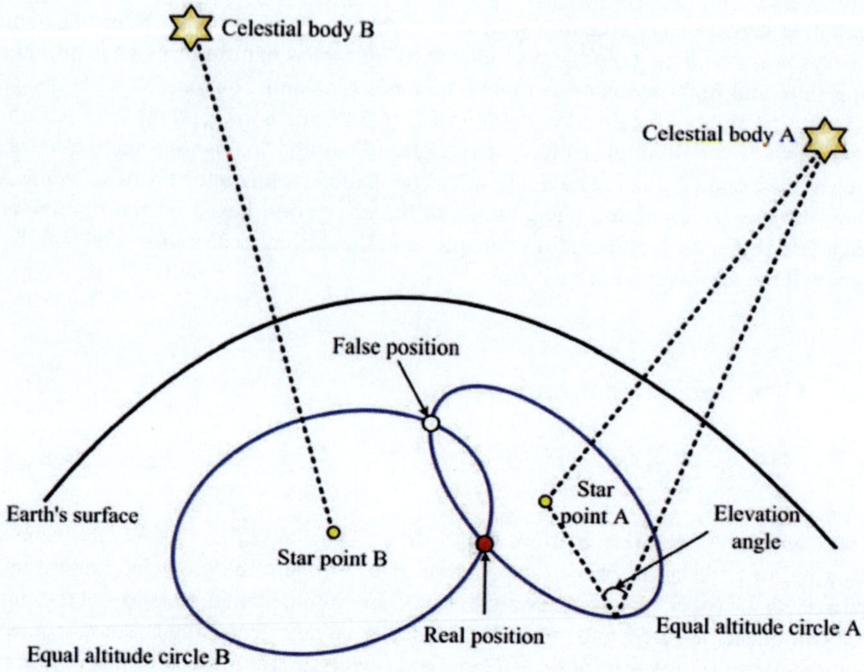

Fig. 4.1 Positioning diagram of ship on sea using CNS

of the Sun or other celestial bodies. The sextant is used to measure the stars or planets to get the position and distance of the movable object.

According to the number of the stars tracked, the celestial navigation can be divided into three ways: single-star navigation, double-star navigation and three-star navigation. For the single-star navigation, the heading reference error is usually great and the positioning accuracy is low. For the double-star navigation, the high positioning accuracy can be achieved by selecting two stars whose azimuth difference is closer to 90°. For the three-star navigation, the third star can be used to check the reliability of the first two measurements, and the three-dimensional position parameters can be determined directly. For the spacecrafts in near-earth orbits, there are usually two methods using the celestial navigation of which one is the direct sensitive horizon method. There are two angles to be measured, respectively, by using the star sensor and the Earth sensor: one is the angle between the star's direction and the spacecraft's vertical line, and another is the angle between the star's direction and the tangent along the Earth's limb. And then, the angle between the star's sight line and the vector direction of the Earth's center can be obtained, also called **angular separation of the star**. Taking the angular separations of the stars as the observables, the measurement equations are formed, and thus by combining the system state equation of orbital mechanics, the spacecraft's position and velocity parameters can be estimated. The other is the indirect sensitive horizon method based on starlight refraction. According to the optical refraction property of the Earth's atmosphere, the measurement equations are established with the angles of the starlight refraction as the observables, and then the orbital parameters of spacecrafts can be solved.

The celestial navigation belongs to a passive navigation system, whose beacons come from the celestial bodies observed directly on the movable objects, without radiating energy to the outside. It has the advantages of autonomy, concealment, anti-interference and complete navigation information, as well as the measurement errors will not be accumulated over time. The observation conditions for celestial bodies are usually ideal in the upper atmosphere or in the outer space, but they are easily limited by the weather in the navigation application near the ground, which requires higher luminosity of the observed objects. In addition, the output of celestial navigation information is not continuous, the positioning time is long and the operation process is relatively complicated. The basic observables of the celestial navigation are angles. If there is the smaller angle error, the greater distance error will be caused. Consequently, the positioning accuracy using the celestial navigation reaches hundreds of meters in the ideal observation condition, and, in general, is thousands of meters and even tens of thousands of meters.

4.3.2 Development and Application

The celestial navigation is an ancient and modern positioning technology, with a very long history. It originates from the application of marine voyage, and then gradually applied to aircrafts, missiles, spacecrafts and other fields. As early as 2000 years ago, there were the written records that the astronomical ways were used for sea voyage in China. An-Liu (179–122 B.C.) who was the king of Huainanin the Western Han Dynasty wrote a book entitled *Huai Nan Zi*. In an article titled Qi Su Xun of this book, he described: "As riding a boat, a man was confused and lost the orientation; as seeing the Big Dipper and Pole Star, he knew right away where to go". It shows that the Big Dipper and Pole Star had been used to guide the ships for voyage during the Western Han Dynasty. During the Yuan Dynasty (1271–1368), with opening up the coastal ocean voyage routes, the celestial navigation technology was rapidly developed on the ocean voyage, and thus the navigation method that the geographic latitude was determined by observing the elevation angles of stars was mastered gradually. The famous navigator in the Ming Dynasty, He Zheng (1371–1433), adopted the celestial navigation method of "identifying the orientation with the rising and falling of the Sun and Moon, as well as measuring the distance with the elevation angles of the Big Dipper", to achieve the real ocean voyage and end the history "of sailing along the coast and with the wind during the day". The technology of "determining the position by leading the stars" was called "**transoceanic leading-star technique**". Between 1405 and 1433, the fleets led by Zheng made seven voyages to the Western Ocean, and sailed on the vast Indian Ocean, using the "transoceanic leading-star map" with the constellation names indicated on the nautical chart and combining the celestial positioning with the guiding device (celestial compass) to improve the positioning and heading accuracy of the ships. By using the celestial navigation techniques, the Zheng's fleets smoothly arrived in the Gulf of Aden, the Red Sea and the east coast of Africa, while European ocean voyage industry was then still in infancy.

In 1434, for the first time, the Portuguese fleets crossed the Cape Bojadorcalled "death limit" on the west coast of Africa, which started the adventure of European ocean voyage and colonial plunder. In 1492, an Italian navigator and explorer, C. Columbus, led his fleet across through the Atlantic Ocean to the American continent. In 1497, a Portuguese, V. da Gama, bypassed the Cape of Good Hope at the southern tip of Africa and opened up a route to India. In 1519, a Portuguese explorer F. Magellan led a fleet of about 270 crews, which took 3 years to achieve the first ocean voyage round the world. With the expansion of European capitalism and the increasing demand of ocean shipping, the celestial navigation technology had developed rapidly. By the eighteenth century, the European fleets had commonly been equipped with the measuring-angle instruments like sextants and the chronometers like astronomical clocks to further improve the ship positioning accuracy based on the celestial navigation technology.

The **sextant** is an optical instrument used to measure the angular separation between two distant visible objects, with a shape of fan. It is composed of a small telescope, a semi-transparent fixed plane mirror (also known as the horizon mirror) and a movably reflecting mirror (also known as the index mirror), with a measuring arc of one-sixth of the circumference. Generally, the sextant is used to measure the angle between the Sun or other celestial bodies and the horizon at a certain time, so as to determine the longitude and latitude of ships or aircrafts. As early as 1699, the sextant's principle was first proposed by Isaac Newton (1643–1727), but his notes were not published until 1742. Around 1730, a British mathematician John Hadley and an American inventor Thomas Godfrey independently invented the octant, a double-reflecting quadrant instrument with the measuring arc of one-eighth of the circumference. By 1757, a British naval officer John Campbell altered the octant's double-reflecting quadrant into the circular arc of one-sixth of its circumference to attain a measuring range of 120°, which was renamed sextant. Such simple and convenient stellar measuring instrument has widely been applied to the ocean voyage since the middle eighteenth century.

In 1735, John Harrison, a British clock mechanist, invented a mechanical marine chronometer with temperature compensation, which was the first practical astronomical clock on board of ships and a revolution of the celestial navigation technology, enabling navigators to accurately compute their longitudes at sea. Before the invention of astronomical clock, the latitude could be estimated but the estimates of longitude were usually very inaccurate. The marine chronometers had been used for more than 200 years by continual improvement, and later were replaced by electronic astronomical quartz clocks. In 1837, the American Captain Thomas H. Sumner proposed the concept of celestial position lines, the circle of equal altitude, as he was approaching the Smalls Lighthouse in thick weather. The longitude and latitude of the ship's position could be estimated by using the circle of equal altitude. In 1875, on the basis of the circle of equal altitude, the French Naval Commander Marcq de Saint-Hilaire proposed the astronomical drawing way and conversion method to determine the ship's position line by using the altitude difference, to simplify the operational process of the celestial navigation. The simplified navigation way laid the theoretical and practical foundation for the modern maritime ship's celestial navigation, which has been used up to now.

Since the 1950s, with the rapid development of electronic technology, computer and space technology, the positioning accuracy based on the celestial navigation has reached several hundred meters. Thereby, the celestial navigation was applied to military ships, submarines, medium and long-range bombers and missiles, and its strategic advantages and military utilization values were demonstrated clearly. The American Apollo Program and the Soviet projects of space stations directly promoted the development of celestial navigation technology and its wide application in the field of spacecraft's navigation and control technology. In the early 1960s, the U.S. Air Force initiated the project of studying the autonomous navigation of satellite, called Program 283. Taking the Earth's circle disk as the measurement reference in the Program 283, the navigation system configuration includes attitude reference system, horizon scanner, onboard computer and clock system, among which the attitude reference system consists of three strap-down inertial gyroscopes and one gimbaled star tracker, and thus the satellite's navigation can be carried out by using the horizon scanning data of the inertial platform. The system's navigation accuracy is mainly limited by the error of the horizon scanner, with the estimated orbit accuracy of 2 km. In addition, a standard space guidance system was developed by the U.S. Air Force, which takes the fixed ground landmarks as the measurement reference, and adopts three measurement schemes, including horizon scanning, known landmark tracking and unknown landmark tracking. By using the landmark tracker to measure the direction vector from the spacecraft to the landmarks, and combining the attitude reference system, the orbit and attitude parameters of the spacecraft can be determined, with the estimated navigation accuracy of 0.2–2 km. In 1973, a spacecraft subsystem that has the integrated capability of onboard orbit determination with attitude determination, called Space Sextant–Autonomous Navigation Attitude Reference System (SS-ANARS), began to be developed in the United States. Actually, the Space Sextant is composed of two optical telescopes mounted on the same three-degree-of-freedom gimbaled platform, one of which is used to track the Earth's limb or the Moon's limb while the other is used to track the known brighter stars, and the basic observables composed of the angles between the two lines of sight are used to estimate the navigation parameters by using the Kalman filter with the onboard digital computer. The demonstration system of the SS-ANARS had been flown as a sortie payload on the Space Shuttle to prove its autonomous-integrated orbit and attitude determination capabilities. With the SS-ANARS, the spacecraft's position accuracy is 243–366 m for any Earth orbit, and the attitude determination accuracy relative to the celestial sphere is better than 1 arc-second. In 1979, a low-cost, strap-down and modular attitude reference navigation system, called the Multi-mission Attitude Determination and Autonomous Navigation (MADAN) program, was developed by the U.S. Air Force to provide continuous and real-time inertial attitude and orbit information for advanced spacecraft application, with the property of full autonomy and long life. By replacing software and hardware modules, the system is suitable for the navigation application of three-axis stabilized spacecrafts from the low Earth orbit to five times the altitude of geosynchronous orbit. The measurement unit is composed of three star sensors and one Earth sensor, in which the detector element is a focal plane array of four imaging Charge-Coupled Devices (CCD). The strap-down star sensors will each provide a 7.1-degree Field-of-View (FOV) and an angular measurement accuracy of better than 2 arc-seconds, and the estimated orbit determination accuracy is 0.9 km for the Low Earth Orbit (LEO) and 9 km for the High Earth Orbit (HEO).

In 1989, an onboard navigation sensor system that is specially used to measure the Earth, Moon and Sun was designed and developed by Microcosm Inc. in the City of Torrance, California, USA, to autonomously determine the orbit and attitude of spacecraft in real time. The system is called Microcosm Autonomous Navigation System (MANS), only suitable for the spacecrafts in near-earth orbit. On the basis of the dual cone horizon scanner, a pair of fan-shaped scanning Sun and Earth sensors are added in the MANS. The spacecraft's orbit and attitude parameters can be determined by measuring the angular radius of the Earth's infrared radiation disk and the direction vectors from the Earth's center to the Sun and Moon. The major properties of the MANS are: the spacecraft's orbits and attitudes can be determined with the measured data from a navigation sensor; the sensor system is suitable for the spacecrafts in the low and middle Earth orbits; the navigation sensor is improved from the dual cone horizon scanner that is originally used to detect the Earth's infrared radiation, with the advantages of small size, lightweight, low power consumption and low cost; the latest study achievements of orbital mechanics assisting, sensor system error calibrating, Earth environment modeling and navigation filtering processing are applied to the system. In March 1994, the first Space Test Experiments Platform (STEP) spacecraft, a technology demonstration satellite whose purpose is to test the Technology for Autonomous Operational Survivability (TAOS), was launched from the U.S. Vandenberg AFB. The MANS was one of the TAOS experimental payloads, and unfortunately, due to the failure of the onboard computer, the MANS measurement data in the orbit have to be transmitted to the ground for post-processing, with the orbital determination accuracy of 200–500 m.

In fact, the above study achievements have been only in the stage of system scheme research and critical technology verification, and have not been put into a large number of engineering applications for spacecraft autonomous navigations. By the 1990s, with the completion of the navigation satellite system, it has been widely used in the autonomous navigation and precise orbit determination for the LEO spacecrafts. From applications of highly accurate navigation, the accuracy of traditional celestial navigation system is not comparable to that of the navigation satellite system. Therefore, in the application of spacecraft high-precision navigation, the celestial navigation is only an auxiliary means of navigation parameter determination. Even though, the celestial navigation usually plays a role of "ballast stone", in some special application fields, particularly considering the properties like autonomy and concealment.

4.4 Inertial Navigation System

4.4.1 Basic Principle

The **Inertial Navigation System (INS)** is a system in which the navigation parameters including position, velocity, attitude and heading are solved, by using the Inertial Measurement Units (IMU) to measure the angular velocity and linear acceleration of the movable objects relative to the inertial reference system, and after making the integral calculation, the movable objects can be guided for safe voyage, where the IMU mainly includes the gyroscopes for the angular velocity measurement and the accelerometers for the linear acceleration measurement. The inertia is a basic property of matter, and matter always keeps the state of rest, or uniform motion in a straight line, unless that state is changed by an external force. The principle of the INS is based on the classic Newton's law of mechanics. In 1687, Isaac Newton published **Three Laws of Motion**:

> every body continues in its state of rest, or of uniform motion in a straight line, unless it is compelled to change that state by forces impressed upon it; the acceleration produced by a particular force acting on a body is directly proportional to the magnitude of the force and inversely proportional to the mass of the body; to every action there is always opposed an equal reaction, or, the mutual actions of two bodies upon each other are always equal, and directed to contrary parts.

Newton's three laws of motion reveal the quantitative relationship of macro-mechanics in essence, which lays a theoretical foundation for the inertial navigation. The inertial navigation is a navigation way of dead reckoning, in which by measuring the heading angles and accelerations of the movable object, the current position is deduced from the previously determined position, and thus the current position information of the movable object can be output continuously. The INS generally consists of the gyroscopes, accelerometers, navigation computers and accessory circuit units.

The **gyroscope** is a kind of angular motion-sensing device, consisting of high-speed spinning rotor, inner ring frame, outer ring frame and fixed frame (as shown in Fig. 4.2), and the rotor is mounted on the frames so that it can spin rapidly about an axis which is free to alter in the direction. The orientation of the axis is not affected by tilting of the mounting, so the gyroscopes can be used to provide stability or keep a reference direction, with the performance of the stable-axis-directing and the precession. The **stable-axis-directing** refers to the property that in the case of no external force moment (torque), the direction of the spinning axis of the rotor remains stable in the inertial space. As the external torque acts on the axis of the outer ring frame, the gyroscope with high-speed spinning rotor will rotate around the axis of the inner ring frame, and as the external torque acts on the axis of the inner ring frame, the gyroscope will rotate around the axis of the outer ring frame. The **precession** refers to the property that the direction of spinning angular velocity and that of external torque is perpendicular to each other. There are many kinds of the gyroscopes. According to the different number of the precession degree of freedom, the gyroscopes are divided into single degree-of-freedom gyroscope and double degree-of-freedom gyroscope;

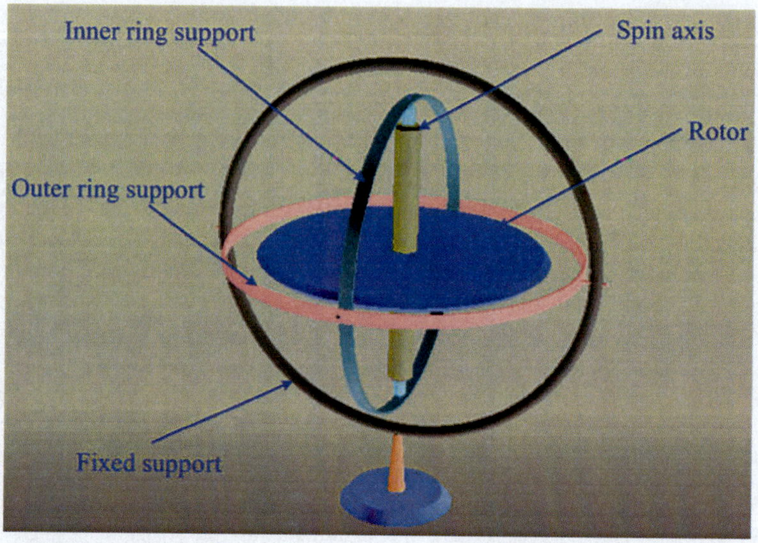

Fig. 4.2 A diagram of simple gyroscope structure

according to the different geometric positions of the gyroscope center of gravity, they are divided into balance gyroscope and gravity gyroscope; according to the different supporting ways, they are divided into frame gyroscope, liquid floating gyroscope, gas floating gyroscope, electrostatic gyroscope and flexible gyroscope; according to the different principles of gyroscope effects, they are divided into rotor gyroscope, vibratory gyroscope, laser gyroscope, fiber optic gyroscope and atomic gyroscope.

The **accelerometer** is a device for measuring linear proper acceleration of the movable object, consisting of sensitive mass, support, potentiometer, spring and shell (as shown in Fig. 4.3), where the sensitive mass is constrained by the support to move along only one axis (also known as input axis). The proper acceleration refers to the acceleration of the movable object in its own instantaneous rest frame, which is not as same as coordinate acceleration. In the terms of the Newton's laws of motion, when the device's shell is accelerated along the input axis with the movable object, the sensitive mass with certain inertia tries to keep its original motion state unchanged, and thus there will be relative motion between the sensitive mass and the shell, so that the spring will deform. Under the action of the spring force, the sensitive mass is accelerated to move. As the spring force is balanced with the inertial force each other, there is no longer relative motion between the sensitive mass and the shell. At the moment, the deformation of the spring represents the magnitude of the measured acceleration, and the acceleration is converted into an electrical signal output by the potentiometer.

In the INS, the spinning axis of the gyroscope rotor, the axis of the inner ring support and the axis of the outer ring support form a mutually orthogonal coordinate system; the heading and attitude angles of the movable object are measured with the

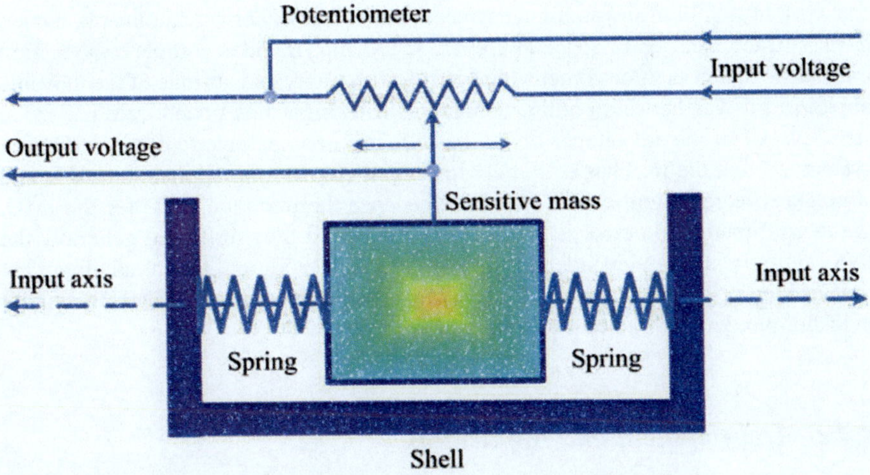

Fig. 4.3 A diagram of accelerometer principle

gyroscope; the input axis of the accelerometer is aligned with the coordinate axis to measure the acceleration of the movable object; finally, the velocity can be obtained by integrating the acceleration about time, and the velocity is integrated about time once again, to get the displacement of the movable object.

According to the different ways by which the IMU is mounted on the movable object, the INS can be divided into two categories, Gimbaled Inertial Navigation System (GINS) and Strap-down Inertial Navigation System (SINS). For the GINS, the real physical platform is utilized, on which the gyroscopes and accelerometers are mounted. The platform coordinate system is used to track the local geographical coordinate system, and thus the three-dimensional position and velocity of the movable object can be solved. Meanwhile, the heading and attitude angles are directly read out from the platform frame. It has the advantages of high navigation accuracy, easy implementation and simple calculation, but it is large in volume, heavy in equipment, expensive in price and high in required technological level. For the SINS, its gyroscopes and accelerations are directly attached to the movable object, and the physical platform is replaced by the mathematical platform established with the navigation computer. Thereby, the navigation parameters can be solved by calculating the attitude matrix and position matrix of the movable object. It has the advantages of small volume, lightweight, low cost, easy installation and maintenance of the IMU, easy redundant configuration, and easy integration with other navigation system hardware, but the IMU is required to have strong anti-vibration and anti-shock performance to adapt to the harsh working environment, and the gyroscopes are also required to have big dynamic measurement range, high-performance rebalancing loop, high stability and corresponding error offset mechanism.

The inertial navigation is a real autonomous and hidden navigation system that does not rely on any external information and not radiate any energy to outside. It can work normally in any medium environment such as underwater, mine pit, tunnel, Earth surface, near-earth space and outer space and provides complete navigation information such as three-dimensional position, velocity and attitude of the movable object in all-weather and continual manner, with high data update rate and good stability. All of the advantages about the INS are incomparable to other navigation system, so that the INS has extremely important engineering application value and strategic research significance. However, because there are the drifts for the IMU, the inertial navigation error is gradually accumulated over time, and generally the INS cannot be used independently for a long time. In addition, the initial alignment time needed is long for the alone INS, and to develop high-precision IMU is difficult in technique, with a high cost and an expensive price.

4.4.2 *Development and Application*

The Swiss mathematician and physicist, Leonhard Euler (1707−1783), spent most of his life in Russia and Germany and was the academician professor at the Saint Petersburg Academy of Science in Russia. He not only made decisive and formative contributions to the subjects of geometry, calculus, mechanics and number theory, but also developed methods for solving problems in observational astronomy and demonstrated useful applications of mathematics in technology and public affairs. In 1765, L. Euler first used analytical method to explain the essential problem of fixed-point rotation of rigid body, and created the fundamental theory of rotor gyroscope mechanics. The Franco-Italian mathematician and astronomer, Joseph-Louis Lagrange (1736−1813), made significant contributions to the field of analysis, number theory, and both classical and celestial mechanics. In 1778, he proposed the differential equations of motion of a rigid body rotating at a fixed point under the action of gravitational moment. In 1851, a French physicist Jean Foucault proved that the Earth rotates about its axis, by interpreting the motion of a heavy iron ball swinging from a wire of 67 m long, which was well known as "Foucault pendulum". The pendulum always swung in the vertical plane, but on a rotating Earth, this vertical plane slowly changed at a rate and direction dependent on the geographic latitude of its location. For this demonstration, the gyroscope was invented by Foucault as part of a two-pronged investigation into the rotation of the Earth. In the 1880s, the technology of using motor to drive gyro rotor appeared to improve the stability of the gyro rotor rotation, and the device for testing the gyrocompass that is a type of non-magnetic compass based on a fast-spinning disk and the rotation of the Earth to automatically find geographical direction was developed. In 1908, Hermann Anschütz-Kaempfe, a German inventor, designed a single-rotor-swinging gyrocompass, which can automatically find the North based on the gravitational moment, solving the problem of long-distance navigation of ships and submarine. Soon afterward, an American inventor Elmer A. Sperry built the first automatic pilot using a gyroscope to maintain an aircraft on course. In 1923, a German engineer Maximilian Schuler pointed out that if the gyrocompass was tuned to have an 84.4-min period of oscillation then it would resist errors due to sideway acceleration of the ship or aircraft carrying it. It is a well-known **Schuler tuning**, a design principle for the INS considering the curvature of the Earth. The Schuler tuning completely solves the problem of tracking the local vertical line with the established artificial vertical line on the movable object, laying a foundation for the practical application of the INS.

In 1942, an elementary inertial navigation device was equipped on the V-2 missile developed by W. von Braun in Germany. As a result, the heading and attitude angles of the missile were provided by the gyroscope, and the initial velocity of the missile into trajectory was provided by the gyro-integral accelerometer installed along the longitudinal axis, so as to guide and control the missile for flight. Although the guidance accuracy of the inertial navigation device was very low, it had created a precedent for the application of the INS. With the improvement of gyro stability and accuracy, inertial technology was gradually applied to the navigation for aircraft and submarine. In 1949, FEBE (named after the Roman Sun god Phoebus), an inertial device with a single degree-of-freedom liquid floating gyroscope, was used for the navigation of the strategic bomber B-29, with a stellar and magnetic coordinate as reference. The FEBE had created a record of 10-h autonomous flight test, which showed the inertial navigation was then feasible over moderate distance without stellar tracking. In 1950, the complete inertial auto-navigator XN-1 was developed by the electromechanical department of North American Aircraft inc. (established as the Autonetics Division in 1955), and its auto-navigational capability was demonstrated in a flight of a C-47 aircraft. Meanwhile, the inertial navigation device N-6 was developed by improving the XN-1, and successfully applied to the navigation of marine ships. In 1958, the U.S. Navy's nuclear submarine Nautilus was equipped with the N-6A inertial navigation system and the MK-19 north-seeking compass, starting from Pearl Harbor near Honolulu of Hawaii, crossing the Arctic icecap and finally arriving at Portland Harbor of Oregon. The entire voyage spent 21 days, and as the Nautilus surfaced near Portland Harbor, its position error was determined to be only 20 nautical miles. It fully demonstrates the INS advantages like the autonomy, concealment and completion, with the unique military application value.

In the 1960s, the technology of liquid floating gyroscope had been improved while the studies on flexible gyroscopes and dynamically tuned gyroscopes had risen. The dynamically tuned gyroscope had the advantages of simple structure, easy manufacture and low cost, with the drift error of 0.01 degrees per hour, and was widely used in the SINS. The inertial navigation system LTN-72 with the dynamically tuned gyroscopes was developed by Litton Industries, Inc. and had widely been used in the civil aircraft at that time. In the 1970s, Honeywell International Inc. and Rockwell Automation Inc. began to develop the electrically suspended gyroscopes with the drift error of less than 0.0001 degrees per hour. The electrically suspended gyroscopes were mainly used for the high-precision reference of strategic weapon throwing platform. Two sets of inertial navigation devices MK-27 with the electrically suspended gyroscopes, developed by the Rockwell, were equipped on the Ohio-class strategic nuclear submarine of the U.S. Navy. Afterward, the laser and fiber optic gyroscopes have developed gradually, whose stability and reliability are far better than the mechanical rotor gyroscopes, with the average failure time of 90,000 h. The laser SINS-LTN-90 series, produced by the Litton, has become a necessary navigation system for American civil and government transport aircrafts. By the 1990s, with the development of the Micro-Electro-Mechanical System (MEMS) technology represented by micro-machines, micro-sensors and micro-actuators, various micro-gyroscopes and accelerometers based on the silicon, which have the advantages of small size, lightweight, easy installation, impact resistance, high reliability and low cost, as well as easy mass production, are emerging constantly. So, with the improvement of manufacture technology and product performance, the IMU based on the MEMS will become the mainstream products in civil and tactical weapon applications.

4.5 Radio Navigation System

The **radio navigation system** is a kind of system that uses radio signal to measure the basic observables including the distances and its change rates, distance differences and orientations between the transmitting and receiving antennas, and that estimates the position and velocity of the movable object and safely guide its voyage. At the beginning of the twentieth century, with the advent of radio beacons, the radio navigation technology started to be applied to the field of sea voyage and aviation. At present, there are about 100 kinds of the radio navigation systems. According to the different positions of signal transmitting antennas, they are classified into two major categories, land-based radio navigation system and space-based radio navigation system. According to the differences of navigation ranges and task types, they are divided into the landing navigation system within less than 100 km, short-range navigation system within 100–500 km, medium range navigation system within 500–1000 km, long-range navigation system within 1000–3000 km and ultra-long-range navigation system with the distance of more than 3000 km. According to the differences of navigation algorithms and basic observables, they are divided into ranging and bearing navigation system, hyperbolic navigation and Doppler navigation system. The radio navigation system has the advantages of using time not limited by climate conditions, simple equipment and high reliability, but the navigation signal is easy to be interfered, and the working range of the land-based navigation system is limited by the coverage area of the ground stations. Generally, the radio navigation accuracy is with the order of several kilometers, for the movable objects on the surface of the Earth. The land-based radio navigation systems, including the ranging and bearing navigation, hyperbolic navigation and Doppler navigation, are introduced in this section, while the space-based radio navigation system, also known as the navigation satellite system, will be discussed in the following section of this chapter.

4.5.1 Ranging and Bearing Navigation System

The **ranging and bearing navigation system** is a kind of radio navigation system which takes the distance and orientation of the movable object relative to the ground stations as the basic observables to determine its position and velocity. It belongs to a kind of short-range navigation system, commonly with the polar coordinate system, and at the beginning is used in the navigation and approach landing of aircrafts. In 1912, the first radio navigation equipment, radio compass, was successfully developed according to the principle of amplitude direction-finding, with the working frequency of 0.1–1.75 MHz. During the Second World War, the direction-finding equipment, such as orientating plane, four-channel beacon and fan-shaped radio-pointing beacon, had been developed. Afterward, a variety of ranging and direction-finding equipment have emerged, which improves the accuracy and safety for aircraft navigating and landing. There are three typical ranging and bearing navigation systems, Very High-Frequency (VHF) Omni-Directional Range (VOR), Distance Measuring Equipment (DME) and Tactical Air Navigation System (TACAN). The **VOR** is a type of short-range radio navigation system with the amplitude direction-finding for aircraft, enabling the aircraft with a receiving unit to determine its position and stay on course by receiving radio signals transmitted by a network of fixed ground radio beacons, with the VHF band of 108.00–117.95 MHz. In 1937, the VOR began to be developed in the United States, and until 1946 it was deployed to use for aircraft navigation. With the wide application of Global Positioning System (GPS), the VOR ground stations are decreased gradually. For example, there were about 3,000 VOR stations operating in 2000, while their number was reduced to 967 by 2013. The **DME** is a radio navigation technology that measures the slant range between an aircraft and a ground station by timing the propagation delay of radio signals in the frequency band between 960 and 1215 MHz. As a result, the line of visibility between the aircraft and the ground station is required. An airborne interrogator initiates an exchange by transmitting a pulse pair, on an assigned channel, to the transponder at the ground station, and the channel assignment specifies the carrier frequency and interval between the pulses. After a known delay, the transponder replies by transmitting a pulse pair on a frequency that is offset from the interrogation frequency by 63 MHz and having specified separation. The **TACAN** is a navigation system for military aircrafts, providing the users with the bearing and slant range to a ground or ship-borne station (as shown in Fig. 4.4), which is a more accurate version of the VOR and DME. The DME portion of the TACAN system is available for civil use, and at the VORTAC facilities where a VOR is combined with a TACAN, civil aircraft can receive VOR/DME readings. The aircrafts equipped with the TACAN avionics can use this system for en-route navigating and non-precision approaches to landing. The navigation system for integrating the VOR and DME has become the standard short-range navigation equipment stipulated by International Civil Aviation Organization (ICAO), and the TACAN is the standard system for military aircrafts.

In addition, there is an aircraft approach and landing guidance system, which uses radio equipment to guide the pilot or autopilot to make the aircraft land safely. The

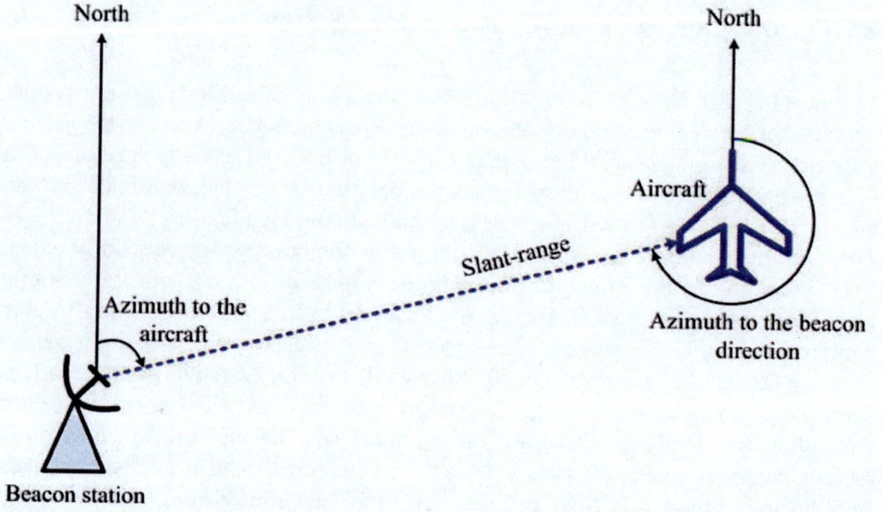

Fig. 4.4 Principle diagram of TACAN

radio landing guidance system provides accurate guidance information for aircrafts, such as landing orientation, glide path and distance. Thereby, the aircraft aligns the runway according to the guidance information, and approaches and lands according to the given glide angle to ensure that the deviation of the ground point is within the specified range. Generally, the landing guidance system is divided into three types, such as Instrument Landing System (ILS), Ground Control Approach System (GCAS) and Microwave Landing System (MLS).

4.5.2 *Hyperbolic Navigation System*

The **hyperbolic navigation system** is a kind of radio navigation system which by taking the ranging differences as the basic observables uses the intersection of hyperbolic position lines to realize positioning for movable objects. Its basic principle is: when the differences between the distances from each point along the trajectory of a movable object to two fixed points are constant, this trajectory is a hyperbola with the two fixed points as the focal points. When there are two hyperbolas intersected in a plane, the intersected point can be positioned. The hyperbolic navigation principle of offshore ship is shown in Fig. 4.5, where P is the ship's position to be determined, A is a major station, both B and C are the sub-stations, and the three stations are deployed in the fixed shore sites and marked in a nautical chart. Given the distance values of AC and AB, when both the distance difference d1 between BP and AP and the distance difference d2 between CP and AP are measured, two hyperbolas can be drawn in the nautical chart by taking A and B as the first focus pair, A and C as

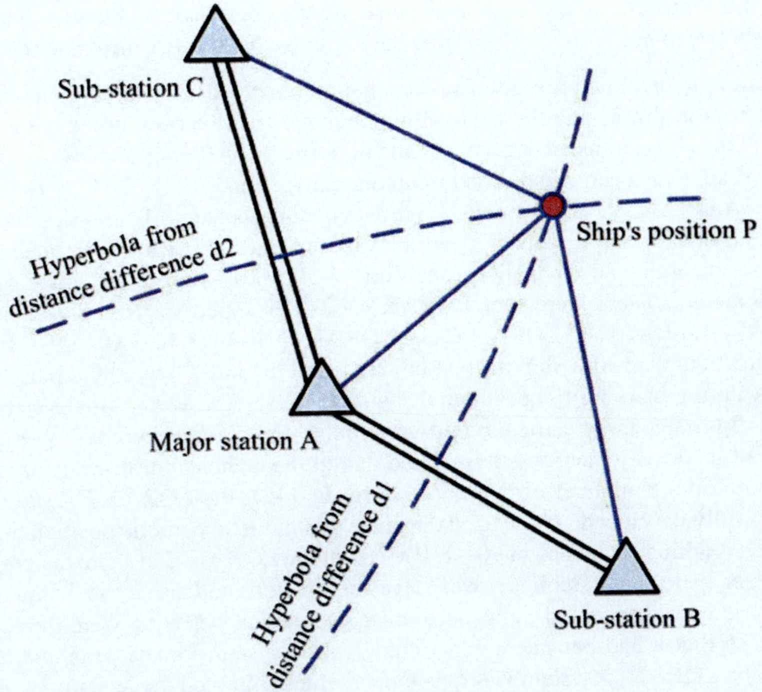

Fig. 4.5 Principle diagram of hyperbolic navigation system

the second focus pair, and d1 and d2 as the distance difference values, respectively. And thus, the intersected point P of the two hyperbolas is the ship's position, which can directly be determined in the nautical chart by using the geometric method. In the application of radio navigation, the distance differences are usually derived from the time differences between the transmitting point and the receiving point, and, thus, sometimes the hyperbolic navigation system is also called the **time-difference navigation system**.

The hyperbolic navigation system belongs to the medium and long-distance navigation system. According to the differently measured radio navigation parameters, it can be divided into three types as follows:

(1) **Pulse hyperbolic navigation system** is a radio navigation system, such as LORAN-A, which can get the distance differences by measuring the time differences between two pulse signals.
(2) **Phase hyperbolic navigation system** is a radio navigation system, such as Decca and Omega, which can get the distance differences by measuring the phase differences between two continuous sine wave signals.
(3) **Pulse and phase hyperbolic navigation system** is a kind of radio navigation system, such as LORAN-C and LORAN-D, which can not only measure the

time differences between two pulse signals, but also the phase differences between both carrier waves, and then obtains the distance differences.

For the hyperbolic navigation system, the number of users is unlimited, the user equipment is simple and the navigation parameters can be determined without the need of the signal transmitting devices and high-precision timing devices on movable objects, so it has been widely used in marine navigation.

The **LORAN**, short for Long-Range Navigation, began to be developed in the United States during the Second World War, and operated at lower frequencies to provide an improved navigation range up to 2415 km, with an accuracy of tens of kilometers. There were four LORAN series developed sequentially, including LORAN-A, LORAN-B, LORAN-C and LORAN-D. In 1943, the LORAN-A system was first equipped with ship navigation, and used the pulse hyperbolic positioning way, with the working frequency of 1.75–1.95 MHz. The LORAN-B system was originally designed by using the mid-wave pulse and phase hyperbolic positioning way, but its development was terminated due to the technical difficulties of identifying the integer number of cycles of phase. In 1957, the LORAN-C system was successfully developed, which used the pulse and phase hyperbolic positioning way with the working frequency of 90–110 MHz. In the LORAN-C system, the problem of integer cycle phase ambiguity was solved completely, and the new techniques like the phase comparison and multi-pulse code correlation detection were then introduced, so that it had become a general navigation system for sea, land and air. In 1968, the LORAN-D system was developed, which inherited the technology of the LORAN-C system and added the mobility of the ground signal transmitting station, so as to become an ideal military tactical navigation system.

The **Decca navigator system** was a hyperbolic navigation system which used the phase comparison of two low-frequency signals between 70 and 129 kHz to determine the positions of ships or aircrafts. In 1930s, the Decca's concept was originally proposed in the United States, but later developed into an operational system by Decca Radio and Television Company Limited of London, the United Kingdom. In 1943, the Decca system began to be equipped with the British Royal Navy. In 1946, the first commercial chain of the Decca's stations was established in southeast England. In the middle 1950s, the problem of the Decca's phase ambiguity was completely removed by introducing the multi-pulse technique. By the 1960s, the Decca system was adopted by the ICAO as the standard airborne navaid in preference of the VOR/DME.

The **Omega navigation system** was a very low-frequency, continuous wave and phase hyperbolic navigation system, which had the capability of global-range navigation, with the working frequency of 10–14 kHz. It was successfully developed in 1967, and operated around 1971 by the United Statesin cooperation with six partner countries. In the Omega system, a radio beacon network with eight fixed terrestrial stations was deployed around the world, so that the aircrafts and ships anywhere could receive the signals from five terrestrial stations at least, of which three stations were selected for positioning calculation, with the positioning accuracy of two nautical miles. Until 1997, the Omega system was shut down in favor of the GPS.

In addition, there are also three hyperbolic navigation systems, including the British Gee, the Soviet (later Russian) Alpha and Chayka, and the Chinese Changhe-2. The Gee was a pulse timing radio navigation system similar to LORAN-A, and began to provide service for the British Royal Air Force in 1942, with the positioning accuracy on the order of a few hundred meters within a range of 560 km. The Alpha (also known as RSDN-20), a hyperbolic navigation system similar to the Omega, was developed by the Soviet Union in 1962, and later evolved into the Russian terrestrial radio navigation and timing system similar to the LORAN-C, called Chayka, using a working frequency of around 100 kHz. The Chayka is available across the entire western portion of Russiaincluding Moscow, the Baltic Sea and Black Sea, as well as in the eastern portion of the country including the Seaof Okhotskand the northern portion of the Sea of Japan. The Changhe-2, a terrestrial medium and long-range radio navigation and timing system, was independently developed by China, with the signal coverage of the China's coastal and related sea areas. In 1990, it was officially opened to a variety of users.

4.5.3 Doppler Navigation System

The **Doppler navigation system** is a radio navigation system which based on the principle of Doppler shift generated by radio signals takes the distance change rates of the movable objects relative to the signal transmitting stations as the observables, to determine the positions of the movable objects. The typical Doppler navigation system can be divided into two types: one is Doppler radar navigation and another is Doppler counting navigation.

The Doppler radar navigation belongs to an autonomous dead reckoning system, mainly used for aircraft navigation. The aircraft's velocity information is measured with the Doppler radar, and its course and attitude are provided by the special attitude reference system, so as to obtain the velocity vector of the aircraft relative to the ground coordinate system. And thus, by integrating the velocity about time once, the position coordinates of the aircraft can be gotten eventually. In 1945, the Doppler radar for measuring speed had occurred. In 1955, the military aircrafts began to use the Doppler radar navigation, crossing through the development stages from electronic tube to transistor, combination and integration. At present, the Doppler radar navigation is a quite mature technique and has become the general navigation equipment for all kinds of aircrafts.

The Doppler counting navigation is mainly used for satellite's orbit determination. Generally, in order to improve the accuracy of measuring Doppler shift, the Doppler shift at an epoch is not measured directly, but the cumulative value of the Doppler shift within a given time interval is measured indirectly, which is called **Doppler counts**. The measurement equations can be established by using the Doppler counts from multiple time intervals, and then by combining the system state equations, the navigation parameters of the movable objects can be estimated with the Kalman filtering. Doppler Orbitography and Radiopositioning Integrated by

Satellite (DORIS), a typical Doppler counting navigation system, was developed in France by Centre National d'Etudes Spatiales (CNES) with cooperation of Grope de Recherche en Géodésie Spatiale (GRGS) and Institut Géographique National (IGN) in the mid-1980s. The system consists of more than 50 ground-based radio beacon stations, a DORIS control center and a number of the onboard DORISreceivers. The ground-based stations emit the radio signals to satellites at two central frequencies 401.25 and 2036.25 MHz, and then the double-frequency Doppler counts are obtained with the DORISreceivers on board of the satellites. Thereby, the satellites can autonomously determine its orbital parameters by using the Doppler counts. The best known satellites equipped with the DORIS receivers are the ocean surface topography mapping satellite TOPEX/Poseidon, and the global ocean circulation monitoring satellites Jason-1 and Jason-2, which are used to observe the ocean surface as well as currents or wave heights. The DORIScontributes to the two orbit accuracies of about 2 cm. Moreover, the DORIS is also used for the orbit determinations of other satellites, such as the Earth observation satellite series SPOT-2, SPOT-4 and SPOT-5; the large geostationary Earth-observing satellite Envisat; the polar icecap monitoring CryoSat-2; the ocean-observing satellite HY-2A; and so on. Apart from the orbit determinations, the DORISobservations are used for positioning of ground stations. Although the positioning accuracy of the DORIS is a bit lower than with that of the GPS, it still contributes to maintain the International Terrestrial Reference Frame (ITRF).

In the early 1980s, the European Space Agency (ESA) initiated a project for testing extremely accurate orbit determination performance, called the Precision Range and Range-rate Equipment (PRARE), which was developed in Germany. The PRARE is a compact, space-borne, two-way, two-frequency microwave satellite tracking system based on the measurement principle of determining the signal travel times and Doppler shifts. The system is composed of the space-borne microwave transceivers, ground microwave transmitting stations, ground master station and ground calibration stations. Its basic operation process is: firstly, the space-borne microwave transmitter sends the downlink signals of modulating the pseudo-random noise ranging codes and data codes to the ground stations at two frequencies 8489 MHz and 2248 MHz, respectively; secondly, the ground stations forward the uplink signals of carrying the modulated codes at the frequency of 7225.29 MHz; thirdly, the space-borne receivers receive the uplink signals, getting the delays of the ranging codes with Delay Locked Loop (DLL) and the Doppler shifts with Phase Locked Loop (PLL), and demodulate the navigation data; finally, the satellites can autonomously determine their orbital parameters by using the observables like the pseudoranges and Doppler shifts, and the navigation data. Since 1991, the PRARE system has been used for the precision orbit determination of the European Sensing Satellites ERS-1 and ERS-2, and the weather observation satellites from Meteor-3 to Meteor-7.

4.6 Navigation Satellite System

The traditional land-based radio navigation system has many disadvantages, such as limited signal coverage area, outdated equipment, backward technology and low positioning accuracy (generally the order of several kilometers), so that it is difficult to meet the requirements for highly precise, highly dynamic and real-time navigations, including the navigations for land vehicles, ships, aircrafts and spacecrafts. In the late 1950s, human beings entered a new era of exploration, development and utilization of space, bringing vitality and power to the traditional radio navigation system. The ground-based radio navigation station equipment is installed on the satellites, and the satellites are used as the dynamic known points. By receiving the signals from the satellites, the movable objects can get the basic observables like distances and Doppler shifts, and thus the navigation parameters including position, velocity and time can be solved in real time. Generally, the kind of space-based radio positioning, navigating and time-transferring system is called **navigation satellite system**. The navigation satellite system is a space infrastructure for real-time acquisition of high-precision navigation information, which can provide all-weather, full-time, high-precision navigation parameters for the users on the Earth's surface and in near-earth space. It has extremely important value for civil and military uses, and has attracted the attention of all aerospace institutions in the world.

4.6.1 Origin of Navigation Satellite Idea

In 1951, two young physics graduates, George C. Weiffenbach (1921–2003) and Willian H. Guier (1926–2011), came to work at the Research Center of the Applied Physics Laboratory (APL), Johns Hopkins University. Both were members of the Research Centergroup, whose assignment was to apply basic mathematics and physics method to tasking problems of the Laboratory. In those early days, their task was to investigate the methods of signal processing to improve the beam rider performance of Terrier and Talos missiles for defense of the Fleets against air attacks, especially by enemy planes at very low altitude. This problem was referred to as the low-angle problem and was of particular concern then.

On October 4, 1957, the world's first artificial Earth satellite, Sputink-1, was successfully launched by the USSR. The Sputink-1 was equipped with two radio transmitters, which broadcasted radio signals to the world at the frequencies 20.005 MHz and 40.002 MHz, respectively. Its purpose is to publicize the Soviet launch of the first satellite, and its implication for the International Geophysical Year. Fortunately, Weiffenbach had a good 20-MHz receiver, and there was the National Bureau of Standards radio stations WWV (a shortwave radio station), only 19.3 kmaway from the APL. The WWV broadcasted the best available frequency and time standards, and it was so close from the APL that a 0.6-m wire hanging from the receiver was an adequate antenna. Therefore, the receiver tuned to 20 MHz using the

WWV could yield a superb reference for a microwave spectrometer. Furthermore, the Soviet designers had set the Sputnik-1 frequency about 1 kHz from an exact 20 MHz so that any receiver would produce an audio tone of 1 kHz plus or minus the Doppler shift generated by the satellite motion. This offset of the satellite's frequency ensured that the received audio tone never went through zero, varying from about 1500 Hz to about 500 Hz, clearly audible throughout an entire pass.

Weiffenbach was responsible for receiving the signals transmitted by the Sputnik-1 while Guier was responsible for the work of the signal processing. By extracting the Doppler shifts, Guier could estimate the orbital parameters of the Sputnik-1. The Sputnik-1 stopped transmitting signals on October 26, 1957. Right after, the Sputnik-2 was launched by the USSR on November 3, 1957. Weiffenbach and Guier also received the dual-frequency signals from the Sputnik-2 and used the dual-frequency Doppler shift data to determine the satellite's orbit parameters. From the culmination of this study using the archived Sputnik-1 data and some of the Sputnik-2 data added, it was demonstrated that a complete set of orbit parameters for a near-earth satellite could be inferred to useful accuracy from a single set of Doppler shift data. The demonstration established that the whole Doppler curve was needed, nearly from the horizon to horizon, and that a first-order correction for ionosphere refraction was required, as well as an inferred correction for the satellite's oscillator frequency and frequency drift. As a result, there was the total number of nine unknown parameters, where the six is the orbital parameters, the one is the ionosphere correction parameter and the other two are, respectively, the bias and drift correction parameters of the satellite's oscillator frequency. Included in this demonstration was a single-parameter model of the ionosphere electron density, which was also inferred along with the six orbital parameters. The measurement equations were established by using the Doppler shifts of a complete observation pass, and thus the nine unknown parameters could be estimated with the method of least squares.

On March 17, 1958, Frank T. McClure (1916–1973), who was then the director of the APL's Research Center, called Weiffenbach and Guier to his office. After confirming that their study on determining an approximate orbit from a single pass of Doppler data had not been exaggerated, McClure presented an inverse solution problem to them, determining the station's position while assuming the orbit is known. This became the well-known "**navigation satellite problem**", later. "Go do an error analysis and let me know the answer ASAP." said McClure [7]. Clearly, the number of the unknown parameters needed to be solved for the problem had been reduced to five, including two plane coordinate parameters of a ground observation station and three system parameters. The study quickly evolved to the assumptions that the satellite is cooperative and emits the signals at two frequencies that are very stable and sufficiently high to effectively eliminate ionosphere refraction errors. And thus, the number of the system parameters was reduced to only one, the onboard clock bias. In other words, finally, there were only three unknown parameters needed to be solved. From the simulation results, it was indicated that the location coordinates of the ground stations could reach high accuracy, which was exciting and even unbelievable for Weiffenbach and Guier. When they reported excitedly the results back to McClure, he, of course with his consistently penetrating insight, was not surprised.

Shortly afterward, aiming at the requirement for navigating Polaris ballistic missile submarines, in which a launch location at sea would be accurately known, ideally within the orders of hundred meters, McClure presented a competitive proposal on the Polaris Doppler navigation system to the U.S. Navy's Special Project Office, finally accepted as the navigation satellite system, which ushered in a new era of the development of space-based radio navigation system.

4.6.2 First-Generation Navigation Satellite System

In December 1958, under the ARPA (renamed DARPA in 1972) sponsorship, the APL began to build the world's first navigation satellite system for the U.S. Navy, called the **Navy Navigation Satellite System (NNSS)**. Since the NNSS satellite orbits are designed to pass over the Earth's north and south poles, the NNSS is also known as **Transit navigation system**. Its basic idea was: the multiple polar orbit satellites emitted the signals encoded with their orbit parameters at two ultra-stable frequencies; a satellite tracking system received the signals with the two frequencies to solve the "direct problem"; an injection station transmitted the resulting orbit parameters to each satellite; the satellites would continue to orbit the Earth so that the submarines with the navigation receivers could determine their positions about once an hour anywhere on the Earth.

In September 1959, the first prototype satellite Transit-1A was launched, but was not sent to the expected orbit due to the failure of the third stage rocket. In April 1960, the Transit-1B satellite was successfully launched, and the test of Doppler navigation technology was first carried out actually, which demonstrated the first engine restart in space and the feasibility of using satellite as navigational aids. In June 1963, the Transit 5A-3 was successfully launched into the orbit, but could not be used for navigation due to the memory malfunction and the oscillator stability degraded. For all that, the satellite was the first to achieve gravity-gradient stabilization, and its other subsystems performed well. By 1964, the Transit system was officially completed and provided navigation services for the U.S. Navy's Polaris submarines. In 1967, the U.S. government announced that the Transit's navigation message would be decrypted to offer to civilian community for use, and thus the Transit system was quickly spread to many countries over the world and applied to many professional fields, such as maritime navigation, hydrographic survey and geodetic survey. Up to August 1988, as the last satellite, Transit O-31, was successfully launched into the orbit, the total number of the launched Transit satellites got to 37. Until 1996, the Transit system ceased service for users, replaced by the GPS.

The Transit system is a space-based, passive radio navigation system using the Doppler shift counts, and consists of three major parts: space segment, control segment and user segment. The space segment is a constellation composed of five Transit satellites, besides which five in-space spare satellites are also added. These satellites are evenly deployed on five low polar orbit planes, spaced 72 degrees apart at right ascension of the ascending node, with the orbital altitudes of 1100 km, the inclinations of 90° and the periods of 106 min, and each satellite transmits navigation signals at two central frequencies of 400 MHz and 150 MHz, respectively. The control segment is a ground network of tracking, measuring and controlling the satellites, composed of the ground stations with known position coordinates. Its tasks are to track each Transit satellite and measure its Doppler shift counts, estimate and predict the satellite's orbit and time parameters, and inject navigation message to each satellite twice a day with the uplinks. The user segment consists of a variety of the Doppler receivers, each of which measures the Transit satellite Doppler counts in a given time interval and establishes the measurement equations, and then the location coordinates of user receivers can be estimated by using the method of least squares. The orbits of the Transit satellites were designed to pass over the North and South Poles of the Earth, and evenly distributed on the equator. One Transit satellite at least can always be observed at any site in the world for a given time, and the site's coordinates can be determined. The time interval between twice positioning is several hours near the equator, 1–2 h in the middle latitude and less than 1 h in the polar regions of the Earth. The goal of building the Transit system is to regularly calibrate the inertial navigation system on board of the Polaris submarine, so the positioning interval, even with several hours, can also meet the system-built requirements. In other words, the Transit system is not applied to the real-time navigation for highly dynamic users. At any site, by measuring the Doppler shift counts of a single pass of a Transit satellite and using the dual-frequency signals to weaken the ionosphere delays, the navigation parameters can be obtained with the positioning accuracy of 200 m and the timing accuracy of 50 microseconds [8].

In 1960s, the Soviet Union began to develop the navigation satellite system, called **Tsiklon**, in order to provide the Soviet ballistic missile submarines with accurate position fixed. As early as 1962, the draft project for a Tsiklon experiment satellite was completed, using the Doppler navigation method, and the satellite would be placed in 800–1000 km altitude orbits, but its development was prolonged due to software problems and lack of accurate geodetic data. Until May 1967, the first Tsiklon satellite Cosmos-158 with a weight of 750 kg was launched into the orbit, with an apogee of 812 km, a perigee of 731 km and an inclination of 74 degrees to carry out the flight tests. Subsequently, the first positioning test was conducted with a Project-680 vessel of the Black Sea fleets, demonstrating a position error of 3 km, an intolerable test result. In 1969, after improving the models of predicting the satellite's ephemerides and incorporating new information on the gravitational anomalies and geoids of the Earth, an average position error reached 100 m over a 5-day period. On the basis of further improving a mapping of the Earth's gravitational anomalies, the Tsiklon system's tests had continued, and until 1972, the system was finally accepted by the Soviet Naval Forces for service as an interim measure, to pend deployment of the improved Tsiklon-B system, known as **Parus**. Between 1967 and 1978, there were a total of 29 Tsiklon satellites launched for test and application, preliminarily building the Soviet first generation of navigation satellite system.

Using the basic KAUR-1 bus (which was a drum covered with solar cells, with a weighted mast on top for gravity-gradient stabilization, and an antenna hung off one side near the bottom), the Parus satellite was a 2.0-m-diameter cylindrical spacecraft body, with solar cells and radiators of thermostatic temperature regulating system mounted on the exterior. Its orientation was by a single-axis magneto-gravitational passive system. The hermetically sealed compartment had the equipment mounted in cruciform bays, with the chemical batteries protecting the radio and guidance equipment mounted at the center. The operational constellation was composed of a minimum of six Parus satellites, each deployed in one of six orbital planes, spaced 30° apart, and one active satellite and one in-space spare was usually maintained in each plane, with a typical lifetime of 1.5–2 years. In 1974, the first Parus satellite Cosmos-700 was launched into the orbit, with a weight of 810 kg, an apogee of 994 km, a perigee of 960 km and an inclination of 83 degrees. The satellites transmitted the Doppler-shifted signals at around 150 and 400 MHz, which carried the satellites' position and orbital characteristics. In 1976, the Parus system was built, with the positioning accuracy of better than 100 m by using the data from several satellites, officially providing military services. Up to 2010, with the last Parus satellite Cosmos-2463 put into the orbit, there were a total of 99 Parus satellites launched. Since then, the Parus system had completed its historical mission, gradually replaced by Global Navigation Satellite System (GLONASS).

In addition, a complementary civilian version of the Parus military navigation system, known as **Tsikada**, was designed for the Soviet Merchant Marine and Academy of Sciences in 1974. The first civilian navigation satellite Cosmos-883 was launched in 1976 to test its performance, and the launch of Cosmos-1000 in 1978 marked the beginning of deployment of the operational system. The Tsikada constellation consisted of a minimum of four operational satellites, each deployed in one of four orbital planes, spaced 45° apart, whose signals were supplemented for military users by the very similar Parus system. The Tsikada system provided global navigation for both the Soviet (later Russian) Navy and commercial shipping. Up to 1995, as the last civilian navigation satellite Cosmos-2315 was sent into the orbit, a total of the launched Tsikada satellites reached 20.

The first generation of navigation satellite system based on the Doppler-shifted principle, represented by the Transit system, has the following defects:

(1) Due to small total of the satellites, it is difficult to continuously navigate and locate. There are only five active satellites and five spare satellites in the orbits in the Transit navigation constellation, so that the time interval between two observations of the Transit satellites at any point on the ground is long, generally 0.8–1.6 h, and varies with time. In the low latitude area, the satellite passes 15 times a day, and in the high latitude area, the satellite passes 30 times a day, and each passing observation time is 10–18 min. Apparently, it is difficult for the Transit system to achieve the continuous and real-time navigating and positioning.

(2) Due to the low Earth orbits, it is difficult to precisely determine the satellite's orbit parameters. The orbital altitudes of the Transit satellites are only 1100 km, belonging to the LEO, which are greatly affected by atmospheric drag. Due to uncertainty of the atmospheric density, satellite's surface-to-mass ratio and atmospheric drag coefficient, it is difficult to accurately model the atmospheric

drag. By using the Doppler-shifted counts from the ground stations tracking the satellites, it is also so difficult to precisely determine the satellite's orbits that the navigating and positioning accuracies are not high finally. In general, the positioning accuracy is 200 m by using the Doppler data during a single pass, and can only reach 10 m by using 15 passes. Thereby, the orbit determination accuracies are so low that the requirements for high-precision navigation and precise geodetic positioning cannot be met.

(3) Due to the low frequencies used by the satellites to transmit the signals, it is difficult to effectively calibrate the ionosphere delay. The Transit satellites transmit the navigation signals, respectively, at two frequencies 150 and 400 MHz, where the two frequencies are in the VHF band. In the dual-frequency Doppler navigating and positioning, the low-order term of the ionosphere delay effect could be weakened by using the two frequencies, but it is difficult to eliminate the errors from the high-order terms. From the studied results, it is demonstrated that the high-order terms of the ionosphere delay effect will cause the geodetic altitude error of more than 1 m near the geomagnetic equator. As long as high microwave frequencies are used as the carrier frequencies of navigation satellite's signals, the ionosphere delay effects will effectively be eliminated to improve the navigating and positioning accuracies.

For the development of follow-up navigation satellite system GPS, the Transit's contributions were mainly to two aspects. One is the dual-frequency technique, which is used to calibrate the time delay of the radio signal induced by the ionosphere, and which is incorporated into the GPS to attain the highest positioning accuracy. The other is the accurate prediction of satellite orbits, another essential GPS technique, as one of the five fundamental challenges that faced the GPS system designers.

4.6.3 Second-Generation Navigation Satellite System

For the first generation of navigation satellite system based on the Doppler-shifted navigation principle, it is difficult to meet the requirements for continuous, highly dynamic and highly accurate navigating and positioning applications. For this reason, using the passive navigation mechanism with single-way timing and ranging, the second generation of navigation satellite systems is designed and developed, and their basic principles are: the user receiver tracks more than four navigation satellites simultaneously, and the pseudorange observables between the satellites and the receiver are obtained by measuring the arrival-times of the pseudo-random noise codes; more than four measurement equations are established; finally four unknown parameters, including three position coordinate components and one clock bias of the receiver, can be estimated by using the least squares method. The built and building second navigation satellite systems, including the American Global Positioning System (GPS), Russian Global Navigation Satellite System (GLONASS),

Chinese BeiDou navigation satellite system and European Galileo navigation satellite system, are four core providers designated by the International Committee on Global navigation satellite system (ICG), and will be highlighted for their introductions in this section. In addition, the Japanese Quasi Zenith Satellite System (QZSS) and the Indian Region Navigation Satellite System (IRNSS) are also developed by using the single-way timing and ranging ways. Both the QZSS and IRNSS, which are the regional navigation systems, can usually be found from the related references for the interested readers, and are not going to be further described in this book.

4.6.3.1 GPS System

In 1960s, besides the Transit system, there were also three navigation satellite test programs developed by the U.S. Forces. One was the **Project 621B** initiated by the Space Division of the U.S. Air Force in 1963; the other was the **TIMATION program**, which was conceived and developed by the Naval Center for Space Technology (NCST) at Naval Research Laboratory (NRL) in Washington D. C. beginning in 1964; the third was the **Sequential Collation of Range (SECOR)** system developed by the U.S. Army, and the first SECOR satellite was launched for military geodetic surveying in 1964.

The Project 621B's proposal was presented originally by Ivan A. Getting (1912–2003) who was then-president of the Aerospace Corporation. Seeing the need for a new satellite-based navigation system, Ivan Getting envisioned a more accurate positioning system that would be available in three dimensions, 24 h a day, 7 days a week. At that time, Getting's energy and foresight were essential to gaining the Air Force support to study the system alternatives. In the Project 621B, the passive-ranging technique was used to eliminate the need for an extremely stable clock in the user equipment and provide three-dimensional positioning. Moreover, the Code Division Multiple Access (CDMA) signal structure was also designed for the satellites, and the prototype versions of CDMA navigation receivers were developed. Between 1968 and 1972, using the receiver on board of aircrafts to receive from the ground-based pseudo-satellite signals, the Project's tests were carried out at White Sands Missile Range, New Mexico, to verify the feasibility of choosing CDMA as the passive-ranging signal and other related techniques.

The core concept of TIMATION was a time-based navigation system with passive ranging, circular orbits and space-borne high-precision clocks synchronized to a master clock, presented by Roger L. Easton (1921–2014), a scientist and engineer at the NRL. In May 1967, the first test satellite TIMATION-1 was launched successfully, which was a small satellite with a weight of 38.3 kg and a power of 6 W, and its key feature was to carry a very stable quartz clock. In September 1969, the TIMATION-2 satellite was launched, which broadcasted on two frequencies to calibrate ionosphere group delay. It was proved from the TIMATION testing results that a system using a passive-ranging technique, combined with highly accurate clocks, could provide the basis for a new and revolutionary navigation system with three-dimensional coverage throughout the world.

The SECOR was an all-weather geodetic survey system, consisting of a satellite and four ground stations, of which three stations' coordinates had been surveyed accurately and the fourth station's coordinates were to be undetermined. By using the successive positions of the satellites in orbits, the unknown locations on the Earth's surface can be determined with exactness over long distances. When radio waves are flashed from the three stations with the known coordinates to the satellite and returned, the satellite's position at any time can be determined by measuring the ranges between the satellite and the three stations. Using these precisely established satellite positions as a base, ranges from the satellite to the unknown stations are used to compute the position of the unknown station. Each ground station was entirely portable, containing three units: a radio-frequency shelter, a data handling shelter and a storage shelter. Later, the initial units were replaced by the lighter weight, solid-state equipment. Between 1964 and 1969, there were a total of 13 SECOR satellites launched. Each satellite is with a mass of 18 kg and spherical shape or cubic shape, carrying a transponder, a telemetry system to monitor temperature and operating voltages, and a power unit comprised of solar panels and batteries. The SECOR system was in operational use for several years, establishing a global survey network, and especially allowing continents and islands to be brought within the same global geodetic grid.

In the early 1970s, although the above four space-based navigation programs had been developed parallel in the United Sates, even including the built land-based worldwide radio navigation system Omega, all of the systems were unable to meet the wide needs for accurate navigation in military and civilian sectors due to their own limitations. Thereby, it was realized that a superior system would be developed by synthesizing the best technologies from the multiple systems. However, almost none of those needs were seen as justification for billions of dollars that would cost in research, development, deployment and operation for a constellation of navigation satellites. In the atmosphere of Cold War arms race, the nuclear threat to the existence of the US was one need that did justify this cost in the view of the U.S. Congress. In addition, it is also the reason for the ultra-secrecy at that time. The nuclear triad consisted of the U.S. Navy's Submarine-Launched Ballistic Missiles (SLBM) along with the U.S. Air Force strategic bombers and Inter-Continental Ballistic Missiles (ICBM). Considered vital to the nuclear deterrence posture, accurate determination of the SLBM launch position was a force multiplier. Meanwhile, the U.S. Air Force with two-thirds of the nuclear triad also had requirements for a more accurate and reliable navigation system. In order to increase the survivability of ICBM, the requirements for using mobile launch platforms were proposed, with the need to fix the launch position similar to the SLBM situation. Furthermore, the U.S. Army also had the urgent needs for precisely geodetic surveying by using navigation satellites. As a result, a meeting of about dozen military officers from the Army, Navy, Marine Corps and Air Force was held at the Pentagon by the U.S. Department of Defense (DoD) to discuss the creation of a Defense Navigation Satellite System (DNSS) in May 1973, and a Joint Program Office (JPO) led by the then-Air Force Colonel Branford Parkinson (1935–) was formed to execute the DNSS program. Later that year, the DNSS program was named **Navigation Satellite Time and Ranging (NAVSTAR)**, or Navigation System Using Timing and Ranging. With the individual satellite being associated with the name of NAVSTAR, a more fully encompassing name was used to identify the constellation of NAVSTAR satellites, **NAVSTAR-Global Positioning System (GPS)**, often simply referred to as **GPS**.

The GPS consists of three major parts: space segment, control segment and user segment. The space segment is composed of 24–32 satellites in Medium Earth Orbit (MEO), and also includes the payload adapters to the boosters required to launch them into orbit. The control segment consists of a master control station, an alternate master control station, four dedicated ground antennas and six dedicated monitor stations. The user segment consists of hundreds of thousands of the U.S. and allied military users, with secure GPS Precise Positioning Service (PPS), and tens of millions of civilian, commercial and scientific users, with Standard Positioning Service (SPS). The GPS satellites broadcast signals to the Earth, and the user receivers receive the signals and calculate their three-dimensional locations including longitude, latitude and altitude, as well as their positioning time. The implementation of GPS program is divided into the following three phases:

(1) From 1973 to 1977 was the scheme demonstration phase. With the Air Force named the Executive Service, then the TIMATION program merged with the Project 621B to form the NAVSTAR–GPS program, the TIMATION-3 satellite originally developed by the NRL was re-designated Navigation Technology Satellite 1 (NTS-1). In the summer of 1974, the NTS-1 with two rubidium-vapor frequency clocks onboard was launched to test the technologies, such as the onboard atomic clocks, time-based ranging, range rate and 12-h orbit, and to demonstrate the capability for pinpointing longitude, latitude and altitude throughout the world, leading directly to the first GPS technology. Three years later, the NTS-2 was successfully sent into the orbit, to verify the earlier technologies, such as cesium frequency standards, nickel–hydrogen battery, radiation dosimeters and a worldwide network for data acquisition, and also considered the first GPS prototype test satellite.
(2) From 1978 to 1985 was the phase that the GPS prototype satellites were deployed. In February 1978, the first GPS prototype satellite, known as Block I-1, was launched into the orbit, with an apogee of 20,308 km, a perigee of 20,095 km and an inclination of 63.3°. During this phase, there were a total of 11 Block I satellites launched, of which one was destroyed in a launch failure and other ten formed a test constellation to validate the feasibility of the GPS system. The Block I satellites weigh 845 kg each in final orbit, and with three rechargeable nickel–cadmium batteries for power and 7.25 m^2 of single-degree solar panels, and themselves, they could only store sufficient information for 3.5 days of independent operation. The Block I satellites emitted the navigation signals at two frequencies L1 (1575.42 MHz) and L2 (1227.60 MHz), where the Coarse Acquisition Code (C/A-Code) signals were only emitted at the L1 for civilian uses and the Precision Code (P-Code) signals emitted at the L1 and L2 for military uses. The uploaded information from the Control Segment was not encrypted for the Block I. The design life of the satellites

was 4.5 years, but their actual lifetime was 8.8 years. The last Block I satellite was decommissioned in the late 1995.

(3) From 1986 to 1994 was the rapidly deployed constellation phase. The second-generation GPS satellites are known as Block II, with approximately twice as heavy as the Block I and a design life of 7.5 years. Between 1989 and 1990, a total of nine Block II satellites were successfully launched into the orbits to rapidly form an operational constellation. The Block II satellites could operate up to 14 days without an upload from the control segment and the uploaded information had been encrypted. The satellites themselves were radiation hardened, and their signals were subject to the Selective Availability (SA). Meanwhile, the advanced Block II satellites, known as Block IIA, were developed during the phase. The Block IIA satellites could store more of the navigation message, and operate without contact with the control segment for 6 months. If it were actually done, of course, their broadcast ephemerides and clock bias corrections would degrade. The first Block IIA satellite was launched in 1990, and there were a total of 19 Block IIA satellites put into the orbits between 1990 and 1997. In December 1993, the GPS system achieved initial operational capability, which indicated a full constellation with 24 GPS satellites available to provide the SPS. The GPS program cost up to 1994, not including the cost of the user equipment, had been estimated at about $5 billion.

The GPS Space and Control Segments had been developed, maintained and operated by the U.S. Air Force. In April 1995, the full operational capability of the GPS system was declared by the U.S. Air Force Space Command, signifying full availability of the military secure PPS. The GPS constellation design originally called for 24 satellites, eight each in the three approximately circular orbits, but later it was modified into six orbital planes with four satellites each. The six orbit planes have approximately 55° inclination, separated by 60° right ascension of the ascending node, with an orbital altitude of 20,184 km, which is the repeated orbit with two circles per sidereal day. The constellation allows that at least six satellites are always within line of sight from almost everywhere on the Earth's surface. For this reason, the four satellites are not evenly spaced 90° apart within each orbit. In general terms, the angular differences between two close-to-satellites in each orbit are 30°, 105°, 120° and 105°, respectively, with a sum of 360°.

In the wake of the Block II satellites, the highest quality signals of the GPS system were reserved for military use while the signals available for civilian use were intentionally degraded, which was so-called the SA. Afterward, with the widespread growth of differential GPS services to improve civilian accuracy, the U.S. military advantage getting the better positioning accuracy was being eliminated. Moreover, the U.S. military was actively developing technologies to deny GPS service to potential adversaries on a regional basis. And thus, in 1996, the then U.S. President Bill Clinton signed a policy directive declaring that the SA would be turned off in May 2000 to provide the same precision to civilians that was afforded to the military, and establishing an Interagency GPS Executive Board (IGEB) to manage it as a national asset. At midnight on May 1, 2000, as the SA software was switched off eventually, allowing civilian users to receive a non-degraded GPS signals globally, the SPS accuracy had been improved from 110 to 20 m available. The GPS is owned and operated by the U.S. Government as the national asset, while the DoD is the steward of GPS. From 1996 to 2004, the IGEB oversaw the GPS policy matters. In 2004, the U.S. President George W. Bush updated the national policy directive, and then the IGEB was replaced by the national executive committee for space-based Positioning, Navigation and Timing (PNT) to advise and coordinate federal departments and agencies on matters concerning the GPS and related navigation systems. The PNT executive committee is chaired jointly by the deputy secretaries of defense and transportation, and its membership includes equivalent-level officials from the departments of state, commerce and homeland security, as well as the joint chief of staff and the NASA. Furthermore, components of the executive office of the president participate as observers of the executive committee, and the chairman of Federal Communication Commission (FCC) participates as a liaison.

Since the GPS deployment, the improvements of the system performance have been implemented consistently, including the added new signals; updated ground stations; improved navigation accuracy and increased continuity, integrity, availability, anti-jam and autonomous survivability. Meanwhile, all of the improvements maintained compatibility with existing GPS equipment. Modernization of the GPS satellites has been an ongoing by the DoD through a series of satellite acquisitions to meet the growing needs of the military, civilian and commercial markets. Between 1997 and 2004, a total of 12 GPS replenishment satellites called Block IIR were launched, of which the first launch was unsuccessful due to the rocket's failure and the other 11 satellites successfully entered into the orbits. There are two significant advancements for the Block IIR satellites. One is with the autonomous navigation capability by using the inter-satellite crosslink ranging, known as AutoNav, and the other is to use the onboard reprogrammable processors to do their own fixes in flight. Moreover, the Block IIR also demonstrated the concept of Distress Alerting Satellite System (DASS). Between 2005 and 2009, a total of eight GPS replenished and modernized satellites were successfully sent into the orbits, known as Block IIR-M. The Block IIR-M satellites emitted the added signals of two codes, the M-code as a new military code and the L2C code as a new civilian code. Between 2010 and 2016, a total of 12 GPS follow-on satellites called Block IIF were put into the constellation. Compared with the Block IIR-M, the improvements of the Block IIF include the third civil signal L5 (1176.45 MHz), 12-year design lifetime, improved rubidium frequency standards, direct orbit insertion and operational DASS repeaters. The third civil signal and the L2C signal, with double carrier frequencies, are very helpful for ionosphere modeling, to further improve the positioning accuracy for civilian users. It was reported that the SPS horizontal accuracy was better than 3.5 m in 2015, using the high-quality GPS receivers with Federal Aviation Administration (FAA) grade. Of course, there would be many factors affecting this accuracy, such as receiver noise, atmospheric issues and multi-path delay.

In December 2018, the new generation GPS satellite Block III-1 was launched successfully, with the design lifetime of 15 years, and by 2023, a total of ten Block III will be deployed to the constellation to meet higher demands of both military and civilian users. According to the different performance improvements, the GPS III satellites are divided into three types, namely, A, B and C, where the GPS IIIA delivers significant improvements over current GPS satellites, including a new international civil signal (L1C) and increased M-code anti-jam power with full Earth coverage for military users; the GPS IIIB will enable a cross-linked command and control architecture, allowing the satellites to be updated from a single ground station instead of waiting for each satellite to orbit in view of a ground antenna; the GPS IIIC will include a high-powered spot beam to deliver greater M-code power for increased resistance to hostile jamming. The final GPS III satellite proves three times better positioning accuracy and up to eight times improved anti-jamming capabilities, with the navigation payload of 70% digital. As of April 2020, there are a total of 32 satellites in orbits for the GPS constellation, including 10 IIRs, 8 IIR-Ms, 12 IIFs and 2 IIIs.

As deploying GPS III satellites to more resistant to jamming and spoofing, the U.S. Air Force is currently looking for ways to shore up the PNT capabilities provided by the GPS. One approach the Air Force believes feasible is to supplement its medium Earth-orbiting GPS constellation with an additional layer of smaller satellites in geosynchronous Earth orbit. To test that idea, the Air Force Research Laboratory in 2022 will launch an experimental PNT satellite, called NTS-3. Like its predecessors the NTS-1 and NTS-2, the NTS-3 is revolutionary in its approach, addressing new capabilities and paving the way to more resilient PNT. In the next few decades, it will blaze the trail to unfettered access to the vital data that the GPS provides, even in the midst of a wide variety of new threats [9].

4.6.3.2 GLONASS System

On the basis of implementing the Tsiklon, Parus and Tsikada navigation system in practice, the Soviet government made a decision to launch the development of the unified space navigation system, its second-generation navigation system GLONASS, in 1976. The design proposal of the GLONASS was completed in the late 1970s, using the constellation Walker24/3/2, with an orbital altitude of 19,130 km and an inclination of 64.8°. The GLONASS satellite's orbits are the repeated orbits, with a period of 17 circles per 8 days. The first-generation GLONASS satellites were 7.8-m tall, with the width of 7.2 m, measured across their solar panels, and the weight of 1260 kg. By using the heavy-lift Proton rocket, three GLONASS satellites would be launched at a time. The GLONASS satellites transmit two types of navigation signals: one is an open standard-precision signal L1OF, and the other is an obfuscated high-precision signal L1SF/L2SF.

In October 1982, the first three GLONASS satellites, respectively, designated Cosmos-1413, Cosmos-1414 and Comos-1415, were successfully launched aboard a Proton rocket. As only one GLONASS satellite (real Cosmos-1414) was ready in time for the launch instead of the expected three, it was decided to launch it along with two mock-ups, the dummies Cosmos-1413 and Cosmos-1415. Between 1982 and 1991, there were 43 GLONASS-related satellites plus 5 test satellites launched.

When the Soviet Unionwas disintegrated in 1991, the GLONASS constellation had 12 operational satellites in two orbital planes, allowing limited usage. Subsequently, the Russian Federationtook over control of the GLONASS and continued its development. In 1993, the GLONASS system was officially declared operational, actually with the then constellation consisting of 12 satellites, and the constellation was finally brought to its optimal status of 24 operational satellites in December 1995. The system was initially designed to have a positioning accuracy of 65 m, but in reality it had an accuracy of 20 min the civilian signal and 10 min the military signal.

Although both the GLONASS and the GPS have similar system architecture, positioning principle and service ways, there are differences between both, such as time reference system, coordinate reference frame, broadcast ephemeris format and satellite-identified ways.

(1) The time reference systems are different between both. Both belong to the atomic time system . The GPS time system is consistent with the Coordinated Universal Time (UTC) maintained by the U.S. Naval Observatory at 0:00 on 6 January 1980, known as UTC(USNO), which is a continuous-time system without leap second. The GLONASS time system is based on the Moscow's Coordinated Universal Time, known as UTC(SU), which is a synchronously leap second time system, so that there is only a system difference of 3 h plus less than 1 μs between the GLONASS time and the UTC(SU), while there is not the leap second difference.

(2) The coordinate reference frames are different between both. The GLONASS uses a coordinate datum named the Earth Parameters 1990, Paramtry Zemli 1990 (PZ-90), in which the precise location of the Earth's North Pole is given as an average of its positions between 1990 and 1995. The GPS's coordinate datum is the World Geodetic System 1984 (WGS-84), which uses the location of the North Pole in 1984. As of September 17, 2007, the PZ-90 datum has been updated to version PZ-90.02, which differs from the WGS-84 less than 40 cm in any given direction. Since December 31, 2013, the updated version PZ-90.11 is being broadcasted, which is aligned to the International Terrestrial Reference System (ITRS) at epoch 2011.0, getting to the centimeter level.

(3) The broadcast ephemeris formats are different between both. The GPS satellite's ephemerides are broadcasted with a format of Keplerian orbit elements, with an interval of 2 h. By using the Keplerian orbit equations and considering the satellite's perturbation motion, the instantaneous positions of the GPS satellites can be calculated in the WGS-84. For the GLONASS system, at a given epoch the satellite's position and velocity and its perturbation acceleration from the Sun and Moon are broadcasted directly as the ephemerides, with an interval of 30 min. Using the four-order Runge–Kutta method for numerical integration, the instantaneous positions of the GLONASS satellites can be calculated in the PZ-90.

(4) The satellite-identified ways are different between both. The CDMA technique is used to identify the different GPS satellite signals, and the navigation signals for all of the GPS satellites are transmitted at the multiple same frequencies. The

Frequency Division Multiple Access (FDMA) technique is used to identify the different GLONASS satellite signals, and each GLONASS satellite transmits the navigation signals within its given frequency band. All GLONASS satellites transmit the same code as their standard-precision signals on a different frequency using a 15-channel FDMA spanning either side from 1602.0 MHz, known as the L1 band. The L2 band signals use the same FDMA as the L1 band signals, but transmit spanning the center frequency 1246 MHz. Since 2008, new CDMA signals have been researched for use with the follow-on GLONASS satellite series, considering the interoperability with other navigation satellite systems.

In the wake of the achievement of full GLONASS constellation in 1995, the replenished satellites could not be launched in time to the constellation due to the Russian financial difficulty. As a result, the constellation had only six operational satellites in 2001, reaching its lowest point. With the Russian economy recovery and state finances improvement, the GLONASS system's restoration was made one of the government's top priorities. For this purpose, the Federal Targeted Program, "Global Navigation System 2002—2011", was launched by the Russian Government Decision No. 587. Meanwhile, responsibility of the program was transferred from the Russian Ministry of Defense to the Russian Federal Space Agency, commonly known as Roscosmos. In December 2003, the second-generation GLONASS satellite, GLONASS-M, was launched for the first time, with a weight of 1415 kg and a design lifetime of 7 years. In the original GLONASS design, only the obfuscated high-precision signal was broadcasted in the L2 band, but starting with the GLONASS-M, an additional civil reference signal L2OF was broadcasted with an identical standard-precision code of the L1OF signal, and a new signal L3OC was also added in the GLONASS-M series launched since 2014. In May 2007, all restrictions of the GLONASS system were lifted by the Russian government, and thus the positioning accuracy, formerly military-only signal with a precision of 10 m, has since then been freely available to civilian users. In June 2008, the GLONASS constellation was composed of 16 satellites, 12 of which were fully operational at the time. The Russia's aim of finishing the constellation in 2010 suffered a setback as a December 2010 launch of three GLONASS-M satellites failed. Following the mishap, the Roscosmos activated two reserve satellites, and decided to make the first improved GLONASS-K satellite, as a part of the operational constellation instead of mainly for testing as was originally planned. The GLONASS-K is with a mass of 935 kg and a design life of 10 years, and broadcasts the new civil signal, designated L3 centered at 1202.025 MHz, using the CDMA to ease interoperability with the GPS and Galileo. The three satellites were launched together into the orbits in February 2011, which would bring a total of satellites in the GLONASS constellation to 23 and achieve almost complete worldwide coverage.

Although the GLONASS constellation had reached a global coverage, its commercialization, especially development of the user segment, had been lacking compared to the GPS system. For example, the first commercial Russian-made GLONASS navigation device for cars, Glospace SGK-70, was introduced in 2007, but it was much bigger and costlier than similar GPS receivers. In the late 2010, there were only a handful of GLONASS receivers on the market, and few of them were meant for ordinary consumers. To improve the situation, the Russian government had been actively promoting the GLONASS for civilian uses, taking a series of measures, such as increasing the import duty on all GPS-capable devices including mobile phones, demanding the car manufacturers in Russia to use the GLONASS starting from 2011, funding the development of the GLONASS navigation base-band chips, developing new supplemental technologies, and maintaining navigational maps to attract more GLONASS users.

As of April 2020, there are 27 satellites in the GLONASS constellation, where 24 are normally operational, 2 are in-orbit spares and 1 is in flight test phase. In addition, a revised version of the GLONASS-K satellite, GLONASS-K2, is being expected to place into the constellation in 2020, with a mass of about 1800 kg and a design life of 10 years. The GLONASS-K2 will transmit eight navigation signals, L1OF, L1SF, L1OC, L1SC, L2OF, L2SF, L2SC and L3OC, where "O" stands for the open signal (standard precision), "S" for the obfuscated signal (high precision), "F" for the FDMA and "C" for the CDMA. Furthermore, a modernized GLONASS-K satellite, known as GLONASS-KM, is being designed at present, which will transmit the legacy FDMA signals on L1 and L2 and CDMA signals on L1, L2 and L3, and also do the CDMA signals on the GPS L5 frequency at 1176.45 MHz. According to the current schedule, the GLONASS-KM series will not be deployed to the constellation until 2025.

4.6.3.3 BeiDou System

In the early 1980s, China began to actively explore the space-based navigation system suitable for its own national conditions, known as **BeiDou Navigation Satellite System (BDS)**. In accordance with the development idea of "first region, then global" [10, 11], the BDS is implemented on the following three phases.

The first phase is from 1994 to 2003, to build BeiDou navigation satellite test system. In 1994, the Chinese government initiated the construction of the BeiDou test system project, known as BDS-1. Two BDS-1 satellites were successfully launched in October and December 2000, respectively, which marked the BeiDou test system was built, and thus China became the third country in the world with independent navigation satellite system. In May 2003, the third BDS-1 satellite was sent into the predetermined orbit to further enhance the performance of the BeiDou test system.

The BeiDou test system is a regional navigation and timing service system for China's territory and surrounding areas, and consists of three parts, space segment, ground segment and user segment. The space segment is a constellation with three Geostationary Orbit (GEO) satellites, which are located, respectively, at 80°E, 110.5°E and 140°E over the equator. The ground segment consists of one control center and many calibration stations, which are used to determine the BDS-1 satellite orbits, measure the distances between the satellites and ground stations, correct ionosphere, determine the BDS-1 user's locations, exchange the user's short messages, etc. The user segment consists of various types of BDS-1 user terminals, such as handheld terminal, vehicle-onboard terminal, ship-onboard terminal and command terminal, which are used to submit the positioning application to the control center, correct time, receive location data, short message, etc. The active navigation mechanism with two-way timing and ranging is used in the BeiDou test system, which can provide location determination, one-way or two-way time service, short-message communication and other services for medium and low dynamic users and static users within China and its surrounding areas, with the positioning accuracy of 20 m, the timing accuracy of 100 ns for one way and 20 ns for two way, and the number of single short-message communication words of 20 Chinese characters. The BeiDou test system has played an important role in promoting Chinese economic construction, social development and national security.

The second phase is from 2004 to 2012, to build the BeiDou navigation satellite system for regional service. In 2004, the Chinese government launched the project of building the BeiDou navigation satellite system, known as BDS-2. In April 2007, the first BDS-2 satellite was launched successfully, and as of October 2012, a total of 16 BDS-2 satellites had been sent into the orbits, forming the BDS-2 constellation with five GEO satellites, five Inclination Geo-Synchronization Orbit (IGSO) satellites and four Medium Earth Orbit (MEO) satellites. The five GEO satellites were located at 58.75°E, 80°E, 110.5°E, 140°E and 160°E over the equator. The five IGSO satellites were with an orbital altitude of 35,786 km and an inclination of 55° to the equatorial plane, where the three satellite sub-point tracks were coincident, with the equatorial intersection at 118°E, and the other two satellites sub-point tracks were coincident, with the equatorial intersection at 95°E. By using the constellation Walker 24/3/1, two of the four MEO satellites were put into the first orbital plane and the other two into the second orbital plane. The MEO satellites were with the orbital altitude of 21,528 km, the inclination 55° to the equatorial plane and the repeated period of 13 circles per 7 days. Moreover, the ground segment of the BDS-2 is composed of monitoring stations, master control stations, time synchronization and injection stations, which are used for the BDS-2 satellites tracking and monitoring, data processing, task planning, operation management, navigation message preparation, navigation message injected to the satellites, etc. The user segment consists of various types of BDS-2 user terminal equipment. The passive navigation mechanism with one-way timing and ranging is used in the BDS-2 system. In 2012, the BDS-2 system had the capability of affording regional continuous navigation, positioning and timing service, with the service area ranges from 55°S to 55°N and from 70°E to 150°E, and the horizontal and vertical positioning accuracies are better than 10 m (2-sigma).

The third phase is from 2013 to 2020, to build the BeiDou navigation satellite system for global service. After the BDS regional system built in 2012, the construction of the BDS project in the third phase was pushed forward continuously to achieve the global coverage of the BDS signals, which is also known as BDS-3. By inheriting the technical schemes of both active and passive services, the BDS-3 system can provide the positioning, navigation and timing, global short-message communication, and international search and rescue services for global users. Meanwhile, for regional users within China and its surrounding areas, it can also offer the satellite-based augmentation, ground-based augmentation, precise point positioning and regional short-message communication services. In March 2015, the first BDS-3 satellite (the 17th BDS satellite) was launched successfully. On June 23, 2020, the 55th BDS satellite was successfully launched into the orbit, which marks the BeiDou global navigation satellite system is officially built, forming the BDS constellation with 45 operational satellites, with the visible satellite number anywhere on the Earth's surface of more than 6.

The time reference for the BDS is called the BDS Time (BDT), which is a continuous timing system with a second defined in the International System of Units (SI) as a basic unit and without the leap seconds, and its initial epoch was defined at 0:00:00 UTC on January 1, 2006. The BDT is traced to the UTC through the UTC (NTSC), which is maintained by the National Time Service Center (NTSC) of China. The offset between the BDT and UTC is less than 50 ns (modulo 1 s). The leap second information between the BDT and UTC is broadcasted in the navigation message.

The BeiDou Coordinate System (BDCS) is an Earth-centered, Earth-fixed terrestrial reference system, which is defined in accordance with the specifications of the International Earth Rotation and Reference System Service (IERS) and aligned with the latest International Terrestrial Reference Frame (ITRF). The definition of the BDCS frame is: the origin is the center of Earth's mass; the *Z*-axis points to the direction of the IERS reference pole; the *X*-axis points to the intersection of the IERS reference meridian with the plane passing through the origin, and is normal to the *Z*-axis; the *Y*-axis together with the *Z*-axis and *X*-axis constitutes a right-handed orthogonal coordinate system; the length unit is with a meter of the SI; the orientation is given by the Bureau International de l'Heure (BIH) orientation of 1984.0; the time evolution in orientation will create no residual global rotation with regard to the crust. As of June 2020, the latest version of the BDCS is 2019V01, and the differences between the BDCS (2019V01) and ITRF2014 are with the Root Mean Square (RMS) of 0.74 mm on three coordinate axes [12].

The BDS has achieved the global signal coverage, with the horizontal positioning accuracy of 10 m (2-sigma), the vertical positioning accuracy of 10 m (2-sigma) and the timing accuracy of 20 ns (2-sigma). In the Asia–Pacific Region, the BDS positioning accuracies are 5 m (2-sigma) horizontally and vertically. With the full navigation capability of the BDS, a relatively complete application industry has been formed, including fundamental products, application terminals, system applications and operation services. The BDS application is gradually extended to various fields, such as transportation, marine fishery, hydrological monitoring, meteorological forecast, forest fire prevention, communication time system, power dispatching, disaster relief and reduction, which is bringing up a wide range social, economic and military benefits.

4.6.3.4 Galileo System

In March 2002, the European Union (EU) member states and the European Space Agency (ESA) finally reached an agreement after many discussions to launch the

European civil navigation satellite program, known as **Galileo navigation satellite system** [13]. In fact, the Galileo program was not officially funded by the EU and ESA until May 2003. The constellation Walker 27/3/1 was used in the Galileo system, with an orbital altitude of 23,616 km, an inclination of 56° to the equatorial plane and an orbital period of 14 h and 22 min, and three in-space spare satellites were added to the constellation. The implementation of the Galileo system was divided into the following three phases.

The first phase was from 2003 to 2010, to test the algorithms of the Galileo satellite's orbit determination and time synchronization. In 2004, the on-ground algorithms for the satellite's orbit determination and time synchronization had been validated by the project on the Galileo system test bed (version 1.0), which provided fundament knowledge to develop the space segment of the Galileo positioning system. In December 2005, the first Galileo In-Orbit Validation Element (GIOVE) test satellite, known as GIOVE-A, was successfully launched into the orbit with an altitude of 23,222 km, to ensure that the Galileo system meets the frequency-filling allocation and reservation requirements for the International Telecommunication Union (ITU), a process that was required to be completed by June 2006. In April 2008, the second Galileo test satellite GIOVE-B was launched, with a more advanced payload than the GIOVE-A. The Galileo navigation technique and equipment, such as navigation signals and algorithms, onboard atomic clocks, Satellite Laser Ranging (SLR) and space radiation environment, were tested and validated in orbits with the GIOVE-A/B. By analyzing the data from the GIOVE-A/B, it was confirmed that the Galileo signals could successfully be operated with the tracking performance as expected. The experimental results based on real data from the GIOVE-A/B were used to mitigate the risks for the In-Orbit Validation (IOV) satellites that would follow from the ground test beds.

The second phase is from 2011 to 2013, to launch the IOV Galileo satellites. The GIOVE-A/B satellites were followed by four IOV Galileo satellites that were much closer to the final Galileo satellite design, and the Search and Rescue feature was installed in the IOV satellites. In October 2011, the first two IOV satellites were launched from Guiana Space center using a Soyuz rocket, and the other two in October 2012. With launching the four IOV satellites, the key validation tests were enabled since ground-based receivers like those in cars and phones need to simultaneously track a minimum of four satellites in order to calculate their positions in three dimensions. On March 12, 2013, a first positioning was performed using the four IOV satellites. Once the IOV phase has been completed, the remaining satellites would be deployed to reach the Full Operational Capability (FOC).

The third phase is from 2014 to form the Galileo constellation composed of the FOC satellites. In August 2014, the first two FOC satellites were launched, each with a mass of about 710 kg, but were not injected into the intended orbit due to the rocket anomaly. In March 2015, the second two FOC satellites were launched successfully into the orbit with an altitude of 23,616 km. As of April 2020, there were a total of 22 FOC satellites launched, and the Galileo constellation has 22 operational satellites, including 19 FOC satellites and 3 IOV satellites.

The FOC Galileo satellites transmit 10 right-hand circularly polarized navigation signals and 1 search and rescue signal, where the navigation signals are broadcasted, respectively, in three carrier frequency bands, namely, 1164–1215 MHz, 1260–1300 MHz and 1559–1591 MHz, and the search and rescue signal is broadcasted in the reserved frequency band of emergency service, 1544–1545 MHz. When the Galileo constellation provides global continuous signal coverage, the ground user receiver can track up to 13 satellites at the same time. The Galileo system affords five types of basic services: Open Service (OS), Commercial Service (CS), Safety-of-Life (SoL), Public Regulated Service (PRS), and Search and Rescue (SAR) [14].

The OS is to provide free positioning, speed measurement and time service information to global users, with the horizontal positioning accuracy of 4 m (2-sigma) and the vertical accuracy of 8 m (2-sigma) for dual-frequency receivers, the timing accuracy of 30 ns (2-sigma) for the triple frequency receivers and the service availability of 99.5%, but the OS users cannot get the integrity information from the Galileo system, and their safeties using the Galileo system are not protected. The CS, which provides the development for professional application, can broadcast commercial encrypted information with high data rate, and obtains high-precision positioning and good system performance service, and its foreseeable applications are to the fields such as wireless communication network integration and indoor navigation. The SoL is with a navigation availability of 99.8% to ensure the safety of life, and its applications are to the fields endangering users' life and safety, such as land vehicles, sea voyage and aviation, which require rapid, timely and comprehensive system integrity information as well as high-level positioning, navigation and timing services. The PRS is to use the special frequency bands and signals to encrypt data information, and is able to get all information of the Galileo system, protecting the space signals of the Galileo system from being destroyed. The PRS is managed and controlled by specialized agency approved by the EU. The Galileo satellites carry the SAR payloads. Once distress alerting information is detected, it will be transmitted to the ground SAR center through the special frequency band to provide the disaster location and relevant information, so as to take rescue measures in time.

4.7 The Navigations for Spacecrafts

It is well known that the motion states of a spacecraft can be expressed by the position and velocity parameters of its center of mass, as well as the parameters of the rotational angles and their rates, in which the spacecraft rotates around its center of mass. The position and velocity parameters are used to express the moving trajectory of the spacecraft, commonly known as **orbital parameters**, and the rotational angle and its rates are used to express the spatial orientation of the spacecraft, commonly known as **attitude parameters**. That is to say, the spacecraft, regarded as a rigid body moving, has six Degrees of Freedom (DOF), three of which represent its orbital motion states, and the other three represent its attitude motion states. The spacecraft moving in space is always affected by various forces and moments, in which the orbital perturbations are brought up by the forces and the attitude disturbances by the moments. The **navigation for spacecraft** refers to a technique and method to guide the spacecraft along the designed orbit to fly to the target orbits or target celestial- bodies in accordance with certain laws, also known as **space navigation**. In order to effectively conduct the space navigation, safely guide the spacecraft for the flight along the designed orbits and directions, and complete the scheduled space flight missions in time, the top priority is to measure the orbit and attitude parameters and carry out the orbit and attitude controls for spacecrafts [15, 16].

4.7.1 Orbit and Attitude Determinations

The **spacecraft's orbit determination** refers to a technique and method to measure and estimate the three-dimensional position and velocity parameters of the spacecraft in a selected reference coordinate system. In general, the spacecraft's position and velocity parameters can be converted into six Keplerian orbital elements as shown in Fig. 4.6, including the right ascension (Ω) of ascending node, the inclination (i) relative to the equatorial plane, the semi-major axis (a), the eccentricity (e), the argument (ω) of perigee and the time (t) when the spacecraft passes through the perigee, where the parameters Ω and i are used to represent the spatial position of the orbital plane; the parameters a and e are used to represent the shape of the

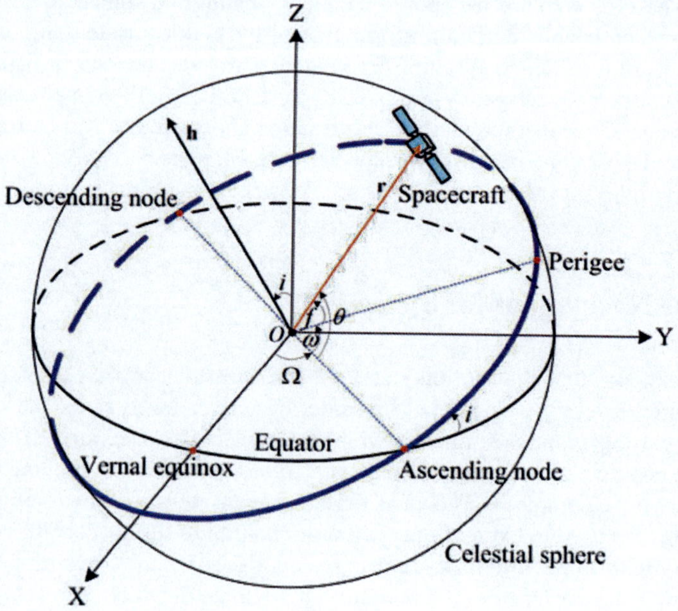

Fig. 4.6 Keplerian orbital elements of spacecraft

elliptical orbit; the parameter ω is used to represent the position of the elliptical orbit; the parameter t is used to represent the starting position of the spacecraft in the elliptical orbit. In fact, the time for the spacecraft to pass through the perigee contains two meanings: one is the moment and another is the corresponding position, the perigee. In practical application, the passing perigee time (t) is often replaced by the mean anomaly (M) or true anomaly (f). So, for a given moment, the spacecraft's orbit states can be determined completely by using the six orbital elements.

According to the dependence of spacecraft on the ground tracking and telemetry stations, the spacecraft's orbit determination can be divided into two major categories: one is **autonomous orbit determination** that the spacecraft autonomously estimates the position and velocity parameters by using the data from its own measurement equipment and that there are not two-way information exchange between the measurement equipment and the Earth or other celestial bodies, also known as **autonomous navigation**, such as the celestial navigation, inertial navigation and pulsar navigation; the other is the **non-autonomous orbit determination** that by tracking and measuring the spacecraft with the radio and optical facilities at ground TT&C stations, the spacecraft's position and velocity parameters can be obtained through data processing. For the spacecraft's navigation depending on the ground TT&C stations, if it is required to track and measure the spacecraft for an entire pass, the need number of the deployed ground stations is

$$\mathrm{N} = \frac{14400}{h} + 2, \tag{4.1}$$

where h is the spacecraft's orbital altitude, at a unit of km. Moreover, the ideal distribution of the ground stations is required, which will inevitably be beyond the territory of a country, or the ground stations have to be deployed on the sea. Obviously, it is uneconomical and unrealistic to continuously track the LEO spacecrafts by increasing the number of ground stations.

The spacecraft's attitudes are to express the transformation relationship between two coordinate systems with each other: one is the spatial reference coordinate system, referred to as r-system; the other is the spacecraft's body coordinate system, referred to as b-system. When the r-system is taken as the inertial coordinate system, the attitude of the spacecraft relative to the reference system is called **inertial attitude**, which is mainly used in astronomical observation satellites, and interplanetary and interstellar flight missions. When the r-system is taken as the orbital coordinate system of the spacecraft, the attitude of the spacecraft relative to the reference system is called **attitude to the Earth**, which is mainly used in navigation satellites, communication satellites, Earth remote-sensing satellites, etc. Certainly, for spacecrafts orbiting other celestial bodies, such as the Moon's probes and Mars' probes, the attitudes of the spacecrafts relative to the orbital coordinate system are the attitudes orientated to the corresponding celestial bodies. Therefore, the **attitude determination** is to study the attitude measurement method of the spacecraft relative to a certain reference system, such as the inertial reference system and the reference system to the Earth. Usually, using the attitude sensors, such as the Sun sensors, Earth sensors, star sensors and inertial gyroscopes, carried with the spacecrafts, the attitude parameters can be measured and estimated autonomously. The attitude estimation accuracy is closely related to the performances of the sensors and the methods of data processing. In general, there are three ways for expressing the spacecraft's attitudes: attitude matrix, Euler angles and quaternion.

When the spatial reference coordinate system (*r*-system) and the spacecraft's body coordinate system (*b*-system) are all right-handed orthogonal coordinate systems, there are nine direction cosines between the two sets of coordinate axes. Using the nine direction cosines, a direction cosine matrix can be formed to determine the geometric direction of the *b*-system relative to the *r*-system. The direction cosine matrix represents the coordinate rotation transformation relationship from the *r*-system to the *b*-system, called **attitude matrix**. Through the attitude matrix, the same direction vector can be represented, respectively, in the spacecraft body coordinate system and the reference coordinate system, connected with each other, and thus the basic orientation relationship of the *b*-system relative to the *r*-system can be determined completely. The attitude matrix is an orthogonal matrix, that is, there are six constraints among the elements of the attitude matrix. In other words, only three of the nine direction cosine elements in the attitude matrix are independent.

It is well known that the orientation of the *b*-system relative to the *r*-system has nothing to do with the relative displacements of their coordinate origins. Thereby, supposed that the origins of two coordinate systems coincide, the motion of the *b*-system relative to the *r*-system can be regarded as the fixed-point rotation of a rigid body. In the terms of **Euler's rotation theorem**, the general displacement of a rigid body (or a coordinate frame) with a fixed point is equivalent to a single rotation about a certain axis that passes through the fixed point, where this axis is called **Euler axis**. It also means that the composition of two rotations is also one rotation, and any rotation may be described using three angles, which are called **Euler angles**. Therefore, as long as the *r*-system has continuously made three rotations about its coordinate axes, it can be transformed into the *b*-system, and the angle of rotation each time is an Euler angle. Furthermore, the attitude matrix determined by the Euler angles is the product of the matrixes from three rotations. Apparently, the attitude matrix is related to the order of three rotations in the coordinate system. The rotation sequence of the coordinate system can usually be divided into two categories: one is the first and third rotations are about the same coordinate axis, the second rotation is about one of the other two coordinate axes, and the other is that each rotation is about a different coordinate axis. In this way, there are 12 kinds of combinations of the Euler angle rotation sequence for the two categories of rotations. The most commonly used Euler angles are obtained by the two kinds of rotation sequences, *Z-X-Z* and *Z-X-Y*, and correspondingly two sets of the Euler angles are labeled as ψ, θ and φ, as well as ψ, φ and θ, respectively.

In Fig. 4.7, both the *r*-system $O\text{-}X_rY_rZ_r$ and the *b*-system $O\text{-}X_bY_bZ_b$ are established with the same fixed point O as the coordinate origins, and their Euler angles are represented in accordance with the rotation sequence of *Z-X-Z*. When $t = 0$, the two coordinate systems are completely coincident. According to the right-hand

Fig. 4.7 Euler angles obtained by the rotation sequence of Z-X-Z

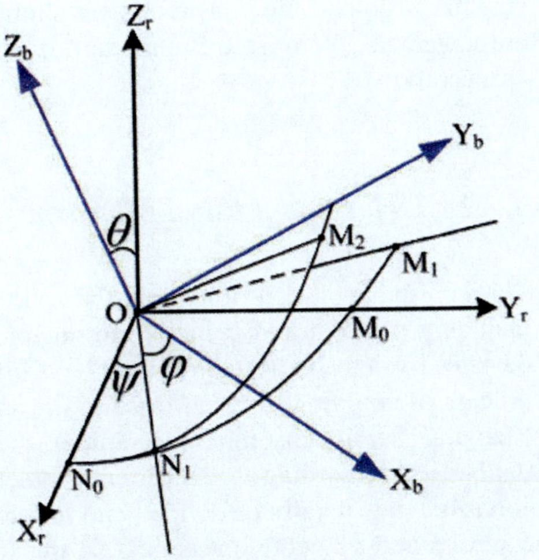

rule, firstly, the b-system $O\text{-}X_bY_bZ_b$ rotates an angle ψ (called **precession**) about the coordinate axis OZ_r, so that point N_0 on the coordinate axis OX_r arrives at point N_1, and the axis ON_1 is usually called **nodal line**. Secondly, the b-system $O\text{-}X_bY_bZ_b$ rotates an angle θ again (known as **nutation**) about axis ON_1, and at this time, two coordinate axes OZ_b and OZ_r coincide no longer, with the angle of θ between the two axes. Finally, the b-system $O\text{-}X_bY_bZ_b$ rotates an angle φ (known as **spin**) about the coordinate OZ_b, and thus the coordinate axis OX_b has left the nodal line of ON_1, rotating the angle of φ. So, at any moment, as long as the three Euler angles (ψ, θ and φ) are known, the spacecraft's space attitude can be uniquely determined.

Moreover, according to Euler's rotation theorem, it is also shown that a rigid body (or a coordinate system) with a fixed point has arbitrary finite motion, which can be achieved with a rotation around the axis of passing through the fixed point. Supposed the b-system is a result that the r-system rotates an Euler angle α about the Euler axis $\mathbf{u}$ (direction vector), the orientation relationship between the two coordinate systems can be expressed by a quaternion vector, namely,

$$\mathbf{q} = \begin{bmatrix} q_1 & q_2 & q_3 & q_4 \end{bmatrix}^T = \mathbf{u}\sin\frac{\alpha}{2} + \cos\frac{\alpha}{2}, \tag{4.2}$$

where $\mathbf{u} = u_x\mathbf{x}_r + u_y\mathbf{y}_r + u_z\mathbf{z}_r$ and $\mathbf{q}$ is the attitude quaternion vector.

In summary, among the above three expressions on the spacecraft's attitude, the attitude matrix is suitable for algebraic calculation in coordinate transformation, which is simple in form, but lack of geometric intuition and easy to generate multiple solutions. The geometric concept of the Euler angles to represent the coordinate system rotation is clear, but the rotation matrix contains trigonometric functions, which has nonlinear transformation relationship and is easy to cause the problem of singularity. The quaternion does not contain any trigonometric function, there is no singular problem, the constraints are simple and the attitude kinematics equations are linear. So, the quaternion method is usually used to determine the attitudes of spacecrafts.

4.7.2 *Orbit and Attitude Controls*

Through the orbit determination, the spacecraft's real-time navigation parameters including position and velocity are obtained to determine whether the spacecraft deviates from the expected orbits. The **orbit control** for the spacecraft refers to the process of applying external force to the spacecraft's center of mass, purposefully changing its trajectory, making it return to the expected orbits or fly to the target celestial bodies. According to the different practical applications, the spacecraft's orbit controls can generally be divided into four categories. One is the **orbit maneuver**, a control process that the spacecrafts are transferred from one free-flight orbit segment to another. For example, when the geostationary orbit satellite is launched, the orbit maneuver will be carried out at the apogee where the orbit is transferred. The other is the **orbit keeping**, a control process of overcoming the orbit perturbation to keep some orbital parameters unchanged for the spacecrafts, such as the station keeping for the GEO satellites, the formation keeping for navigation constellations, the orbital inclination keeping for the Sun synchronous orbit satellites and the repeated period keeping for the orbit-repeated satellites. The third is the **orbit rendezvous**, a control process that two spacecrafts reach the same position in space at the same time and at the same speed, such as the rendezvous and docking between spaceship and space station, etc. The fourth is the **re-entry**, a control process that the spacecraft leaves its original flight orbit and returns to the Earth's atmosphere, such as the re-entry for the spaceship back to the ground, etc.

Through the autonomous attitude determination, the spacecraft's real-time attitude parameters are obtained to determine whether their errors are within a tolerance range. The **attitude control** for the spacecraft refers to the process of applying the moment of a force to the spacecraft around the center of mass to maintain or change its spatial orientation, including attitude stabilization and attitude maneuver. The **attitude stabilization** refers to the control process of keeping the attitudes of spacecrafts in the specified direction, which requires the control system to work continuously during the lifetime of the spacecrafts, with the long-term and frequent characteristics and the small control moments required. The **attitude maneuver** refers to the reorientation process of the spacecraft from one attitude to another, which generally lasts for a short time and uses a great control moment to make the spacecraft's attitude change obviously. According to the different motion ways of spacecrafts, the attitude stabilization can be roughly classified into two categories: one is the **spin stabilization**, a control process that the spacecraft rotates about the spin axis and maintains the inertial spatial direction of the spin axis by using the spinning momentum torque; the other is the **three-axis stabilization**, the control process of keeping the three orthogonal axes of the spacecraft in a certain reference direction based on the active attitude control or using the environmental torque. For a dual-spin stabilized satellite composed of the spinning wheel (the rotor) and the de-spun platform, the spin axis is orientated by the rotor rotating rapidly about it, and connected with the bearing of the de-spun platform each other. With the spin axis orientating, the de-spun platform is controlled by the bearing motors to obtain a three-axis stabilization effect. For this reason, the payloads are usually assembled on the de-spun platform in practical applications.

In general, the spacecraft's orbit and attitude controls are associated with each other. Before implementing the orbit control, the spacecraft's attitude should meet the requirements for its orbit control. On the contrary, the implementation of the attitude maneuver for the spacecraft requires corresponding orbit conditions. Of course, for some specific case or special space missions, the orbit and attitude controls can be carried out independently. For example, the Earth space environment exploration satellite, which operates in the Kepler orbit, can meet the requirements for space exploration, without the need of implementing the orbit control, and only with the requirements for the attitude control. According to the different sources of force and torque for the spacecraft's orbit and attitude controls, the controls for the spacecraft are divided into two categories: one is the **active control**, the control of consuming the spacecraft's own energy to generate the control force and moment, which is generally realized by the closed-loop control system, composed of the spacecrafts and the ground facilities; another is the **passive control**, using space environment and spacecraft dynamics characteristics to provide the control force and moment, without consuming the energy for the spacecraft itself, such as using aerodynamic force and moment, solar radiation pressure, gravity-gradient moment and magnetic moment, all of which are also able to achieve the orbit and attitude controls for spacecrafts.

From practical applications, the active controls are the major means for the spacecraft controls, and usually include two kinds of control systems: autonomous control system and space-ground large loop control system. The autonomous control system is composed of the sensors, computers, controllers, actuators, and orbit and attitude dynamics units, carried by the spacecrafts, not relying on the ground TT&C stations, the spacecrafts are able to autonomously complete the control tasks, as shown in Fig. 4.8. For example, the autonomous controls are used in the de-spun controls for the dual-spin stabilized satellites, the attitude controls for the three-axis stabilized satellites and the east–west station keeping controls for the GEO satellites. The space-ground large loop control system is composed of the onboard remote control receiving units, actuators, orbit and attitude dynamics units, sensors and telemetry sending units, as well as the ground telemetry receiving equipment, tracking and measuring facilities for the orbits and attitudes, control center and remote control sending devices, as shown in Fig. 4.9. For example, the space-ground large loop controls are used in the attitude controls for the single-spin and dual-spin stabilized satellites, and the orbit controls for most of the spacecrafts. It is mainly because all of the spacecrafts have not achieved really autonomous navigation at present.

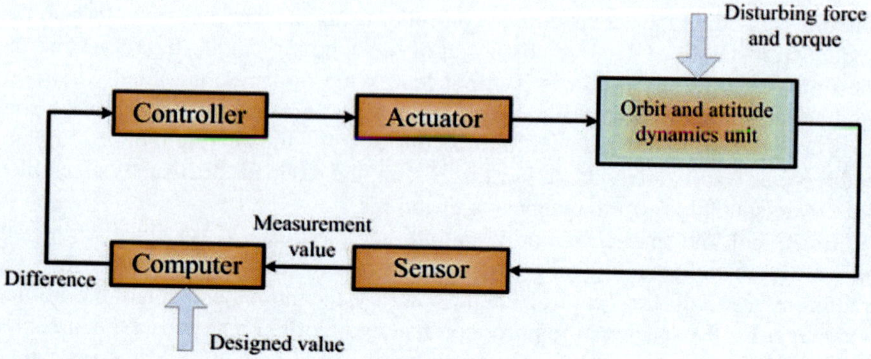

Fig. 4.8 Diagram of autonomous control system for spacecrafts

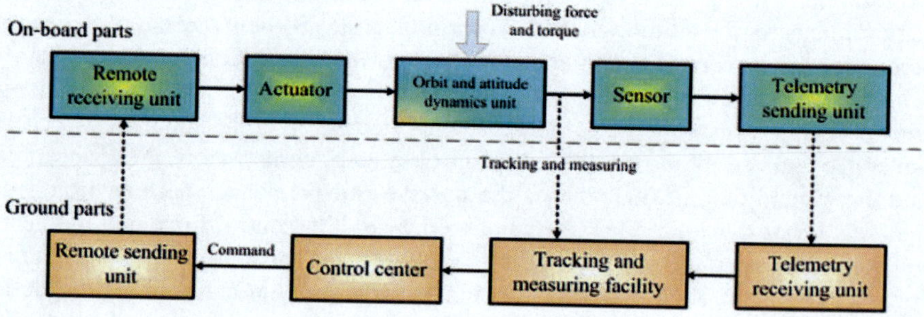

Fig. 4.9 Diagram of space-ground large loop control system

4.8 Ground Navigation Networks for Spacecrafts

4.8.1 Relation Between TT&C System and Navigation Network

4.8.1.1 Components and Functions of TT&C System

The **TT&C system** refers to a ground support system of tracking, monitoring, controlling and managing the spacecraft's motion and working states for a long time, which is an important part of large-scale space program. Its mission is to ensure the spacecraft to fly and work according to the pre-designed states, and to achieve the scheduled objectives of space flight. Moreover, the TT&C system also has voice, television and user data transmission and other communication functions, and sometimes, it is also called **TT&C and communication system**. From the perspective of system functions, the TT&C system consists of many subsystems, such as tracking and measurement, telemetry, remote control, space-ground communication and data transmission, data processing, monitoring and display, ground network communication, time reference and auxiliary support [17]. The tracking and measurement subsystem takes charges of making the TT&C antennas (or telescopes) point to spacecrafts, establishing the space-ground links, getting the spacecrafts' motion parameters like direction angles, distances and radial velocities, and finally determining the operational states of spacecrafts. The telemetry subsystem takes charges of receiving the measurement data from spacecrafts, collecting the data transmitted through the radio downlinks, storing and processing the data, and finally sending the data to the ground control center as required. The remote control subsystem takes charges of forming the injecting commands and data codes, and sending them to the spacecrafts via the radio uplinks, so as to conduct the real-time control or the program control for the spacecrafts. The space-ground communication and data transmission subsystem takes charge of transmitting the voice, message, television, image and special data between the spacecrafts and the ground TT&C stations, which is mainly used in manned spaceflights and remote-sensing satellites. The data processing subsystem takes charges of processing, analyzing and calculating the measurement data, and forming the injecting commands and data codes, so as to achieve the information exchange and equipment manage for the TT&C system. The monitoring and display system takes charges of collecting, processing, classifying and displaying the information sent from the data processing subsystem and concerned by the command and control personnel, so as to provide the reference basis for analyzing, deciding, commanding and controlling. The ground network communication subsystem takes charge of connecting the control center with the TT&C stations, and transmitting the information such as data, voice and image. The time reference subsystem takes charge of providing the unified standard time and frequency signals. The auxiliary support subsystem takes charge of providing the technical supports and logistics services, such as the meteorological data, geodetic data, power supply and air-conditioning environment for the control center and TT&C stations; the position, heading and attitude parameters for the tracking ships; the stable directing beams for the tracking and measurement antennas; and so on.

Tracking and measuring the orbits and attitudes of spacecrafts have always been the core tasks of the TT&C system, and thus the tracking and measurement subsystem is essentially the ground navigation system for spacecrafts, which usually adopts two kinds of tracking and measuring techniques, optical and radio. For the optical equipment, due to the disadvantages like strict observation condition and limited operating distance, it is unable to meet the requirements for the spacecraft's orbit determination for long-term operation, and thus the optical observations are usually used only in the stage of launching the spacecrafts into orbits. The radio tracking and measuring equipment, with the advantages of all-weather, long operating distance, easy information transmission and real-time data processing, usually plays a leading role in the TT&C system. According to the different working ways, the radio tracking and measuring equipment can be divided into two types, pulse and continuous wave. The pulse-type equipment is the measurement radar of transmitting radio-frequency pulse signals, with the advantages of simple device, as well as single station, single relay and multi-station working at the same time. The continuous wave-type equipment adopts the ways of continuously transmitting radio-frequency signals, which is easy to realize the comprehensive utilization of velocity measurements and carrier signals, mainly including Doppler velocity measurement unit, range and its rate measurement unit and microwave-unified unit. The microwave-unified units share the same microwave carrier channel and a set of antennas, functionally integrating the tracking and measurement together with the telemetry, remote control and communication, and thus it is widely used in the TT&C and communication for spacecrafts.

4.8.1.2 Framework of TT&C System

The tracks of projecting the spacecraft's flight orbits on the Earth's surface, usually known as **sub-satellite point tracks**, are mostly with the shapes round the globe, and only the sub-satellite point tracks of the geosynchronous orbit satellites are "8" shape. Usually, the TT&C equipment is deployed along the sub-satellite point tracks on the surface of the Earth. In order to achieve the tracking measurement and communication coverage with a large arc-segment for a spacecraft, it is necessary to consider the deployment of the TT&C stations within the global range and even build the space-based relay stations. From the perspective of system architecture, the TT&C system is composed of many TT&C units. The so-called **TT&C unit** refers to an organic combination of several functional TT&C subsystems, and as a relatively independent unit, it is usually placed in the appropriate site near the sub-satellite points to perform the space TT&C tasks with special functions. In general, there are two basic types of the TT&C units, control center and TT&C stations. In other words, the TT&C system is composed of one control center and a number of widely distributed TT&C stations. Each scattered node unit can be connected into an organic whole through the time-unified and communication, forming an information exchange network under the same time reference, which is called **TT&C network**, as shown in Fig. 4.10.

The **control center** is the central measurement and control unit for the TT&C network, consisting of the data processing, monitoring display, command communication, voice communication, data transmission, television monitoring and other equipment. According to the different division of tasks, the control center is also divided into the operation center of TT&C network, and the mission command and control center, where the former takes charges of the organization, coordination, operation and management of the TT&C network, and the latter takes charges of planning and organizing the spaceflight missions, as well as monitoring and operating the spacecraft's flight states. Currently, the TT&C network has achieved automatic operation control. The control center, TT&C stations and their devices are equipped with remote monitoring and controlling system. According to the requirements for the practical flight missions, the control center can dispatch and allocate the TT&C stations and their devices through the remote commands, and set the working parameters for the equipment. Meanwhile, the state parameters of TT&C devices can also be collected by the remote monitoring system of the TT&C stations, and sent to the

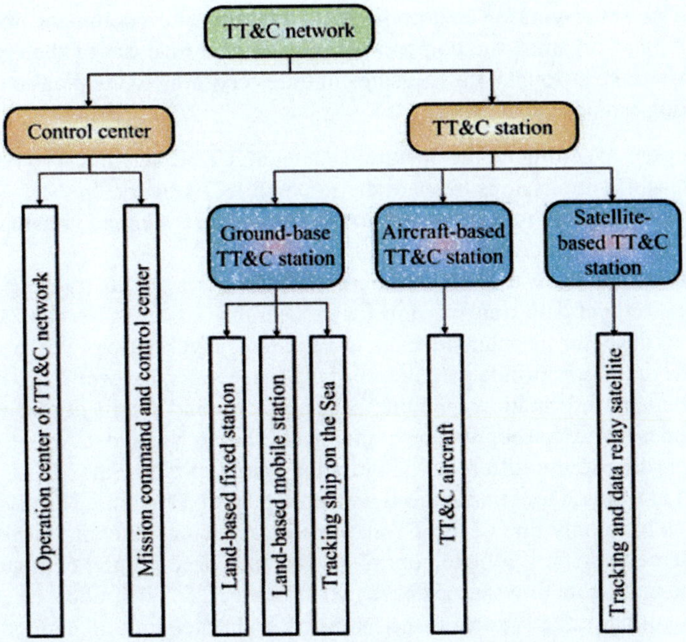

Fig. 4.10 Framework of TT&C network for spacecrafts

control center for analysis and processing, so as to check and monitor the status of the devices and ensure that the devices can normally perform the TT&C missions.

For the mission command and control center, its specific tasks are as follows:

(1) To make the TT&C plans in accordance with the spaceflight missions.
(2) To command and dispatch the spaceflight missions and the TT&C network.
(3) To monitor the operation status of spacecrafts, the working status of their instruments and devices, and the physiological status of astronauts.
(4) To communicate in voice with the astronauts and guide the astronauts for carrying out relevant works.
(5) To determine the orbit and attitude parameters of spacecrafts and the working state parameters of the onboard instruments and devices, by processing and analyzing the measurement data.
(6) To generate the injecting commands and data codes for the orbit maneuver, attitude adjustment, flight program-revised, parameter setting of the devices, so as to achieve the TT&C and management for spacecrafts.
(7) To improve the support service capability for the spacecrafts with the needs of special TT&C, according to the requirements for the users or the payload subsystems.
(8) To collect and send TT&C information, organizing the information exchange within the space engineering system and the TT&C network.

(9) To provide the working environmental conditions for the command and control staffs, and organize the comprehensive technical analysis of the spaceflight missions, so as to take the measures in time according to the plan or decision-making in the case of failure.

The purpose of setting up the operation center of TT&C network is to coordinate multiple spaceflight missions to share the same TT&C network. In the case of less TT&C stations and missions, the operation center can be combined with the mission command and control center.

The **TT&C station** is a TT&C unit that directly carries out the tracking, control, communication and data transmission for spacecrafts. Under the organization and command of the control center, the main tasks of the TT&C stations are: to track and measure the orbit and attitude parameters of the spacecrafts; to receive and demodulate the telemetry information from the spacecrafts; to send the injecting commands and data codes to the spacecrafts; to communicate and exchange data with the spacecrafts. According to the different configurations and functions, the TT&C stations are divided into single function stations and multi-function stations. The single function stations have only one of the TT&C functions, such as tracking measurement station, telemetry station, remote control station and space-ground communication station. The multi-function stations have two or more TT&C functions, usually with the typical configurations, such as the telemetry and space-ground communication station, the tracking measurement and telemetry and remote control station, and the integrated station with four TT&C functions.

According to the different deployment locations, the TT&C stations are divided into ground-based station, aircraft-based station and satellite-based station, where the ground-based station includes land-based fixed station, land-based mobile station and tracking ship on the sea. The **land-based fixed station** refers to the ground station with the fixed location on the land, which is usually equipped with large-scale infrastructure facilities, and the selection of its site location required with small shielding angle, good electromagnetic environment and convenient living condition (such as power supply, communication, transportation and water source), as well as considering suitable for various types of TT&C missions and observation arc-segments. The **land-based mobile station** is a kind of the TT&C station on board of the vehicle, that is, usually installing the TT&C equipment on its chassis or in its carriage, and the vehicle is driven to the designated place to perform the TT&C tasks. The mobile stations are mainly used for temporary supplementary measurements and covering some TT&C arc-segment gaps. It is very necessary to carry out the TT&C at the gaps for some spaceflight keypoints, such as the braking points for the recoverable satellites, the re-entry points for the manned spacecrafts and the arc-segments for the manned spacecrafts flying out of the ionization blackout. The mobile stations are usually deployed at the sites where the spaceflight missions are few frequent while it is unnecessary to set up the long-term stations, or which are located in the remote mountainous areas where it is not suitable to build the fixed stations. The **tracking ship** is a mobile TT&C station, which can freely be deployed in the most favorable TT&C sea area in accordance with the sub-satellite points and the requirements for

TT&C, so as to make up for the limitations of the land area, with the advantages of large range of activity, flexible deployment and high coverage of TT&C. The **aircraft-based TT&C station** is a kind of mobile TT&C station that is achieved by the special aircraft, mainly using for the TT&C tasks of the manned spacecraft in the section launching into the orbit and the returning section, to ensure the two-way voice communication between the control center and the spacecraft, to receive and record telemetry information, and to send the commands to the spacecraft when necessary. The **satellite-based TT&C station** is a mobile TT&C station achieved by Tracking and Data Relay Satellite (TDRS) in the GEO, and its space-ground TT&C link consists of the user spacecraft, relay satellite and ground TT&C terminals, of which the TT&C terminals are connected with the control center through the ground communication link to exchange the TT&C information. Generally, for the LEO spacecrafts with an inclination of 45° or so, the TT&C coverage of each orbit can be obtained by deploying three tracking ships on the oceans at three appropriate locations, respectively, which have the different geographic longitudes while the values of their latitudes nearly equal to the inclination value. Moreover, for the spacecrafts in the medium and low Earth orbits, only using two relay GEO satellites, with the located-longitude spaced 130° apart, whose information is transmitted by the same ground terminal station, the TT&C coverage for more than 85% of the orbital arc-segments can be achieved.

4.8.1.3 Classifications of Ground Navigation Networks

The TT&C network is a special network for tracking, measuring and controlling the spacecrafts after launching into orbits, and focuses on the navigation, control and management for the spacecraft's platform and the support services for the astronauts. Therefore, from the perspective of general navigation concept, the TT&C network is essentially the ground navigation network for spacecrafts, and from the perspective of special navigation concept, the ground navigation network is the TT&C network only with the function of tracking and measuring spacecrafts. In this book, although the concept of general navigation is used to define the ground navigation network, only the tracking measurement and control for spacecrafts are discussed in detail.

According to the different spaceflight missions, the ground navigation network can be divided into three basic types, respectively, providing services for the spacecrafts in near-earth space, the navigation constellations and the deep-space explorations. The navigation network for the spacecrafts in near-earth space is usually a tracking and measuring network for the spacecrafts in the LEO, MEO, GEO and HEO, except for the navigation satellites. Due to the restriction of the straight-line propagation of electromagnetic wave and the curvature of the Earth, using single TT&C station, the tracking duration of each circle is generally about 4–7 min for the LEO spacecrafts, about 15 min for the MEO ones and more than 1 h for the GEO or HEO ones. The TT&C frequencies are L-band, S-band and C-band, and the diameter of microwave TT&C antennas is generally no more than 10 m. Based on the design idea of "one network and multi-purpose", the navigation network for the spacecrafts is planned and designed to meet the needs of various spaceflight missions in near-Earth space.

The ground navigation network for the navigation constellations is a special network for tracking, measuring and operating the navigation constellations, and generally does not undertake the TT&C tasks for other spacecrafts. For example, the GPS, GLONASS, BeiDou and Galileo constellations have their own navigation networks, commonly referred to as the **control segments** or the **ground control networks**. The special navigation frequency L-band is used to track and measure the navigation satellites, while the TT&C frequency is S-band or C-band.

The ground navigation network for the deep-space explorations is a special network designed for exploring the Moon, interplanetary space and interstellar space, with the characteristics of continuous tracking, long navigation duration and far distance for the measurement equipment to work. Consequently, it usually needs to use a large diameter antenna of more than 25 m. Furthermore, considering the influence of background noise in the Galaxy, the TT&C frequency should be transformed from L-band to S-band. For the navigation constellations and the deep-space explorations, the deployments and tracking measurements of their ground navigation networks will be discussed in the last two parts of this section, respectively.

4.8.2 *Ground Control Networks of GNSS*

It is well known that the Global Navigation Satellite System (GNSS) is composed of the space constellation segment, ground control segment and user terminal segment, where the ground control segment is implemented by a ground control network worldwide, taking charges of tracking and measuring the operation status of the navigation satellites, monitoring their health status, generating the navigation message and injecting the message to the navigation satellites. The network is essentially to provide a navigation function for the space constellation. From the above section, it is known that there are four major GNSSs, the GPS, GLONASS, BeiDou and Galileo, whose control segments all have similar capability and function. Thereby, as examples, the ground control networks represented by the GPS and GLONASS system are introduced here in detail.

4.8.2.1 Ground Control Segment of GPS Constellation

The ground control network (or the ground segment) takes charge of the proper operation for the GPS system, and its main tasks are: to monitor and control the orbital parameters of the satellites, and their health and subsystem status, such as the solar arrays, the battery power and the level of propellant used for maneuvers; to activate the spare satellites; to update the navigation message like ephemeris, almanac and clock correction parameters; to treat the satellite's anomalies; to control the SA and Anti-Spoofing (A/S); to track passively the satellites. The network consists of the monitor stations, master control stations and injection stations (also known as the dedicated ground antennas). The monitor stations take charges of tracking and measuring all GPS satellites passing over the stations, obtaining the pseudorange, carrier phase and navigation message, measuring the atmospheric parameters like temperature, humidity and air pressure, and finally transmitting all data to the master station for processing. Using the measurement data from the monitor stations, the master station takes charges of calculating the precise orbits and clock biases of the GPS satellites, forming the navigation messages and commands for the GPS satellites, and finally transmitting them to the injection stations. Through the S-band antennas, the injection stations take charges of injecting the navigation messages and commands to the GPS satellites, and receiving the telemetry information sent by the satellites.

Since beginning to form the GPS constellation in 1978, as to April 2020, there are a total of 74 GPS satellites with 7 types launched, including 11 Block I, 9 Block II, 19 Block IIA, 13 Block IIR, 8 Block IIR-M, 12 Block IIF and 2 Block III. With the GPS satellites' modernizing, the optimized constellation operating and the User Ranging Error (URE) decreasing, the performance of the GPS system is improved gradually. In order to support the evolution of the platform and payload technologies of GPS satellites, it is required to continuously improve the ground control network. The development of ground control network for GPS constellation is divided into the following four phases.

(1) From 1978 to 1985 was the demonstration phase. The ground closed-loop control network was initially set up to support the on-orbit operation of GPS Block I satellite series mainly through manual operation.
(2) From 1985 to 2007 was the full-constellation operation phase. The first generation of ground control segment with basic network architecture was built, which has the capability to support the coordinated operation of 10 network nodes to ensure the on-orbit operation of GPS Block II, IIA, IIR and IIR-M satellite series.
(3) From 2007 to 2016 was the capability-improved phase. The second-generation ground control segment with a network architecture facing users was built, which can support more than 10 users of network nodes, to ensure the on-orbit operation of the modernized GPS satellites, such as Block IIR-M and IIF satellite series.
(4) From 2016 entered the new capability-upgraded phase. The Next-Generation Operational Control System (OCX) will be built, with the network architecture of central interface, modularization, flexibility and expandability, and the capability of supporting more than 100 network node users, to ensure the on-orbit operation of current GPS satellites and transit to the TT&C tasks of the GPS III satellites.

The GPS OCX will dramatically enhance GPS command, control and mission capabilities and provide significant benefits to military and civilian users around the world, and its development is following an incremental approach, with three versions of Block-0, Block-1 and Block-2. The Block-0 is the launch and control system intended to control launch and early orbit operations and the on-orbit checkout of all GPS III satellites. It is a subset of the Block-1 providing the hardware, software and cyber-security base for the Block-1. The Block-1 fields the operational capability to control all legacy satellites, civil signals and military signals, as well as the modernized military signals, the modernized civil signal, the aviation safety-of-flight signal, the GPS III satellites and the globally compatible signals, to fully meet information assurance and cyber-defense requirements. The Block-2 fields the advanced operational capability to control the advanced features of the modernized military signals. The OCX program represents a critical part of the GPS modernization, and its specifically final capabilities are as follows:

(1) To control and manage the legacy GPS satellites and the next generation of GPS III satellites, while enabling the fully array of military signals.
(2) To build on a flexible architecture that can rapidly adapt to the changing needs of current and future GPS users allowing immediate access to the GPS data and constellation status through secure, accurate and reliable information.
(3) To provide the war fighter with more secure, actionable and predictive information to enhance situational awareness.
(4) To enable new modernized signals like L1C, L2C and L5, and has M-code capability, which the legacy system is unable to do.
(5) To provide significant information assurance improvements over the current program including detecting and preventing cyber-attacks, while isolating, containing and operating during such attacks.
(6) To support higher volume near real-time command and control capabilities.

At present, the GPS control network is composed of 2 master control stations, 16 monitor stations and 12 injection stations, as shown in Fig. 4.11. The master control station, located at Schriever Air Force Base (AFB) in Colorado Springs, is the central control node for the GPS constellation, which is operated and managed by the 2nd Space Operation Squadron (2SOPS) of the U.S. Air Force, while the alternate master control station is located at Vandenberg AFB in California. Using the information from the monitor stations, the master control station computes the orbit and clock parameters of GPS satellites, generates the navigation message, monitors the integrity of navigation message and system, and evaluates the health status of GPS constellation. The 2SOPS uses the master control station to perform the satellite's maintenance and anomaly resolving. In the event of a satellite failure, the master control station can reposition the satellites to maintain an optimal GPS constellation. Six of the sixteen monitor stations are specially deployed by the U.S. Air Force, respectively, located at Schriever AFB in Colorado Spring, Hawaii and Cape Canaveral in Florida within the US, as well as Ascension Island in the South Atlantic Ocean, Diego Garcia Island in the central Indian Ocean and Kwajalein Atoll in the west-central Pacific Ocean, for tracking the GPS constellation. The other ten monitor stations are deployed by the U.S. National Geospatial-intelligence Agency (NGA), respectively, located at Fairbanks in Alaska and USNO Washington within the US, as well as Quito in Ecuador, Buenos Aires in Argentina, Hermitage in the United Kingdom, Manama in Bahrain, Pretoria in South Africa, Osan in South Korea, Adelaide in Australia and Wellington

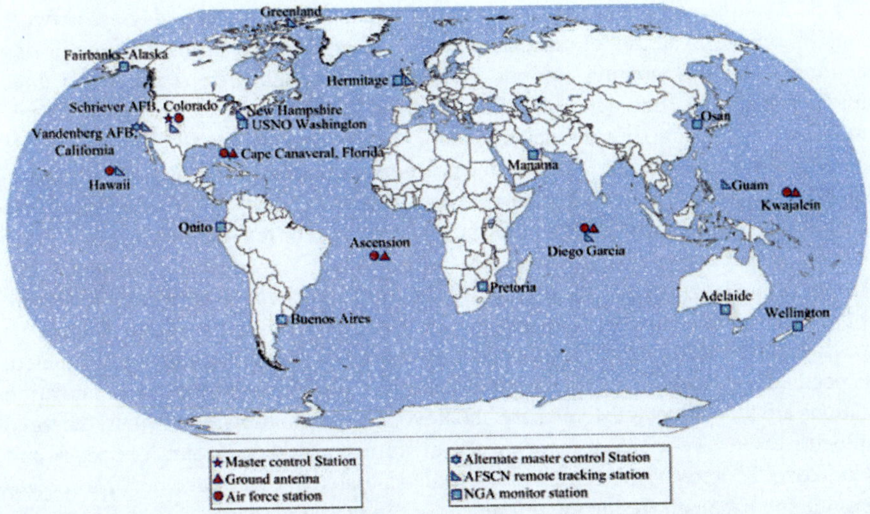

Fig. 4.11 Ground control network of GPS constellation

in New Zealand. The purpose of deploying the 16 monitoring stations is to ensure that each GPS satellite can be tracked and measured by more than three stations at the same time, to compute more precise orbit and ephemeris parameters and further improve the system accuracy. There are four of the twelve injection stations co-located with monitor stations, respectively, at Cape Canaveral, Ascension Island, Diego Garcia Island and Kwajalei Atoll, which are specially designed and configured for operating the GPS constellation. The other eight injection stations are the remote tracking stations worldwide, which are subordinated to the Air Force Satellite Control Network (AFSCN), respectively, located at Schriever AFB in Colorado Springs, Vandenberg AFB in California, Hawaii and New Hampshire within the US, as well as Greenland Island in the North Pole region, Hermitage in the UK, Diego Garcia Island in the central Indian Ocean and Guam in the western Pacific Ocean, to increase the visibility, flexibility and robustness for commanding and controlling the GPS satellites. The ground antennas equipped at the monitor stations uplink the commands and data codes to the GPS satellites via S-band radio signals and collect the telemetry. The information can be uploaded to each satellite three times per day, i.e., every 8 h, and nevertheless, it is usually updated just once a day. Moreover, using the S-band ranging data from the injection stations, the 2SOPS is allowed to provide anomaly resolution and early orbit support.

4.8.2.2 Ground Control Segment of GLONASS Constellation

Like the GPS, the GLONASS ground segment takes charges of tracking, monitoring, operating and managing its constellation, determining the satellite ephemeris and clock parameters, and uploading them to the satellites twice a day. Its network consists of one System Control Center (SCC), five TT&C stations, ten monitoring stations, two laser ranging stations, two time synchronization stations and three upload stations, as shown in Fig. 4.12, where the SCC is located at the town of Krasnoznamensk (Краснознаменск) in Moscow Oblast, Russia; the five TT&C stations are located at Schelkovo (Щёлково), Komsomolsk-on-Amur (Комсомольск-на-Амуре), Saint Petersburg (Санкт-Петербу́рг), Ussuriysk (Уссур´ийск) and Yeniseysk (Енисе́йский) within Russia, respectively; the ten monitoring stations are located, respectively, at Schelkovo, Krasnoznamensk, Yeniseysk, Komsomolsk-on-Amur, Yakutsk (Якутск), Ulan-Ude (Улан-Удэ), Vorkuta (Воркута́), Murmansk (Му́рманск) and Bolshoy Zelenchuk (Большой Зеленчук) within Russia, as well as Nurek (Норак) in Tajikistan; the two laser ranging stations are located, respectively, at Schelkovo and Komsomolsk-on-Amur; the two time synchronization stations are located, respectively, at Schelkovo and Komsomolsk-on-Amur; the three upload stations are located, respectively, at Komsomolsk-on-Amur, Yeniseysk and Schelkovo. Moreover, the central clock is also situated at Schelkovo near Moscowto provide the time and frequency reference for the whole network.

The SCC takes charges of controlling and managing the GLONASS constellation, providing the telemetry, command and control function for the satellites, and coordinating all functions and operations at the system level. Its core tasks are to process the information and measurement data from the TT&C stations, the monitor stations and the laser ranging stations; monitor the satellite's states; compute the orbit and clock parameters; generate the navigation message and commands; send them to each

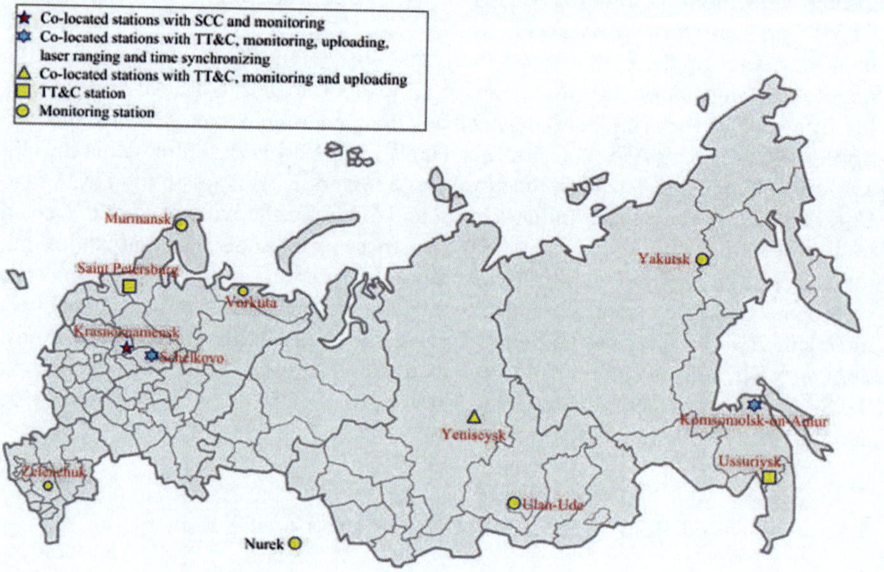

Fig. 4.12 Ground Control Network of GLONASS Constellation

GLONASS satellite via the upload stations. The central synchronizer implements the GLONASS system timescale by the means of a high-precise hydrogen atomic clock. The Russian State Etalon UTC allows the synchronization of the cesium atomic clocks onboard the GLONASS satellites.

The TT&C stations track the GLONASS satellites in view, and collect ranging data and telemetry from the satellites. All of the ranging data and telemetry are transmitted to the SCC for processing and analyzing. Moreover, both the monitor stations and the laser ranging stations track and measure the satellites, whose measurement data is used to precisely determine the orbit and clock parameters of each GLONASS satellite at the SCC. As a single source of calibration data for the GLONASS ephemeris determination, the advantages of the laser ranging stations are: to provide the means of accuracy estimating and radio-frequency calibrating for the GLONASS satellite's orbit measurements; to monitor the onboard clocks and allow use of the data for operational control of the GLONASS time and ephemeris data, with geodetic-class RF navigation receivers connected to hydrogen maser frequency standards; to use their site coordinates as a geodetic base for the GLONASS reference frame; to present the declared value of ephemeris precision for the satellites using the laser ranging data.

4.8.3 Ground Deep Space Networks

According to the definition of "Deep Space" by the ITU, the space range of more than 2×10^6 km away from the Earth's surface is called **deep space**. In the terms of this definition, the deep-space exploration frequency band specified by the ITU would not be used in the exploring activities for the Moon and even the Lagrange Points L1 and L2 in the Sun–Earth space. From the current situation of space activities, the spacecrafts that explore the Moon, planets and interplanetary space are called **deep-space probes**, and the ground networks that track, navigate and communicate for the deep-space probes are called **ground deep-space networks** or **ground navigation networks** for the deep-space probes. The networks have the functions of tracking and measuring the probe's orbits, receiving and processing the telemetry information and scientific data from the probes, and monitoring and controlling the operational states of the probes. Generally, according to the difference of basic working units, the deep-space TT&C network can be divided into deep-space stations, communication links and computing control center in its architecture. According to the different uses of the TT&C units, the network is composed of many subsystems, such as tracking, downlink telemetry and demodulation, digital detection information demodulation, uplink command modulation, uplink data modulation, communication, data processing, operational control and simulation. According to the different using ranges and exploring objects, the networks are divided into the TT&C networks for lunar exploration and for planets and interstellar space explorations, and between the two networks, there are discrepancies in acting range, principle of measuring angle and operational control. Compared with the TT&C for the spacecrafts in the near-Earth orbits, the difficulty of deep-space exploration is to solve the TT&C

of ultra-long distance, and to carry out the transmission of telemetry information and scientific data, and the tracking and control task for the probes. Since 1957, the United Statesand the Soviet Union (later Russia) have been the leading forces of planetary exploration. More than 200 unmanned planetary probes have been launched to explore planets and their interplanetary space in the solar system, so as to demonstrate the technical strength for space exploration. As a result, their own deep-space TT&C networks have been established successively. In the past 20 years, the deep-space exploration has once again become a research hotspot in the field of space science and technology. Besides the American and Russian deep-space TT&C networks, ESA, China, France, India and Italy are also vigorously promoting the constructions of their deep-space networks to meet the tracking, navigation and communication task requirements for deep-space probes. As a represented example, the NASA Deep Space Network (DSN) is introduced in the following.

The DSN is a worldwide network of large antennas and communication facilities, composed of three deep-space communication complexes, which are located, respectively, at California in the Unites States, Madrid in Spain and Canberra in Australia, to support interplanetary spacecraft missions. The antennas at the three DSN complexes communicate directly with the Deep Space Operations Center located at the JPL facilities in Pasadena, California. Moreover, the DSN also performs radio and radar astronomy observations for the exploration of the solar system and the universe, and supports selected Earth-orbiting missions. The three DSN stations are situated in semi-mountainous, bowl-shaped terrain to help shield against radio-frequency interference, and placed approximately 120° apart in the geographic longitude around the Earth, to permit constant observation for spacecrafts as the Earth rotates, and make the network the largest and most sensitive scientific telecommunication system in the world. Viewing from the North Pole of the Earth based on the three-site configuration, the spacecraft with a distance of more than 30,000 km away from the Earth's surface is always tracked by one of the stations at least. Unlike tracking LEO spacecrafts, the deep-space missions are usually visible for long periods of time from a large portion of the Earth's surface, and thus only require a few of TT&C stations, but the huge aperture antennas, ultra-sensitive receivers and powerful transmitters need to be equipped at the stations, in order to transmit and receive the commands, telemetry information and scientific data over the vast distance involved. Each of the three DSN stations has many deep-space terminals equipped with ultra-sensitive receiving system and large parabolic-dish antennas, including one 70-m antenna, one 34-m high-efficiency antenna, one 26-m antenna, as well as two or more 34-m beam waveguide antenna, where there are three beam waveguide antennas at the Goldstone Station in California, two at the Robledo de Chavela Station in Madrid and two at the Tidbinbilla Station in Canberra.

The DSN is a NASA facility, managed and operated for the NASA by the JPL, which is a part of the Cal Tech. The Interplanetary Network Directorate (IND) manages the program within the JPL and is charged with its development and operation. The IND is considered to be the JPL's focal point for all matters relating to telecommunications, interplanetary navigation, information system, information technology, computing, software engineering and other relevant technologies. The

IND is not only best known for its duties relating to the DSN, but also maintains the advanced multi-mission operation system and the institutional computing and information services for the JPL. For the DSN, the specific TT&C tasks are: to acquire the telemetry data from spacecrafts, to send TT&C commands to the spacecrafts, to upload the updated software to the spacecrafts, to carry out the VLBI, to conduct radio scientific tests, to detect radio wave anomalies, to collect scientific detection data, to monitor and control the network's performance, etc.

Since the beginning of the Voyager Interstellar Missions in the early 1990s, the general capabilities of the DSN have not substantially changed. Certainly, the advancement in digital signal processing, arraying and error correcting has been adopted by the DSN. The capability to array several antennas was incorporated to improve the data returned from the Voyager-2 encountering the Neptune. Later, the capability was extensively used in the Galileo exploration mission for the Jupiter, when the high-gain antenna was deployed incorrectly. Since performing the Galileo mission, the DSN array is currently available, in which the 70-m dish-shaped antenna at the Goldstone Station in California can link with an identical antenna located in Australia, and two 34-m antennas at the Tidbinbilla Station in Canberra. The California and Australia sites had been used concurrently to pick up communications with the Galileo spacecraft, which was an American automatic space probe that studied the Jupiter and its moons, as well as other celestial bodies in the solar system, sent into the orbit in 1990. Arraying of antennas within the three DSN stations is also used. For example, a 70-m dish-shaped antenna can be arrayed with a 34-m dish-shaped antenna. For especially vital missions, like the Voyager probes, the Canberra 70-m dish can be arrayed with the Parkes Radio Telescope in Australia, and the Goldstone 70-m dish is arrayed with the very large array of antennas in New Mexico. In addition, two or more 34 m dishes at the same DSN station are commonly arrayed together. All of the stations are remotely operated from a centralized signal processing center at each complex. These centers house the electronic subsystems that point and control the antennas, receive and process the telemetry data, transmit commands and generate the navigation data for spacecrafts. Once the data is processed at the complexes, it is transmitted to the JPL for further processing and for distribution to the relevant science teams over modern communication network.

There are a number of limitations to the current DSN, and a number of challenges going forward. It is noted that the Deep Space Network is something of misnomer, as there are no current plans, nor future plans, for exclusive communication satellites anywhere in space to handle multiparty, multi-mission uses. All of the transmitting and receiving equipment are the ground based. Thereby, data transmission rates from and to any and all spacecrafts and space probes are severely constrained due to the distances from the Earth. The DSN is required to support the legacy missions that have remained operational beyond their original lifetime but are still returning scientific data. For example, the Voyager missions have been operating long past their original mission termination date. The programs like the Voyager also need some of the largest antennas. Replacing major components can cause problems as it will leave an antenna out of service for months at a time. The older 70-m antennas are reaching the end of their lives. At some point these will need to be replaced.

The leading candidate for 70-m replacement had been an array of smaller dishes, but more recently the decision was taken to expand the provision of 34-m beam waveguide antennas at each station to a total of four. As of 2020, the DSN is required to support twice the number of missions compared with that in 2005. However, due to the limited finance support, and the decay and lack of replacement of the existing antennas, the support for the increased spaceflight missions will continue to be an ongoing problem. For this reason, the new spacecrafts intended for missions beyond geocentric orbits are being equipped to use the beacon mode service, which allows such missions to operate without the DSN most of the time.

For the radio navigation in ultra-deep space, the navigation error increases linearly with the increase of TT&C distance. As increasing one Astronomical Unit in the TT&C distance each, the ranging error will increase by 4 km. Furthermore, the TT&C delay for the DSN is so great that the DSN lacks the capability of real-time navigation and control for deep-space crafts. Before 2030, the navigation accuracy requirements for the NASA deep-space exploration missions are shown in Table 4.1 [18]. It can be seen from the table that in the next 10 years, the NASA requires the orbit determination accuracy of deep-space crafts to reach the level of 100 m. The JPL has successively upgraded the antennas of the DSN stations. For example, the reconstruction of the Tidbinbilla TT&C station was completed in 2014, and then the Goldstone and Robledo de Chavela stations were reconstructed in 2016, to further improve the TT&C capability of the DSN. Obviously, it is difficult to meet the requirements of navigation accuracy indicators for the future deep-space missions, by using the existing DSN, or the traditional navigation means like the CNS and

Table 4.1 Requirements of navigation accuracies for NASA's deep-space explorations (**Notes*: The landing accuracy is expressed by the error ellipse)

Accuracy \ Year	2005	2010	2020	2030
Near-earth space or Earth-Mars mid-course trajectory (km)	2	2	1	0.5
Interplanetary flight trajectory (km)	20	20	10	2
Orbits around the planets (km)	6.75	1.50	1.00	0.25
Landing on the inner planets* (km)	21×5	7×7	1×1	0.1×0.1
Orbit maneuver (m, on the radial)	330	0.5	0.1	0.1
Landing on asteroid's surface (m)	—	3×3	2.5×2.5	2.5×2.5
Lander positioning (m)	10	10	1	1

INS. So, the NASA considers using the X-ray pulsar-based navigation technology to enhance the autonomous navigation ability for deep-space crafts, so as to greatly improve the navigation accuracy using the DSN in the direction perpendicular to the line of sight. In addition, the X-ray pulsar-based navigation provides an independent TT&C way that does not relay on the Earth-based DSN, so that the ultra-deep space crafts can finally achieve the autonomous navigation.

4.9 Autonomous Navigation Technology for Spacecrafts

4.9.1 Concepts of Autonomous Navigation

For a spacecraft, the **autonomous navigation** refers to the process in which the spacecraft without the support from the ground-based TT&C system for a long time, relying on its onboard devices, obtains all kinds of measurement data; determines the navigation parameters like orbit, time and attitude; guides automatically itself operation; completes finally the scheduled missions. In general, the spacecraft's autonomous navigation should have five basic characteristics: navigation information integrity, real-time operation, no signals radiated, un-relying on the Earth-based stations and long-time automatic operation. The autonomous navigation of spacecraft has extremely important practical application value and theoretic research significance. On the one hand, it can reduce the workload of ground TT&C network, the number of the ground TT&C stations required, the frequency of information injection from the ground stations to spacecrafts and the cost of building the spacecraft (or constellation) systems and keeping them operation for a long time. On the other hand, it can reduce the dependence of spacecrafts on the ground TT&C network, and enhance the capability of the system's anti-jam and autonomous survivability. Therefore, the spacecraft's autonomous navigation has been always regarded as the power and goal for the world's major space countries to promote the development of astronautic technology.

Since the 1960s, based on traditional celestial navigation methods, the United States, the Soviet Union (later Russia) and other countries have proposed many autonomous navigation systems for spacecrafts, and developed a variety of the onboard measurement devices, such as the horizon scanning, landmark tracker, tar sensor, space sextant, MANS and microwave radar altimeter. However, most of the research results have been in the stage of system simulation and flight test, and have not been put into practical application. Although the DORIS system developed by France and the PRARE system developed by Germany can determine the spacecraft's orbit with high accuracy, both need to exchange information with the ground stations, and hence do not belong to the above-defined autonomous navigation for spacecrafts. In the past 20 years, with the development and wide application of the GPS system, the GPS has become the main means to determine the precise orbits of LEO satellites, but the GPS constellation itself needs the ground control network to upload the commands and navigation message every day to maintain operation for a long time. So, the GPS system is usually considered to be a kind of semi-autonomous navigation way for the LEO satellites. From the basic characteristics of the autonomous navigation for spacecrafts, the spacecrafts have not achieved the autonomous navigation in the real sense at present, and most spacecrafts still rely on the ground TT&C system to complete the navigation tasks. It fully shows the complexity and difficulty of autonomous navigation technology for spacecrafts.

For the navigation constellations, the requirements for their orbit determination and time synchronization accuracies are higher than other spacecrafts. Therefore, it is very significant to achieve the autonomous navigation for the navigation constellations, and its benefits are: to reduce the number of ground control stations; to be able to keep the normal operation of the constellation without the worldwide monitor stations; to reduce the frequency of upload information from the injection stations to the navigation satellites; to reduce the cost of the construction and long-term maintenance of the constellation; to monitor the integrity of navigation information in real time and enhance the availability of the system; to reduce the dependence of the navigation satellites on the ground control network, and enhance the anti-jam and autonomous survivability of the system. Furthermore, with the support of the ground control network, the crosslink measurements within the constellation provide an independent means to examine the ephemeris and clock parameters of the satellites, and through joint data processing, further improve the system performance and navigation accuracy.

Generally, there are two key indicators used to evaluate the performance of the autonomous navigation for navigation constellations: one is autonomous operational time and the other is User Ranging Error (URE). The URE refers to the root mean square of the broadcast ephemeris errors and clock biases of the navigation satellites projected along the line-of-sight direction. The URE dynamically changes over time, whose reasons are: on the one hand, the elevation angles of the satellites are always changing; on the other hand, the predicted ephemeris errors and clock biases increase over time. As a result, the predicted ephemeris and clock parameter accuracies are closely associated with the Age of Data (AOD) for the navigation satellites.

4.9.2 Autonomous Navigation of GPS Constellation

In the early 1980s, the GPS system was still in the phase of engineering test. On the basis of launching the GPS Block I satellites and conducting the navigation test, the JPO proposed specific requirements for the development of payloads of the new generation GPS Block II satellites: they could provide high-precision navigation and positioning services, and also have the real-time monitoring capability of global nuclear test. As a result, researchers from the University of Texas, Aerospace Corporation, National Bureau of Standards, International Business Machines Corporation (IBM) and Rockwell Automation Inc. soon proposed a nuclear explosion monitoring proposal based on GPS inter-satellite communication link. The basic process of the nuclear explosion monitoring through the GPS satellites is shown in Fig. 4.13, from which it can be seen that the inter-satellite link is the core technology of the nuclear explosion monitoring, and the abnormal event information is quickly transmitted to the ground information collection center through the inter-satellite link.

In the mean time, the researchers also discussed how to improve the orbit determination accuracy of navigation satellites by using the information from the two-way ranging and communication crosslink between the satellites, and theoretically demonstrated the feasibility of autonomous navigation for the GPS constellation. In 1984, M. P. Ananda, et al. from the IMB gave the theoretical demonstration results of GPS satellites' autonomous navigation for the first time [19]. Moreover, the JPO entrusted Rockwell Automation Inc. to study the engineering feasibility for the GPS Block II satellites to realize autonomous navigation. It was shown for the studied results that the GPS satellites could achieve the autonomous navigation as long as the onboard data processing units were properly modified and the corresponding payloads similar to the nuclear explosion monitoring process were equipped with. The basic principle of the GPS satellites' autonomous navigation is

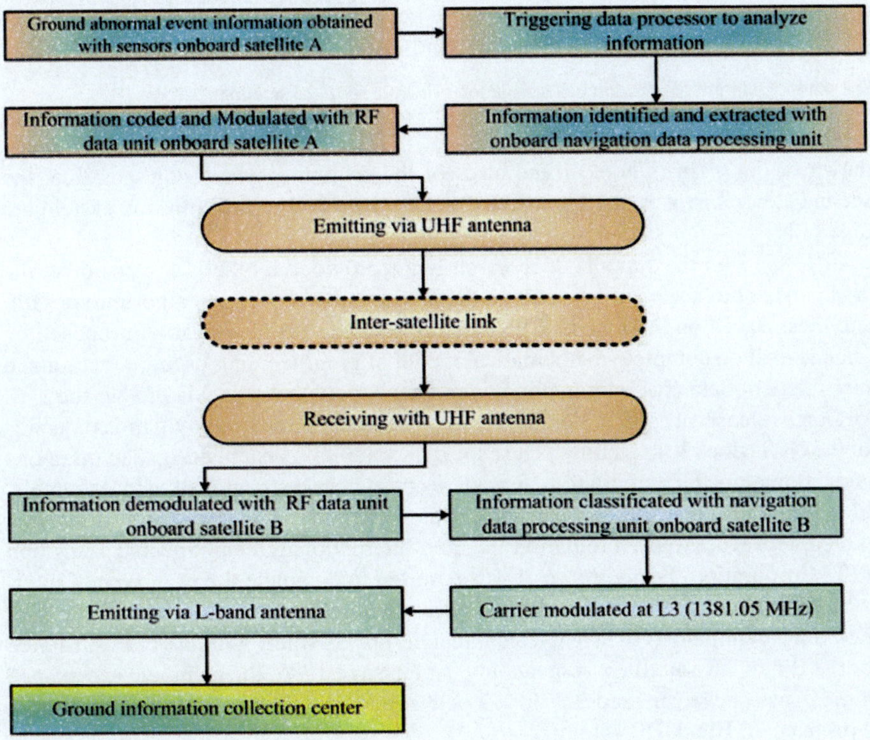

Fig. 4.13 Flowchart of nuclear explosion monitoring of GPS

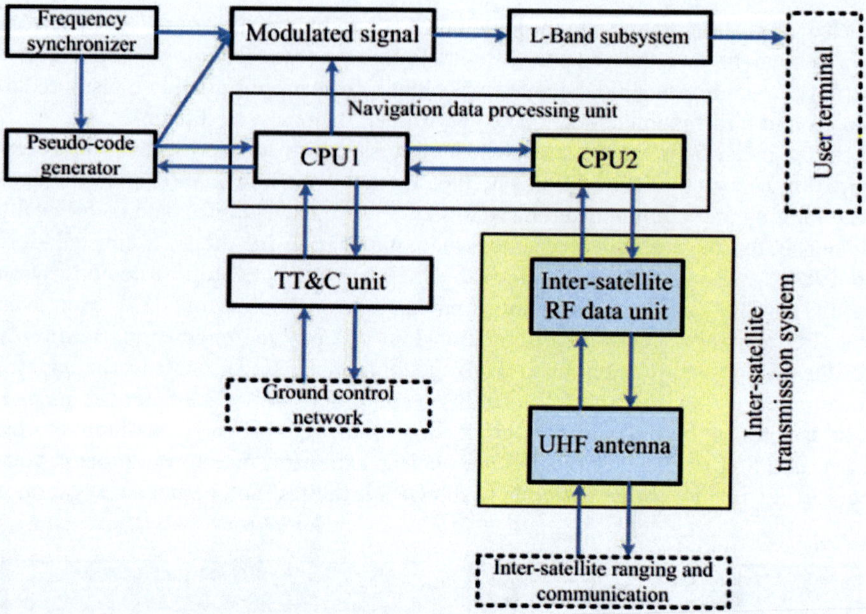

Fig. 4.14 Diagram of autonomous navigation principle for GPS satellites

shown in Fig. 4.14, where on the basis of the original payload configuration, the Second Central Processing Unit (CPU2) and inter-satellite transmission system are added [20].

In 1985, with the support of research funds provided by the U.S. Air Force, the IBM carried out a special research on the autonomous navigation algorithm of GPS satellites. Based on the idea of GPS satellite's autonomous navigation proposed by Ananda et al., a complete mathematical model of preprocessing pseudoranges and a measurement data checkout method were further developed. In May 1987, the U.S. Air Force released a contract for the development of autonomous navigation payloads for the GPS Block IIR satellites. The Block IIR satellite's system design, autonomous navigation algorithm simulation, hardware equipment development, and Assemble, Integration and Test (AIT) were carried out by Stanford Telecommunication Inc. jointed with Rockwell Automation Inc. and International Telephone and Telegraph (ITT) Corporation. Furthermore, it was required to complete the prototype of inter-satellite transmission system within 1 year. In July 1988, Rockwell and General Electronic Company (GE) signed technical design contracts with the U.S. Air Force for the Block IIR satellites, respectively, and in June 1990, the technical acceptance of the contracts was passed. In 1991, Lockheed Martin Corporation won a contract to produce 20 Block IIR satellites.

Without the support of ground control network, the GPS Block IIR satellites would start the inter-satellite crosslink with ranging and communication function.

The so-called **inter-satellite crosslink** is essentially a set of radio-frequency transmitting and receiving devices, which used Time Division Multiple Access (TDMA) to realize the two-way ranging and data exchange between the GPS satellites, so as to autonomously generate the satellites' ephemeris and clock parameters by processing the inter-satellite measurement data with the onboard filters. The original design goal of autonomous navigation of the GPS constellation was: during 180 days, the URE of GPS constellation was kept within 6 m, the navigation and positioning accuracy would not degrade significantly and the integrity of the satellite's signals could be monitored, so as to enhance the GPS system's availability, continuity and reliability. The autonomous navigation mode and technical features of GPS satellites contain the inter-satellite crosslink antenna, topological link architecture, communication mechanism and operational mode of autonomous navigation.

The inter-satellite crosslink is composed of inter-satellite signal transmitter, signal receiver, feed network, transmitting antenna unit, receiving antenna unit, and signal and data processing unit, with the functions of inter-satellite two-way ranging, data communication and instruction release. The transmitting antenna of the inter-satellite crosslink is an isolated unit antenna with the double-cone helix shape, while the receiving antenna adopts the direct plane array composed of nine radiating units, of which one is located at the center of the array, and the other eight are evenly arranged around the central unit, whose feed phases are opposite to the phase of the central unit and whose feed amplitudes are configured in proportion. The phase center of the receiving antenna beam is stable, and the beam edge gain can reach 7 dBi or so.

According to the design of GPS constellation formation and inter-satellite crosslink antenna beam-forming, the distance of GPS inter-satellite crosslink can reach 49,465 km. For a constellation with 24 satellites, there are a total of 8–16 links, including the forward and backward links in the same orbit, as well as lateral links between adjacent orbits.

The TDMA spread spectrum technique is adopted for the inter-satellite communication, with the communication frequency band of UHF (250–290 MHz). Each satellite is allocated a time interval of 1.5 s for dual-frequency ranging and communication. Thereby, for the constellation with 24 satellites, a sub-frame is 36 s and master frame is 900 s. The sub-frame zero is the ranging frame; the 1st sub-frame is used to correct the inter-satellite ranging errors; the 2nd sub-frame is used for the inter-satellite measurement and data communication; the 3rd sub-frame is reserved time, which can be used for data processing and communication preparation; the 4th–9th sub-frames are used for the satellites' ephemeris and clock parameters to exchange each other; the 10th–24th sub-frames are reserved time for inter-satellite crosslink signal transceiver, which can be used for the Kalman filtering and navigation message compiling, etc. For the GPS Block IIR and IIR-M satellite series, their inter-satellite crosslink ranging period can be set, respectively, as 15 min, 1 h, 2 h, 3 h, 4 h and 6 h, among which 1 h is the default value [21].

There are four operational modes for the GPS satellite autonomous navigation: one is the shut-off mode, only with the inter-satellite ranging and data storage; the other is the slip mode, in which the inter-satellite ranging, data storage, orbit propagation and navigation message generation, as well as the updated navigation message that is broadcast to users, can be carried out; the third is the isolated mode, in which without satellite data used, single satellite itself carries out information processing; the fourth is the normal mode, in which all autonomous navigation operations can be performed. The autonomous navigation function for the GPS satellites can be turned on or off by the ground control network. When the autonomous navigation payload is suspended, the GPS Block IIR satellites also have the functions and working modes similar to the Block IIA satellites.

In January 1997, the development and test of the first GPS Block IIR satellite with the autonomous navigation was completed in accordance with the requirement of the contract, and was launched on the schedule. However, due to the failure of solid rocket engines, the satellite took off from the launch site for 13 s and then crashed with the rocket. In July 1997, the second Block IIR satellite was successfully launched into orbit [22], which is the first autonomous navigation satellite deployed to the GPS constellation. As of April 2020, there are 32 satellites in the GPS constellation, including four satellite series of the Block IIR, IIR-M, IIF and III. All of these GPS satellites have the autonomous navigation capability.

4.9.3 Whole Rotation of Navigation Constellation

For the autonomous navigation of navigation constellation based on the inter-satellite crosslink, the constellation's space orientation cannot be determined due to the lack of external space–time reference information. As a result, the constellation regarded as a rigid body will drift relative to the Earth-Centered Inertial (ECI) coordinate frame, which is referred to as **whole rotation of navigation constellation**. In other words, among six Keplerian elements of each navigation satellite, only the semi-major axis (a), eccentricity (e) and mean anomaly (M) are observable, while the right ascension (Ω) of ascending node, inclination (i) to the equatorial plane and argument (ω) of the perigee are unobservable. Obviously, with the two-way measurement distances between the satellites as the observables, it is impossible to eliminate or suppress the whole rotation error of the constellation, so that it is very difficult for the navigation constellation to operate autonomously for a long time [23].

The whole rotation error modeling is one of the key technologies to realize the autonomous navigation of navigation constellation, and there are two problems: one is always with the modeling error, which makes the real rotation error unable to be separated correctly, resulting in a false rotation phenomenon of the constellation and then the divergence of the onboard filter and the other is that it is difficult to correctly allocate the rotation error between the ground control center and the onboard filter [24, 25]. If the total rotation of the constellation can directly be measured by using the star sensors to observe the changes of the background of stars, the absolute orientation of the constellation relative to the ECI frame is gotten to correct the drift errors of the satellite's orbits. However, it requires a very high accuracy of constellation orientation measurement. In view of the current technical level of star sensors, it cannot meet the practical application requirements yet. If the ground stations regularly transmit the ranging signals and the orientation parameters to the constellation, and the satellites autonomously carry out the information processing, the whole rotation error accumulated with time will be eliminated or restrained, which is known as the "**ground-based anchor**" technique [26]. However, the solution of establishing satellite-ground links violated the principle that the autonomous navigation satellites do not rely on the ground stations to operation for a long time.

In addition, the Earth's rotation rate and polar wander parameters need to be known in the navigation and positioning process, so as to calculate the rotation angle of the Earth-Centered Earth-Fixed (ECEF) coordinate system relative to the ECI, and add it to the navigation message. Therefore, the user receivers can obtain the position and velocity components of the satellites in the ECEF in real time to solve the navigation parameters. Due to the non-uniformity error of the Earth's rotation rate and the residual of polar wander, the ECEF will drift relative to the ECI, resulting in user's navigation accuracy to degrade. The autonomous navigation of the constellation based on the inter-satellite crosslink is also unable to solve the problem that the non-uniformity error and polar residual are accumulated with time. In fact, the error of Earth's rotation rate is the difference between the UT1 and UTC, and its drift is approximately 1 s per year, where the UT1 is the Universal Time (UT) corrected by the polar wandering error, and the UTC is a Coordinated Universal Time using the length of a second defined by the International Atomic Time (TAI) and synchronizing with the UT approximately. In order to keep the consistency between the UT1 and UTC time system, it is necessary to make a leap second correct for the UTC each year. The global monitoring data of the Earth rotation is analyzed and processed by the Bureau International des Poids et Measures (BIPM), and the difference between the UT1 and UTC is published regularly. Moreover, the non-uniformity error and polar residual of the Earth rotation can be predicted for a long time, and the users on the Earth's surface can also use the modern communication networks to directly download relevant parameters from the BIPM website for correction.

According to the design specifications for the GPS Block IIR and IIR-M satellites' autonomous navigation, it was required that the URE is less than 6 m during 180 days, and the navigation and positioning accuracy of users will not be degraded significantly. However, from the analysis of the post-processing results of the actual inter-satellite ranging data, it was shown that the whole rotation error of the constellation increases rapidly over time. It is difficult to meet the original design requirements. For this reason, in the design of the GPS Block IIF satellite, the autonomous navigation requirement had been changed, that is, the URE is less than 2 m, with the autonomous navigation capability of 60 days [27].

In summary, it is seen that the ground-based anchor technique does not meet the definition of spacecraft's autonomous navigation. From the perspective of the development of GPS constellation autonomous navigation, its technical vein is gradually formed, from the inter-satellite links to the ground-based anchor, and finally "space-based anchor". The so-called **space-based anchor** is a technology that is taking the X-ray signals radiated by pulsars as natural beacons, and the navigation satellites can autonomously determine the orbit and time parameters, so as to realize the long-time and high-precision autonomous operation and control of the constellation. In this way, the future development of X-ray pulsar-based navigation technology is to achieve the long-time and high-precision autonomous navigation for the navigation constellation, as an ultimate goal. The autonomous navigation constellation based on the X-ray pulsars is to use the X-ray signals radiated from the pulsars or the ephemeris parameters determined by the dedicated pulsar-based LEO satellites, as the external reference information, without the whole rotation error accumulated with time, so as to achieve the autonomous operation and control of the constellation for a long time and with high accuracy. So, the X-ray pulsar-based navigation provides a new approach for studying the autonomous navigation technology for the spacecrafts and navigation constellations.

References

1. Zhou QH (2003) The explanation of the term [Navigation]. Navigation 39(4):103–106 (Chinese)
2. Hofmann-Wellenhof B, Legat K, Wieser M (2011) Navigation: principles of positioning and guidance. Springer Science & Business Media, New York
3. Zhang HZ (1954) The collection of ancient mines. Geological Publishing House, Beijing (Chinese)
4. Wang ZG (2013) Discussions on fully using the historical lighthouse functions. Sci Technol Innov Herald 24:252–253 (Chinese)
5. Rice H, Kelmenson S,Mendelsonhn L (2004) Geophysical navigation technologies and applications. In: Proceedings of IEEE PLANS, pp 618−624
6. Zhou J, Liu YY (2009) Research progress on the new autonomous navigation methods based on polarized skylight. J Astronaut 30(2):409–414 (Chinese)
7. Guier WH, Weiffenbach GC (1998) Genesis of satellite navigation. J Hopkins APL Tech Dig 19(1):14–17
8. Kaplan ED (1996) Understanding GPS: principles and application. Artech House, Inc., Boston
9. ErwinS (2019) Air force experiment NTS-3 could point the way to the next generation of GPS. Space News, March 16, 2019. https://www.spacenews.com/
10. China Satellite Navigation Office (2013) Development of the BeiDou Navigation Satellite System (Version 2.2), December 2013. https://www.beidou.gov.cn
11. China Satellite Navigation Office (2019) Development of the BeiDou Navigation Satellite System (Version 4.0), December 2019. https://www.beidou.gov.cn
12. China Satellite Navigation Office (2019) BeiDou Coordinate System Template. December 2019. https://www.beidou.gov.cn
13. OnidiO (2002) GALILEO Is Launched. In: Proceedings of the 15th international technology meeting of the satellite division of the Institute of Navigation, ION GPS/GNSS 2002, Portland, OR, 24–27 Sept 2002, pp 245–248
14. DuttonL, Rumens D et al (2002) GALILEO's services. In: Proceedings of the 15th international technology meeting of the satellite division of the Institute of Navigation, ION GPS/GNSS 2002, Portland, OR, 24–27 Sept 2002, pp 249–257
15. Yang JC (1995) Orbital mechanics and control for the spacecrafts (Volume 1). China Astronautic Publishing House, Beijing (Chinese)
16. Yang JC (2002) Orbital mechanics and control for the spacecrafts (Volume 2). China Astronautic Publishing House, Beijing (Chinese)

17. Xia NY (2002) Space telemetry, tracking and command system. National Defense Industry Press, Beijing (Chinese)
18. Graven P, Collins J et al (2007) XNAV beyond the Moon. In: Proceedings of the 63rd annual meeting of the Institute of Navigation, 23–25 April 2007, pp 423–431
19. Ananda MP, Bernstein H et al (1990) Global Positioning System (GPS) autonomous navigation. In: Proceedings of IEEE position, location and navigation symposium, Las Vegas, Nevada. 20–23 March 1990, pp 497–508
20. Parkinson BW, Spilker JJ (1996) Global Positioning System: theory and applications (Volume I). Published by the American Institute of Aeronautics and Astronautics Inc., Washington, D.C., pp 149–159
21. Rajan JA (2002)Highlights of GPS IIR autonomous navigation. In: Proceedings of the ION 58th annual meeting, Albuquerque, NM. 24–26 June 2002, pp 345–363
22. KiserK, Vaughan SH (1998) GPS IIR joins the GPS constellation. In: Proceedings of the 11th international technology meeting of the satellite division of the institute of navigation, Nashville, Tennessee, 15–18 Sept 1998, pp 1915–1923
23. Shuai P, Qu GJ (2005) The time synchronization technology for autonomous navigation constellation. J Astronaut 26(6):768–772 (Chinese)
24. Shuai P, Qu GJ (2006) The ephemeris updating technology for autonomous navigation constellation. J Astronaut 27(2):187–191 (Chinese)
25. Shuai P, Qu GJ, Chen ZG (2006) The autonomous navigation technology for navigation constellation. China Eng Sci 8(3):22–33 (Chinese)
26. Rajan JA, Brodie P, Rawicz H (2003) Modernization GPS autonomous navigation with anchor capability. In: Proceedings of the 16th international technology meeting of the satellite division of the Institute of Navigation, ION GPS/GNSS 2003, Portland, OR. 9–12 Sept 2003, pp 1534–1542
27. Fisher SC, Ghassem K (1999) GPS IIF–the next generation. Proc IEEE 87(1):24–47

Chapter 5
X-ray Pulsar-Based Navigation: Theories and Experiments

The pulsar is a high-speed rotating neutron star, with extremely stable periodicity. Especially, the stability of pulse periods of millisecond pulsars gets to 10^{-19}–10^{-21} s/s, which is known as the most stable astronomical clock in nature. **Pulsar navigation** refers to a process that taking the X-ray photons radiated by the pulsars as natural beacons, spacecrafts autonomously determine their orbit, time and attitude parameters to guide them to fly to the target obits or celestial bodies along the designed trajectories safely, effectively and just in time, and its full name is **X-ray Pulsar-based Navigation** (XPNAV). Because X-ray photons radiated from the pulsars are unable to pass through the dense atmosphere of the Earth, the pulsar navigation cannot directly be used for the movable objects on the Earth's surface, and it is essentially a kind of space navigation, which provides navigation information services for the spacecrafts in near-earth orbit, deep space and interstellar space, as well as the rovers on the surfaces of the planets without dense atmosphere. So, the pulsars are also known as **lighthouses for space flights**.

In this chapter, the theories and experiments are attempting to explain the architecture and logical relations of the pulsar navigation from the perspective of top-level design of a system, including basic concepts and principles, fundamental theories and methods, mathematical models and algorithms, numerical experiments and results, flight tests and verifications, and typical scenarios and applications. The so-called **system** refers to an organic whole with specific functions, composed of a number of related things or some subsystems with the connection in consciousness one another. The purpose of combing the pulsar navigation system is to demonstrate its overall picture, and further sketch the future of navigation satellite system.

P. Shuai, *Understanding Pulsars and Space Navigations*, Navigation: Science and Technology 11, https://doi.org/10.1007/978-981-16-1067-7_5

5.1 Principle of Pulsar Navigation

Generally, the center of mass of solar system, also called the Solar System's Barycenter (SSB), is selected as a space-time reference point for the XPNAV. The SSB is the center of mass of the whole solar system, which is the most stable point in the solar system, located near the Sun's surface and on the side of the Jupiter, and the Sun and its planets move around this point. The SSB is constantly shifting since the planets are always in motion. The positions of the planets determine those of the barycenter. The pulse profile and its pulse Time of Arrival (TOA) of a pulsar relative to the SSB can be accurately measured and predicted. Therefore, the mathematical models and basic observables of the XPNAV are usually expressed in the SSB coordinate system.

With the SSB as the coordinate origin O_{SSB}, and the X_{SSB}-axis pointing to the vernal equinox defined by the J2000.0 reference epoch, the SSB coordinate system $O_{SSB}-X_{SSB}Y_{SSB}Z_{SSB}$ can be established in the equatorial plane of the celestial sphere, as shown in Fig. 5.1, where PSR, SC, O_E and O_S, respectively, denote the pulsar, spacecraft, Earth's center of mass and Sun's center of mass; λ and α, respectively,

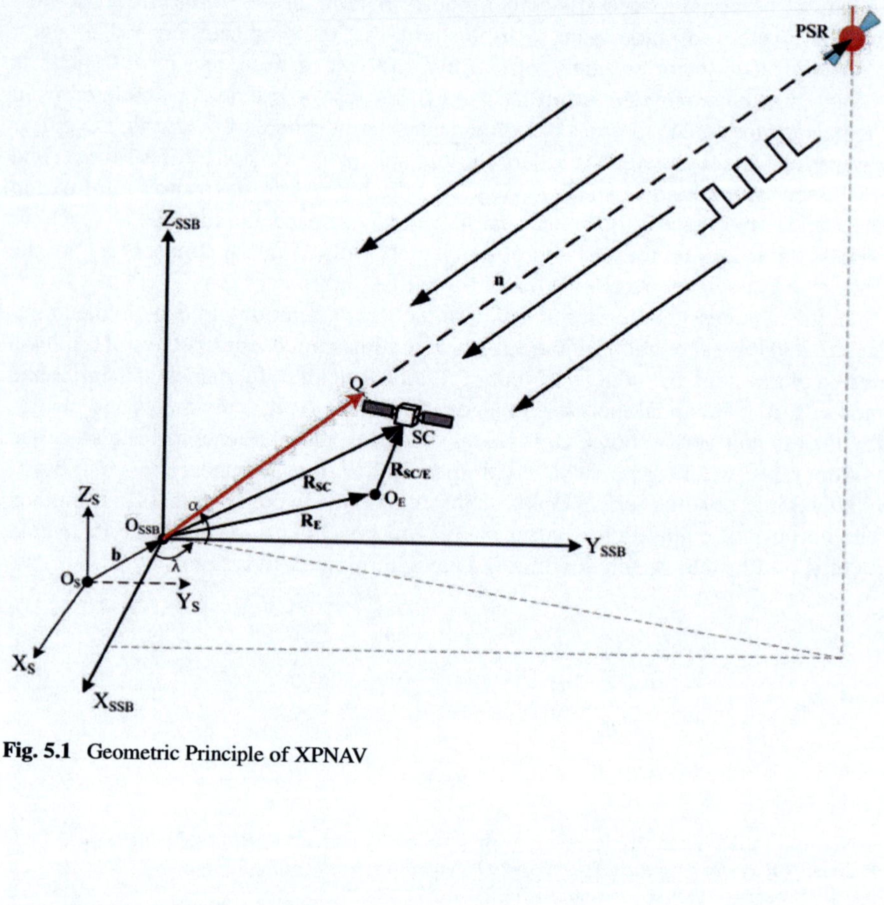

Fig. 5.1 Geometric Principle of XPNAV

denote the right ascension and declination of the pulsar PSR in the SSB coordinate system, also called **pulsar's angular position**; $\mathbf{R}_{\text{SC}}$, $\mathbf{R}_{\text{E}}$ and $\mathbf{R}_{\text{SC/E}}$, in the SSB coordinate system, respectively, denote the position vectors of the spacecraft, Earth and spacecraft relative to the Earth; b denotes the position vector of the SSB in the Sun-mass-centric coordinate system $O_S{-}X_S Y_S Z_S$; Q denotes the projection of the spacecraft's position along the direction from the SSB to the pulsar; **n** denotes the position unit vector of the pulsar, also called **angular position vector**, namely,

$$\mathbf{n} = \left[\cos\alpha\cos\lambda\ \cos\alpha\sin\lambda\ \sin\alpha\right]^T \tag{5.1}$$

Because pulsars are extremely far away from the solar system, reaching thousands of light-years or even tens of thousands of light-years, the **n** can generally be regarded as a constant vector for the whole solar system. In the SSB coordinate system, the right ascensions and declinations of the pulsars can also be measured accurately [1]. In other words, the direction vectors of the pulsars in the SSB coordinate system are known.

For a spacecraft equipped with X-ray detection device and atomic clock, as a pulsar rotates and its magnetic pole beam sweep over the detection device, the spacecraft will receive an X-ray pulse signal and the arrival-time of pulse signal, t_{SC}, is recorded with the atomic clock. Meanwhile, the time of the same pulse signal arriving at the SSB, t_{SSB}, can be accurately predicted by pulsar timing model. For the whole solar system, the beams emitted from pulsars can be regarded as plane waves. Thereby, in Fig. 5.1, the time when the same pulse signal arrives at spacecraft SC is same as that when it arrives at point Q. According to the trigonometric projection relation, and considering a bias of the onboard clock, the measurement equation of the XPNAV can be expressed as

$$c(t_{SSB} - t_{SC}) = \mathbf{n}\cdot\mathbf{R}_{SC} + c\cdot\delta t_{\varphi}, \tag{5.2}$$

where $\mathbf{R}_{SC} = \left[x\ y\ z\right]^T$, c is the light velocity, and δt_{φ} is the bias of the onboard clock.

There are four unknowns in Eq. (5.2), including three-position coordinate components of the spacecraft and one clock bias. When four pulsars are observed at the same time, then four measurement equations can be established to solve the four unknowns. It is referred to as **geometric orbit determination**. Combining with the model of orbital mechanics of the spacecraft, it is only needed to observe one pulsar during each observation arc-segment, and then the position and time parameters of the spacecraft can be determined, which is referred to as **dynamics orbit determination**.

Obviously, the principle of the XPNAV is very similar to that of the GNSS, and their basic observables are distances. The distance measurement of the XPNAV is determined by the time differences between the same X-ray pulse signal arriving, respectively, at the SSB and spacecrafts, while that of the GNSS is determined by the arrival-time from the pseudo-random noise signals broadcasted by the navigation satellites to the user terminals. At present, the navigation accuracy using the GNSS is better than 10 m, and the ultimate accuracy of the XPNAV can also reach 10 m level in the future [2]. Furthermore, for the traditional celestial navigation based on star optical sensors, its basic observables are angles. Since a small angle measurement error will bring up a great orbit error for spacecrafts, the accuracy using the celestial navigation is typically the level of several kilometers, even tens of kilometers.

Considered that the X-ray signals radiated by the pulsars are extremely weak, the signal integration time is long in order to get the pulse profiles with enough high Signal-to-Noise Ratio (SNR). Moreover, the measurement accuracy of the pulse arrival-time for the XPNAV is not as high as that of the pseudo-code arrival-time for the GNSS, and currently in practice, it is still difficult for the spacecraft to carry multiple X-ray detection devices and track multiple pulsars at the same time. Therefore, using the geometric orbit determination method to determine the orbit accuracies of the spacecrafts can only reach the level of thousands of meters, or even lower. From the perspective of practical engineering application, the dynamics method is used for the spacecraft's orbit determination, which is helpful to improving the accuracy of estimating the navigation parameters, to make the orbit propagation in real time, and to need not to simultaneously observe more than four pulsars.

The XPNAV is a feasible approach to achieve long-term and high-precision autonomous navigation for spacecrafts, with the performance advantages that the traditional navigation technologies cannot be comparable to. There are five significant technical-features for the XPNAV. The first feature is the full information, that is, the XPNAV can provide ten-dimensional navigation information for spacecrafts, including three-dimensional position, three-dimensional velocity, three-dimensional attitude and one-dimensional time. The second feature is the all space, that is, the XPNAV can be applied to the whole solar system, even the whole universe, from the near-earth orbit to deep space, and interstellar flight spacecrafts, as well as the rovers on the planetary surfaces without dense atmosphere, so as to achieve seamlessly autonomous navigation for all space flight missions. The third feature is the long term, that is, the XPNAV using the SSB as the space-time reference point can provide the navigation information service for the navigation constellations under an "absolute" reference frame, so as to solve the problem of the whole rotation of navigation constellations and realize the autonomous operation of the constellation during a long time. The fourth feature is the high accuracy, that is, the XPNAV can ultimately achieve the orbit determination accuracy of 10 m, the time synchronization accuracy of 1 ns, and the attitude measurement accuracy of 3 arc-seconds, which are incomparable to the traditional celestial navigation technology. The fifth feature is the autonomy, that is, due to the X-ray signals radiated by pulsars as natural beacons, the XPNAV has the characteristics of information integrity, operating in real time, no emitting signals required by spacecrafts, no relying on the ground stations and long-term automatic operation. So, the pulsar navigation is an effective approach to realize the autonomous navigation for spacecrafts.

5.2 Progress on Pulsar Navigation

5.2.1 The Proposed Concepts

In 1967, the first radio pulsar was discovered by a doctoral student Jocelyn Bell and her mentor Professor Antony Hewish at the University of Cambridge, the United Kingdom. In 1972, the Pioneer-10 spacecraft was launched by the NASA to explore the planets of the outer solar system and the interstellar space. The Pioneer-10 carried the gold-anodized aluminum plate design by Carl E. Sagan et al., on which the pattern using 14 radio pulsars to determine the positions of the Sun and Earth in the Galactic System was drawn. One day, if the Pioneer-10 is intercepted by the intelligent extraterrestrial civilization, then its launching location and epoch will be calculated by observing the signals radiated from these pulsars. It is the first attempt to obtain the position information by using the trigonometric intersection measurement based on the pulsars and to get the time information based on the change rates of pulse periods of the pulsars. Afterward, another gold-anodized aluminum plate and two golden records with the 14-pulsar pattern were also mounted, respectively, on the Pioneer-11, and Voyager-1 and Voyager-2, to transmit the Earth's human civilization information to space out of the solar system.

In 1974, George S. Downs working then at the NASA's JPL proposed an autonomous orbit determination method for the spacecrafts in interplanetary space flight by using the radio pulsars, and estimated the orbit determination accuracy, about 150 km [3]. There are three defects in this method: firstly, because the radiated signals from pulsars in radio frequency are extremely weak, a high gain and large aperture radio telescope are required in order to detect the radio pulse signals, and typically, the antenna is with a diameter of more than 25 m; secondly, the radio signal intensity radiated by pulsars are so low that the signal integration time is required to be as long as 24 h to get enough high SNR; thirdly, most of the celestial radiation and galactic diffusion signals are distributed in the entire radio frequency band, which submerges the extremely weak signals radiated by the pulsars, so that it is disadvantageous to detect and extract the useful signals and eliminate the noises. Downs's method is difficult in engineering application, and by using the method, the orbit determination accuracy is so low that cannot meet the requirements of high-precision autonomous navigation for spacecrafts. For all that, Downs is recognized as the pioneer conducting the studies on the pulsar's application. For infrared, visible and ultraviolet pulsars, due to the small number and low luminosity, the large aperture telescope, high pointing accuracy and advanced signal processing technology are required, so these types of pulsars are not suitable for the autonomous navigations of spacecrafts [4, 5].

In 1981, the idea of using the X-ray signals radiated from pulsars to navigate for spacecrafts was first proposed by Chester and Butman [6]. Most of the energy radiation from the pulsars is concentrated on the X-ray band, which is advantageous to detect and process, and reduces the time of signal integration. Especially, it is conductive to the design of miniaturized detection devices, usually with the detection area of less than 1 square meter, so as to make it possible that the spacecrafts carry the detection devices to observe the X-ray pulsars for navigation application.

In 1993, based on the traditional celestial navigation method that is the Earth's atmosphere occultation method, a comprehensive test project, called Unconventional Stellar Aspect experiment (USA), was proposed by Kent S. Wood working at the U.S. NRL [7–9]. The USA project was a part of the "X-ray Astronomy Program" of the NRL, and the Stanford Linear Accelerator Center was responsible for the thermal control design, equipment development and test assembly of the USA. A doctoral student at the University of Stanford, John E. Hanson, had done a lot of research work on the USA project and proposed a design scheme of spacecraft's attitude measurement and time-keeping phase-locked loop by using the X-ray sources from celestial bodies, including X-ray pulsars. From his numerical analysis results, it is shown that the attitude determination accuracy for the spacecrafts reaches 0.01° and the time-keeping accuracy is better than 1.5 ms [10]. In February 1999, the Advanced Research and Global Observation Satellite (ARGOS) developed by the U.S. Air Force, carrying the USA experimental devices, was launched into the solar synchronous orbit with an orbital altitude of 840 km and an inclination of 98°, to carry out a total of nine space-science experiments including the USA. A large amount of test data were gotten by using the X-ray detection devices of the USA experiment and used to study four special subjects: X-ray astrophysics; X-ray navigation; X-ray detection on the upper atmosphere of the Earth; advanced signal and data processing technology under space environment. In the USA experiments, two sets of double-axis gimbaled collimated gas proportional counters were used, with the detection energy spectrum range of 1−15 keV, the effective detection area of 0.1 m^2 and the Field of View (FOV) of 1.2° by 1.2°. In November 2000, due to the counter's inert gas mixture leaking, the USA experiments had to be terminated. It should be pointed out that the spacecraft's navigation using the X-ray sources in the USA experiments is a traditional celestial navigation method based on the atmospheric occultation, so the orbit determination accuracy of the ARGOS mainly depends on the upper atmosphere model, and is typically the level of tens of kilometers. In other words, the way of pulse arrival-time ranging had not been adopted in the USA experiments, so the experiments are not the space flight tests of pulsar navigation in real sense.

In 2004, based on the intensive studies on the physical characteristics of pulsars and the passive navigation mechanism of one-way arrival-time ranging used in the GNSS, Suneel I. Sheikh who was then a doctoral student at the University of Maryland theoretically demonstrated the feasibility of the autonomous navigation for spacecrafts using the X-ray pulsars [11, 12]. In addition, taking the time difference that the same X-ray signals radiated from pulsars arrive respectively at two GPS satellites, as aiding observables, Dennis W. Woodfork who was then a graduate student at the U.S. Air Force Institute of Technology studied the method improving the estimation accuracies of the orbit and clock parameters of the GPS satellites. From his numerical experimental results, it was shown that the accuracies of estimating the GPS satellite's orbit and clock parameters could be improved by 30.8% and 21.4%, respectively, by adding the measurement data of the pulse arrival-time difference compared with the inter-satellite cross-link pseudoranges used alone [13].

In summary, although the navigation concept using the radio pulsars was presented in the 1970s, the XPNAV technology was proposed only 15 years ago, using the idea of arrival-time ranging of the navigation satellite system, and increasingly attracted the attention to the major space countries in the world. At present, the XPNAV is in the stage of key technology research and space flight test. From the in-orbit flight tests, the feasibility of the XPNAV system has been demonstrated preliminarily.

5.2.2 Relevant Development Programs

In 2004, the U.S. Defense Advanced Research Projects Agency (DAPAR) was delving into creation of a "Pulsar Network" for satellite to know their exact whereabouts in space and time, also known as "military eyes pulsar navigation network". The idea was to create a celestial seasoning of sorts, adding a little pizzazz to the GPS constellation of satellites in the Earth orbit. The effort was dubbed X-ray Source-based Navigation for Autonomous Position Determination (XNAV) Program and was being spearheaded under the DAPAR's tactical technology work [14]. "We're on our way… basically creating a pulsar network", said Tony Tether, Director of the then DAPAR during opening remarks at the Small Satellites: Complimentary or Disruptive Technology Conference, being held at Utah Sate University in August 2004. In the GPS world there is a term called "Dilution of Precision (DOP)", a measure of the quality of the GPS data being received from the satellites. "I don't think we'll have that problem with pulsars", Tether said. "First of all, there's lots of pulsars", Tether said. "What hasn't been done is that they haven't been characterized. It would be useful for anywhere in the solar system… a great way to navigate around, all over space". The XNAV is a research and development to look into an autonomous position, attitude and time determination system using celestial sources in the X-ray band of the electromagnetic spectrum. The goal of the Program is to prove the feasibility and viability of the use of celestial sources, pulsars as well as neutron stars for position, attitude and time determination of low Earth orbiting spacecrafts. Such a capability would provide an autonomous backup system to military navigation and communication satellites. Using the XNAV, a satellite in space would be able to find where it is using pulsars.

The XNAV project was contracted to two research teams: one is a research team led by Ball Aerospace & Technologies Corporation, in collaboration with Los Alamos National Laboratory, Applied Physics Laboratory of Johns Hopkins University, and National Institute of Standards and Technology (NIST), carried out the research work in the project's overall design, device development and system integration; the other is a research team led by the NRL, which cooperated with Massachusetts Institute of Technology (MIT) and Brookhaven National Laboratory, carried out the studies on navigation database and X-ray detector technology. The ultimate goal of the XNAV Program is to establish a pulsar navigation network that can provide the orbit determination accuracy of 10 m, time synchronization accuracy of 1 ns and attitude determination accuracy of 3 arc-seconds, so as to meet the requirements of high-precision autonomous navigation for the future spacecrafts from the near-earth orbit to deep space, interstellar space.

By 2030, the orbit determination accuracy of deep-space probes of the NASA will be required to reach the order of 100 m. However, it is difficult to meet the navigation accuracy requirements for the deep-space exploration missions by using the ground deep space TT&C network or traditional celestial navigation methods. For this reason, a research program of deep-space navigation technology using X-ray pulsars was initiated by the NASA in February 2006. The general contractor of this program was the Microcosm Inc., in cooperation with the ASTER Labs Inc., CrossTrac Engineering Inc., NRL and Goddard Space Flight Center (GSFC), to carry out the related research work [15–17].

In 2009, on the basis of the XNAV Program, X-ray Timing (XTIM) Program was also funded by the DARPA. The XTIM was led by Lockheed Martin Corporation, cooperated with Ball Aerospace & Technologies Corporation, to together develop a large-area array collimation detector and finally build an independent and stable space-ground pulsar timing reference system. In 2011, the GSFC together with the U.S. Universities Space Research Association (USRA) proposed a technology demonstration enhancement mission of "Neutron star Interior Composition Explorer (NICER)", called "Station Experiment for X-ray Timing and Navigation Technology (SEXTANT)". The NICER is an International Space Station (ISS) payload devoted to the study on neutron stars through soft X-ray timing, and its scientific objectives are to resolve the nature of ultra-dense matter at the threshold of collapse to black hole and to reveal the interior composition, dynamic processes and radiation mechanisms of neutron stars. The SEXTANT is to demonstrate the feasibility of XNAV in real time through the sequential observation of multiple millisecond pulsars, which is an important milestone for the technical enhancement test of the NICER mission. The SEXTANT system is composed of four parts, including X-ray Timing Instrument (XTI), in-orbit flight software, ground data support and ground test bed [18–20]. The XTI as the core device of the NICER represents an innovative configuration of high-heritage components, and its heart is an aligned collection array of 56 X-ray concentrator optics and Silicon Drift Detector (SDD) pairs. Each of the X-ray concentrator optics collects X-rays over a large geometric area from a roughly 30 square arc-minutes region of the sky and focuses them onto a small SDD. The XTI is a cubic shape, with the volume of 778 mm × 1245 mm × 800 mm, the detection energy spectrum range of 0.2–12 keV, the effective detection area of more than 2000 cm^2 at 1.5 keV and 600 cm^2 at 6 keV, the energy resolution of 85 eV at 1 keV and 137 eV at 6 keV and the time-tagging resolution of less than 300 ns. In June 2017, the NICER was launched to the ISS by a SpaceX Falcon-9 rocket, to carry out the experiments of X-ray astronomy and the tests of SEXTANT, where a flight navigation experiment was conducted in November 2017, with an initial orbit determination accuracy of better than 10 km [21].

Since 2005, China Academy of Space Technology (CAST) has continued to carry out studies on the theory and method of XPNAV system and made the breakthroughs in X-ray detector technology, pulsar database construction method, basic theory of large-scale navigation and ground test method. Meanwhile, the CAST has also developed the multiple types of X-ray detector prototypes and products, the ground test system and the dedicated test satellite, known as XPNAV-1. In November 2016, the world's first dedicated pulsar navigation test satellite XPNAV-1 was successfully launched from Jiuquan Satellite Launch Center of China, to verify the performance of X-ray detectors and the feasibility of the XPNAV system. Using the observation data from the XPNAV-1 satellite in the orbit, the improved orbit determination accuracy initially gets to 38.4 km [22]. The XPNAV-1 satellite and its observation results are introduced in detail in Sect. 5.8 of this chapter. Afterward, the first X-ray astronomy of China, known as Hard X-ray Modulation Telescope (HXMT), was launched into orbit in June 2017, to observe and investigate the high-energy objects such as black holes, neutron stars and active galactic nucleus through high-precision orientation in space. The HXMT satellite was manufactured by the CAST, and the Institute of High-Energy Physics of the Chinese Academy of Science as the user system was responsible for developing the payload of the satellite. With the launches of the XPNAV-1 and HXMT, a large amount of X-ray photons radiated from pulsars are collected to speed up the development of the XPNAV technology.

In addition, a project of deep-space navigation based on pulsars, known as Project ARIANDA 03/4202, was released by the ESA in 2004. The ARIANDA is a mechanism for collaborative joint research projects between the ESA's internal research think tank, the Advanced Concepts Team and academia, to establish closer link with the European academic community in advanced space technology research. The Project ARIANDA 03/4202 was undertaken by a joint research team composed of two universities from Spain: one is the Department of Signal Theory and Communications at University Politecnica de Catalunya, and another is the Department of Astronomy and Meteorology at University de Barcelona. The joint research team carried out the research on theoretical methods and numerical experiments of pulsar navigation, and finally formed a report, entitled "Feasibility Study for a Spacecraft Navigation System Relying on Pulsar Timing Information" [23]. In 2009, the Space Research Institute of the Russian Academy of Science carried out the investigation into the XPNAV technology. In 2012, an interstellar positioning system for spacecrafts far from the Earth using pulsars was developed by Max Planck Institute for Extraterrestrial Physics in Munich, Germany. Furthermore, during the past 15 years, the studies on the XPNAV technology have also been conducted by some researchers in other countries, such as Britain, India, Japan, Iran and Italy.

5.3 Space-Time Reference Based on Newtonian Mechanics

5.3.1 Description for Particle Motion

The Newtonian mechanics takes particles as research objects, and the motion state of the entire particle system can be inferred by studying the forces on each particle. The theoretical framework of classic mechanics is composed of Newton's three laws of motion and its law of universal gravitation, which is applicable to the study on the laws of motion of the macroscopic objects whose moving speeds are far less than the velocity of light. In the framework of classic Newtonian mechanics, time is a parameter to describe the movement and change of substance, which is independent of space; the coordinate values of the same space-time point given by different reference systems may be different, while the time and space intervals between two space-time points are unchanged, which is known as **Newton's absolute space-time view**. In other words, time is completely independent of space. The clock of the time-keeping system is an ideal standard clock, and the scale of the length measurement system is an ideal standard measurement bar. Both have nothing to do with the motion form of specific substance and the selected reference system and are with absoluteness.

In order to describe the motion state of a particle, a three-dimensional spatial reference system K can be selected, and the time t is taken as the independent variable, that is, the equation of particle motion path can be expressed as

$$x^i = x^i(t), \tag{5.3}$$

where the value of superscript i is, respectively, 1, 2 and 3.

Correspondingly, the velocity and acceleration of the particle motion can be expressed in component form as

$$\left.\begin{aligned} v^i &= \frac{dx^i(t)}{dt} \\ a^i &= \frac{d^2x^i(t)}{dt^2} \end{aligned}\right\}. \tag{5.4}$$

In the reference system K, the motion law of the particle is described by Eqs. (5.3) and (5.4). If the particle is not affected by the external force, it will remain at rest or move in a straight line at a constant speed, then the reference system K is an inertial system. Letting another inertial system K' move in a straight line at a constant speed relative to the inertial system K, the transformation of the particle motion equation satisfies the Galilean transformation, i.e.,

$$\mathbf{r}' = \mathbf{r} - \mathbf{u}t, \tag{5.5}$$

where $\mathbf{r}$ and $\mathbf{r}'$ stand for the position vectors of the particle at the K-system and K'-system, respectively; $\mathbf{u}$ stands for the velocity of the coordinate origin of the K'-system relative to the K-system, with an uniform motion.

When $t = 0$, the coordinate origins of the two inertial systems are coincident, the coordinate axes of the K'-system are always parallel to the corresponding coordinate axes of the K-system, that is, the K'-system is making the translational motion. Similarly, the velocity and acceleration of the particle in the K'-system are expressed respectively in vector form as

$$\left.\begin{aligned} \mathbf{v}' &= \frac{d\mathbf{r}'}{dt} = \mathbf{v} - \mathbf{u} \\ \mathbf{a}' &= \frac{d^2\mathbf{r}'}{dt^2} = \mathbf{a} \end{aligned}\right\}. \tag{5.6}$$

In two inertial systems which move uniformly in a straight line, it can be seen that the acceleration of a particle is an invariant.

In general, the origin O' of the K'-system not only makes a curve motion relative to the K-system, but also the K'-system itself rotates relative to the K-system. That is to say, the coordinate axes between the two reference systems keep parallel no longer. Thereby, it is needed to introduce a middle reference system K_m, whose coordinate origin O_m coincides with the origin O' of the K'-system, and the K_m-system only makes the translational motion relative to the K-system. In this way, the motion law of the particle is first transformed from the inertial system K to the middle reference system K_m, and then from the K_m-system to the K'-system. So, it is only needed to consider the rotation transformation between two reference systems, and then combining the translation relationship between the reference systems, the general motion law of a particle can be gotten in different reference systems.

For single particle, its rotation motions between two different reference systems have been discussed in the above. For a particle system, its fixed-point rotation is investigated, which is the problem of fixed-point rotation of a rigid body, as described in Sect. 4.7 of Chap. 4. In general, a free rigid body has six Degrees of Freedom (DOF), which needs six generalized coordinates to describe its position, and the position of the rigid body can be determined completely by the coordinates of any three non-collinear points. When a point on the free rigid body is selected as a fixed point, three constraints are added. It becomes the problem of fixed-point rotation of a rigid body with three DOFs, and thus the motion of the rigid body can be described by using the generalized coordinates composed of three Euler angles, which is known as Euler's kinematical equation. For the problem of fixed-point motion of rigid body under external force moment, Euler's dynamical equation is used to express. Consequently, the nonlinear second-order differential equations can be obtained, composed of Euler's kinematical and dynamical equations for the fixed-point motion of a rigid body [24]. When the external moment is known, the integral calculation is carried out for the differential equations, and the relationship of three Euler angles changing over time can be obtained, so as to master the fixed-point motion law of the rigid body.

5.3.2 *Concepts and Classifications of Time*

The time contains two basic concepts, the moment and the time interval. The so-called **moment** refers to the time when a phenomenon or an event occurs, and the so-called **time interval** refers to the process experienced by a phenomenon or an event, which is numerically equal to the difference between the start and end moments of the phenomenon or event. In astronomy and navigation application, the moment of recording an event is called the **epoch**; the unit used to measure or meter the time interval is called **timescale**; the starting point of measuring the moment is called **time origin** or **initial epoch**. The **time system** is a time measurement reference standard composed of the time origin and timescale, also known as **time reference**. Generally, the timescale is the core of establishing the time reference, and the time origin can be chosen according to the practical application.

If a continuous changing physical quantity x is expressed as a function taking the time t as an independent variable, and by measuring the change of the x, the change characteristics of the physical quantity are recorded accurately, then the x can be used as a kind of means of measuring time. Usually, it is expected that the time function of the selected physical quantity is a simple and linear function form, namely,

$$x = f(t) = a + bt, \tag{5.7}$$

where a is the time origin; when $t = 0$, $x = a$; b is the timescale.

Generally, formula (5.7) is called **time measurement equation**. From the practice of establishing the time system, it is shown that it is most convenient to choose the physical quantity with observable periodic motion as the time reference. In this way, after a certain time interval or motion period, all states of the physical quantity can be reproduced, which is advantage to the accurate measurement of the moment and time interval. Therefore, the basic conditions for selecting an observable physical quantity as a time reference are as follows:

(1) The motion should be continuous and periodic.
(2) The period of motion should have enough stability, and at any time, the period of the motion should be constant, seldom affected by the change of external environment conditions.
(3) The periodic phenomenon of the motion should be reproducible. That is to say, the periodic motion can be reproduced by the observations and experiments at any time and any place.

Due to the difference of the selected periodic motion phenomena, there are different time systems. Actually, there is not an ideal time reference that fully meets the above basic conditions in the objective world. Generally, it can only be approximately consistent, and thus the actual time measurement equation should be expressed as

$$x = f(t) = a + bt + \phi(t), \tag{5.8}$$

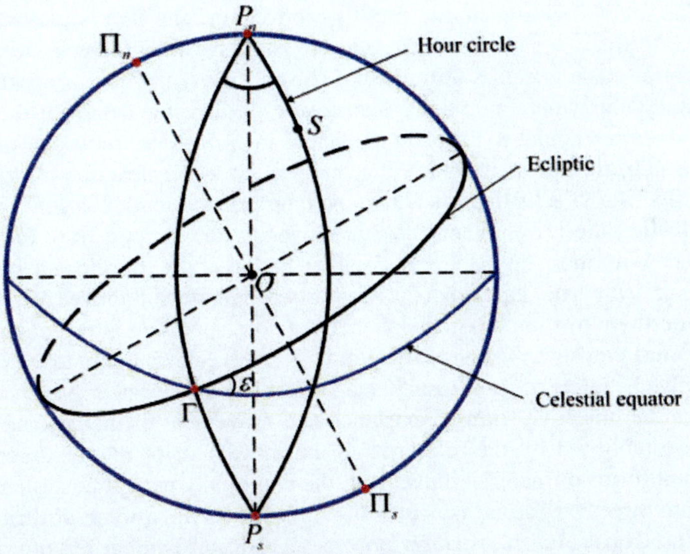

Fig. 5.2 Basic elements and their relations of celestial sphere

where $\phi(t)$ denotes the nonlinear items affected by external and internal conditions, including the quadratic terms, and the polynomials and noise terms of higher than them. As a time reference, it is required that the terms $\phi(t)$ should be as small as possible and easy to be modeled, so as to improve the precision of the time reference.

In order to explain the concept of time system, the definitions of the celestial sphere and its point, line, surface and circle, and their mutual relationship are given first of all. The so-called **celestial sphere** refers to an imaginary sphere with the SSB or the Earth's center of mass as its central point O, and with any length R as its radius, as shown in Fig. 5.2. In astronomy and astrometry, celestial bodies are usually projected on the spherical surface of the celestial sphere, and the spherical coordinate system is used to express and study the positions of celestial bodies and their relations. In this section, the Earth's center of mass was selected as the center of the celestial sphere, as an example to explain the celestial sphere and its coordinate system. The extension line of the Earth's rotation axis is called **celestial axis**, and two intersection points of the celestial axis with the celestial sphere, P_n and P_s, are called **north celestial pole** and **south celestial pole**, respectively. The plane that passed through the Earth's center of mass and is perpendicular to the celestial axis is called **celestial equatorial plane**, coinciding with the equatorial plane of the Earth. The great circle where the celestial equatorial plane is intersected with the celestial sphere is called **celestial equator**. The plane containing the celestial axis and passing through any point on the surface of the Earth is called **celestial meridian plane**, and the great circle where the celestial meridian plane is intersected with the celestial sphere is called **celestial meridian circle**. Half of the great circle where the plane containing the celestial axis is intersected with the celestial sphere is called **hour**

circle. The great circle where the Earth's orbit around the Sun is intersected with the celestial sphere is called **ecliptic**, and the angle between the ecliptic plane and the equatorial plane is called **obliquity of the ecliptic** (ε), which is approximately 23.5°. Two points where a straight line passing through the center of the celestial sphere and perpendicular to the ecliptic plane is intersected with the celestial sphere are called **ecliptic poles**, in which the intersection point near the north celestial pole is called **north ecliptic pole** (Π_n), and that near the south celestial pole called **south ecliptic pole** (Π_s). When the Sun moves on the ecliptic from the southern hemisphere to northern hemispheres, the intersection of the ecliptic and the celestial equator is called **vernal equinox** (Γ). Similarly, when the Sun moves on the ecliptic from the northern to southern hemispheres, the corresponding intersection is called the **autumnal equinox**. The vernal equinox and the celestial equatorial plane are, respectively, the reference point and plane to establish the celestial reference system.

Time is the objective form of existence and movement of matter, and the time system is established by the reference to the movement of matter. According to the different forms of material movement, the commonly used time systems can be divided into three categories, as shown Fig. 5.2. One is the **universal time system**, which is based on the Earth's rotation movement, being also the earliest time metering system established by human beings; the other is the **ephemeris time system**, which is based on the orbital revolution of the Earth around the Sun; the third is the **atomic time system**, which is based on the movement characteristics of the atoms in the matter. According to the different space reference points selected for observing the Earth's rotation, the universal time system is divided into sidereal time, apparent solar time, mean solar time and universal time. According to the different definitions and application scenarios, the atomic time system is divided into International Atomic Time (TAI), Coordinated Universal Time (UTC) and navigation satellite time. The classifications of these commomly used time sytems are further shown in Fig. 5.3.

5.3.2.1 Sidereal Time

The **sidereal time** refers to a timescale that is the time measured by the apparent motion about the Earth, of the distant, so-called fixed points, such as the vernal equinox and stars. The time interval between two successive passages of the local meridian circle (the upper culmination) at the vernal equinox is called a **sidereal day**, which is subdivided into 24 sidereal hours, 1,440 sidereal minutes and 86,400 sidereal seconds. The sidereal time is numerically equal to the hour angle of the vernal equinox relative to the local meridian circle, and the hour angle of the vernal equinox is equal to the sum of the hour angle of any star and its right ascension. For the same moment, the sidereal time gotten at different observation stations is different. That is to say, the sidereal time is with locality, and the local sidereal time at a moment can be obtained by observing the stars. Due to the influence of the axial-precession and nutation, the direction of the Earth's rotation axis in space is constantly changing, so that the position of the vernal equinox on the celestial sphere is not fixed. For the same epoch, the north celestial pole can be divided into the apparent north celestial

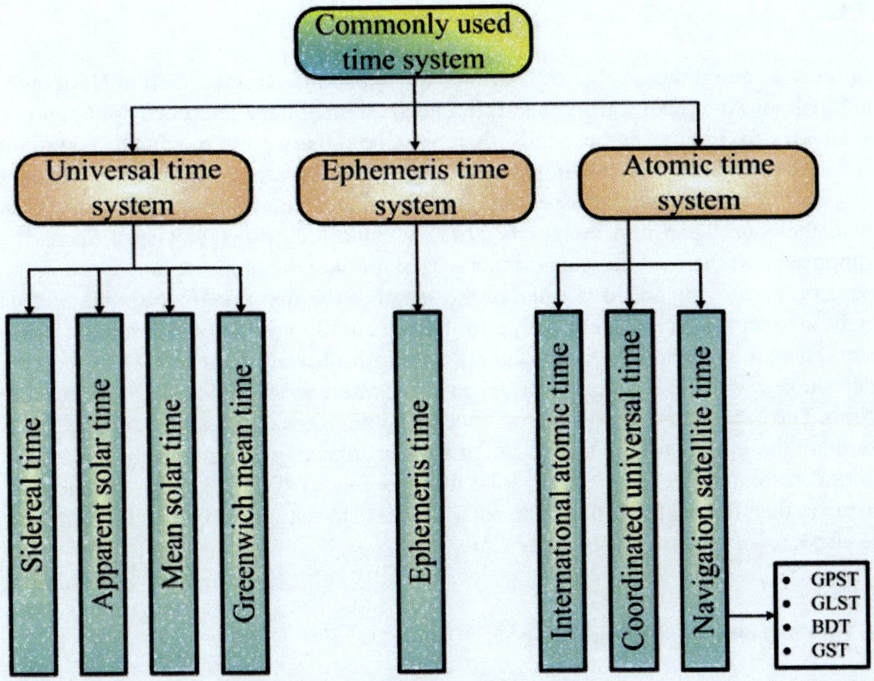

Fig. 5.3 Classifications of commonly used time systems

pole and the mean one, and the corresponding vernal equinox can also be divided into the apparent vernal equinox and the mean one. So, the sidereal time can also be divided into the apparent sidereal time and the mean one.

5.3.2.2 Apparent Solar Time

The **apparent solar time** refers to the time measured by the apparent motion of the Sun, taking the Sun's center as a reference point, also known as **true solar time**. The time interval between two successive passages of the Sun's center through the upper culmination is called **true solar day**. The apparent motion of the Sun on the celestial sphere results from the joint action of the rotation and revolution of the Earth. Because the Earth's orbit (ecliptic) is an ellipse with the perturbation from the Moon and planets, hence its revolution speed is non-uniform, and the ecliptic does not coincide with the celestial equator, resulting in the change of the length of the true solar day. It is shown from the observation results that the difference between the longest and shortest true solar days reaches 51 s. Obviously, the apparent motion of the Sun does not meet the basic conditions as a time measurement reference.

5.3.2.3 Mean Solar Time

In order to avoid the change of length of the apparent solar day, Simon Newcomb, a Canadian-American astronomer, introduced an imaginary reference point, known as **mean sun**. Firstly, supposed that there is an imaginary point moving at a constant speed in the ecliptic, its moving speed is equal to the average speed of the apparent motion of the Sun, and it passes through the perigee and apogee at the same time with the Sun. Then, the imaginary point is called **ecliptic mean sun**. Secondly, supposed that there is the second imaginary point moving at a constant speed in the equator, its moving speed is equal to the speed of the first imaginary point, and its right ascension as close as possible to the celestial longitude of the ecliptic mean sun. Then, the second imaginary point is called **equatorial mean sun**. The timescale determined with the equatorial mean sun as the reference point is called **mean solar time**. The time interval between two successive passages of the equatorial mean sun through the upper culmination is called a mean solar day, which is subdivided into 24 mean solar hours, 1,440 mean solar minutes and 86,400 mean solar seconds. As same as the sidereal time, the mean solar time is also with locality, and sometimes it is also known as **local mean solar time**.

5.3.2.4 Greenwich Mean Time

The **Greenwich Mean Time (GMT)** refers to the mean solar time at the Royal Observatory in Greenwich, London, taking the mean midnight (lower culmination) as a starting point of zero hour, also known as **Universal Time (UT)**. It can be seen that the universal time is based on the timescale of the mean solar time, and both discrepancy is only due to the difference of the time origin. The mean solar time is to take the Earth's rotation as a reference. However, it has been found from the modern astronomical observations that the rotation rate of the Earth is not uniform, and the average length of a solar day caused by tidal friction increases by 0.0016 s per hundred years. Moreover, there are various kinds of periodic, seasonal and irregular changes for the UT. Since 1965, the seasonal correction terms of the polar wander and rotation rate of the Earth have been introduced into the UT, and the resulting UT is expressed as UT1 and UT2, respectively, while the uncorrected UT directly calculated by observatories using their measurement data is expressed as UT0. Obviously, after the correction of the polar wander and rotation seasonal changes, the resulting UT2 is still affected by the long-term and irregular changes of the Earth's rotation rate. Therefore, the UT2 is not a strictly uniform time system yet.

5.3.2.5 Ephemeris Time

Celestial mechanics is the theory of celestial motion in the solar system on the basis of the classic Newtonian mechanics. Usually, the time in the motion equations is used as an argument to calculate the celestial position ephemeris. Obviously, the time here should mathematically be an evenly independent variable, known as the **Newtonian time**. When the position of the celestial body at a certain moment is gotten from astronomical observation, then Newton's time can easily be obtained from the ephemeris. However, as solving the differential equations of motion, due to the errors of the integral constants determined by the actual measurements, as well as those of the theoretical models of celestial motions, only the approximate Newton time can be given by any celestial position ephemeris. Generally, the time given by the celestial position catalogue is called **Ephemeris Time (ET)**. The ET could be obtained by observing the orbital position of any planet or satellite and then using an ephemeris, which lists the calculated orbital positions as a function of time. Due to the small difference between the ETs given by the position ephemerides of the Moon, Venus, Mars and Sun (actually the Earth), a uniform time system, Newcomb's tables of the Sun, is selected as the standard to define the numerical measure of the ET. In fact, Newcomb's tables are the orbital positions of the Earth about the Sun, developed mathematically by Simon Newcomb in 1898. That is to say, the ET is the time variable of Newcomb's tables of the Sun. The timescale of the ET is the tropical year, the length of which is equal to the time interval between two successive passages of the mean sun through the vernal equinox, that is, every 360° increase of the geometric mean of the Sun's ecliptic longitude is a tropical year. In 1960, the IAU decided that the position ephemerides of the celestial bodies in the solar system are reckoned by the ET, and the ephemeris second is used as the definition of a second in the International System of Units (SI). The ET based on the Earth's revolution around the Sun is not perfect either in theory or in practice. The annual apparent motion of the Sun on the celestial sphere is relatively slow. When the error of observing the position of the Sun is 0.1 arc-seconds, then the error of the reckoned ET will reach 2.5 s. In fact, the ET is determined by observing the Moon, but its accuracy is not better, and the needed observation time is so long that the results cannot be obtained in time. Furthermore, there are some astronomical constants in the definition of the ET. Once the astronomical constant system is redefined or modified, the discontinuity of the ET will be caused. So, another new time metering system, the atomic time system, has been adopted since 1967, and the ET has completely been replaced by the atomic time since 1984.

5.3.2.6 International Atomic Time

The **atomic time** is a timescale generated by atomic clocks, which furnish time more accurately than was possible with previous astronomical means, such as the sidereal time, mean solar time and ephemeris time. In 1967, at the 13th General Conference on Weights and Measures, a new definition of second length in the SI units was adopted: an **atomic second** refers to the duration of 9,192,631,770 periods of the oscillating radiation, based on the transition between two hyperfine levels of the ground state of Cesium 133 atom on the sea level and in zero magnetic field. The atomic time has high precision and stability, so it has been developed and widely used. Many countries have established their own local atomic time systems, but there are always differences between the atomic times at different places. For this reason, a lot of the atomic clocks distributed at the laboratories of various countries are compared with one another, and then the global unified atomic time system is calculated by data processing, which is called **International Atomic Time (TAI)**. The TAI was initially established and maintained by the Bureau International de l'Heure (BIH), and maintained by the Bureau International des Poids et Measures (BIPM) since 1988. The time origin of the TAI is at zero hour of the UT on January 1, 1958, and the timescale is the atomic second defined by the SI units. It was initially hoped that the UT and TAI had the same time origin, but for technical reasons, there was a small difference of 0.0039 s between both. At present, the TAI is based on a system consisting of about 270 laboratory-constructed atomic clocks. Signals from these atomic clocks are transmitted to the BIPM and used to generate the TAI. The Cesium fountain clocks now provide the SI unit's second to an unprecedented level of accuracy. These clocks are predicted to be offset by less than one second for more than 50 million years.

5.3.2.7 Coordinated Universal Time

Although the implementation of the atomic time makes a qualitative leap in the accuracy of time measurement, the atomic time cannot replace the application of the UT. The UT based on the rotation of the Earth conforms to people's daily habits and still has a wide range of uses. However, due to the trend that the Earth's rotation rate slows down for a long time, the UT is about 1 s slower than the atomic time every year, and the difference between the two is accumulating year by year. In order to avoid excessive deviation between the published atomic time and the UT, a compromised time system based on the atomic second length has been adopted since 1972, which is well known as **Coordinated Universal Time (UTC)**. The length of a second of the UTC is exactly equal to that of the atomic time, and the leap second method is used to make the UTC as close as possible to the moment of the UT. When the time difference between the UTC and UT exceeds ± 0.9 s, a leap second with the positive or negative is introduced into the UTC, which is generally added at the end of June 30 or December 31 of each year. The need for additions of leap seconds is determined by the International Earth Rotation and Reference Systems Service (IERS), located at Paris Observatory. Except for the introduction of the leap seconds, the UTC is essentially the TAI, so the establishment, measurement and comparison of the UTC systems are basically as same as the atomic time. The UTC is calculated by the weighted average of the BIPM's global atomic clocks, with the highest long-term stability and precision.

Currently, there are 72 laboratories that have established and maintained the UTC systems around the world. According to the international agreement, the UTC of each laboratory is expressed as UTC (k), where k is the abbreviation of the laboratory name. For example, UTC (NIM) stands for the UTC maintained by the National Institute of Metrology in Beijing, China; UTC (NTSC) stands for the UTC maintained by the National Time Service Center at Lintong, Xi'an, China; UTC (USNO) stands for the UTC maintained by the U.S. Naval Observatory in Washington D.C.; UTC (APL) stands for the UTC maintained by the Applied Physics Laboratory in Maryland, the United States; UTC (SU) stands for the UTC maintained by the Institute of Metrology for Time and Space in Moscow, Russia; UTC (PTB) stands for the UTC maintained by the Federal Physical Institute (Physikalisch-Technische Bundesanstalt), the national metrology institute with scientific and technical service tasks, in Braunschweig and Berlin, Germany. The TAI can be provided in the form of "paper clock", while the UTC is broadcasted by the time service stations to meet the needs of national economic construction and scientific research applications. Therefore, the UTC (k) everywhere is achieved by the atomic clocks, which would be a physical atomic clock or a group of atomic clocks controlled by computers, and its signals are outputted by the computers.

5.3.2.8 Navigation Satellite Time

The navigation satellite system is the space-based radio navigation and time transmission system, and thus the generation, measurement and maintenance of time information is one of its core tasks. The navigation satellite time is with the atomic time as the timescale, selecting the corresponding UTC (k) as the time origin, and realized and maintained by the atomic clocks at the ground control stations together with those of the onboard navigation satellites. The **GPS Time (GPST)** is based on the UTC (USNO) time maintained by the U.S. Naval Observatory, and its time maintenance is achieved together by the atomic clocks deployed at the master control stations, monitor stations and on-orbit satellites. The origin of the GPST is at 0:00 UTC (USNO) on January 6, 1980, and its timescale is length of a second of the TAI. The GPST is a continuous time system without the leap seconds, which is expressed by the GPS weeks and second counts of a week. There is a constant deviation of 19 s between the GPST and TAI. Besides the leap seconds, the GPS ground control stations keep the system difference between the GPST and UTC (USNO) within 1 μs, which can be corrected by using the corresponding parameters of the navigation message. The **GLONASS Time (GLST)** is based on the UTC (SU) maintained by the Institute of Metrology for Time and Space in Moscow, achieved together by the atomic clocks deployed at the GLONASS system control center, TT&C stations, monitor stations, Laser ranging stations and time synchronization stations. The GLST is a time system with synchronous leap seconds, with the system difference of 3 h and less than 1 μs between the GLST and UTC (SU) and without the leap second differences between the two. The **BeiDou Navigation Satellite System Time (BDT)** is continuous timing system based on the second length of the SI units, and its initial epoch is at 0:00:0.0 UTC on January 1, 2006, which is expressed by the BeiDou weeks and second counts of a week. The BDT is connected with the international UTC time through the UTC (NTSC) maintained by National Time Service Center of China, and the deviation between the BDT and UTC is kept within 100 ns (module 1 s). The **Galileo Navigation Satellite System Time (GST)** is generated and maintained by the two active hydrogen atomic clocks and four cesium atomic clocks installed at the master control station, and is compared with the UTC times maintained by the Italian National Institute of Metrological Research (IT), the British National Physical Laboratory (NPL), the French National Metrology and Testing Laboratory (OP) and the Federal Physical Institute of Germany (PTB), to correct the time from the hydrogen atomic clocks. The GST is similar to the GPST, also a continuous timing system, with the timescale of the TAI, and its daily stability is better than 4.3×10^{-15} s/s relative to the TAI, with the time bias of less than 28 ns.

5.3.3 *International Terrestrial Reference System and Its Realization*

The terrestrial reference system is a non-inertial coordinate system which is fixed on the Earth, rotating together with the Earth. If the Earth is an ideal rigid body, any three-axis coordinate system fixed on the Earth is suitable, and the choice of the terrestrial reference system only depends on the convenience of the ground observation station. However, the Earth is a discontinuous deformation body, and due to the coupling motions among various Earth spheres and the polar wander, it is complicated to fix the reference system in the Earth using the ideal way. The ideal terrestrial reference system is a kind of reference system which only has deformation, no whole rotation and translation relative to the Earth, while relative to the inertial reference system, only contains the whole motion of the Earth (i.e., the rotation and revolution of the Earth). For the ideal terrestrial reference system, besides defining its fixed way relative to the Earth, its coordinate origin, scale and orientation must also be defined in practical application. Therefore, the establishment of geocentric coordinate system should meet the following four basic conditions:

(1) The coordinate origin is located at the center of mass of the whole Earth, including the solid Earth, ocean and atmosphere.
(2) The coordinate scale is the scale within a local terrestrial frame based on the general relativity (seen in Sect. 5.4).
(3) The orientation parameters are determined by the Earth rotation parameters at an epoch. For example, the Conventional Terrestrial Pole (CTP) and the zero meridian line are determined at an epoch by the BIH. The *Z*-axis and *X*-axis in the BIH 1984.0 coordinate system are, respectively, the CTP and Zero Meridian determined at the 1984.0 epoch by the BIH.
(4) The orientating evolution over time is subject to the constraint that there are not the whole rotation and translation relative to the Earth.

Generally, the physical frame composed of a series of observation stations on the Earth's surface is the best approximate and concrete form of the ideal terrestrial reference system, and its motion and deformation must meet the demand that there are not the whole rotation and translation relative to the Earth. On the basis of selecting the physical frame, the parameters and models to describe the frame need to be determined. The purpose of establishing the physical frame model based on the ground stations is to determine its geometric formation (the stations' coordinates) and internal deformation (time-varying model). The connotation of the conventional terrestrial reference system contains the theoretical methods and basic physical frame to determine these parameters. The definitions and uses of the parameters and models related to the terrestrial reference system are usually consistent with the international convention, and so is the determination of the mean pole of the Earth.

The polar wander makes orientating of the terrestrial coordinate system change, which brings much inconvenience in practice. Thereby, in 1967, the International Astronomical Union (IAU) and the International Association of Geodesy (IAG) proposed to use five International Latitude Service observing stations, including Carloforte (in Italy), Kitab (in USSR/Russia), Mizusawa (in Japan), Gaithersburg (in USA) and Ukaih (in USA), using the average latitudes between 1900 and 1905, and to determine the position of the mean terrestrial pole, as a reference origin. That is to say, the mean terrestrial pole is to use the average position of the Earth's rotation axis on the Earth's surface during this period, commonly known as **Conventional International Origin (CIO)**, and the corresponding equatorial plane of the Earth is called **mean equatorial plane**, or **conventional equatorial plane**. In practice, the CIO is still widely used as the CTP. The terrestrial coordinate system established with the CTP as the reference point is called **Conventional Terrestrial System (CTS)**, and the coordinate system corresponding to the instantaneous terrestrial pole is called **Instantaneous Terrestrial System (ITS)**, as shown in Fig. 5.4.

The International Terrestrial Reference System (ITRS) is an internationally recognized conventional terrestrial reference system with the best accuracy and stability. In 1911, the International Union of Geodesy and Geophysics (IUGG) decided that the IERS would take charge of the definition, realization and maintenance of the ITRS. The basic definition of the ITRS is as follows:

(1) The coordinate origin is located at the center of mass of the whole Earth, in which the mass of the whole Earth is the total sum of mass of the solid Earth, ocean and atmosphere.
(2) The length of a meter defined in the SI units is used as the coordinate scale, which is a local terrestrial reference frame scale defined based on the general relativity. This coordinate scale is consistent with the timescale of Geocentric Coordinate Time (TCG) defined by the IAG and IUGG, and it can be calculated by the theoretical models of relativity.
(3) The orientation of coordinate system is consistent with the CTP and BIH Zero Meridian at the 1984.0 epoch.
(4) The time evolution of the orientating parameters does not bring up residual global rotation relative to the crust. In other words, there is not a whole residual rotation for the horizontal motion of the reference system relative to the whole Earth's plates.

The ITRS is provided in the form of International Terrestrial Reference Frame (ITRF) through the IERS ITRS Product Center, including the independent terrestrial reference frames established by using space geodetic technologies, such as the VLBI, Lunar Laser Ranging (LLR), SLR, GNSS and DORIS. The specific contents of

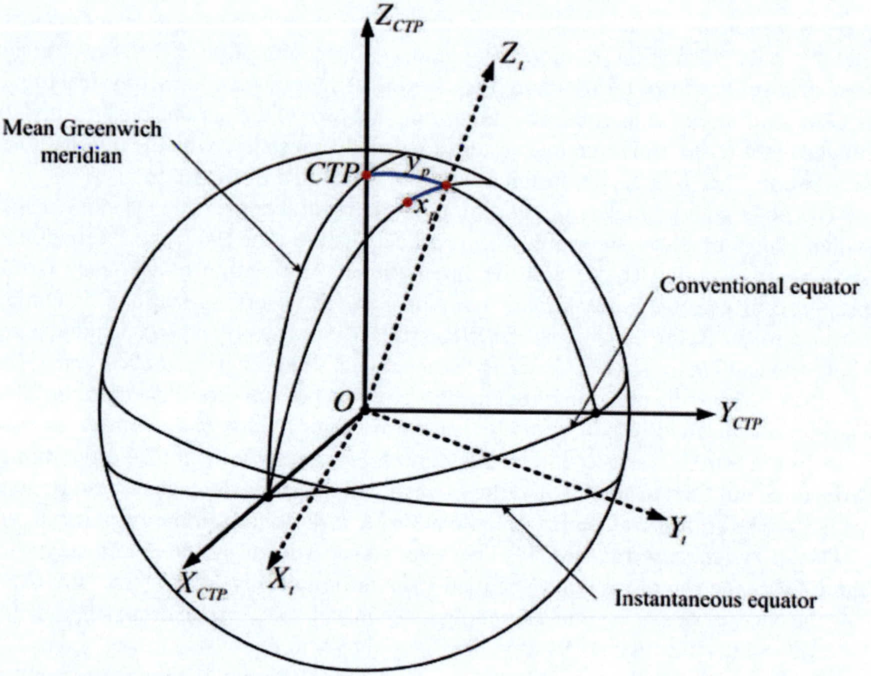

Fig. 5.4 Relationship between CTS and ITS

each reference frame include the sites' positions, velocities and complete covariance matrices provided in the form of Solution Independent Exchange (SINEX) format, and using the integrated data processing way, the unified ITRF results can be gotten finally. The integrated data processing is achieved by the local connection of the co-location sites, which are the observation stations with more than two kinds of same or different geodetic techniques near the same station. The realization methods of co-location observation include the simultaneous observation of the same technique, different time observation of the same technique, observation of changing position and co-location observation with different techniques. Consequently, the ITRS has high accuracy and long-term stability.

The ITRF is the specific realization form of the ITRS, composed of the position coordinates of many geodetic sites and their rates, and these geodetic sites fixed on the Earth's surface only make to move linearly. The establishment of the ITRF network sites should meet the following four basic conditions:

(1) The continuous observation times of the sites are generally at least 3 years.
(2) The sites are away from the plate boundaries and the deformation areas.
(3) The velocity accuracies of the sites are better than 3 mm per year.
(4) The velocity residuals of at least three different solutions are less than 3 mm per year.

The orientation of ITRS is determined by the CTP and Zero Meridian of the BIH at epoch 1984.0, and its specific position on the Earth is reflected by the coordinates of the sites distributed globally. Theoretically, as long as the position coordinates of any three sites on the Earth are known, a conventional terrestrial coordinate system can be determined uniquely. However, due to the observation errors, there will be errors for the initial meridian plane, coordinate origin and Z-axis direction, and thus more than three reference sites are usually used for orientating the terrestrial coordinate system. Moreover, due to the influence of the Earth's plate motion and local crustal deformation, the coordinates of the reference sites always change dynamically. The ITRF is a geocentric coordinate system derived from the observation data of the global sites, a four-dimensional geocentric coordinate reference frame. Besides giving a high-precision space rectangular coordinate form of the sites, it also provides the drift rates of the sites. At present, the site's coordinate accuracy gets to the level of millimeter. By using multiple observation means like the VLBI, SLR, LLR, GNSS and DORIS, and combining with the solution results from multiple data analysis center, the ITRF composed of a series of the sites' coordinates and velocities relative to a reference epoch can be obtained. After establishing the ITRF, in order to get the ITRF coordinates of an undetermined point, it is only necessary to measure the position coordinates of the point relative to the ITRF sites.

As early as 1984, an initial terrestrial reference frame, known as the BIH Terrestrial System (BTS84), was established by the BIH using the observation data from the VLBI, LLR, SLR and Transit, and then the series of BTS85, BTS86 and BTS87 were completed sequentially. In 1988, the IUGG and IAU jointly created the IERS, and then the ITRF is established by the IERS. At present, there are thirteen ITRF versions released by the IERS, including ITRF88, ITRF89, ITRF90, ITRF91, ITRF92, ITRF93, ITRF94, ITRF96, ITRF97, ITRF2000, ITRF2005, ITRF2008 and ITRF2014, in which the number of year following "ITRF" denotes the data of the next year used for establishing the reference system. For example, the ITRF97 denotes the reference frame with the sites' positions and velocities established in 1999, by using all available IERS data up to 1998. The observation techniques, reference epochs and plate motion models used in the versions from the ITRF88 to ITRF2014 are listed in Table 5.1.

With the increasing number of the observation sites and co-location observation sites distributed all over the world, as well as the improvement of performance of various measurement devices, the ITRF will be refined constantly. For example, the ITRF88 has about 100 stations and 22 VLBI/SLR/LLR co-location sites, the ITRF2000 includes 500 stations and 101 co-location sites, and the ITRF2014 includes 1499 stations and 975 co-location sites. Meanwhile, with constant improvement of the data processing ways for various IERS analysis centers, the difference between the coordinate and velocity solutions of ITRF sites will become smaller and smaller. Currently, the ITRF accuracy has reached the millimeter level. The transformation parameters between the ITRF versions can be obtained from the websites of the IERS and International GNSS Service (IGS).

Table 5.1 Versions from ITRF88 to ITRF2014

Number of version	Observation technique	Epoch	Plate motion model	Available year
ITRF88	VLBI/SLR/LLR	1988.0	AM0-2/AM1-2	1989
ITRF89	VLBI/SLR/LLR	1988.0	AM0-2/AM1-2	1990
ITRF90	VLBI/SLR/LLR	1988.0	AM0-2/AM1-2	1991
ITRF91	VLBI/SLR/LLR/GPS	1988.0	AM0-2/NNR-NUVEL-1	1992
ITRF92	VLBI/SLR/LLR/GPS	1988.0	AM0-2/NNR-NUVEL-1	1994
ITRF93	VLBI/SLR/GPS	1993.0	NNR-NUVEL-1A	1995
ITRF94	VLBI/SLR/GPS	1993.0	NNR-NUVEL-1A	1996
ITRF96	VLBI/SLR/GPS/DORIS	1997.0	NNR-NUVEL-1A	1998
ITRF97	VLBI/SLR/GPS/DORIS	1997.0	NNR-NUVEL-1A	1999
ITRF2000	VLBI/SLR/GPS/DORIS/LLR	1997.0	NNR-NUVEL-1A	2001
ITRF2005	VLBI/SLR/GNSS/DORIS	2000.0	NNR-NUVEL-1A	2006
ITRF2008	VLBI/SLR/GPS/DORIS	2005.0	NNR-NUVEL-1A	2009
ITRF2014	VLBI/SLR/GNSS/DORIS	2010.0	NNR-NUVEL-1A	2015

5.3.4 Defects on the Theory

In the Newtonian mechanics, two concepts of mass are introduced: one is the inertial mass introduced from Newton's Second Law of Motion; another is the gravitational mass introduced from Newton's Law of Universal Gravitation.

According to Newton's Second Law of Motion, there is

$$\mathbf{F} = m\mathbf{a}, \tag{5.9}$$

where m is the inertial mass of an object, representing the ability for the object to resist acceleration; $\mathbf{a}$ is the acceleration of the object relative to a reference system; $\mathbf{F}$ is the total sum of external forces acting on the object.

Only in the inertial coordinate system, formula (5.9) can be established, and its coordinate transformation must satisfy the Galileo space-time transformation relation. Meanwhile, from Newton's Law of Universal Gravitation, there is

$$\mathbf{F} = -\sum_{i=1}^{n} \frac{Gmm_i}{\|\mathbf{r} - \mathbf{r}_i\|^3}(\mathbf{r} - \mathbf{r}_i), \tag{5.10}$$

where G is the universal gravitational constant; m is the gravitational mass of the object, representing the ability for the object to generate and accept gravity; m_i is the mass of object i, generating the gravitational force on the object with mass of m; $\mathbf{r}$ is the position vector of object m; $\mathbf{r}_i$ is the position vector of object m_i; $\mathbf{F}$ is the total sum of the gravitational forces acting on the object with the mass of m, resulting from the objects with the mass of $m_1, m_2, \ldots, m_n$, respectively.

The gravity is considered as a conservative force. At the position vector $\mathbf{r}$, if the strength $\mathbf{g(r)}$ of the gravitational field is defined as the gravitational force acting on the object with a unit of mass, then

$$\mathbf{g(r)} = \frac{\mathbf{F}}{m} = -\nabla\Phi(\mathbf{r}), \tag{5.11}$$

where $\Phi(\mathbf{r}) = -G\sum_{i=1}^{n}\frac{m_i}{\|\mathbf{r}-\mathbf{r}_i\|}$, known as **gravitational potential**;

$\nabla = \frac{\partial}{\partial x}\mathbf{i} + \frac{\partial}{\partial y}\mathbf{j} + \frac{\partial}{\partial z}\mathbf{k}$, known as **Hamiltonian operator**.

For continuously distributed matter, the gravitational potential function is

$$\Phi(\mathbf{r}) = -G\int\frac{\rho(\mathbf{r}')}{\|\mathbf{r}-\mathbf{r}'\|}d^3\mathbf{r}', \tag{5.12}$$

where $\rho(\mathbf{r}')$ is the matter density at position vector $\mathbf{r}'$.

By using Gauss's theorem, it can be proved that the differential expression of Eq. (5.12) satisfies Poisson's equation, namely,

$$\Delta\Phi = \nabla^2\Phi = 4\pi G\rho, \tag{5.13}$$

where $\Delta = \nabla^2 = \frac{\partial^2}{\partial x^2} + \frac{\partial^2}{\partial y^2} + \frac{\partial^2}{\partial z^2}$, known as **Laplace operator**.

For the gravitational potential of a celestial body like the Earth, its interior satisfies Poisson's equation, while its exterior space does Laplace's equation. Conceptually speaking, the inertial mass and gravitational mass are essentially two different physical quantities. If the ratio of both is the same for all substances, then it can be treated as the same quantity in practice. It is the problem of the equivalence between the inertial mass and the gravitational mass. For this reason, Isaac Newton (1687), Roland Eötvös (1889), Robert H. Dicke (1964) and Vladimir Braginsky (1971), et al., successively designed the special experiments, to measure the small deviation of the ratio of the inertial mass to the gravitational mass relative to 1. Generally, when the measurement accuracy is on the level of 10^{-12}, the inertial is equal to the gravitational mass. In other words, the ratio of the inertial mass to the gravitational mass has nothing to do with the size and material of the object, and the two are equal within the range of experimental accuracy, for which the reason needs further theoretical research to explain.

For an object, if only gravity acts on it, substituting formula (5.10) into formula (5.9), then the physical quantity m used to describe the inherent physical properties of the object will be eliminated. That is to say, no matter what the nature of the object itself is, as long as the initial velocity is same, any object will move along the same trajectory. So, there is such question about the Newtonian mechanics: the inertial mass m of an object was originally introduced as an important dynamics quantity, but it was lost as discussing the dynamical problem on the gravitational field. It is the first issue that Newton's theory of gravitation cannot explain itself.

Secondly, the real difficulty of the Newtonian gravity is that the Mercury's orbital precession calculated by Newton's theory is not consistent with the actual observation results, and the difference between both is 43 arc-seconds per 100 years.

Thirdly, if it is admitted that the universe is infinite and supposed that the matter is evenly distributed in universe, then the field strength at any point in the gravitational field will be infinite by using the Newtonian gravity. It is the so-called **Neumann–Seeliger puzzle**.

Finally, Newton's theory of gravitation cannot meet the Lorentz transformation, and the universal gravitation is a kind of acting force cross over distance, whose attraction to objects reaches instantaneously. In other words, the propagation of its field energy is infinite, which does not conform to the idea of the special relativity.

It should be pointed out that Newton's theory of gravitation is successful in describing the motion of objects in the solar system and around the Earth, and its calculation accuracy is also quite high. In particular, the Neptune in the solar system was successfully predicted and discovered by using Newton's theory of gravitation. As the famous physicist Albert Einstein said: a principle with such a wide range of universality is effective and with such a high precision, for physical phenomena in one field, but it is ineffective in another field; it is impossible from a priori point of view.

In fact, Newton's theory of gravitation is completely applicable to weak gravitational field. The basic criterion for the strength of the gravitational field of a celestial body is expressed as

$$\alpha = \frac{2GM}{rc^2} = \frac{r_g}{r}, \tag{5.14}$$

where $r_g = \frac{2GM}{c^2}$, called the gravitational radius of celestial body; G is the gravitational constant; M is the mass of celestial body; r is the physical radius of celestial body; c is the velocity of light.

When $\alpha \ll 1$, the corresponding gravitational field is regarded as the weak gravitational field. The strengths of the gravitation fields of the Universe, Galactic System and typical celestial bodies are listed in Table 5.2, from which it can be seen most of the celestial bodies belong to the weak gravitational field, and thus Newton's theory of gravitation is applicable completely.

Table 5.2 Strengths of the Universe, Galactic System and typical celestial bodies

Name	α	Name	α
Moon	$10^{-10.1}$	White dwarf	$10^{-4.0}$
Earth	$10^{-8.9}$	Neutron star	$10^{-1.0}$
Sun	$10^{-6.0}$	Universe	1
Galactic System	$10^{-5.4}$		

5.4 Space-Time Reference Based on General Relativity

Any physical quantity is represented by a set of ordinal numbers, which are usually associated with the selection of coordinates. In order to study the relationship between the ordered array and the coordinate transformation in n-dimensional affine space, the concept of tensor is introduced. The tensor analysis is gradually developed in studies on differential geometry. In the eighteenth and nineteenth century, Karl F. Gauss, Georg F. Bernhard Riemann and Edwin B. Christoffel, et al., introduced the concept of tensor to study the differential geometry. Between 1887 and 1901, a fundamental framework of tensor analysis was established by Gregorio Ricci-Curbastro and Tullio Levi-Civita, et al. In 1916, the tensor analysis and Riemann geometry were used by Albert Einstein to describe the basic principle of general relativity, which greatly promoted the development of differential geometry. In this way, the tensor analysis becomes a powerful mathematical tool, used widely.

The XPNAV is carried out in the large-scale space-time frame, and its theoretical basis is the space-time metric relation of the general relativity. In the studies on the space-time reference theory and photon arrival-time transformation model for the XPNAV, some fundamental theories, such as the tensor analysis, Riemann space, gravity and space-time bending, relativistic timescale, international celestial reference system and its realization, will be involved. Thereby, the related concepts and basic methods are introduced in this section.

5.4.1 *Concept of Riemann Space*

5.4.1.1 Tensor Foundation and Metric Tensor

In three-dimensional affine space, let $x^{i'}$ and x^j, respectively, stand for the coordinates of the new and old coordinate systems, and the basis vectors of the two coordinate systems are, respectively, $\mathbf{g}_{i'}$ and $\mathbf{g}^{i'}$ as well as $\mathbf{g}_j$ and $\mathbf{g}^j$, the coordinate transformation from the old coordinate system x^j to the new one $x^{i'}$ can be expressed as

$$x^{i'} = x^{i'}\left(x^j\right), \left(i', j = 1, 2, 3\right). \tag{5.15}$$

When $x^{i'}\left(x^j\right)$ is a monotone continuous differential-function within the definition domain of x^j, then there is the Jacobian determinant that does not equal to zero, namely,

$$\left.\begin{array}{l} \left|\dfrac{\partial x^{i'}}{\partial x^j}\right| \neq 0 \\ \left|\dfrac{\partial x^j}{\partial x^{i'}}\right| \neq 0 \end{array}\right\}. \tag{5.16}$$

Meanwhile, the corresponding inverse transformation relation of expression (5.15) is expressed as

$$x^{j} = x^{j}\left(x^{i'}\right). \tag{5.17}$$

In this way, the functional relation of mutual transformation between the two coordinate systems is established.

According to the definition of covariant basis vectors, the transformational relations of the covariant basis vectors between two coordinate systems are

$$\mathbf{g}_{i'} = \beta_{i'}^{j}\mathbf{g}_{j}. \tag{5.18}$$

$$\mathbf{g}^{i'} = \beta_{j}^{i'}\mathbf{g}^{j}, \tag{5.19}$$

where $\beta_{i'}^{j} = \frac{\partial x^{j}}{\partial x^{i'}}$, known as **covariant transformation coefficients;** $\beta_{j}^{i'} = \frac{\partial x^{i'}}{\partial x^{j}}$, known as **contravariant transformation coefficients.**

For any vector **a** in three-dimensional affine space, the different basis vectors can be decomposed in different coordinate systems. In the new and old coordinate systems, the different covariant components of the same vector are obtained by decomposing the contravariant basis, while the different contravariant components of the same vector are obtained by decomposing the covariant basis. Although the components of different coordinate systems are different, the sum of decomposition vectors will not change due to the different coordinate systems. Therefore, in the different coordinate systems, the relations between the components of the same vector can be determined by the transformations of the basis vector, that is

$$\left.\begin{aligned} a_{i'} &= \beta_{i'}^{j}a_{j} \\ a^{i'} &= \beta_{j}^{i'}a^{j} \end{aligned}\right\}. \tag{5.20}$$

Under the condition of satisfying expression (5.15), the components of any vector in three-dimensional affine space have the transformational law described in expression (5.20). This law of coordinate transformation is extended to n-dimensional affine space, to study the coordinate transformation of an ordered array. In n-dimensional affine space, for a quantity with n^{N} components, each component can be transformed by the following expression:

$$x^{i'} = x^{i'}\left(x^{j}\right). \left(i', j = 1, 2, \ldots, n\right) \tag{5.21}$$

According to this transformation law, there is

$$T_{i'_1 \cdots i'_m}^{j'_1 \cdots j'_l} = \beta_{j_1}^{j'_1} \cdots \beta_{j_l}^{j'_l}\beta_{i'_1}^{i_1} \cdots \beta_{i'_m}^{i_m}T_{i_1 \cdots i_m}^{j_1 \cdots j_l}, \tag{5.22}$$

where $T_{i_1\cdots i_m}^{j_1\cdots j_l}$ is the function about x^j, $T_{i'_1\cdots i'_m}^{j'_1\cdots j'_l}$ is the function about $x^{i'}$, and $N = l+m$.Thereby, the quantity with n^N components is called ***N*-order tensor**, and $T_{i_1\cdots i_m}^{j_1\cdots j_l}$ (a total of n^N) is called the **mixed-variant component** with the m-order covariant and l-order contravariant for the N-order tensor.

Obviously, the concept of tensor is the generalization of vector. The scalar is the zero-order tensor; the vector is the first-order tensor; the matrix is the second-order tensor; the three-dimensional array is the third-order tensor, with an arrangement form of "cubic matrix"; the tensors with the order number of more than three are usually difficult to express by using a graph.

For a given affine space coordinate system x^i, there is always a set of the determined covariant basis vectors $\mathbf{g}_i$ and a set of the determined contravariant basis vectors $\mathbf{g}^i$. Any vector in the affine space can be decomposed, respectively, according to the two sets of the basis. As a result, the two sets of the basis can also be decomposed each other, namely,

$$\mathbf{g}_i = g_{ij}\mathbf{g}^j, \tag{5.23}$$

$$\mathbf{g}^i = g^{ij}\mathbf{g}_j, \tag{5.24}$$

where g_{ij} is the covariant component of the basis vector $\mathbf{g}_i$, and g^{ij} is the contravariant component of the basis vector $\mathbf{g}^i$.

Supposed that an entity of the tensor g_{ij} is expressed as

$$\mathbf{G} = g_{ij}\mathbf{g}^i\mathbf{g}^j. \tag{5.25}$$

Moreover, it can also be proved that

$$\mathbf{G} = g^{ij}\mathbf{g}_i\mathbf{g}_j. \tag{5.26}$$

Apparently, g_{ij} and g^{ij} are, respectively, the covariant and contravariant component of the same second-order tensor **G**, representing the scale relation in the affine-space coordinate system, and thus the tensor **G** is known as **metric tensor**.

5.4.1.2 Riemannian Space and Minkowski Space

If there is a metric tensor, $g_{ij}(x^i) = g_{ji}(x^i)$, in n-dimensional vector space, and the distance between the adjacent two points in the space is determined by the following positive-definite quadratic form,

$$ds^2 = g_{ij}dx^i dx^j. \tag{5.27}$$

Then the vector space is known as **Riemannian space**. The quadratic form ds^2 is the line element of the Riemannian space, and g_{ij} is the metric tensor of the Riemannian space. In this way, the differential form of the arc length of a curve can be defined as

$$ds = \sqrt{g_{ij}dx^i dx^j}. \tag{5.28}$$

Both the three-dimensional Euclidean space that is well known, and the four-dimensional Murkowski space that is discussed in the special relativity, are special cases of the Riemannian space. In three-dimensional Euclidean space, using Cartesian rectangular coordinate system, let $x^1 = x$, $x^2 = y$ and $x^3 = z$, then the distance formula between two adjacent points is expressed as

$$ds^2 = dx^2 + dy^2 + dz^2. \tag{5.29}$$

Correspondingly, the metric tensor is

$$g_{ij} = \begin{bmatrix} 1 & 0 & 0 \\ 0 & 1 & 0 \\ 0 & 0 & 1 \end{bmatrix}. \tag{5.30}$$

It is shown that in the Cartesian coordinate system, the components of the metric tensor have nothing to do with the space-point's position. When the spherical coordinate system is used, and let $x^1 = r$, $x^2 = \theta$ and $x^3 = \varphi$, then the distance formula between two adjacent points is expressed as

$$ds^2 = dr^2 + r^2 d\theta^2 + r^2 \sin^2\theta d\varphi^2. \tag{5.31}$$

And thus, the corresponding metric tensor is

$$g_{ij} = \begin{bmatrix} 1 & 0 & 0 \\ 0 & r^2 & 0 \\ 0 & 0 & r^2\sin^2\theta \end{bmatrix}. \tag{5.32}$$

It can be seen that in the three-dimensional Euclidean space, when the spherical coordinate system is selected, the metric tensor is a function of the point's position coordinates, which will change with the point's position.

For the Minkowski space, a four-dimensional space-time relation is established by using the one-dimensional time coordinate and three-dimensional space coordinates in the coordinate system. Let $x^0 = ct$, $x^1 = x$, $x^2 = y$ and $x^3 = z$, then the invariant distance is expressed as

$$ds^2 = -c^2dt^2 + dx^2 + dy^2 + dz^2, \tag{5.33}$$

where c is the velocity of light. With the unit of length as the dimension of the coordinate components, the metric tensor is

$$g_{ij} = \begin{bmatrix} -1 & 0 & 0 & 0 \\ 0 & 1 & 0 & 0 \\ 0 & 0 & 1 & 0 \\ 0 & 0 & 0 & 1 \end{bmatrix} = \eta_{ij}, \quad (5.34)$$

where η_{ij} is the four-dimensional Minkowski metric of space-time.

In the Riemannian space, when a proper coordinate system is selected, and the metric tensor has the following basic forms,

$$g_{ij} = \begin{cases} \pm 1 & i = j \\ 0 & i \neq j \end{cases}. \quad (5.35)$$

And then, this space is called **flat Riemannian space.**

5.4.1.3 Differential Equation of Geodesic Line

In the Riemannian space, the affine connection coefficients can always be determined in a unique way to meet the following conditions:

(1) The affine connection is torsion-free, i.e., $\Gamma_{ij}^k = \Gamma_{ji}^k$.
(2) The translation generated by the affine connection keeps the vector's length unchanged.

This affine connection is called **Riemannian connection.**

In the Riemannian space, investigating the contravariant vector $\mathbf{a}^i(M)$ of the point M, the vector is made the translation to the adjacent point N by using the Riemannian connection, and the connection equation, with the first-order small quantity and keeping its length constant, can be expressed as

$$g_{ij,k} - g_{lj}\Gamma_{ik}^l - g_{il}\Gamma_{kj}^l = 0. \quad (5.36)$$

The connection equation with respect to the subscripts i and j is symmetric, and it is considered that the Riemannian connection space is torsion-free. As a result, the number of independent connection equations is equal to that of independent components of the Riemannian connection.

In the Riemannian space, a scalar integral is introduced for any curve as follows:

$$s = \int_{M_0}^{M} ds, \quad (5.37)$$

where ds is the invariant distance between two adjacent points on the curve; M_0 is a fixed point on the curve; M is any point on the curve; s is the proper length from M_0 to M, which is a natural scalar argument to mark the point on the curve.

Here, taking s as an affinor, the tangent vector is defined as

$$a^i = \frac{dx^i}{ds}. \tag{5.38}$$

Considering that the tangent vector **a** is always the unit vector, the geodesic differential equation with s as the affinor can be derived in the Riemannian space.

$$\frac{da^i}{ds} + \Gamma^i_{jk} a^j a^k = 0. \tag{5.39}$$

5.4.2 *Gravity and Space-Time Bending*

5.4.2.1 Einstein Field Equations

At the end of nineteenth century, James C. Maxwell established the electrodynamics theory, in which the basic laws of electromagnetism, well known as Maxwell's equations, was expressed in detail, and predicted the existence of electromagnetic waves. The velocities of the waves are c, which is the velocity of light, but what is the c calculated relative to? At first, it was thought that the c was relative to a kind of the medium propagating electromagnetic waves, called ether, like the sound wave propagated by air. However, it is shown through experiments that there is no the so-called ether. That is to say, absolute velocity cannot be given by electromagnetism or optics.

In 1905, on the basis of previous studies, Einstein concluded a general principle that all physical laws should have the same form in any inertial system, which is known as the **special relativity**, including two basic principles: one is that any inertial system is equivalent; another is that the velocity of light is constant. According to the special relativity, a new space-time transformation between inertial coordinate systems, Lorentz transformation, can be derived as follows:

$$\left.\begin{aligned} x' &= \frac{x-vt}{\sqrt{1-\frac{v^2}{c^2}}} \\ y' &= y \\ z' &= z \\ t' &= \frac{t-\frac{xv}{c^2}}{\sqrt{1-\frac{v^2}{c^2}}} \end{aligned}\right\}. \tag{5.40}$$

When $v \ll c$, the Lorentz transformation degenerated into the Galilean transformation. In other words, the Galilean transformation is a special case of the Lorentz transformation, and Maxwell's equations are covariant for the Lorentz transformation, so two theoretical systems, the electromagnetism and Newtonian mechanics, have been unified by the special relativity. The discrepancy between the special relativity and Newtonian mechanics is: the time in the Newtonian mechanics is absolute, and if two events occur simultaneously in an inertial system, then they must also do at the same time in other inertial systems; in Einstein's space-time view, the absoluteness of the simultaneity is discarded.

However, the special relativity did not solve completely the problem of Newton's theory of gravitation. For this reason, in 1915 Einstein published a covariant theory of gravitation, known as the **general relativity**, which is a basic theoretical system composed of equivalence principle and general relativity principle. The general relativity generalizes the special relativity and refines Newton's law of universal gravitation, providing a unified description of gravity as a geometric property of space and time. In particular, the space-time bending is directly related to the energy and momentum of whatever matter and radiation are present. This relation is expressed as the Einstein field equations which represent the dependence of the spatial metric on the distribution of matter.

Both the Newtonian mechanics and special relativity fail to solve the essential problem of inertial system and inertial force. Inspired by the equality of the inertial mass with the gravitational mass, Einstein proposed the first basic principle of general relativity that a gravitational field is locally indistinguishable from the dynamical effects of an inertial force. The equivalence of inertial force and gravity obtained from mechanical experiments is called **weak equivalence principle**, which has been proved by more and more precise experiments. Furthermore, it is assumed for any physical experiments, such as mechanical, electromagnetic and other ones, that the effects of both inertial force and gravity are indistinguishable in a local region. In other words, the dynamical effects in the weak equivalence principle are extended to the deeper assumption for any physical effects, which is known as **strong equivalence principle**. It should be pointed out that the strong equivalence principle does not have sufficient experimental supports and needs to be verified in practice.

From the equivalence principle, it is known that there is no strict inertial system in the universe, so it is usually inevitable for people to investigate the physical laws in the non-inertial system. For the equivalence principle, the inertial force in the non-inertial system is regarded as the gravitational field, and considering the influence of the gravitation field on the physical laws, the gravitational effect is merged into space-time background. Therefore, another basic principle was generalized by Einstein, which all reference systems are equally weighted. In other words, the real- objective physical laws should have the same form under any coordinate transformation and be generally covariant, which is called the **general relativity principle**.

Obviously, the superiority of the inertial frame has been removed for the equivalence principle and general relativity principle, so that all reference frames are equally weighted, and the influence of gravitational field must be considered to get a correct physical law. For the equivalence principle, it is allowed to describe the physical process by using the non-inertial system and treat the permanent and non-permanent gravitational fields as the same. The equivalence principle is the basis and premise for the general relativity principle, but it is uncertain that the general relativity principle is derived from the equivalence principle. Both are independent of each other and constitute together the fundamental theoretical system of the general relativity.

The equivalence principle extends the concept of gravity and implies that the space-time of gravitational field is a curved Riemannian space. The physical effect of gravitational field can be reflected by the metric tensor of the Riemannian space. Similar to Newton's gravitational theory, the distribution of gravitational potential depends on that of the density of static matter, so the metric field should depend on the energy-momentum tensor of matter. By the tensor analysis of Riemannian space, Einstein derived that the differential equation of metric field is as follows:

$$G_{ij} + \Lambda g_{ij} = -\kappa T_{ij}, \tag{5.41}$$

where $G_{ij} \equiv R_{ij} - \frac{1}{2} g_{ij} R$, known as **Einstein Tensor**; $R \equiv g^{ji} R_{ij} = R^{j}_{\bullet j}$, known as **scalar curvature** or **Ricci scalar**; Λ is the cosmological constant; g_{ij} is the metric tensor; $\kappa = \frac{8\pi G}{c^4}$, the relativity constant, in which c is the velocity of light and G is the gravitational constant; T_{ij} is the energy-momentum tensor.

In general, when the cosmological constant is very small, it will not conflict with the differential equation of Newton's gravitational field theory. In the simplest case, let $\Lambda = 0$, and then

$$R_{ij} - \frac{1}{2} g_{ij} R = -\frac{8\pi G}{c^4} T_{ij}. \tag{5.42}$$

Usually, both formulas (5.41) and (5.42) are known as the **Einstein field equations**, where the former is with the cosmological constant and the latter not.

The left side of the Einstein field equation represents the geometric properties of space-time, and the right side does the basic properties of matter distribution, indicating the influence of matter and motion on the background of space-time and the dependence of gravitational field on the distribution of matter. In the general relativity, the importance of Einstein field equation is equivalent to Poisson's Eq. (5.13) in Newton's theory of gravitational field. The Einstein field equation is a more accurate description of gravitational field, and in the case of weak gravitational field, it will degenerate to Poisson's equation.

5.4.2.2 The Gravity as Geometry

In any gravitational field, the dynamical equation of a free particle is the geodesic equation. In the Minkowski space, when a particle moves freely and its connection coefficient is zero, the following differential equation is easily obtained from formulas (5.38) and (5.39).

$$\frac{d^2 x^\mu}{ds^2} = 0, \ (\mu = 0, 1, 2, 3), \tag{5.43}$$

where $x^{\mu} = (ct, x, y, z)$.

In order to distinguish in writing conveniently, it is ruled in this book that, under the general theory of relativity, the superscripts or subscripts for three-dimensional space are marked with English letters like *i, j, k* and *l*, with the values of 1, 2 and 3; the indicators for four-dimensional space-time are marked with Greek letters like α, β, μ and ν, with the values of 0, 1, 2 and 3.

The non-inertial system is introduced by a generalized coordinate transformation, and its four-dimensional space-time coordinate is expressed as

$$x^{\mu} = x^{\mu}\left(x^{\nu'}\right). \tag{5.44}$$

Correspondingly, its inverse transformation is

$$x^{\nu'} = x^{\nu'}(x^{\mu}). \tag{5.45}$$

Therefore, the dynamical equation that a particle moves freely can be expressed as

$$\frac{d^2x^{\mu}}{ds^2} = -\Gamma^{\mu}_{\nu\alpha}\frac{dx^{\nu}}{ds}\frac{dx^{\alpha}}{ds}, \tag{5.46}$$

where $\Gamma^{\mu}_{\nu\alpha} = \frac{\partial^2 x^{\mu'}}{\partial x^{\nu}\partial x^{\alpha}}\frac{\partial x^{\mu}}{\partial x^{\mu'}} + \beta^{\nu'}_{\nu}\beta^{\alpha'}_{\alpha}\beta^{\mu}_{\mu'}\Gamma^{\mu'}_{\nu'\alpha'}$.

In Eq. (5.46), its left side denotes the particle's acceleration, and its right side denotes the gravitational force of the particle per a unit of mass. It can be seen that the strength of inertial force field is represented by the Riemannian connection. According to the principle of equivalence, both the gravitation field and inertial force have the same status in physical laws, so the strength of gravitational field is usually described by the connection of space, which in turn describes the geometric structure of space. The basic idea of expressing the gravity through space geometry is called the **gravity as geometry**.

When the strength of gravitational field is described by the Riemannian connection, which is the gravitational acceleration, then the metric tensor is equivalent to the gravitational potential. In Newton's theory of gravity, the gravitational potential is a scalar field, and in the general relativity, it is a second-order symmetric tensor. Supposed that the space-time is flat, a set of Mikowski coordinates whose connection coefficients are constantly equal to zero can always be found, so that the effects of the gravitational field are completely disappeared. In fact, the real physical space-time is a curved Riemannian space, and the curvature tensor must not be zero. As a result, it is impossible to eliminate all gravitational effects. Certainly, according to the principle of equivalence, it is possible to eliminate the connection of a point in bending space-time. It means that within an infinitesimal neighborhood of any space-time point, the gravitational effects can be approximately eliminated to establish a local inertial frame.

5.4.2.3 The Space-Time Metric

The four-dimensional space-time is a set of one-dimensional time coordinate (x^0) and three-dimensional space coordinates (x^1, x^2, x^3), which is called the "**world**" by H. Minkowski. Any point $M(x^i)$ in the four-dimensional space-time represents an event, called **world-point**, and the motion of a particle in space-time is a curve formed by a series of events, called **world-line**. In the Newtonian mechanics, the motion of particles is described by $\boldsymbol{r}(t)$. Because of the absoluteness of the time t in the Newtonian mechanics, the t is an invariable physical quantity, and thus it can be used as a basic variable to describe the motion of particles. However, in the general relativity, the t no longer has this property, and it is conceptually divided into proper time and coordinate time.

The time τ of a clock moving with a particle is used as the basic quantity to describe the motion of the particle. Commonly, the τ is known as the **proper time**, which is the real record of the time interval between any two events by the same standard clock in the gravitational field. The τ is an invariant, whose definition does not rely on the physical process of coordinates, not changed with the change of coordinate system. Considered that the clocks at rest are not synchronous in the gravitational field, it is imagined that a reference clock placed in the local inertial frame flies from the infinity to each point in the gravitation field to calibrate the frequencies of these clocks, and then the time t recorded by the calibrated clocks is known as the **coordinate time**. The coordinate time is related to the selected coordinates, and it is a variable that changes with the coordinate system. However, in the same coordinate system, each point in the gravitational field has only a coordinate time, which is a measure of the time passing at the point.

The world-line of a particle moving in space-time is generally expressed as the function of the proper time, that is

$$x^i = x^i(\tau). \tag{5.47}$$

In the reference frame (or the local inertial frame) which falls freely at point $M(x^i)$, the space-time interval between any two world-points $M(x^i)$ and $N(x^i+dx^i)$ is the invariant that has nothing to do with the reference frame, i.e.,

$$ds^2 = -c^2 d\tau^2 = \eta_{ij} dx^i dx^j = -c^2 dt^2 + \delta_{ij} dx^i dx^j, \tag{5.48}$$

where ds is the space-time interval between two world-points, c is the velocity of light, $d\tau$ is the proper time interval between two world-points, dt is the coordinate time interval between two world-points, η_{ij} is the four-dimensional Minkowski metric, and δ_{ij} is the three-dimensional Euclidean metric in Cartesian coordinate system, an identity matrix with the order of 3 by 3.

In general, it is impossible that the coordinate transformation from the inertial system to the non-inertial system is a linear transformation. Let x^i and x^j denote the coordinates in the inertial and non-inertial systems, respectively, and thus according to the generalized coordinate transformation, the differential forms of coordinate transformation are expressed as follows.

$$\left.\begin{aligned} dx^{i'} &= \beta_i^{i'} dx^i \\ dx^i &= \beta_{i'}^i dx^{i'} \end{aligned}\right\}, \tag{5.49}$$

where $\beta_i^{i'} = \frac{\partial x^{i'}}{\partial x^i}$ and $\beta_{i'}^i = \frac{\partial x^i}{\partial x^{i'}}$, which are the coordinate transformation coefficients. Apparently, $\beta_i^{i'} \beta_{i'}^i = 1$.

From the coordinate transformation condition of formula (5.49), the space-time interval invariant in the inertial system can be expressed in the non-inertial system as

$$ds^2 = \eta_{ij} dx^i dx^j = g_{i'j'} dx^{i'} dx^{j'}, \tag{5.50}$$

where $g_{i'j'} = \beta_{i'}^i \beta_{j'}^j \eta_{ij}$, which is a basic quantity to describe the gravitational field, denoting the space-time metric relation in the non-inertial system. Obviously, the space-time metric is a function of time and space with the symmetric subscripts and has ten independent components. Moreover, the function form of space-time metric is related to the selection of specific coordinates.

For the local space-time, the time and space can not only be separated, but also be regarded as a flat space-time. In this way, the Cartesian coordinate system can be established within the local space-time range near an observer, which is a rigid frame, which is at rest relative to the observer and whose three axes are perpendicular to each other. The local Cartesian coordinate system and the "clock" carried by the observer form together with a local reference frame, in which the observer can measure the time, distance and orientation of the happened events in its neighborhood. However, in order to describe the motion of matter and its relationship in the large-scale space-time, it is necessary to establish a global coordinate system. Due to the non-Euclidean property of space-time, the large-scale global coordinate system cannot meet the conditions of Cartesian coordinate system. In addition, considering the unity of space-time, the time and space cannot be absolutely separated.

In other words, under the general theory of relativity, two kinds of "time" with different properties are brought up: one is the proper time, the time that is measured by the ideal clock carried by the observer, and it is suitable for the local reference frame where the observer is located; another is the coordinate time, which is determined by the space-time metric, and it is used as a time-like variable of time coordinate. Obviously, the proper time is an objective physical quantity, which has nothing to do with the choice of coordinate system. In any coordinate system, the proper time of the observer is uniquely determined. The proper time has a clear physical meaning, which according to the definition of the length of a second can be directly realized by a physical clock or some measurement means. However, the proper time varies with the observer's space-time position and velocity, and it can only be used in the local space near the observer. It is impossible to use the same proper time in the global space. Although the coordinate time has no clear physical meaning and is a coordinate quantity with only mathematical meaning, whose mathematical relation can only be given by the space-time metric, it has the clear definition within the whole space range of the reference system. The relationship between the "local" proper time and the "global" coordinate time can be given by the space-time metric. The coordinate time of any "space point" in the reference system can be calculated by the proper time measured by the local clock.

When the time axis of a coordinate system is orthogonal to its space-axis, then the coordinate system is called **time-axis orthogonal coordinate system**, i.e., $g_{0i} = 0$. Furthermore, the axes of the selected three-dimensional space coordinate system are also orthogonal to one another, i.e., $g_{ij} = 0$, then the coordinate system is called **space-time orthogonal coordinate system**. In the space-time orthogonal coordinate system, there are only four independent components for the space-time metric, and the corresponding space-time interval is expressed as

$$ds^2 = -c^2 d\tau^2 = g_{00}(cdt)^2 + g_{11}\left(dx^1\right)^2 + g_{22}\left(dx^2\right)^2 + g_{33}\left(dx^3\right)^2. \quad (5.51)$$

In the three-dimensional Euclidean space, the distance between any two space points must be greater than zero. However, in the general relativity, the space-time interval ds^2 (or the proper time interval $d\tau^2$) between any two world-points may be greater than, less than or equal to zero. When $ds^2 > 0$, then the space-time interval is called the **space-like interval**, which means that there is no physical connection between two events and there is no causal relationship between them; when $ds^2 < 0$, then the space-time interval is called the **time-like interval**, which means that there is the connection between two events by using certain physical signal and there is the causal relationship between them; when $ds^2 = 0$, then the space-time interval is called the **light-like interval**, which means that only particles moving at the velocity of light can meet this condition.

In the four-dimensional space-time, any space-time point's coordinate transformation, $x^i = x^i\left(x^{j'}\right)$, is carried out. In order to make the first variable in the four variables (x^0, x^1, x^2, x^3) represent the time coordinate, and to do the last three represent the spatial coordinates, the space-time metric must meet the following necessary and sufficient conditions:

(1) $g_{00} > 0$;

(2) $\begin{vmatrix} g_{00} & g_{01} \\ g_{10} & g_{11} \end{vmatrix} < 0$;

(3) $\begin{vmatrix} g_{00} & g_{01} & g_{02} \\ g_{10} & g_{11} & g_{12} \\ g_{20} & g_{21} & g_{22} \end{vmatrix} > 0$;

(4) $\begin{vmatrix} g_{00} & g_{01} & g_{02} & g_{03} \\ g_{10} & g_{11} & g_{12} & g_{13} \\ g_{20} & g_{21} & g_{22} & g_{23} \\ g_{30} & g_{31} & g_{32} & g_{33} \end{vmatrix} < 0.$

This is known as **physical coordinate conditions of space-time metric**. If a space-time metric cannot meet one of the above four conditions, the selected coordinate system is non-physical, relative to which a physical observer cannot be at rest locally.

In 1916, the distribution expression of the vacuum gravitational field outside the spherically symmetric and stationary celestial body with the mass of M was derived from the Einstein field equations by Karl Schwarzschild (1873—1916), a German physicist and astronomer, which is known as **Schwarzschild metric**. Taking the center of mass of the celestial body as the coordinate origin, the Schwarzschild metric in the spherical coordinate system is expressed as

$$\begin{aligned} ds^2 = -c^2 d\tau^2 = &-\left(1 - \frac{2GM}{rc^2} - \frac{\Lambda}{3}r^2\right)c^2 dt^2 + \left(1 - \frac{2GM}{rc^2} - \frac{\Lambda}{3}r^2\right)^{-1} dr^2 \\ &+ r^2 d\theta^2 + r^2 \sin^2\theta d\varphi^2, \end{aligned} \tag{5.52}$$

where GM is the product of the gravitational constant and the mass of the celestial body; r, θ and φ are, respectively, three components of the spherical coordinates; τ and t are, respectively, the proper time and coordinate time; c is the velocity of light in vacuum; Λ is the cosmological constant.

The Schwarzschild metric space-time is a static space-time of the spherically symmetric celestial body. The Schwarzschild metric components are independent of time, with the property of orthogonal time axis. If the term of the cosmological constant is ignored, then there is

$$\begin{aligned} ds^2 = -c^2 d\tau^2 = &-\left(1 - \frac{2GM}{rc^2}\right)c^2 dt^2 + \left(1 - \frac{2GM}{rc^2}\right)^{-1} dr^2 \\ &+ r^2 d\theta^2 + r^2 \sin^2\theta d\phi^2. \end{aligned} \tag{5.53}$$

5.4.2.4 Bending of Space-Time

Would there be no gravitational field in space, the four-dimensional space-time becomes into the Minkowski space with maximum symmetry. Once there is a gravitational field, the symmetry will be destroyed, and the four-dimensional space-time is no longer the Minkowski space, but the curved Lobachevsky space or Riemannian space. In order to conveniently explain the influence of gravitational field on time and space, Einstein's Disc model is quoted here.

Let $S(T, R, \Theta)$ denote the inertial system of one-dimensional time and two-dimensional polar coordinates, and $s(t, r, \theta)$ denote the corresponding non-inertial system with the disk rotating at a constant velocity. At the initial moment, the inertial system S coincides with the non-inertial system s, and all clocks and scales are calibrated and assigned to each point in the S-system and s-system. Using the coordinate transformation relation between the rectangular and polar coordinate systems, the space-time interval of two world-points in the inertial system can be expressed by formula (5.48) as follows.

$$ds^2 = -c^2dT^2 + dR^2 + R^2d\Theta^2. \tag{5.54}$$

When the s-system rotates uniformly at angular velocity ω relative to the S-system, the space-time coordinate transformation relations between the two coordinate systems are expressed as

$$\left.\begin{array}{l} T = t \\ R = r \\ \Theta = \theta + \omega t \end{array}\right\}. \tag{5.55}$$

By differentiating the coordinates of the above formulas, the space-time interval in the s-system is expressed as

$$ds^2 = -\left(1 - \frac{r^2\omega^2}{c^2}\right)c^2dt^2 + dr^2 + \left(\frac{2r^2\omega}{c}\right)cdtd\theta + r^2d\theta^2. \tag{5.56}$$

And thus, the space-time metric in the s-system is

$$g_{ij} = \begin{bmatrix} g_{00} & g_{01} & g_{02} \\ g_{10} & g_{11} & g_{12} \\ g_{20} & g_{21} & g_{22} \end{bmatrix} = \begin{bmatrix} -\left(1 - \frac{r^2\omega^2}{c^2}\right) & 0 & \frac{2r^2\omega}{c} \\ 0 & 1 & 0 \\ \frac{2r^2\omega}{c} & 0 & r^2 \end{bmatrix}. \tag{5.57}$$

Investigating the change of the scale and the disk's circumference in the S-system and s-system at any time, since the disk's radius and scale are always perpendicular to the direction of the linear velocity of the motion during the entire process of the disk's rotation, there is no the Lorentz contraction, which is the phenomenon that a moving object's length is measured to be shorter than its proper length, so that the radial coordinates in the two coordinate systems are consistent with the measurement results. In the s-system, when the disk's circumference is measured by the calibrated unit scale, there is a total of n standard scale length, but from the view of the S-system, there is the Lorentz contraction for the scale. In other words, the length of the scale in the S-system is $1 \cdot \sqrt{1 - \frac{r^2\omega^2}{c^2}}$,and the disk's circumference measured by the scale is

$$L = \oint \left(\frac{dl}{\sqrt{1 - \frac{r^2\omega^2}{c^2}}}\right) = \int_0^{2\pi} \left(\frac{rd\theta}{\sqrt{1 - \frac{r^2\omega^2}{c^2}}}\right) = \frac{2\pi r}{\sqrt{1 - \frac{r^2\omega^2}{c^2}}} = n. \tag{5.58}$$

Therefore, in the s-system, the ratio of the circumference L of a circle to its diameter D can be expressed as

$$\frac{L}{D} = \frac{2\pi r}{2r\sqrt{1 - \frac{r^2\omega^2}{c^2}}} = \frac{\pi}{\sqrt{1 - \frac{r^2\omega^2}{c^2}}} > \pi. \tag{5.59}$$

According to the Riemannian geometry, for a geometric space, when its ratio of the circumference of a circle to its diameter is equal to π, it is the Euclidean space; when its ratio is greater than π, it is the Lobachevsky space, which is a hyperbolic space; when its ratio is less than π, it is the Riemannian space, which is an elliptic space. Among them, the last two kinds of spaces are called the non-Euclidean space, which are both the bending space-times. Apparently, the non-inertial system established on Einstein's Disc belongs to the Lobachevsky space.

It is shown from Einstein's Disc that for the observers in the inertial and non-inertial systems, their views of space-time are different. From the view of the observer in the inertial system, since the disk is accelerating at the acceleration $r\omega^2$ relative to the inertial system, the observer is subjected to the inertial force $r\omega^2$ with a unit of mass. From the view of the observer in the non-inertial system, the observer does not know that it is being accelerated, and thereby the measurement result that the ratio of the circumference of a circle to its diameter is greater than π is explained as that caused by the gravitational field. Meanwhile, the change of the clock on the disk is investigated. According to the special theory of relativity, once the clocks placed statically everywhere are calibrated and synchronized in the inertial system, they will not be affected by the existence and motion of the surrounding matter, and they will always keep in the synchronization. However, according to the general theory of relativity, even if the standard clocks in a reference system are calibrated and synchronized at a certain time, they will not be able to keep synchronization forever due to the effect of gravitational field. It is clearly shown that the speeds of the clocks will be affected by the existence and motion of the surrounding matter. In the view of the S-system, for the clock placed statically at the position (r, θ) on the disk, there is Einstein's delay, that is,

$$t = T\sqrt{1 - \frac{r^2\omega^2}{c^2}} \approx T\left(1 - \frac{r^2\omega^2}{2c^2}\right). \tag{5.60}$$

Obviously, it is impossible to synchronize the clocks that are placed all over the disk. Only at the center of the disk will the clocks be synchronized. The space-time metric not only describes the gravitational field, but also the bending of space-time. The bending degree of space-time and the strength of gravitational field are two kinds of statements of the same problem. The greater is the curvature of space-time, the stronger the gravitational field. Conversely, the stronger is the gravitational field, the greater the curvature of space-time also.

5.4.3 *Physical Interpretation of Space-Time Measurement*

5.4.3.1 Reference System and Coordinate System

In order to describe an event or a particle's motion, the corresponding reference system and coordinate system must be selected. It is of no practical significance to discuss the motion law of an event or a particle without the reference system or coordinate system. According to the general theory of relativity, the four-dimensional physical space-time is a curved Riemannian space, and generally, there is no global unified time system and spatial coordinate system. Therefore, the local reference system and coordinate system are only discussed in this section. The reference system and coordinate system are two different concepts, which should be distinguished to better describe the regularity of objective events.

The so-called **reference system** refers to the clock and local rigid frame (scale) carried by an observer moving in a certain way in space-time, and focuses on the reference object adopted by the system, rather than the specific mathematical expression. The so-called **coordinate system** refers to the establishment of a certain one-to-one correspondence between the space-time position of a specific physical event and an array composed of four parameters. In this way, selecting a space-time coordinate system, any definite value of the array represents a certain space-time point. A reference system can contain many coordinate systems which are essentially the same, such as the rectangular coordinate system and spherical coordinate system. The description of objective laws will not be affected by the selection of any coordinate system. In other words, it is essentially equivalent in the same reference system to select two coordinate systems for the description of the same event. Obviously, for the objective description of an event, the selection of reference system is very important. Once a reference system is selected, it does not matter what form of coordinate system is used. When a coordinate system is selected, its specific form is clearly determined to provide a definite description of the space-time position, but it may not necessarily represent a physically meaningful reference system. When a reference system is selected, it is shown that the specific form of coordinate system is not focused on.

It is well known that any physical quantity must be observable, and any measurement is always related to a certain observer's local reference system. The theoretical studies of physical problems in the general relativity need the coordinate system, but all measurements depend on the reference system. However, the coordinate system is usually different from the reference system, which leads to a question: what is the relationship between the tensor's physical quantities obtained in a coordinate system and the physical quantities measured by an observer who is observing the same phenomenon? For a specific measurement problem, there are usually four factors needed to be considered: the first is the definite observation object, which refers to the physical quantity in a certain physical phenomenon; the second is the space-time geometric properties of the event; the third is the selected reference system, which refers to the observer moving in a certain way; the fourth is the selected coordinate system, to describe the movement of the observation object and the observer. In fact, any observation result only depends on the first three factors, namely, observation object, space-time structure and observer, while has nothing to do with the fourth factor. For example, for the measurement of the space-time interval between two events, once the first three factors are determined, the observer can measure a definite number, the selection of the coordinate system is arbitrary and the space-time interval will not be affected due to the transformation of the coordinate system.

In summary, since the space-time coordinate system is arbitrarily selected in the general relativity, the space-time coordinates are only a marker of an event, used to distinguish from other events, generally without direct measurement significance. With the measurement significance is the proper time measured by the standard clock and the proper length measured by the standard scale, which are the basic observables of the general relativity. From the perspective of tensor analysis, the basic observable of the general relativity must be a scalar.

5.4.3.2 Expression of Space-Time Reference

From Einstein's general relativity, there is a new understanding of the space where matter moves and the time through which it passes, and the time and space cannot exist independently without matter. The structure and property of space-time depend on the distribution of matter, and the motion of matter in the gravitational field is limited by the geometric characteristics of space-time. Under the general theory of relativity, the space-time reference (or the reference system) is defined by the local space-time metric, that is

$$ds^2 = g_{\mu\nu} dx^\mu dx^\nu. \tag{5.61}$$

The space-time metric $g_{\mu\nu}$ is a function of the observer's space-time position. Selecting the different space-time metric is equivalent to selecting the different space-time coordinate system, and thus an array of four parameters, representing a determined space-time point, is gotten.

For the flat space, when the Euclidean metric δ_{ij} is selected, the commonly used rectangular coordinate system will be obtained; when the spherical metric is selected, the spherical coordinate system is obtained; when the cylindrical metric is selected, the cylindrical coordinate system is obtained. These coordinate systems can be transformed one another and be extended to the whole space to describe the events in the flat space. For the curved Riemannian space, the coordinate system represented by arbitrary space-time metric has only local meaning, and generally it cannot be extended to the whole space. Because under the general theory of relativity, the timescale and length-scale change with the change of spatial position, there is not a unified coordinate system. According to the theory of relativity, the changes of the timescale and length-scale mainly come from two factors: one is the kinematical effect (also known as the special relativity effect), the moving clock is slowed down, and the moving scale is shortened, called length contraction; another is the gravitational field effect (also known as the general relativity effect), which slows down the speed of the clock and shortens the scale. So, the clocks and scales placed at different space points are different completely, which is the locality of the clocks and scales.

The clock and scale are the fundamental elements of establishing the space-time reference, and hereby it is required that the space-time scale standards at different space positions are unified, so as to establish a unified space-time scale standard system. The so-called **space-time scale standard system** refers to selecting a relatively-stationary reference system or frame within a small-enough area, in which the relatively-stationary time-keeping system and length measuring system are set as the standard time and the standard scale, respectively, and then the clocks and scales in other regions or reference systems are calibrated by taking those in standard reference system as the datum. For example, the clock and scale of the BIPM are used as the datum to calibrate the clocks and scales deployed in the other places. In general, the clock's calibration can be achieved by the Two-Way Satellite Time and Frequency Transfer (TWSTFT) method, and the scale's calibration is more complicated because the contraction effects of the scale placed in different directions are different.

Supposed that a free particle moves along the world-line in the gravitational field, a point on the world-line as the coordinate origin is selected to establish the locally orthogonal coordinate frame which moves with the particle. According to the orientations of different coordinate axes, the different names of the coordinate frames are used. For example, if the coordinate axes point to distant stars, the coordinate system is called stellar orientation local frame; if the three coordinate axes always point to the directions of the tangent, main-normal and sub-normal lines of the particle's world-line, respectively, the coordinate system is called natural local frame; and so on. The local coordinate systems established in the general theory of relativity cannot be extended but can be connected one another through a series of coordinate transformations.

Although the local coordinate systems are equally weighted and have no advantages or disadvantages, there is always the "superior" coordinate system in practical application. Generally, a dynamical self-consistent local coordinate system should be established, which meets two basic properties: one is that the external gravitational potential appears in the form of tidal potential; another is that the internal gravitational potential is the gravitational potential of the celestial body under investigation. In fact, the local reference system is established under the specific space-time background, so the reference system in the general relativity can be divided into background and local reference systems.

In summary, the complete processes of establishing the space-time reference under the general relativity are: to define the coordinate origin of reference system; to select the world-line of the coordinate origin; to select the rotation of the local reference system relative to the spatial axis of the background reference system; to select the coordinate conditions; to finally determine the space-time metric. The IAU recommends that the celestial mechanics of the general relativity is used to deal with the mutual transformation between the background and the local reference systems for the first-order post-Newtonian approximation.

5.4.3.3 Time Synchronization Measuring

In the four-dimensional space-time, the coordinate system $S(x^i)$ is selected. When the observer's coordinate intervals and four-dimensional velocities are, respectively,

$$\left.\begin{aligned} dx^0 &= -cdt \neq 0 \\ dx^i &= 0 \end{aligned}\right\}. \tag{5.62}$$

$$\left.\begin{aligned} u^0 &= -\frac{1}{\sqrt{-g_{00}}}c \\ u^i &= 0 \end{aligned}\right\}, \tag{5.63}$$

then the observer is locally at rest relative to the coordinate system, called **local-rest observer**. Therefore, with the passing time of the clock carried by the local-rest observer as the investigated object, the time interval measured by the observer is expressed as

$$dT = -\frac{1}{c^2}g_{00}u^0 dx^0 = \sqrt{-g_{00}}dt. \tag{5.64}$$

From the above formula, it can be seen that the time interval recorded by the observer's clock is directly proportional to the coordinate time interval, and its proportional coefficient $\sqrt{-g_{00}}$ is a function of the observer's position. Even if the coordinate time intervals dt between the observers at different points in space are the same, the proper time intervals are also different. In other words, the time measured by the different observers is different, which is the effect of gravitational field on the clocks. Would the gravitational field disappear, the proper time recorded by the standard clock at different positions is the coordinate time.

The time synchronization refers to the mutual calibration of two remote clocks, with the simultaneity of time measurement, and belongs to the basic concept of time system keeping. In the special relativity, the time synchronization is related to the choice of coordinates, but in the same coordinate system, it can always be established. In the general relativity, it is generally difficult to synchronize two remote clocks, or only when certain conditions are satisfied, the simultaneity of two remote clocks can be discussed. In the bending space-time, the alignment of the at-rest clocks at two adjacent points M and N can be achieved by optical signal propagation. When the coordinate time indicated by the clock's face is $x^0_{M(1)}$, the at-rest observer at point M, sends out a light signal; when the signal arrives at point N, its coordinate time is x^0_N; finally, when the signal reflects back to point M, its coordinate time is $x^0_{M(2)}$, as shown in Fig. 5.5, where l_M and l_N represent the world-lines of the at-rest observers at points M and N, respectively, and both are parallel to the time axis. As a result, the coordinate differences between the light signals going and returning are, respectively,

$$dx^0_{(1)} = x^0_N - x^0_{M(1)}, \tag{5.65}$$

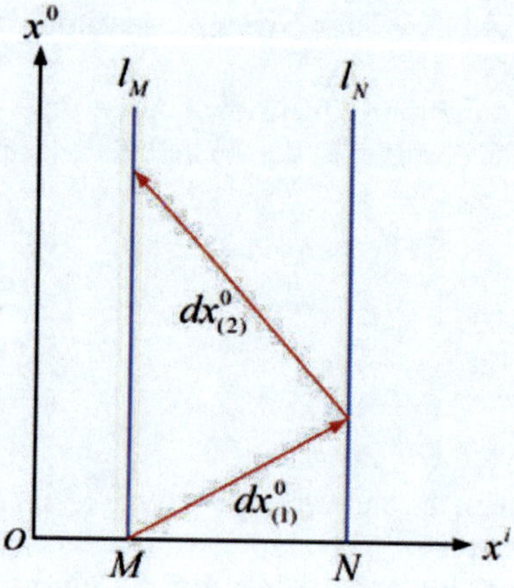

Fig. 5.5 At-rest clocks at remote positions synchronized by light signal

$$dx^0_{(2)} = x^0_{M\,(2)} - x^0_N. \tag{5.66}$$

Let $x^0_M = \frac{1}{2}\left(x^0_{M\,(1)} + x^0_{M\,(2)}\right)$ denotes the coordinate time at point M, which has the simultaneity of time measurement with the coordinate time x^0_N at point N. In general, the velocity of light is not necessarily isotropic, and $dx^0_{(1)}$ is not necessarily equal to $dx^0_{(2)}$, so there is the time synchronization error between two remote clocks, that is,

$$\Delta x^0 = x^0_N - x^0_M = \frac{1}{2}\left(dx^0_{(2)} - dx^0_{(1)}\right). \tag{5.67}$$

Since the light-like interval is zero, the coordinate difference between the light signal going and returning should meet the following equation:

$$g_{00}\left(dx^0\right)^2 + 2g_{0i}dx^0dx^i + g_{ij}dx^idx^j = 0. \tag{5.68}$$

By solving Eq. (5.68), the following solutions can be obtained.

$$dx^0_{(1)} = \frac{-g_{0i}dx^i \pm \sqrt{\left(g_{0i}g_{0j} - g_{00}g_{ij}\right)dx^idx^j}}{g_{00}}, \tag{5.69}$$

$$dx^0_{(2)} = \frac{g_{0i}dx^i \pm \sqrt{\left(g_{0i}g_{0j} - g_{00}g_{ij}\right)dx^idx^j}}{g_{00}}. \tag{5.70}$$

Therefore,

$$\Delta x^0 = \frac{1}{2}\left(dx^0_{(2)} - dx^0_{(1)}\right) = \frac{g_{0i}}{g_{00}}dx^i. \tag{5.71}$$

By using the formula (5.71), the coordinate clocks everywhere along any open path can be calibrated to synchronize. However, Δx^0 is not generally a total differential, so the integral along a closed path will not be equal to zero. Obviously, it is impossibly achieved that two coordinate clocks at different positions are calibrated along different paths. In other words, although the clocks at points M and N_1 can be calibrated to synchronize, and the clocks at point N_1 and N_2 can also be calibrated to do, the clocks at points M and N_2 are not necessarily synchronized. Even if in the same coordinate system, the at-rest coordinate clocks are not synchronized, it is also difficult that the at-rest standard clocks are calibrated to synchronize. As a result, in the general relativity, it is so difficult to realize the time synchronization of two remote clocks that the concept of time cannot be established.

However, when the time-axis orthogonal coordinate system is used, i.e., $g_{0i} = 0$, then $\Delta x^0 = 0$, which shows that two remote clocks can be calibrated to synchronize and the time process can be described by the space-time coordinate system. For all stationary gravitational fields, a time-axis orthogonal coordinate system can always be found. Therefore, in the general relativity, the time-axis orthogonal coordinate system is of great significance, which is the fundamental condition for establishing the simultaneity of the clocks at different places. In this way, the time of coordinate clocks at every space point can be calibrated to synchronize, so as to achieve the simultaneous transfer of time.

5.4.3.4 Spatial Distance Measuring

Investigating the problem of the spatial distance measuring of the local-rest observer in its own neighborhood, the ordinary space metric can be calculated from formula (5.63), that is,

$$h_{\mu\nu} = \frac{1}{c^2}u_\mu u_\nu + g_{\mu\nu} = \frac{1}{c^2}g_{\mu\alpha}g_{\nu\beta}u^\alpha u^\beta + g_{\mu\nu} = g_{\mu\nu} - \frac{g_{\mu 0}g_{\nu 0}}{g_{00}}. \tag{5.72}$$

From the above formula, the time and its cross-terms of metric $h_{\mu\nu}$ can be expressed as

$$h_{00} = h_{0i} = h_{i0} = 0. \tag{5.73}$$

And thus, the distance between two adjacent points measured by the local-rest observer is

$$dL = \sqrt{h_{ij}dx^i dx^j}. \tag{5.74}$$

For the measurement of the space-time interval of the local-rest observer, the intuitively physical explanation can be given from another perspective. The spatial distance between two adjacent points M and N is explained as that half of the traveling time of the light signal, emitted by the at-rest observer at point M and reflected back to point M through point N, multiplied by the velocity of light. Thereby, from formulas (5.69) and (5.70), half of the going and returning time between the M and N is expressed as

$$dt = \frac{1}{2c}\left(dx^0_{(1)} + dx^0_{(2)}\right) = \frac{\sqrt{\left(g_{0i}g_{0j} - g_{00}g_{ij}\right)dx^i dx^j}}{cg_{00}}. \tag{5.75}$$

Considering formula (5.64), the spatial distance between two points is expressed as

$$dL = cdT = c\sqrt{-g_{00}}dt = \sqrt{\left(g_{ij} - \frac{g_{0i}g_{0j}}{g_{00}}\right)dx^i dx^j}. \tag{5.76}$$

Apparently, formula (5.76) is completely coincident with formula (5.74). It should be pointed out that formula (5.76) is derived from the precondition that there is the local-rest observer. If the precondition is physical realizable, the dL measured by the observer must be real. It is required that the $h_{ij}dx^i dx^j$ in formula (5.74) is a positive-definite quadratic form. In other words, the space-time metric h_{ij} is required to meet the physical coordinate conditions.

For the local region of gravitational field, the Euclidean geometric distance between two adjacent points can be defined as

$$dl = \left(dx^2 + dy^2 + dz^2\right)^{\frac{1}{2}} = \left(dr^2 + r^2 d\theta^2 + r^2 \sin^2\theta d\varphi^2\right)^{\frac{1}{2}}. \tag{5.77}$$

The geometric distance dl is called **coordinate length.**

Due to the different gravitational intensity at the different points in the gravitational field, the scale which is not affected by the gravitational field in the at-rest inertial system is taken as the standard of measurement. The standard scale is placed in the local inertial system, and when it passes through each point in the gravitational field, the length of the scale at each point is calibrated to keep the same standard scale length at every point.

Furthermore, the real spatial distance between two adjacent points in the gravitational field measured by the at-rest observer is defined as the proper length, and in the Schwarzschild metric with weak gravitational field, there is the following expression:

$$dL = \left(1 + \frac{2GM}{rc^2}\right)^{\frac{1}{2}} dl, \tag{5.78}$$

which is the relation formula between the coordinate length and the proper length measured by the local-rest observer using the standard scale length.

For the spatial distance within a small area, when the metric is independent of time, the spatial integration of formula (5.78) can be carried out directly. Under the Newtonian limit, the proper length of any two points $M(\mathbf{r}_1)$ and $N(\mathbf{r}_2)$ in space is

$$\begin{aligned}\Delta L &= \int_{r_1}^{r_2} \left(1 + \frac{2GM}{rc^2}\right)^{\frac{1}{2}} dl \approx \int_{r_1}^{r_2} \left(1 + \frac{GM}{rc^2}\right) dl \\ &= \|\mathbf{r}_2 - \mathbf{r}_1\| + \frac{GM}{c^2} \ln \frac{r_2 + \mathbf{r}_2 \cdot \mathbf{n}}{r_1 + \mathbf{r}_1 \cdot \mathbf{n}},\end{aligned} \tag{5.79}$$

where $\mathbf{n} = \frac{\mathbf{r}_2 - \mathbf{r}_1}{\|\mathbf{r}_2 - \mathbf{r}_1\|}$.

In formula (5.79), the first term on the right denotes the Euclidean geometric distance between two points in space, and the second term denotes the distance correction resulted from the space bending. It is clearly demonstrated that the space bending is caused by the action of gravitational field.

5.4.4 Timescale of Relativity

In order to study the dynamics of celestial bodies in the solar system more accurately, the introduction of the timescale of relativity in astronomy was proposed at the 16th IAU general assembly in 1976. Two kinds of the relativistic timescales, **Barycentric Dynamical Time** (TDB) and **Terrestrial Dynamical Time** (TDT), were defined at the 17th IAU general assembly in 1979. The TDB refers to the time variables used in the equations of motion relative to the SSB and in the compiled star catalogue. The TDT refers to the time variables used in the equations of motion relative to the Earth's center of mass and in the catalogues of the apparent positions of the observed celestial bodies at the Earth's center. In 1991, at the 21st IAU general assembly, the space-time forms of the Barycentric Celestial Reference System (BCRS) and Geocentric Celestial Reference System (GCRS) were proposed, and correspondingly, two new timescales, **Barycentric Coordinate Time** (TCB) and **Geocentric Coordinate Time** (TCG), were defined, respectively. Meanwhile, it was decided in the IAU general assembly to change the TDT to **Terrestrial Time** (TT). The TT is the time reference datum of the apparent ephemeris at the Earth's center, its difference with the TDT is a proportional constant, and its measurement-selected unit is consistent with the length of a second in the SI units on the geoid. At 0:0:0.0 TAI on January 1, 1977, the TT's reading is 0:0:32.184 on the same day. The TCB refers to the one-dimensional time component of any space-time coordinate point in the BCRS, and the TCG refers to the one-dimensional time component of any space-time coordinate point in the GCRS.

Strictly speaking, the observer must be in the neighborhood of the clock and remain at rest relative to the clock. Due to the difference of gravitational field, there may be a slight difference between two actual lengths of a second in the SI units, and hence the relativistic effect should be considered in the realization of second of the atomic time. In 1980, the TAI was defined by the International Advisory Committee on the definition of "seconds". The TAI is the coordinate timescale in the geocentric coordinate system, taking the second of the SI units on the rotating geoid as the unit. In fact, the TT is measured along the world-line by using the clock carried with the observer, who is at rest relative to the GCRS on the geoid.

Therefore, there are four astronomical timescales in the general relativity, where the TDB and TCB belong to the BCRS, and the TT and TCG belong to the GCRS. Since 1984, the dynamical time has been used in the compilation of celestial ephemeris in the solar system. However, in order to maintain the continuity of the ephemeris catalogue, it is necessary to use an astronomical timescale to succeed the previous ET, and finally the TDT (later renamed the TT) is introduced to replace the ET as the time variable in the ephemeris catalogue.

The trajectory curve of any observer and his local reference system is a world-line, and the line element between two adjacent points on the world-line can be expressed as

$$ds^2 = -c^2 d\tau^2, \tag{5.80}$$

where τ is the proper time measured by using the ideal clock carried by the observer.

The proper time is connected with the coordinate time through the space-time metric, i.e.,

$$d\tau^2 = (1 - 2K)dt^2, \tag{5.81}$$

where $K = \frac{1}{2}\left(1 + g_{00} + g_{0i}\frac{v^i}{c} + g_{ij}\frac{v^i v^j}{c^2}\right)$; $v^i = \frac{dx^i}{dt}$, the three-dimensional velocity of the observer.

Because the coordinate time is not a physical quantity, it can only be realized indirectly by the proper time for the celestial bodies. It is not difficult to understand that for any physical clock used to measure the proper time, only if its frequency is adjusted or its time corrected, it can be changed into a coordinate clock.

In fact, there are still some limitations in the space-time metric defined by the IAU in 1991. For this reason, the complete expression of the post-Newtonian space-time metric was further proposed at the 24th IAU general assembly in 2000. In the BCRS, the metric tensor of any space-time coordinate point $(ct, \mathbf{x})$, where $t = TCB$, is expressed as

$$\left.\begin{array}{l} g_{00} = -1 + \frac{2}{c^2}[w_0(t, \mathbf{x}) + w_L(t, \mathbf{x})] - \frac{2}{c^4}\left[w_0^2(t, \mathbf{x}) + \Delta(t, \mathbf{x})\right] \\ g_{0i} = -\frac{4}{c^3}w^i(t, \mathbf{x}) \\ g_{ij} = \delta_{ij}\left(1 + \frac{2}{c^2}w_0(t, \mathbf{x})\right) \end{array}\right\}, \tag{5.82}$$

where $w_0(t, \mathbf{x}) = \sum\limits_A \frac{GM_A}{|\mathbf{x}-\mathbf{x}_A|}$;

$$w^i(t, \mathbf{x}) = G\sum_A \left\{ \frac{M_A v_A^i}{|\mathbf{x}-\mathbf{x}_A|} - \frac{[(\mathbf{x}-\mathbf{x}_A)\times \mathbf{S}_A]^i}{2|\mathbf{x}-\mathbf{x}_A|^3} \right\};$$

$$\Delta(t, \mathbf{x}) = \sum_A \left\{ \frac{GM_A}{|\mathbf{x}-\mathbf{x}_A|} \left[-2v_A^2 + \sum_{B\neq A} \frac{GM_B}{|\mathbf{x}_B-\mathbf{x}_A|} + \frac{1}{2}\frac{((\mathbf{x}-\mathbf{x}_A)\cdot \mathbf{v}_A)^2}{|\mathbf{x}-\mathbf{x}_A|^2} \right.\right.$$
$$\left.\left. + (\mathbf{x}-\mathbf{x}_A)\cdot \mathbf{a}_A \right] + \frac{2G\mathbf{v}_A\cdot[(\mathbf{x}-\mathbf{x}_A)\times \mathbf{S}_A]}{|\mathbf{x}-\mathbf{x}_A|^3} \right\},$$

where $\mathbf{x}_A$, $\mathbf{v}_A$, $\mathbf{a}_A$, $\mathbf{S}_A$ and M_A are, respectively, the position vector, velocity vector, acceleration vector, total angular momentum and gravitational mass of a celestial body A in the BCRS; $w_L(t, \mathbf{x})$ is the expansion of the post-Newtonian multiple moments involving all celestial bodies; δ_{ij} is the Kronecker symbol.

In most cases, the accuracy requirements can be satisfied for the single mass approximation metric, and then $w_L(t, \mathbf{x}) = 0$. However, in order to keep the continuity and integrity of the model for all cases, the term $w_L(t, \mathbf{x})$ is still remained in formula (5.82).

In the GCRS, the metric tensor of any space-time coordinate point (cT,$\mathbf{X}$), where $T = TCG$ can be expressed as

$$\left.\begin{array}{l} G_{00} = -1 + \frac{2W(T,\mathbf{X})}{c^2} - \frac{2W^2(T,\mathbf{X})}{c^4} \\ G_{0a} = -\frac{4}{c^3}W^a(T,\mathbf{X}) \\ G_{ab} = \delta_{ab}\left(1 + \frac{2}{c^2}W(T,\mathbf{X})\right) \end{array}\right\}, \quad (5.83)$$

where $W(T, \mathbf{X})$ and $W^a(T, \mathbf{X})$ are, respectively, the scalar and vector gravitational potentials resulted from the tidal force of the Earth and external celestial bodies.

Since the world-lines of any two observers do not coincide, it is impossible to directly compare the proper time measured by the clocks carried by the observers. In the local reference system, the time reading obtained by the observer using the local clock cannot be directly applied to other local reference systems or global (background) reference systems. The time difference caused by the relativistic effect between different reference systems must be considered. It is the problem of timescale transformation in the general relativity, which will be discussed in detail in Sect. 5.6.2 of this chapter.

5.4.5 International Celestial Reference System and Its Realization

5.4.5.1 Celestial Sphere Reference System and Its Frame

An ideal celestial reference system is a three-dimensional inertial reference system established under the Newtonian mechanics. It is a coordinate system which describes the corresponding relationship between the actual space point and three-dimensional array, defined by the origin, orientation and scale of coordinate axis. In order to establish the celestial reference system, some basic elements such as reference points, lines, planes and circles and their relationships must be defined at the first. A celestial coordinate system is defined by coordinate origin, datum point and datum plane. In this coordinate system, the position of any celestial body can be expressed in the form of rectangular or spherical coordinates.

The Earth is an approximately rotating ellipsoid with the equatorial bulge. Under the gravitational action of the Sun, Moon and other celestial bodies on the equatorial bulge, when the Earth moves around the Sun, the spatial orientation of its rotational axis will change, thus resulting in the vernal equinox to drift to the West slowly on the ecliptic. In Astronomy, this phenomenon is called **axial-precession**, as shown in Fig. 5.6a. Under the influence of the axial-precession, observing from above the north ecliptic pole in space, the Earth's rotation axis rotates slowly clockwise around the north ecliptic pole, so that the north celestial pole rotates around the north ecliptic pole in the same way. On the celestial sphere, the trajectory of the north celestial pole approximately forms a small circle with the north ecliptic pole as the center and the ecliptic obliquity ε as the radius. The westward drift of the north celestial pole is about 50.371 arc-seconds per year on this small circle, and its period is about 25,800 years. In general, the north celestial pole with regular motion on the

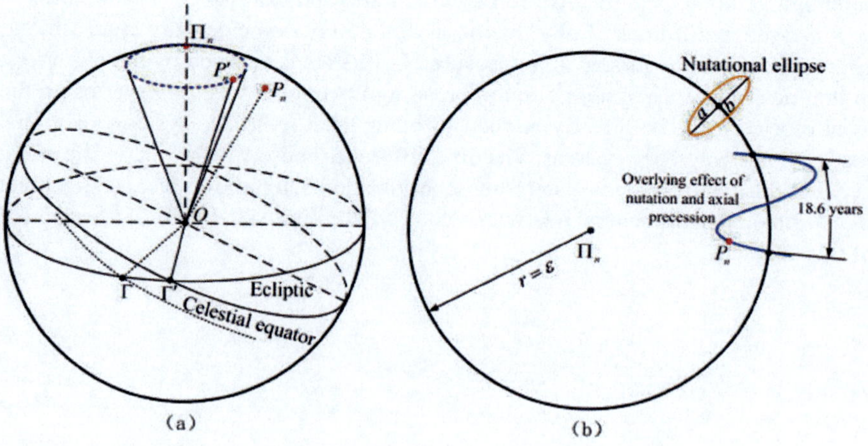

Fig. 5.6 Axial-Precession and Nutation, where **a** is Axial-precession and **b** is Nutation

celestial sphere is called **mean north celestial pole**, and the corresponding celestial equator and vernal equinox are called **mean celestial equator** and **mean equinox**, respectively. Under the influence of the gravity from the Sun, Moon and planets, the instantaneous north celestial pole will rotate around the mean north celestial pole, and its trajectory will be approximately elliptical, with the semi-major axis of about 9.2 arc-seconds and the period of about 18.6 years. In astronomy, this phenomenon is called **nutation**, as shown in Fig. 5.6b. Under the influence of the axial-precession and nutation, the orientation of the axis in the instantaneous celestial coordinate system is always changing, which no longer meets the fundamental condition keeping the datum point and plane unchanged for establishing the inertial system. In the non-inertial system, the Newtonian mechanics cannot be directly used to study the motion laws of celestial bodies and spacecrafts.

In order to establish a reference system close to the inertial system, a certain moment is generally selected as a standard epoch t_O, and at epoch t_O, the instantaneous Earth's rotation axis pointing to the north celestial pole and the direction from the Earth's center to the instantaneous vernal equinox are taken, respectively, as the Z-axis and X-axis after the corrections of the instantaneous axial-precession and nutation, so as to establish a space-fixed coordinate system, which is called the **Conventional Celestial Reference System** (CCRS) for the standard epoch. From January 1, 1984, the IAG and IAU decided to initiate the CCRS, whose coordinate-axial orientations are defined by the equator and equinox at 12:00 TDB on January 1, 2000, with the J2000.0 epoch as the standard one.

The CCRS is defined by the pure theory associated with the space-time and gravitational model, and its specific realization, the Conventional Celestial Reference Frame (CCRF), is usually established by kinematical or dynamical method. The CCRF is achieved through the optical star catalogue, pulsar catalogue containing the radio and X-ray pulsars, solar system planets and lunar ephemeris. Therefore, the establishment and maintenance of the CCRF depend on the natural celestial bodies, such as the millisecond pulsars, extragalactic radio sources, quasars, stars, planets and Moon, as well as the man-made Earth satellites. The observational technologies, such as the optics, radio, X-ray, Very Long Baseline Interferometer (VLBI), Very Large Array (VLA), radar-ranging, Lunar Laser Ranging (LLR), Satellite Laser Ranging (SLR) and satellite-positioning, are usually used to establish and maintain the CCRF, and the observational equipment may be placed on the ground or in space, as shown in Fig. 5.7.

Currently, the establishment and maintenance of CCRF are mainly achieved by the ground-based observations. The astrometry on the ground is conducive to the construction of large-diameter telescopes, and the long-term management and maintenance cost of the equipment are relatively low. Nevertheless, the ground-based observation conditions are greatly affected by atmospheric refraction, and it is difficult to realize the all-sky observation by single equipment. The advantages of space-based astrometry are there is no atmospheric refraction problem, and thus the observed star image is a fixed diffraction image, which is more accurate than the star image observed on the ground; there is not the effect of mechanical moment, so that the position of the star image at the center of the FOV does not change with

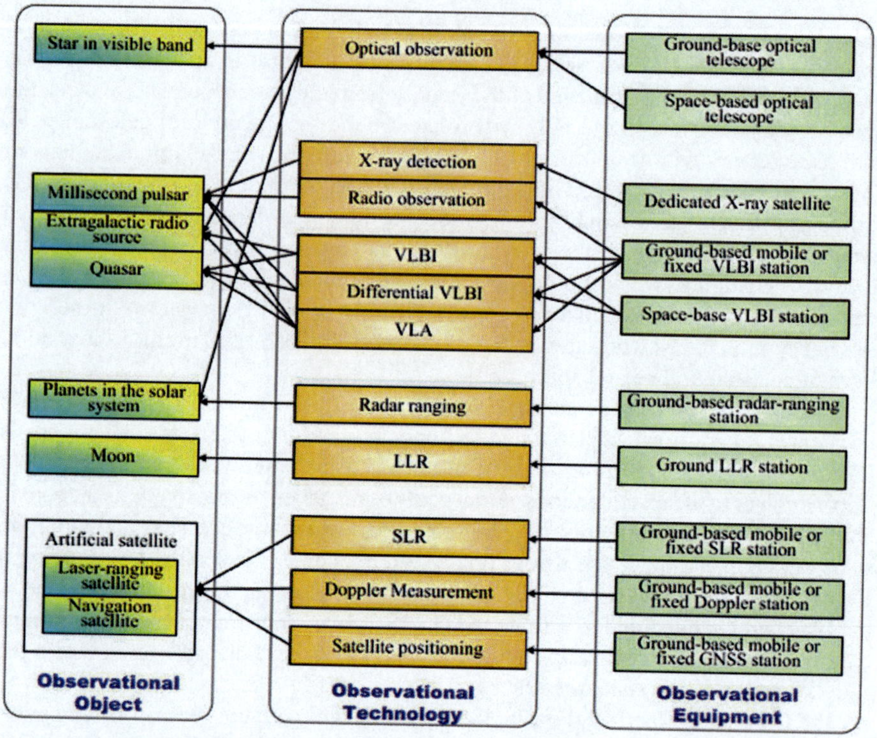

Fig. 5.7 Technological approaches to establishing and maintaining CCRF

the change of telescope's orientation; the observation for the all-sky area can be achieved only with a single set of equipment. Particularly, the observations of X-ray pulsars on a space-based constellation can form a long baseline beyond the Earth's diameter, and by using the X-ray photon correlation measurement, the highly accurate angular positions of the pulsars can be obtained. Furthermore, by accurately measuring the arrival-time of X-ray photons, the pulsar timing system is established, so as to maintain the highly accurate celestial reference frame.

5.4.5.2 International Celestial Reference System

In 1997, at 23rd IAU general assembly, the celestial reference system realized with the extragalactic radio sources was submitted by the Work Group of Reference Frame (WGRF) of IAU, and adopted as the International Celestial Reference System (ICRS). The ICRS is a reference system established by the kinematical method using the angular positions of a group of extragalactic radio sources, its coordinate origin is defined as the SSB, and its datum plane is defined as the mean equatorial plane close to the J2000.0 epoch. The VLBI technology is adopted to establish the celestial reference frame with the extragalactic radio sources. Moreover, the VLBI can also be used to adjust the correction value of the IAU axial-precession and nutation model and the drift of the mean celestial pole at the J2000.0 epoch relative to the celestial reference pole defined by the ICRS. Generally, two nutation models, IERS1996 and MBH2000, are used to evaluate the deviation of the J2000.0 mean celestial pole relative to the ICRS's celestial pole. Using the IERS1996, the deviation of the J2000.0 mean celestial pole is +17.1 milliarcseconds at 12-h direction, and +5.0 milliarcseconds at 18-h direction. Using the MBH2000, the deviation is +16.6 milliarcseconds at 12-h direction and +6.8 milliarcseconds at 18-h direction. It is shown that the distance between the ICRS and J2000.0 poles is within 20 milliarcseconds.

The vernal equinox of the ICRS is determined by the mean value of the right ascensions of 23 extragalactic radio sources at the J2000.0 epoch, selected from three radio source catalogues provided by the GSFC, JPL and NGA. In the three radio source catalogues, the origins of right ascensions are determined by using right ascension 12 h 19 min 6.6997 s of quasar 3C273B in the Fifth Fundamental Catalogue (FK5), and thus the origin of right ascension of ICRS is determined. As a result, it can be deduced that the distance of the origin of ICRS's right ascension relative to the dynamical equinox at epoch J2000.0 is 78 ± 10 milliarcseconds.

5.4.5.3 International Celestial Reference Frame

The ICRS is realized by the International Celestial Reference Frame (ICRF) calculated by the IERS using the global VLBI observations. The ICRF is composed of a group of extragalactic radio sources with precise position coordinates, including defining, candidate and other sources. The **defining sources** refer to the sources with more observation numbers over a long time span, which is mainly used to maintain the stability of the coordinate axes of ICRS. The **candidate sources** refer to the sources with short observation time and less observation numbers, but can be used as the defining sources in the future. The **other sources** refer to the sources with very few observation numbers, but can be used for connection between different reference frames. In 1995, using arc-parameter elimination method, the WGRF carried out the adjustment processing of a total of 1.6 million pairs of delays and their rates from the global VLBI observation data, and then it was obtained that the Root-Mean-Squares (RMS) of the delays and their rates are, respectively, 32.6 picoseconds and 104.2 picoseconds per second. A total of 1305 global parameters and 650,000 arc-parameters were solved, and the WGRF radio source catalogue was gotten finally. Using 133 co-sources, the WGRF radio source catalogue was compared with the IERS integrated-radio catalogue in 1995, i.e., RSC (ICRS) 95C02. In this way, the WGRF radio catalogue was rotated to the RCS (ICRS) 95C02, and finally the ICRF was obtained, which was published at the 23rd IAU general assembly in 1997.

The ICRF has 608 extragalactic radio sources, including 212 defining sources, 294 candidate sources and 102 other sources, and their angular position accuracies reach 0.25 milliarcseconds. In the data processing, the basic principles of selecting the sources are the accuracies of the sources' angular position are better than 1 milliarcsecond; the observation numbers are more than 20; the coordinate deviations in several test calculations are less than 3-sigma or 0.5 milliarcseconds; the structural changes and motions of the sources are very small. Since January 1, 1998, the ICRF has officially replaced the FK5 catalogue, applied to the research fields, such as astrometry, astro-geodynamics and space exploration.

The IERS uses mathematical models to maintain the stability of the ICRF, mainly including the elimination of local deformation (local system error) and the orientation keeping. The ICRF does not depend on the Earth's rotation, or on the ecliptic, but is only affected by the observation accuracy. According to the radio source catalogues provided by every analysis center each year, the IERS synthesizes them by mathematical methods to produce the combined radio source catalogue, which is used to define and realize the ICRF and maintain its stability.

The ICRF is considered to be non-rotational, but it has been found that some selected compact extragalactic radio sources, such as 3C273 and HC39.25, are still the extended sources in fine structure, and their structures vary at the order of milliarcseconds. Therefore, it is necessary to continually observe the extended source's structures and calculate the influence of structural changes on time delay, and then obtain the source's structural indicators according to the influence degree. By adding the correction of source's structural change to the time delay and considering the apparent proper motion of the radio sources resulted from the change of cosmic matter distribution, the stability of the established extragalactic radio reference frame can generally reach the order of microarcseconds.

5.4.5.4 New Instantaneous Reference Frame

In practice, when the ICRF is taken as a space-fixed reference frame, there are still many factors that affect the Earth's motion. For the Earth orientating, it is necessary to establish the instantaneous reference frame of motion, which is a kind of instantaneous Earth's equatorial coordinate system in strict sense. For this reason, the IAU decided to initiate a new instantaneous reference frame of motion from 2003. The new reference frame has nothing to do with the dynamical reference frame, and it is determined by the new precession-nutation model, Celestial Intermediate Pole (CIP) and Celestial Intermediate Origin (CIO). The CIO does not depend on the theoretical model, and its long-term prediction accuracy reaches the order of milliarcseconds. Due to the CIO associated with the non-rotating origin, a new definition of the Universal Time (UT1) measured by the Earth rotational angle (ERA) is introduced. There is only a linear relationship between the ERA and UT1.

Just as the FK5 system realized with the J2000.0 reference frame is replaced by the ICRS realized by the ICRF, the instantaneous reference frame of motion, established on the basis of IAU 1980 nutation theory, Celestial Ephemeris Pole (CEP) and true equinox, is also replaced by the new reference frame. The basic features of the new instantaneous reference frame are as follows:

(1) The CIP is the reference pole determined by the new precession-nutation model (IAU 2000), used to replace the original CEP and taken as the pole of new instantaneous reference frame. The motion that the period of CIP is less than two days is regarded as the polar wander. In this way, for the terrestrial reference frame, the residuals between observations and models, the high-frequency variation of the Earth-orientating parameters, and the diurnal term caused by the defects of nutation model can be absorbed by the polar wander.
(2) Since the ICRF has nothing to do with the vernal equinox, it is unnecessary that the orientation of the X-axis of the instantaneous reference frame of motion is related to the equinox. By introducing the Celestial Ephemeris Origin (CEO) to replace the vernal equinox, the confusion between the star catalogue and dynamical equinoxes can be eliminated. In 2003, the IAU decided to change the name of the CEO to the CIO. The CIO is determined by the non-rotating origin, its specific position on the equator can be calculated by the precession-nutation model and observational data, and its hour angle is the ERA, similar to the sidereal angle. Only by selecting the CIO as the starting point of right ascension will the ERA's formula not relay on the axial-precession and nutation.
(3) The ITRS is defined by the IUGG and realized by the ITRF. At present, ITRF2014, the latest reference frame issued by the IERS, contains the position and velocity vectors of 1499 stations. The non-rotating origin of the Earth is used in the ITRF to give the Terrestrial Ephemeris Origin (TEO) of the CIP on the equator. The 25th IAU general assembly decided to change the name of TEO to Terrestrial Intermediate Origin (TIO). The TIO is an exact definition of longitude starting point to meet the requirement for the UT1's accuracy.
(4) The ERA is an angular distance between the CIO and TIO, around the orbit of the CIP, and replaces the Greenwich Apparent Sidereal Time (GAST), whose starting point is the vernal equinox. Since the equator moves relative to the ecliptic each other, the relation between the GAST and UT1 contains the terms caused by the axial-precession and nutation, and the relation between the ERA and UT1 does not depend on the axial-precession and nutation. The ERA represents the stellar rotation of the Earth around the CIP axis, and thereby it is also called **sidereal angle**. The UT1 can be gotten from its linear relation with the ERA. The non-rotating origin is introduced to make the time derivative of the UT1 proportional to the instantaneous angular velocity of the Earth. The relation between the Greenwich Mean Sidereal Time (GMST) and UT1 can be derived from that between the ERA and UT1.
(5) The new precession-nutation model is used to replace the IAU1980 nutation theory and IERS 1996 precession-nutation model. Using the new precession-nutation model, the new instantaneous reference frame of motion can be determined, and transformed to the ICRF. There are two versions for the new model: IAU2000A and IAU2000B, in which the accuracy of the former is 0.1 milliarcseconds and that of the latter is 1 milliarcsecond. The new model includes the deviation between the reference frames, geodetic precession and geodetic nutation, but does not include free core nutation. The new model is used to avoid

the artificial separation of the axial-precession and nutation, while the axial-precession and nutation models are used separately in the frames based on the vernal equinox.

(6) The astronomical constants of IAU1976 cannot meet the requirements for high-precision astrometry. Therefore, the optimal estimates of the defining constants and primary constants were given by the IAU, IERS and IUGG working groups, and then the latest astronomical constant values obtained. The new and more accurate observational data is used to improve the planetary and lunar ephemerides. For example, the JPL DE/LE405 ephemeris is used to replace the JPL DE/LE200 introduced in 1984. The new ephemeris table is consistent with the ICRS, and the corresponding constant values are also given.

The new instantaneous reference frame of motion is defined and realized on the basis of the ICRF, with three different forms: the first is the old system in which the vernal equinox and CEP, as well as the axial-precession, nutation and constants introduced in 1984 are used, with low accuracy; the second is the system which still adopts the old concept, definition and vernal equinox, but in which the axial-precession, nutation, constants and celestial reference pole are improved; the third is the new system in which the new precession-nutation model and the latest constants are used, with the CIP as the pole and with the CIO as the starting point of the right ascension. The third form of reference frame has the advantage of high precision and does not depend on the vernal equinox. It should be pointed out that the vernal equinox cannot be measured accurately by the modern observation technologies. In order to define the ICRS strictly, the IAU introduces two space-fixed reference systems, the BCRS and the GCRS. Under the general relativity, the coordinate transformation between the two reference systems can be achieved by using the metric tensors and post-Newtonian approximations. The orientations of the two reference systems at their respective origins can be determined through the ICRS, and there is no rotation between them. However, the timescales used in the two reference systems are different and are, respectively, the TCB and TCG.

In the new precession-nutation model, the expansion of the axial-precession is not given, but the IAU1976 precession formula is still used, and only the specific values, such as the precession rate of the Sun and Moon at the J2000.0 ecliptic longitude as well as the obliquity-rate between the moving equator and the fixed ecliptic, are improved. A new and more accurate precession-nutation theory is being studied, whose nutation series will give 678 solar and lunar terms and 687 planetary terms, with the periods of greater than 2 days and the amplitude of greater than 15 μs, and it will provide the celestial polar coordinates with an accuracy of 0.2 milliarcseconds in the GCRS. The GCRS moves around its center of mass and has a fixed direction relative to the BCRS and extragalactic radio sources. However, under the general relativity, the GCRS is originally as a non-rotating geocentric reference system in the sense of dynamics, but with a relative rotation. The long period part of the biggest components of the GCRS is the geodetic precession and the short period is the geodetic nutation. It is completely different in effects from the usual precession and nutation. The new precession-nutation model contains the geodetic precession and nutation. In this way, the coordinate transformation between the ITRF and ICRF can be achieved by using the new precession-nutation model, CIP, CIO, TIO and ERA.

The introduction of the new instantaneous reference system will change the theory and practice for astrometry. The WGRF of IAU is defining a set of systematic theoretical methods and standardized terms. For example, the right ascension is usually calculated from the vernal equinox. If the terms of the right ascension and declination continue to be used in the new instantaneous reference system, their specific meanings must be redefined and explained. Of course, during the transitional period, the ICRS will coexist at the same time with the new reference system by which the theoretical model, basic parameter, observational data and commonly used term will be defined, and gradually introduced into the relative articles in the future.

5.5 Theory of Pulsar Timing System

Space and time are the basic forms of material existence. Space-measuring system depends on the realization, establishment and maintenance of time system. In astrometry research and space technology application, the time system is the datum to accurately describe the positions of celestial bodies and spacecrafts, and the connections between them and ground TT&C stations. One of the core tasks to achieve the XPNAV is to construct and maintain a space-ground integrated time system with pulsar time as datum.

5.5.1 *Expressions of Pulsar Time*

It is well known that time is usually measured according to the motion law of celestial bodies. The UT1 and ET are the timescales established, respectively, by using the rotation and revolution of the Earth, and the corresponding mean solar second and ephemeris second had ever been used to define the length of second in the SI units. The UT1 is obtained by observing the rotation of the Earth, which is proportional to the ERA around its axis. Due to the instability of the Earth's rotation, the uncertainty of the mean solar second is about 10^{-8} s/s. The ET is achieved by calculating and observing the orbit of the Earth's revolution around the Sun, and it is also not a uniform time standard, with the uncertainty of about 10^{-9} s/s. The atomic time using the electronic transition radiation period to define the length of a second has a very stable frequency. The uncertainty of TAI reaches 10^{-15} s/s, which is obviously better than the traditional astronomical timescales. Nevertheless, if it is further expected to improve the long-term stability of the atomic time and establish and maintain large-scale time reference, then there are some difficulties in both theory and practice.

Therefore, in the past thirty years, it has been studied to create a large-scale and long-term stable time system based on the pulsar's rotation.

The pulsars are high-speed rotating neutron stars, with extremely stable periodicity, especially the millisecond pulsars, such as PSR B1937+21, B1855+09 and J1713+074, whose pulse period change rates have reached $10^{-19} - 10^{-21}$ s/s. As a result, the millisecond pulsars are known as the most stable clocks in nature [25]. The time system established by using extremely stable pulsar-rotating period is called **pulsar timing system**, and the corresponding timescale is called **Pulsar Time** (PT). The pulsar timing system has the following basic characteristics:

(1) The pulse periods of pulsars (also called rotating periods) have extremely good uniformity, and its measurement accuracy is expected to be better than 10 ns.
(2) The pulsars can rotate and radiate signals for a long time continuously and naturally. At any time and anywhere in the solar system, stable pulsating signals can be gotten via processing the observation data from pulsars.
(3) Both the pulse-arriving time and PT can be expressed by using precise mathematical models, with the mathematical description characteristics of general disciplines.
(4) With the aid of auxiliary devices, the PT has the features of counting and displaying.

Obviously, the pulsars can satisfy the fundamental conditions for establishing a time system and are the ideal standard sources metering time and frequency. Since the first millisecond pulsar, PSR B1937+21 with a period of 1.557 ms, discovered in 1982, studies on using the millisecond pulsars for timing have been developed rapidly. Currently, the accuracy of measuring the pulse arrival-time based on the pulsar timing has reached the order of 0.1 μs, and the time stability for the pulsars observed continually for about one year is better than 10^{-14} s/s. Moreover, the instability of the TAI has been detected by the PT.

The pulsar timing system is to establish a phase function represented by the TDB or TCB as an independent variable, which is called **pulsar timing model**. Using the pulsar timing model, the pulse-arriving time can be predicted accurately. In the BCRS, the pulsar timing model can be expressed as the Taylor series expansion of the phase of the pulsating signal, that is,

$$\Phi(t) = \Phi(t_0) + \nu \cdot (t - t_0)^2 + \frac{1}{2}\dot{\nu} \cdot (t - t_0)^2 + \frac{1}{6}\ddot{\nu} \cdot (t - t_0)^3 + \cdots, \tag{5.84}$$

where $\Phi(t)$ is the total pulse phase accumulated at epoch t; $\Phi(t_0)$ is the pulse phase at initial epoch t_0; ν, $\dot{\nu}$ and $\ddot{\nu}$ are the pulsating signal frequency, and its first-order derivative and second-order derivative, respectively.

Considering the basic relation between the frequency and period of pulsar's rotation, there are

$$\left.\begin{aligned} \nu &= \frac{1}{P} \\ \dot{\nu} &= -\frac{\dot{P}}{P^2} \\ \ddot{\nu} &= \frac{2\dot{P}}{P^3} - \frac{\ddot{P}}{P^2} \end{aligned}\right\}. \tag{5.85}$$

Thereby, the pulsar timing model can be also expressed as

$$\Phi(t) = \Phi(t_0) + \frac{1}{P} \cdot (t - t_0)^2 - \frac{\dot{P}}{2P^2} \cdot (t - t_0)^2 + \left(\frac{\dot{P}^2}{3P^3} - \frac{\ddot{P}}{6P^2}\right) \cdot (t - t_0)^3 + \cdots, \tag{5.86}$$

where P, $\dot{P}$ and $\ddot{P}$ are the pulse period, and its first-order derivative and second-order derivative, respectively.

Obviously, the establishment of the pulsar timing model is essentially to determine the periods or frequencies of the pulse signals, as well as their derivatives. The diagram of the pulsar timing principle is shown in Fig. 5.8, where the pulsar observation systems, including ground-based radio telescopes and space-based X-ray detectors, receive the radio and X-ray signals radiated by pulsars, obtain radio signal flux density or X-ray photon flux intensity, and record their Time of Arrival (TOA_{obs}); the TOA_{obs} recorded by the atomic clocks at ground stations or on board of spacecrafts will be transformed to the TCB or TDB in the BCRS; using a segment of observation data, epoch folding and pulse-averaging are carried out in the expected

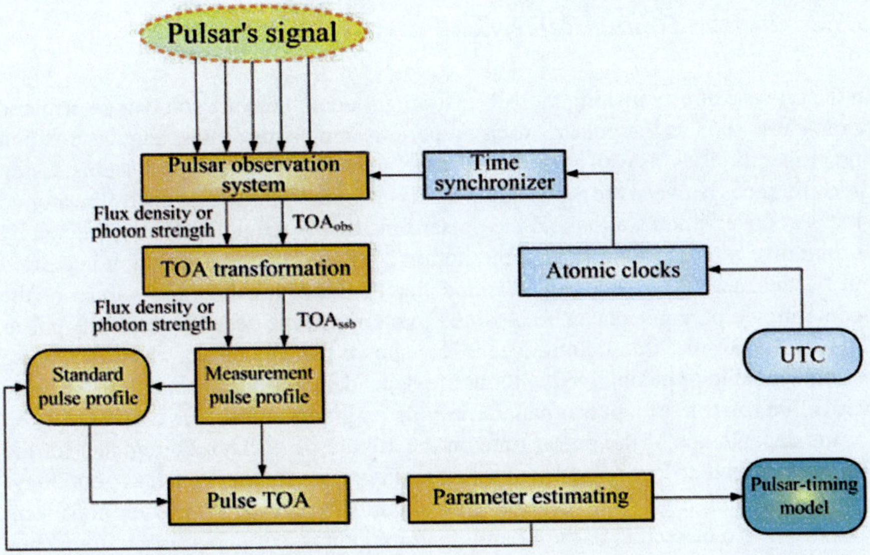

Fig. 5.8 Diagram of pulsar timing principle

pulse cycle to generate the measurement pulse profile; by comparing the measurement pulse profile with the standard pulse profile, the pulse TOA can be measured; by taking a series of the pulse TOA as the basic observables, and using the method of least squares, the parameters, such as the pulsating signal frequency, and its first-order and second-order derivatives, can be estimated. In this way, the pulsar timing model is established, and the standard pulse profile template will be improved further.

For most of the pulsars, the radiation fluxes are very weak, and pulse signals are often submerged in the background noise. For the ground-based observation of pulsars, due to the different signal strength, waveform and polarization of single pulse, it is difficult to obtain a stable pulse signal. However, in an observational frequency band, the average waveform obtained by folding of hundreds or thousands of single pulses is very stable and repeatable. Therefore, the pulsar timing observation usually requires a long signal integration time to obtain the average pulse profile with enough high SNR. The peak on the average pulse profile is selected as the reference point, corresponding to a fixed point in the pulsar's radiation region, which is used to measure the pulse TOA. The time span of pulsar observation is usually several weeks, and continual observation for many years can obtain a high-precision pulsar timing model. TOA measuring is based on the atomic time. The reference atomic clocks are regularly compared with the laboratories of the TAI, so that the measurement of TOA can take the TAI as reference. An observer can measure the TOA from pulsar *i*, which corresponds to a certain moment of pulsar time PT_i. In theory, any event can be compared with the pulse TOA, just as any event can be compared with the second of atomic time.

5.5.2 *Pulsar Timing Reference Frame*

In the process of establishing the pulsar timing model, there are some undetermined parameters, such as the pulsar's rotation periods and change rates, angular position and proper motions, as well as the orbital parameters for X-ray binary systems. Using the differences between the atomic time that is the pulse TOA measured by the atomic clocks at observation station and the pulsar time that is the pulse TOA predicted by an initiative pulsar timing model, the iterative approximation estimation is carried out by the method of least squares, and finally the best estimation values of the undetermined parameters can be obtained. As a result, the dependence of the pulsar time on the atomic time is introduced. It is shown that the rate of change of pulse period is unable to be obtained without the dependence of the atomic time. In other words, the rotation of a pulsar cannot provide an independent time unit. Moreover, for the dependence of the pulsar time on the atomic time, the requirements for the atomic timescale to be adopted are correspondingly put forward, such as good long-term stability, high precision of time interval unit (i.e., it is required to be consistent with the length of second in the SI units), no frequency drift, no seasonal frequency variation and enough readout accuracy. So, the TT is an ideal timescale for the pulsar timing measurement.

At present, the TOA measuring accuracy for the millisecond pulsars is 1 μs, and that for some pulsars reaches the level of 0.1 μs. For the observation time of more than one year, the stability of the difference (AT-PT_i) between the atomic time and the single pulsar time is about 1×10^{-14} s/s. Besides the radio pulse SNR, measurement system error, space-time transformation model error and data processing algorithm error, there are also the atomic timescale error, pulsar radiation mechanism modeling error, interstellar propagation effect uncertainty, solar system planetary ephemeris error and gravitational wave effect. These are all the factors affecting the stability of the time difference (AT-PT_i), and generally they cannot be discriminated only by using single pulsar timing observations. In theory, the Pulsar Timing Array (PTA) can be used to discriminate and detect the factors affecting the timing stability, so as to improve the accuracy of pulsar timing system.

From the point of view of creating a stable timescale and comparing with the atomic time, the fundamental method of establishing the PTA is: for multiple pulsar time (PT_i), except for the errors of atomic time system itself, assuming that other error sources are independent one another, an ensemble-time based on the multiple pulsar time, denoted as PT_{en}, is obtained by weighted average. Obviously, the noise level of the PT_{en} is always lower than that of single PT_i, and its long-term stability will be better than that of the atomic time. The PT_{en} can be expressed as

$$AT - PT_{en} = \sum_{i=1}^{n} W_i \cdot (AT - PT_i), \tag{5.87}$$

where n is the total of the pulsars used in the ensemble, and W_i is the weight of single pulsar time PT_i.

In general, the higher is the stability of a single pulsar, the greater the value of its weight. In order to obtain the best ensemble pulsar timing effect, the following factors should be considered in establishing and maintaining the PTA.

(1) The pulsars used in the ensemble should have good stability of the pulse period.
(2) The radiation signals of the pulsars should have high SNR.
(3) The pulsars are required to evenly distribute in space and to reach a certain number.
(4) The time span of observing the pulsars is long enough.
(5) The reference timescale should be enough stable.
(6) There are the high-precision time and frequency transfer ways between the observation stations and the reference clock laboratories.
(7) A mathematical model with the complete space-time transformation and effect correction is adopted.
(8) An advanced data collecting and processing system is adopted.
(9) The requirements for other applications are also considered. For example, the orthogonal distribution of the pulsars in space is advantage to detecting the gravitational wave.

For data processing of the ensemble pulsar time, the following points should be paid attention to:

(1) The original timing observation data and unified theoretical model are used to ensure the consistency of the residuals. In addition, the residual data and its model parameters are obtained to examine the difference between the theoretical models. Especially for the observation data of the same pulsar obtained from different stations, the influence of the deviation between observation systems should be eliminated at first.
(2) It should be avoided that the system error resulted from residual deterministic trend is mixed in noise.
(3) The equal interval observation data is used to facilitate data processing.
(4) In order to avoid the step of time series caused by the change of the number of ensemble pulsars, a correction value should be added, that is

$$AT(t) - PT_{en} = AT(t + \delta t) - \left(PT'_{en}(t + \delta t) + a\right), \tag{5.88}$$

where a is the added correction value, and PT'_{en} is the ensemble pulsar time calculated after increasing or decreasing the pulsar's number.

The pulsar time is different from the atomic timescale and needs not to adjust the frequency, because in the process of fitting the parameters by using the timing residuals, the deterministic trend part of frequency variation has been eliminated. Since the long-term stability of the ensemble pulsar time is better than that of the atomic time, it is possible to transfer the accuracy of atomic time from one time-segment to another as long as the continuity of observation is maintained. In order words, if the frequency reference fails temporarily, the ensemble pulsar time can be used to maintain the accuracy of atomic time. The future frequency reference has higher accuracy, and after years of stable operation, it is possible to provide a new definition of second. The accuracy of the ensemble pulsar time can be used for a reverse transfer, thus improving the accuracy of the previous atomic timescale. Certainly, this is also limited by the instability of pulsar time itself. In addition, the long-term stability of atomic time can be analyzed by the residual values of AT–PT, and the related parameters of pulsar time are obtained by fitting the AT–PT values. It is shown that the fitting method of least squares can be used to filter out all the short period terms with the period of less than 1 year and the long period terms with the same order of magnitude as the observation time span. Obviously, only when the observation time span is more than two years, the stability of AT–PT can be effectively estimated and the long-term stability of atomic time be analyzed. So, it is an effective approach to improving the stability of timescale by using the long-term accumulated pulsar observation data to establish the ensemble pulsar time.

5.5.3 *Evaluation of Pulsar Timing Stability*

Generally, Allan variance is used to analyze the frequency stability of atomic clocks. For the atomic clocks with linear frequency drift and low-frequency noise (also called **red noise**), Hadamard variance is usually used to analyze the frequency stability. For the pulsar timing, the methods used to analyze the stability of atomic clocks cannot be directly used because of the irregularity of radio pulsar observation data interval. For this reason, using the third-order polynomial fitting of pulsar's timing residuals, a statistical analysis method similar to the improved Hadamard variance was proposed by Joseph H. Taylor in 1991, which is called **Sigma-z variance** (σ_z^2). The Sigma-z variance is not sensitive to the randomness of pulsar observation and is an ideal statistical method to compare the stability of the pulsar time with that of the atomic time, mainly used to characterize the influence of the low-frequency pulsar timing residuals on the stability of frequency. The so-called **timing residuals** refer to the timing errors remained still after the phase, frequency and its rate, angular position and proper motion in the pulsar timing model have been determined, also known as **timing noise**.

By determining a set of polynomial coefficients $p_j (j = 0, 1, 2, 3)$, and utilizing the frequency measurement residual sequence $r(t_i)$ in the pulsar timing at observation time $t_i (i = 1, 2, \ldots, N)$, the power spectral density of the pulsar timing residual can be estimated. The parameters c_j can be fitted by the method of least squares, that is

$$\chi^2 = \sum_{i=1}^{N} \left[r(t_i) - \sum_{j=0}^{3} c_j p_j(t_i) \right] = \min, \tag{5.89}$$

where c_0, c_1 and c_2 are, respectively, the pulsar's phase, period and its rate, and c_3 is the estimating parameter of power spectral density of the residual, that is,

$$S_1 = (c_3)^2. \tag{5.90}$$

The corresponding frequency is $f \approx \frac{1}{T}$, where $T = t_N - t_1$, representing the observational time span.

The frequency measurement residual sequence $r(t_i)$ with the observational time span of T is divided into m subsequences with approximately equal length, and then the every subsequence is divided once again using the same way. As a result, m new parameter-estimating values c_3 will be obtained, and thus the power spectral density is

$$S_m = \sqrt{\frac{1}{m} \sum_{k=1}^{m} (c_3)_k^2}. \tag{5.91}$$

The corresponding frequency is $f \approx \frac{m}{T}$, where $m = 2, 4, 8, \ldots$.

For this way, an infinite dimensional power spectrum estimation S_m is obtained. Similar to the method usually used in the Allan variance, the S_m is divided by the third power exponent of time, and thus an infinite dimensional Sigma-z variance is obtained, i.e.,

$$\sigma_z^2(\tau) = \frac{1}{\tau^3} S_m, \tag{5.92}$$

where $\tau = \frac{1}{m} T$.

This is the original algorithm of the Sigma-z variance proposed by Joseph H. Taylor. The Sigma-z variance $\sigma_z^2(\tau)$ obeys the chi-square distribution with the degree of freedom of m. This variance statistical method is applicable to the processing of non-equal-interval and discontinuous pulsar timing observation data. Of course, it is also applicable to the analyzing and processing of the stability of atomic clock with equal interval measurement data.

In addition, by directly using the pulsar timing residual measurement data to fit a third-order polynomial function, an improved Sigma-z variance statistical algorithm was proposed by Matsakis et al. in 1997 [26]. Similarly, the time residual sequence is divided into m subsequences (m = 1, 2, 4, 8, …), and then the residual data of each subsequence is used for the least squares fitting, so as to meet

$$R^2 = \sum_{i=1}^{N_m} \left[x(t_i) - \frac{X(t_i)}{\sigma_i} \right]^2 = \min, \tag{5.93}$$

where $X(t_i) = c_0 + c_1(t_i - t_0) + c_2(t_i - t_0)^2 + c_3(t_i - t_0)^3$; $x(t_i)$ is the phase measurement residual of time; σ_i is the uncertainty of time measurement; t_i is the observational time; t_0 is the arbitrarily selected deviation time; N_m is the measurement number for the subsequence.

Therefore, an infinite dimensional Sigma-z variance estimation is

$$\sigma_z^2(\tau) = \frac{\tau^2}{2\sqrt{5}} \sqrt{\sum_{j=1}^{m} (c_3)_j^2}, \tag{5.94}$$

where τ is the sampling time for the subsequence.

The Sigma-z variance has the following power-law exponential relation to τ.

$$\sigma_z^2(\tau) \propto \tau^\mu, \tag{5.95}$$

where $\mu = \begin{cases} -\alpha - 1 & \alpha < 3 \\ -4 & \alpha \geq 3 \end{cases}$.

The Sigma-z variance is especially suitable for statistical analysis for the low-frequency phase noises with the "red" or "pink", which are the typical pulsar timing residuals. However, by using the Sigma-z variance, it is difficult to discriminate the "blue" noise process with $\alpha \geq 3$.

5.5.4 Stability of Comparing PT with ATI

The pulsar timing observation is based on Atomic Time (AT). Using the pulse TOA measurement data for many years, the parameters related to pulsars, such as the pulse phases, frequencies and their rates, as well as the angular positions, proper motions and parallaxes of the pulsars, can be accurately estimated, and then the timing residuals are the difference between the reference AT and Pulsar Time (PT), namely, AT–PT. In order to ensure the accuracy of the PT, the TAI or TT is usually taken as the reference to calculate the PT. The millisecond pulsars have the extremely stable rotation frequencies and their signals can be read out cumulatively. Therefore, the PT can be compared with other timescales, and it has the performance of ordinary clocks.

Since 1982, two millisecond pulsars PSR B1937+21 and PSR B1855+09 have been observed at the Arecibo Observatory for the long term, accumulating 16 years of observational data. Using the observation data with a time span of 3300 days and adopting the Sigma-z variance statistical analysis method, the residuals of the PT from B1937+21 and B1855+09 relative to the Terrestrial Time in 1996 (TT96) was calculated by Demetrios N. Matsakis, et al. In addition, the statistical results of the Sigma-z timing residuals on several terrestrial timescales are also presented. From the studied results, it can be shown that:

(1) The slopes of Sigma-z variance curves of the two pulsars are -1.5 or so during the sampling interval of 3 years, and the timing residuals are mainly caused by independent measurement errors. When the sampling interval is more than 1 year, the timing residuals are mainly caused the low-frequency noises, which are generally considered as the inherent noise between the pulsar's and atomic timescales.

(2) With the sampling interval increasing, the frequency stability curve of PSR B1855+09 always demonstrates a downward trend, while the stability of PSR B1937+21 is greatly affected by the higher order derivatives of the rotational period, and thus it gradually shows upward trend during the final segment.

(3) When the sampling interval is more than 1 year, by comparing the time residuals of TT96–PT(1937) with those of TT96–PT(1855) and TA(PTB)–TA(A.1), and considering the equivalence of TT96 and TA(PTB), the frequency stability of the PT has reached or been better than that of TA(A.1). Among them, PT(1937) and PT(1855) denote the pulsar times from PSR B1937+21 and PSR B1855+09, respectively; TA(A.1) denotes the atomic time maintained by the U.S. Naval Observatory; TA(PTB) denotes the atomic time maintained by the Federal Physical Institute of Germany.

(4) The short-term stability of the PT is not as good as that of the AT. For short observation time, the PT accuracy generally gets to the microsecond level, which is relied on the accuracy level of currently measuring the pulse TOA.

Only by observing the millisecond pulsars continually during the long term and accurately measuring their parameters, a precise timing model can be established and its timing stability improved further. As establishing the pulsar timing model, the angular position accuracy of pulsars is required to reach 0.1 milliarcseconds to accurately make the time transformation of the TOA. In addition, the long-term component errors in the solar system ephemeris, as well as the influences of the selected TOA transforming and timing models, should be considered. Generally, the error caused by imperfection of the TOA timing model is 5–10 times larger than that of the fitting of least squares. Thereby, the high-precision PT can be obtained only by processing the cumulatively observed data for many years. Moreover, the angular positions, proper motions, distances and other parameters of some pulsars are obtained by the VLBI observations. It is required that the VLBI observation reaches the accuracy level of 0.1 milliarcseconds, and the connection between the BCRF or GCRF and the VLBI reference frame should also reach the corresponding accuracy level. At present, except for a few millisecond pulsars like PSR B1937+21, whose angular position accuracy using the VLBI reaches the order of 0.1 milliarcseconds, most pulsars are only the order of 10 milliarcseconds due to their low radiation flux density. The VLBI is used to measure the pulsar's angular positions to avoid the fluctuation errors from the atomic time introduced in the pulsar's parameters fitting. With the development of the VLBI, the accuracy of measuring the pulsar's angular positions will be improved further.

For the pulsar timing residuals, besides the intrinsic noises of the atomic time and millisecond pulsar's rotation, there are also the errors of interstellar dispersion, gravitational wave background and solar system planetary ephemeris. Except for the noise of atomic time itself, other error sources are independent of one another for different pulsars, and thus the ensemble pulsar time can be established by weighting every single pulsar time. In this way, the influence of various independent noise sources can be weakened, and the ensemble pulsar time should be more stable than the single pulsar time. With the utilized pulsar number and the observational time increasing, the influence of random errors on the long-term stability of the ensemble pulsar time is weakened, and thus the stability of the ensemble pulsar time will finally reach or be better than the level of the atomic time.

For the short-term stability of pulsar timing, it is far lower than the comparison accuracy of the atomic time. Currently, the comparison accuracy of the atomic time based on the TWSTFT technique of GNSS satellites has been better than 3 ns level, and the short-term noise of pulsar is still larger than the atomic time. Generally speaking, only by using a few months of pulsar's observation data, the pulsar time cannot reach the level of stability of the atomic time yet. On the one hand, the pulsar time is defined by the TOA analysis model, but the model parameters cannot be accurately determined in advance, usually obtained by the least squares fitting with the TOA observables. On the other hand, using the timing residual AT–PT to fit the pulsar's parameters inevitably absorbs part of the instability of atomic time into the fitted parameters, leading to the deviation of the pulsar's parameters, and finally the obtained timing residuals cannot fully reflect the actual difference.

In the past 10 years, the PTA has been developing rapidly over the world, and more than 10 PTA stations have been put into operation. According to the standard operation procedures, the PTA is composed of the radio telescopes, each of which is an array element in the pulsar timing observation and carries out long-term observation and data processing for the selected millisecond pulsars. Using the long-term observation data from the Parkes Pulsar Timing Array (PPTA), an ensemble pulsar time was established by George Hobbs et al., in 2012 [27], and then the clock bias sequences of the ensemble pulsar time compared with the TAI. A total of 19 millisecond-radio pulsars are used for establishing the TT (PPTA11), in which the shortest observation time span is 8.1 years and the longest span is 17.0 years. The TT (PPTA11) denotes the TT achieved by using the data from the PPTA until 2011, which is the ensemble pulsar time. Generally, the BIPM adjusts the reference frequency standard every few years, resulting in the maximum difference between the TT (BIPM11) and TT (TAI) around 1998, where TT (TAI) denotes the TT achieved by the TAI, and TT (BIPM11) denotes the TT achieved by the TAI that was corrected by the BIPM in 2011. From Hobbs' results, it can be seen that from 1994 to 1998, the difference between the TT (PPTA11) and TT (TAI) is large, with the maximum bias of 1300 ns or so; since 1998, the difference between the TT (PPTA11) and TT (TAI) has gradually decreased, with the maximum difference of about 500 ns, and the long-term instability of the TAI has been detected by the ensemble pulsar time; the TT (PPTA11) is well consistent with the TT (BIPM11), and both change trends are same, with the maximum difference of 300 ns after 1998, which indicates the effectiveness of the TAI corrected by the BIPM.

With the standardization and scale of the millisecond pulsar timing observation, the mature application of coherent de-dispersion technology, the improvement and perfection of software and hardware, and the increase of timing observation station and observational time span, the pulsar timing observation accuracy will be improved, so as to refine the short-term and long-term stability of the pulsar time. For the XPNAV, the stability of the X-ray pulsar timing will be better than that of the radio pulsar timing because there is no dispersion effect in the interstellar propagation of X-ray photons. By using the pulsar timing models with high stability, the pulse TOA can be accurately predicted, which is conducive to the high-precision autonomous orbit determination and time synchronization for spacecrafts.

5.6 Methods of Large-Scale Navigation

The pulsars are high-speed rotating neutron stars, far away from the solar system, reaching thousands of light-years, even tens of thousands of light-years. The XPNAV takes the general relativity as the fundamental theory, and by using the XPNAV, the spacecrafts in the Riemannian space can autonomously determine its navigation parameters. It is a new space navigation approach based on four-dimensional space-time transformation, which is defined here as **large-scale navigation**.

5.6.1 Navigational Modes and Basic Observables

According to the different types of navigation parameters and achieving order, the autonomous navigation of spacecrafts based on the X-ray pulsars can be divided into four parameter-determined modes, including time, attitude, velocity and position. Determining the time parameters is helpful to getting the pulse profile, identifying the pulsar's characteristics, measuring the photon's arrival-time and updating the onboard clock parameter estimation; determining the attitude parameters is to get the spacecraft's orientation relative to the orbit coordinate system, so as to judge whether the spacecrafts are rotating or rolling normally; determining the velocity parameters is to get the motion directions and rate parameters of spacecrafts, so as to carry out the orbit maneuver for spacecrafts, and generally the pulsars which are approximately perpendicular to the orbital plane can be selected for navigation to reduce the Doppler frequency shift effect; determining the position parameters is to improve the estimation accuracy of position and velocity parameters, which is closely related to photon arrival-time measurement and time transformation. The parameter-determined order from the time to attitude, velocity and position is suitable for recovering the spacecraft's navigation system in start-up mode or correction mode, and advantage to solving all navigation parameters. Of course, the order of determining the time and attitude parameters may be changed according to the requirement for actual mission, and sometimes the onboard clock itself provides enough accurate time information. In fact, the spacecraft's position, velocity and time parameters are often coupled with one another for the system state estimation based on the Kalman filtering.

According to selecting the navigation reference points, the autonomous navigation of spacecraft based on the X-ray pulsars can be divided into absolute navigation and relative navigation. The **absolute navigation** refers to the process of directly determining the three-dimensional position and velocity components of the spacecrafts relative to the BCRS and guiding them to arrive at the target orbits or in celestial bodies. The **relative navigation** refers to the process of solving the navigation parameters of the spacecraft relative to a reference object frame, and the reference object may be another spacecraft, observation station, Lagrange point or one of the solar system planets.

Using the X-ray pulsars to determine the navigation parameters of spacecrafts, no matter which navigation mode is adopted, they are achieved by using the X-ray detectors installed on the spacecrafts to detect the X-ray signals radiated from pulsars and get the basic observables. The X-ray signals from pulsars contain a variety of navigation information, such as the pulse-arriving time, Doppler frequency shift of pulsating signal and position of pulsar's image. According to the different requirements for solving the navigation parameters, the different observables can be extracted from them.

The pulse-arriving time (or pulse TOA) is a basic observable to determine the position and time parameters of spacecrafts. When the X-ray photons radiated from pulsars enter the field of view of the onboard X-ray detectors, the photon counter will record the number of the incident X-ray photons, whose arriving times are marked by the onboard clock. Consequently, the number and arrival-time of photons are the original observables in practice. However, it is difficult to obtain the basic characteristics of pulsars from the arrival-time of single photon. The arrival-time of photons will be transformed into the BCRS to make the signal processing, so as to obtain the average pulse profile curve during integration time. As a result, the time corresponding to the peak of the pulse profile near the middle segment during the integration time is taken as the TOA of the pulse, and the peak corresponds to a fixed point in the beam area radiated from the pulsar.

The Doppler frequency shifts of pulsating signals are the basic observables to determine the velocity of spacecrafts. The pulsars radiate the periodic pulse signals. When the spacecrafts move toward the pulsars, the observed pulse frequencies are greater than the transmitting ones, and when the spacecrafts move away from the pulsars, the observed pulse frequencies are less than the transmitting ones. The differences between the transmitting and receiving pulse frequencies are the Doppler shifts. By calculating the derivative of pulsar timing model, the pulse frequency can be predicted. The Doppler frequency shifts are determined by comparing the measured pulse signal frequencies with the predicted ones. The velocities of the spacecrafts along the line of sight of pulsars can be computed by the Doppler frequency shifts. In this way, the velocity measurement equations can be established. When the Doppler frequency shifts of more than four pulsars are observed at the same time, the three-dimensional velocities and clock frequency drifts of spacecrafts can be estimated directly by using the method of least squares.

The positions of pulsar's images are the basic observables to determine the attitude parameters of spacecrafts. The positions of the images generated by the pulsars on the detector's plane can be read directly from the plane grid. When the detector is installed on a three-axis gimbaled platform, the rotation angles of the detector coordinate system relative to the spacecraft-body coordinate system can be directly obtained from the rotation of the platform frame. Obviously, it is equivalent to measuring the position coordinates of the pulsar's image in the spacecraft-body coordinate system. The identifiers of the pulsars in the navigation database can be identified by the cross-correlation processing between the measured and standard pulse profiles, and then the angular positions of the pulsars in the BCRS can be looked up in the navigation database using the identifiers. Through the coordinate transformation, the angular positions of the pulsars in the orbital coordinate system are obtained. So, from the rotation transformation matrix between the spacecraft-body coordinate system and the orbital coordinate system, the attitude angle components of spacecrafts can be gotten. The imaging time of the pulsars on the detector's plane relies on the rotation speed of spacecraft and the field of view of the detector. The slower does the spacecraft rotate, and the larger is the field of view of the detector, the shorter the imaging time required.

5.6.2 *Transformation of Photon Arriving Time*

5.6.2.1 Transformation of Timescales

In the XPNAV, the time of the X-ray photons radiated from pulsars arriving at the spacecraft is measured by the onboard atomic clock in the spacecraft-body coordinate system, which is the proper time. The process of transforming the proper time of the X-ray photons arriving at the spacecraft into the TCB or TDB is known as **timescale transformation**. The flow chart of the transformation relations among the proper time and several commonly used timescales is shown in Fig. 5.9.

Without loss of generality, taking the spacecrafts orbiting the Earth as an example, the scale transformation of arrival-time of the photons is discussed here. Supposed that the time of the photons arriving at the spacecrafts is measured with the GPST, the specific transformation steps and formulas are as follows.

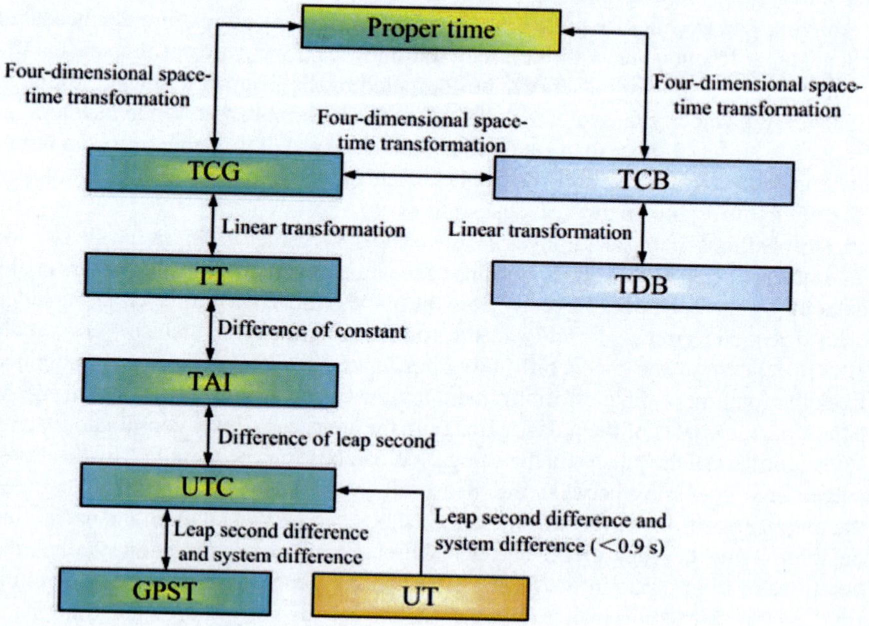

Fig. 5.9 Transformations between proper time and commonly used timescales

(1) Transformation from GPST to UTC

The GPST is traced back to the UTC (USNO), whose transformation formula is

$$UTC(USNO) = GPST - \{Ls + a_0 + a_1[GPST - T_0 + 604800(WN - WN_0)]\}, \tag{5.96}$$

where Ls is the total number of the leap seconds; a_0 and a_1 are, respectively, the coefficients of the constant and first-order terms of time polynomial; T_0 is the reference epoch of UTC (USNO); WN is the number of weeks at current epoch; WN_0 is the number of weeks at the reference epoch of UTC (USNO).

The parameters in formula (5.96) can be obtained from the navigation messages broadcasted from the GPS satellites. The international UTC is obtained by the BIPM-coordinated time comparison and weighted calculation with the UTC maintained by the global laboratories, where the weight of UTC (USNO) accounts for about 60% of the UTC and there is a bias of about 10 ns between them, i.e.,

$$UTC = UTC(USNO) + \Delta t_{USNO}, \tag{5.97}$$

where Δt_{USNO} is the time bias between the UTC (USNO) and UTC, gotten from the periodical BIPM bulletins.

(2) Transformation from UTC to TAI

According to the definition of UTC, except for the introduction of the leap seconds, the UTC is essentially the TAI, and thus the basic relation between both can be expressed as

$$TAI = UTC + n \times 1^s, \tag{5.98}$$

where n is an adjustment parameter of the leap second, released by the IERS.

(3) Transformation from TAI to TT

The TAI clocks used for time keeping on the ground are not located on the geoid, and hence the velocities of ideal clocks are not zero relative to the TT. When the accuracy of time metering of the TT or TAI is required to reach the order of 1×10^{-15}, then the altitude surveying accuracy must be better than 0.1 m. In practical application, all atomic clocks participating in time keeping always have a certain inherent clock velocities relative to the ideal clock. Generally, the clock velocities resulted from the effect of relativity do not need to be corrected separately and can be offset by the integrated average processing of the clocks distributed on the world. However, for the cesium clocks which determine the accuracy of TAI, their relative velocities still need to be corrected.

According to the definition of TAI and TT, the TAI takes the length of second of the SI units on the rotating geoid as the unit, which is regarded as the coordinate timescale in the ECEF reference system, and is also a concrete realization form of the ideal timescale TT. That is to say, in the GCRS, the proper time, metered along the world-line on the geoid and relative to the at-rest observer and its carried clock, is defined as the TAI. The basic relation between the idea TT and TAI is expressed as

$$TT = TAI + 32^s.184. \tag{5.99}$$

However, for the atomic clocks carried by the spacecrafts, there are orbital velocity relative to the GCRS and gravitational potential relative to the geoid, and thus,

$$TT = \left[1 - \frac{1}{c^2}\left(W_G - \frac{3\mu_E}{2a}\right)\right](\tau - \tau_0) + \frac{2}{c^2}\sqrt{\mu_E a}(\sin E - \sin E_0), \tag{5.100}$$

where τ is the proper time in the spacecraft-body coordinate system, which has been transformed to the TAI; τ_0 is the clock reading as the TT is equal to zero; c is the velocity of light in vacuum; W_G is the gravitational potential on the geoid, which is the standardized definition constant of IERS 2000, with the value of 62636856.0 m^2/s^2; μ_E is the gravitational constant of the Earth; a is the semi-major axis of the spacecraft's elliptic orbit; E is the eccentric anomaly of the spacecraft's orbit.

The first term on the right side of formula (5.100) is a long-term one, and the second term is a periodic one. In order to avoid the long-term drift of the onboard clock relative to the TT, the clock's frequency must be adjusted, which is the effect from the long-term term of formula (5.100). Supposed that the reference frequency of the onboard clock is f_0, the frequency before launching the spacecraft should be adjusted to

$$f = (1 - k)f_0, \tag{5.101}$$

where $k = \frac{1}{c^2}\left(W_G - \frac{3\mu_E}{2a}\right)$.

For example, the semi-major axis is 26570 km for the GPS satellite orbits, the reference frequencies of their onboard clocks are 10.23 MHz, and thus $k = 4.472 \times 10^{-10}$. In other words, before launching GPS satellites, the clock reference frequencies are adjusted to 10.22999999543 MHz, to eliminate the long-term effect of the relativity.

(4) Transformation from TT to TCG

According to the resolution of IAU general assembly in 1991, the transformation from the TT to TCG is linear, and the difference between them is a proportional constant, i.e.,

$$TCG = TT + L_G \times (JD - 2443144.5) \times 86400, \tag{5.102}$$

where $L_G = 6.969290134 \times 10^{-10}$, which is an astronomical constant in the standardized definition of IERS 2000; JD is the Julian day at the corresponding observation epoch.

(5) Transformation from TCG to TCB

According to the relevant resolution of IAU2000, the complete post-Newtonian expression for a space-time coordinate point (ct, $\mathbf{x}_E$) in the BCRS has been given, where t =TCB, and thus the four-dimensional space-time transformation from the TCG to TCB is expressed as

$$TCB = TCG + \frac{1}{c^2}\left[\int_{t_0}^{t}\left(\frac{1}{2}v_E^2 + w_{0ext}(\mathbf{x}_E)\right)dt + \mathbf{v}_E \cdot (\mathbf{x} - \mathbf{x}_E)\right]$$
$$- \frac{1}{c^4}\int_{t_0}^{t}\left[-\frac{1}{8}v_E^4 - \frac{3}{2}v_E^2 w_{0ext}(\mathbf{x}_E) + 4v_E^i w_{ext}^i(\mathbf{x}_E) + \frac{1}{2}w_{0ext}^2(\mathbf{x}_E)\right]dt$$
$$+ \frac{1}{c^4}\left[3w_{0ext}(\mathbf{x}_E) + \frac{1}{2}v_E^2\right][\mathbf{v}_E \cdot (\mathbf{x} - \mathbf{x}_E)], \tag{5.103}$$

where $\mathbf{x}_E$ and $\mathbf{v}_E$ are, respectively, the position and velocity vectors of the center of mass of the Earth in BCRS; v_E is the module of vector $\mathbf{v}_E$; $w_{0ext}(\mathbf{x}_E)$ is the total sum of the Newtonian gravity potentials of celestial bodies of the solar system at the Earth-center, except for the Earth; $w_{ext}^i(\mathbf{x}_E)$ is the total sum of the vector potentials of celestial bodies of the solar system at the Earth-center, except for the Earth; t is the upper limit of integration, representing the TCB; t_0 is the lower limit of integration, which is at 0:0:0.00 TAI on January 1, 1977.

(6) Transformation from TCB to TDB

Similarly, according to the resolution of the IAU general assembly in 1991, the transformation from the TCB to TDB is linear, and there is a proportional constant difference between them, that is,

$$TDB = TCB - L_B \times (JD - 2443144.5) \times 86400, \tag{5.104}$$

where $L_B = 1.55051976772 \times 10^{-8}$, an astronomical constant in the standardized definitions of IERS 2000; JD is the Julian day at the observation epoch.

5.6.2.2 Transformation of Time Delays

The pulsar timing model is established in the BCRS using the timescale TCB or TDB. Through the above timescale transformations, the time of the X-ray photons arriving at the spacecraft has been transformed from the GPST to TCB. Under the TCB, the process of transforming the photon arrival-time from the spacecraft to the SSB is called **time delay transformation**. And further, the pulse profile can be gotten, and the pulse-arriving time measured. By comparing the measured pulse-arriving time with the predicted one, the delays from the spacecraft to the SSB along the line of sight of the pulsar are obtained and then are used to establish the measurement equations of determining the spacecraft's orbit and time parameters.

From the point of view of geometric orbit determination, the time transformation model with 1 μs error will result in the orbital error of 300 m for spacecrafts. According the general relativity, for the X-ray photons radiated from the pulsars with thousands or even tens of thousands of light-years away from the solar system, their propagation path is curved, resulting in the delay of arrival-time. For this reason, the high-precision time delay transformation model needs not only to correct the spatial geometric distance, but also consider the effect of the relativity. A single pulse signal is composed of a large number of X-ray photons radiated from the pulsar, and the trajectory curve of each photon in four-dimensional space-time is a world-line. For the X-ray photons radiated from the pulsar's magnetic pole arriving at the spacecraft's detection devices, their passed through world-lines are the zero-geodetic lines. That is to say, the light-like space-time interval is equal to zero. So, the time delay transformation model with high accuracy can be obtained by using the second-order post-Newtonian space-time metric in the BCRS and ignoring the space-time cross-terms [24], and then,

$$
\begin{aligned}
t_{SSB} - t_{SC} = {} & \frac{1}{c}\mathbf{n}_{SSB}\cdot(\mathbf{D}-\mathbf{b}) - \frac{1}{c}\mathbf{n}_{SC}\cdot(\mathbf{D}-\mathbf{p}) \\
& - \sum_{k=1}^{m_P}\frac{2\mu_k}{c^3}\ln\left|\frac{\mathbf{n}_{SSB}\cdot\mathbf{b}_k + b_k}{\mathbf{n}_{SSB}\cdot\mathbf{D}_k + D_k}\right| + \sum_{k=1}^{m_P}\frac{2\mu_k}{c^3}\ln\left|\frac{\mathbf{n}_{SC}\cdot\mathbf{p}_k + p_k}{\mathbf{n}_{SC}\cdot\mathbf{D}_k + D_k}\right| \\
& + \frac{15}{4}\frac{\mu_S^2}{c^5 D_y}\left[\arctan\left(\frac{\mathbf{n}_{SSB}\cdot\mathbf{D}}{D_y}\right) - \arctan\left(\frac{\mathbf{n}_{SSB}\cdot\mathbf{b}}{D_y}\right)\right] \\
& - \frac{15}{4}\frac{\mu_S^2}{c^5 D_y}\left[\arctan\left(\frac{\mathbf{n}_{SC}\cdot\mathbf{D}}{D_y}\right) - \arctan\left(\frac{\mathbf{n}_{SC}\cdot\mathbf{p}}{D_y}\right)\right] \\
& + \frac{2\mu_S^2}{c^5 D_y^2}\left\{\begin{array}{l} \mathbf{n}_{SSB}\cdot(\mathbf{D}-\mathbf{b})\left[1+\left(\dfrac{\mathbf{n}_{SSB}\cdot\mathbf{D}}{D}\right)^2 - \dfrac{2D^2}{bD}\right] \\ +2(\mathbf{n}_{SSB}\cdot\mathbf{D})\left(\dfrac{b}{D}-1\right) \end{array}\right\} \\
& - \frac{2\mu_S^2}{c^5 D_y^2}\left\{\begin{array}{l} \mathbf{n}_{SC}\cdot(\mathbf{D}-\mathbf{p})\left[1+\left(\dfrac{\mathbf{n}_{SC}\cdot\mathbf{D}}{D}\right)^2 - \dfrac{2D^2}{pD}\right] \\ +2(\mathbf{n}_{SC}\cdot\mathbf{D})\left(\dfrac{p}{D}-1\right) \end{array}\right\},
\end{aligned}
\tag{5.105}
$$

where t_{SSB} and t_{SC} are the time of the same pulse signal emitted from the pulsar, respectively, arriving at the SSB and the spacecraft's detection device; c is the velocity of light in vacuum; $\mathbf{n}_{SSB}$ is the direction vector from the SSB to pulsar; $\mathbf{n}_{SC}$ is the direction vector from the spacecraft to pulsar; **b**, **p** and **D** are, respectively, the position vectors of the SSB, spacecraft and pulsar in the coordinate system at the center of mass of the Sun, $O_S - X_S Y_S Z_S$, as shown in Fig. 5.1; $b = \|\mathbf{b}\|$; $p = \|\mathbf{p}\|$; $D = \|\mathbf{D}\|$; p_x and D_x are, respectively, the components of vectors **p** and **D** on the X-axis in the $O_S - X_S Y_S Z_S$; D_y is the component of vector **D** on the Y-axis in the $O_S - X_S Y_S Z_S$; $\mathbf{b}_k$, $\mathbf{p}_k$ and $\mathbf{D}_k$ are the vectors from the center of mass of the solar system's planet k to the SSB, spacecraft and pulsar, respectively; μ_S and μ_k are, respectively, the gravitational constants of the Sun and solar system's planet k; m_P is the total number of the Sun and the solar system's planets.

On the right of formula (5.105), the first and second terms denote the time delay produced by the geometric distance from the pulsar to spacecraft; the third and fourth terms denote the sum of the Shapiro delay effect resulted from the planets in the solar system; the fifth, sixth, seventh and eighth terms denote the light deflected in the gravitation field of the Sun, whose time delay is generally less than 1 ns. Theoretically, using formula (5.105) under timescale the TCB, the coordinate time of X-ray photons arriving at the spacecraft is transformed to the SSB, with the transformation accuracy of 0.1 ns level. Although the time delay transformation accuracy using formula (5.105) is high, the computing process is extremely cumbersome. In practical application, it is necessary to simplify the processing, and of course, it will degrade the accuracy of time transformation. If the deflection of light produced by the gravitational field of the Sun and the gravitational effects produced by the planets may be ignored, and it is supposed that the velocity of proper motion of the pulsar is constant, as well as the time differences emitting the photons approximately equal to those receiving them, and the direction of the line of sight of the pulsar is approximately a constant vector, then a simplified time delay model can be expressed as

$$\begin{aligned} t_{SSB} - t_{SC} = {} & \frac{\mathbf{n} \cdot \mathbf{r}}{c} + \frac{1}{2cD_0}\left[(\mathbf{n} \cdot \mathbf{r})^2 - r^2 + 2(\mathbf{n} \cdot \mathbf{b})(\mathbf{n} \cdot \mathbf{r}) - 2(\mathbf{b} \cdot \mathbf{r})\right] \\ & + \frac{2\mu_s}{c^3} \ln\left|\frac{\mathbf{n} \cdot \mathbf{r} + \mathbf{n} \cdot \mathbf{b} + \|\mathbf{r} + \mathbf{b}\|}{\mathbf{n} \cdot \mathbf{b} + b}\right|, \end{aligned} \tag{5.106}$$

where **n** is the pulsar's direction vector relative to the solar system; **r** is the position vector of the spacecraft relative to the SSB; D_0 is the distance from the pulsar to the SSB at initial observation epoch.

On the right side of formula (5.106), the first term denotes the time delay produced by the geometric distance from the spacecraft to the SSB along the line of sight of the pulsar, known as **Doppler delay**; the second term denotes the time delay resulted from the X-ray photons parallel arriving at the SSB, and the first and second terms are collectively referred to as **Roemer delay**; the third term denotes the time delay resulted from the deflection of light under the gravitational field of the Sun, known as **Shapiro delay**.

The should be pointed out that for a certain observation epoch, the pulsar timing model can be expressed at any known points, such as the Earth's center of mass, the

Lagrange points between the Earth and the Sun, the Mar's center of mass and the position of anther spacecraft. And hence, the photon's arrival-time can also be transformed to a known reference point rather than the SSB. In this case, it is only needed that the SSB's position vector **b** in the above time delay transformation formula is replaced by the position vector of the corresponding reference point. Moreover, in the process of photon's arrival-time transformation, the parameters like the positions and velocities of spacecrafts are needed to use, which are exactly the navigation parameters required to solve. The usually adopted approach is firstly, the approximate state parameters of the spacecrafts are used for the time delay transformation to get the basic observables, and then the more accurate state parameters are solved; secondly, the solved state parameters are used for the time delay transformation again; finally, the iterative approximation way is used until the accuracy requirements are met. Generally, only will a few iterations needed meet the accuracy requirements by using the short-term orbit propagation of spacecrafts.

5.6.3 *Methods of Getting Pulse Profiles*

5.6.3.1 Pulse Profile Folding

The pulse profiles of X-ray pulsars are composed of a sequence of the photons. A two-dimensional array grid detector is used to measure the number, position and arrival-time of the X-ray photons entering the field of view, in which the photon arriving time is measured by the onboard clock. Each photon provides a quantified unit of energy within the detector grid. According to the number and energy level of the photons received per unit time, it can be judged whether the pulse signals are detected. In order to observe the selected pulsars, the detectors are required to aim at the lines of sight of the pulsars. As a photon is detected, the onboard clock records the time when the photon arrives at the detector, which means that each photon is marked with a time-stamp. For a pulsar, during the given observation time, when there are n X-ray photons recorded, the time sequence from τ_0 to τ_{n-1} is composed of the measuring time of each photon arriving at the detector coordinate frame, and through the timescale and time delay transformations, the time sequence of the photons arriving at the SSB in the TCB or TDB is obtained. Thus, the arrival-time of the n photons is transformed to the quasi-inertial coordinate system which is almost at rest relative to the pulsar. Usually, the given observation time is much longer than single pulse period, so the n detected photons span over countless pulse cycles. Each photon is only an energy point of a single pulse, so it is difficult to extract the pulse profile information from single photon. Obviously, for the n photons, the photon measuring data is essentially their arrival-time sequence.

In order to get the pulse profile, all of the measured photons must be aligned according to the time of arrival and arranged within a pulse cycle. The process of data processing in which the measured photons are arranged into a pulse cycle is called **pulse profile folding**, or **epoch folding**. In other words, the epoch folding is a process in which all photon energies are averaged synchronously in a desired pulse cycle [0, 2π] or [0, 1]. If the pulse period is divided into *m* equal time intervals, then each interval stands for a phase box of the pulse profile curve within a cycle, usually known as a "**bin**". The *n* photons are allocated to the corresponding bins according to the folding time sequence. As a result, some bins will have more photons, indicating strong photon radiation flux, while the number of photons in other bins is relatively scarce, indicating that they have weak photon radiation flux. Finally, a recognizable pulse profile curve can be produced [24, 28], which is similar to the photon histogram drawn during a pulse cycle. The smaller is the bin's value, the narrower the histogram block and the higher the resolution of pulse profile curve. Otherwise, the resolution is lower. Certainly, it is possible that the selected value of bin is too small to get more photons and clear profile.

The pulse profile is the unique identifier of pulsar, which is obtained by the synchronous average of a large number of photons. The relevant pulse parameters, such as the amplitude, peak number, period and its rate, as well as noise characteristics, can be measured through the pulse profile. In this sense, the pulse profiles are also the basic observables for the XPNAV. Only by extracting the pulse profile can the observed pulsar be identified, and its relevant parameters in the onboard database gotten to carry out the autonomous navigation computation for the spacecraft. Generally, the average pulse profile curve obtained during a short observation time, such as tens of minutes, is called **measuring pulse profile**, while the pulse profile obtained by long-term observation data processing is called **standard pulse profile**. Although the measuring pulse profile is with low SNR, it has the inherent characteristics of the pulse profile for identifying the pulsar. The standard pulse profile has a high SNR, which is a reference template of recognizing the pulsar for the XPNAV. In the XPNAV, the soft X-ray radiation spectrum range of 1−15 keV is generally selected to obtain more X-ray photons per unit time per unit detection area, which is beneficial to shorten the update time of measurement of the autonomous navigation filtering for spacecrafts.

5.6.3.2 Pulse Arrival-Time Measuring

The pulse arrival-time is a basic observable to determine the position and time parameters of spacecrafts. Only by comparing the measured pulse arrival-time with the predicted one can the time delay from the SSB to the spacecraft along the line of sight of the pulsar be obtained, and then the measurement equations are established to solve the navigation parameters. Supposed that the measuring pulse profile is obtained by synchronous average of a large number of single pulses within enough long observation time, which can characterize the inherent properties of the pulse profile generated within the same energy spectrum range, thus the pulse arrival-time can be measured by the cross-correlation processing between the measuring and standard pulse profiles.

From the studied results, it is shown that for a pulsar, the measuring pulse profile is the best approximation of the standard pulse profile, but there are the differences between both, including the constant deviation, scale transformation factor, time delay and random noise. In other words, for an observation epoch t, the measuring pulse profile $p(t)$ can be expressed by the standard pulse profile $s(t\text{-}d)$ with a time delay of d, i.e.,

$$p(t) = a + b \cdot s(t - d) + g(t), \tag{5.107}$$

where a is the constant deviation between the two pulse profiles; b is the scale transformation factor of the pulse profile; d is the time delay or displacement of the origin of time; $g(t)$ is the background noises of X-ray pulse signals, such as space background noise, high-energy particle noise, nebula noise and detector self-noise, all of which are assumed as the additive random noise process.

The purpose of the cross-correlation processing between the measuring and standard pulse profiles is to determine the values of three parameters a, b and d. Within the given observation time, the d will gradually be increased to the starting observation time of getting the first pulse arrival-time, so that the peak of the measuring pulse profile is aligned with that of the standard pulse profile. In the process of aligning the peak of pulse profile, the time delay from the SSB to spacecraft along the line of sight of the pulsar is obtained at the same time, so as to get the pulse arrival-time [28]. In this way, the measurement of the pulse arrival-time has been converted into the problem of solving the time delay d of two pulse profiles, so the d is sometimes referred to as the pulse arrival-time. The Discrete Fourier Transform (DFT) is a commonly used cross-correlation processing method for the pulse profiles. By comparing the two pulse profiles in the frequency domain, the accuracy of parameter estimation can be evaluated directly and has nothing to do with the photon sampling interval. The time-domain cross-correlation method can also be used to estimate the pulse arrival-time, but its accuracy depends on the photon sampling interval.

In the process of time transformation, the prior knowledge of positions and velocities of the spacecraft is required. However, for the absolute navigation, these parameters are often unknown or approximate estimates. If the proper time series of the photons arriving at the onboard detector within the given observation time are folded directly in the spacecraft-body coordinate system, the shape of the pulse profile will distort, and its SNR is so low that it is difficult to discriminate the pulse peaks from the photon flux intensity. In addition, the Doppler effects have a great influence on the pulse profiles. When the spacecrafts relative to the pulsars move rapidly, it is required to use the velocity parameters as accurate as possible in the time transformation to reduce the influence of the Doppler effects on the extracted pulse profiles.

5.6.4 Integer Ambiguity of Pulse Phase

The pulse periods of X-ray pulsars are extremely stable, which is conducive to extracting the pulse profile, and through the cross-correlation processing of the pulse profiles to obtain the time of arrival of the pulses. Meanwhile, this kind of stable

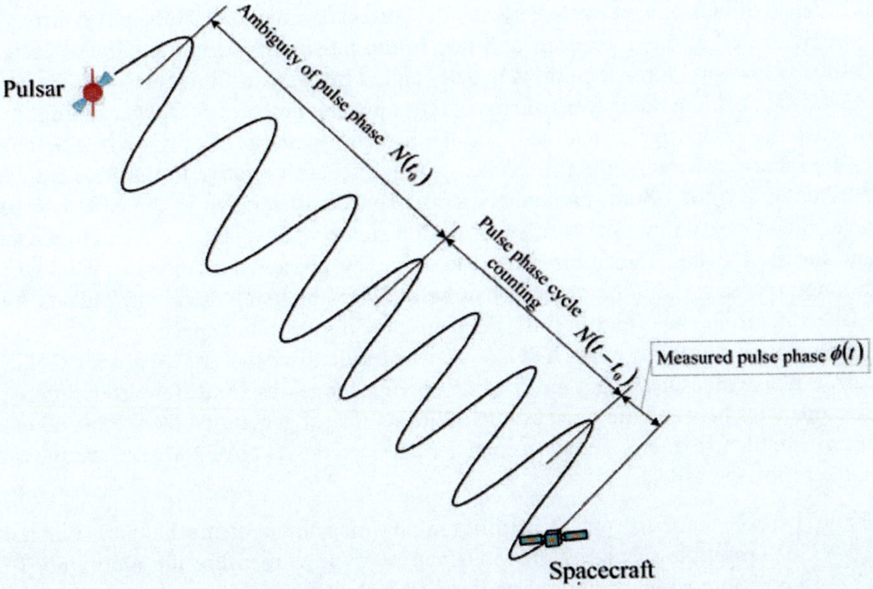

Fig. 5.10 Schematic diagram of pulse phase ambiguity between pulsar and spacecraft

pulse period signal makes the measurement of the pulse arrival-time produce the problem of phase ambiguity. Which pulse cycle radiated from a pulsar is arriving at the spacecraft? The ambiguity of pulse phase from the pulsar to spacecraft is shown in Fig. 5.10, from which for an epoch t, the total pulse phase can be expressed as

$$\Phi(t) = N(t_0) + N(t - t_0) + \phi(t), \tag{5.108}$$

where t_0 is the initial observation epoch; $N(t_0)$ is the integer ambiguity of pulse phase at the initial observation epoch; $N(t\text{-}t_0)$ is the integer of pulse phase cycle counts from t_0 to t; $\phi(t)$ is the fractional part of the phase within a pulse cycle at epoch t.

If the pulse signal emitted from the pulsar is detected at the initial observation epoch t_0, and always tracked and measured continuously, then the $N(t\text{-}t_0)$ can automatically be counted by the detector system. Moreover, the phase $\phi(t)$ within a pulse cycle is also observable. Consequently, only $N(t_0)$ is an unknown number related to the initial observation epoch, pulsar's position and spacecraft's position, with integer property. In the entire observation process, as long as the pulsar is continuously tracked and measured, the $N(t_0)$ is always a constant.

In fact, the number of the integer cycles from the pulsar to spacecraft need not to be investigated in the XPNAV, but the difference value of integer cycles of the spacecraft relative to the SSB or certain known reference point. Therefore, the integer difference of pulse phase cycle between the corresponding point of the pulse arrival-time measured at the spacecraft and that of the pulse arrival-time predicted at the SSB or a known reference point is usually called **integer ambiguity**.

For the absolute navigation of spacecraft, once the problem of integer ambiguity of the pulse phase cycle is solved, and the measuring and predicting pulses belong to the same pulse cycle, the time delay of the spacecraft relative to the SSB can be measured, and the orbital parameters of the spacecraft relative to the SSB can be determined in real time. In other words, as long as there are enough observation data and the appropriate algorithm is used to solve the unknown of integer cycles, the autonomous navigation parameters can be provided by using the X-ray pulsars for spacecrafts in the near-earth orbits, deep space and interstellar space.

The integer ambiguity of XPNAV can be studied by the analogy with GNSS carrier phase measuring, but the specific solving processes are different, due to the discrepancies between the two systems. Without loss of generality, the GPS is taking as an example to explain. The discrepancies between the XPNAV and GPS are shown in the following aspects:

(1) There is the discrepancy in differencing measurement mechanism. For the GPS, multiple receiving antennas can be used to measure the ambiguity of two differencing stations. For the XPNAV, its basic observables are gotten by comparing the pulse arrival-time measured by the detector onboard the spacecraft and that predicted by the pulsar timing model defined at the SSB, so as to determine the absolute position coordinates of the spacecraft. In fact, there is not any physical detector placed at the SSB. Of course, in the relative navigation application of two spacecrafts, multiple detectors can be installed at different positions to determine the relative position.
(2) There are the discrepancies in periods and wavelengths. Currently, each satellite of the GPS constellation mainly broadcasts the navigations signals on three central frequencies L1 (1575.42 MHz), L2 (1227.60 MHz) and L5 (1176.45 MHz), with wavelengths of 19.0 cm, 24.0 cm and 25.5 cm, respectively, and thus single carrier or multiple carrier observables can be used to solve the ambiguity. For the XPNAV, the pulse profile of each X-ray pulsar is unique, and the pulse periods are completely different, typically from a few milliseconds to a few seconds, even a few minutes, which is advantageous to solving the integer ambiguity. The wavelengths of the pulse cycles are more than several hundred kilometers, which is an easily-detected "large cycle slip".
(3) There is the discrepancy in system mechanism. The GPS is a man-made system, its satellites orbit around the Earth, and their signal strengths mainly meet the needs of the users on the ground and in the near-earth space. The pulsars belong to natural celestial system. Although their distance away from the solar system is very far distant and the intensities of their radiation signals are also not controlled artificially, the stably periodic radiation signals can be detected in the whole solar system or Galactic System.
(4) There is the discrepancy in electromagnetic wave band for emitting the signals. The navigation signals of the GPS satellites are modulated in the L-band. The

pulsars generate the signal radiations in the electromagnetic wave bands from the radio to the infrared, visible, ultraviolet, X-ray and gamma-ray. The radio signals can be observed on the ground and in space, but require the larger aperture telescopes. Because the X-ray signals are unable to pass through the Earth's dense atmosphere, they can only be detected on the spacecrafts in the outer atmosphere of the Earth, or the rovers on the surface of the planets without dense atmosphere.

(5) There is the discrepancy in ranging way. For the GPS, the distances between the GPS satellites and the receiver can be determined by using the pseudo-random noise codes with the time-stamped, so as to achieve the receiver positioning. The pulsars are very far away from the solar system and emit the pulse signals without the time-stamped message, so it is difficult to accurately measure their distances. Obviously, the distances between the pulsars and spacecrafts cannot be used as the observables to carry out the positioning.

(6) There is the discrepancy in phase-measured way. For the GPS, the integer ambiguity can be determined by the combination of carrier phase and ranging code signals. For the XPNAV, the pulsars emit the X-ray signals, and the integer ambiguity is solved by measuring the pulse phase signals rather than the carrier signals.

(7) There is the discrepancy in phase change rate. The GPS receivers can accurately track the carrier signals transmitted by the satellites and measure the carrier phase change rates, i.e., the Doppler shifts. However, the carrier signals emitted from the pulsars cannot be directly tracked and measured, so it is difficult to measure the carrier phase change rates. Even though, the motions of spacecrafts relative to the pulsars will produce the Doppler shifts of pulse phases, and thus the change rates of pulse phases can be measured.

For the GPS carrier phase measurements, the commonly used fast ambiguity resolution methods include exchanging antenna, P-code dual-frequency technique, ambiguity function method, least squares searching and ambiguity covariance, among which the latter three methods are referred to as the space-searching method of the ambiguity, which are also the basic method to solve the integer ambiguity for the XPNAV. The so-called **ambiguity space-searching method** is a way to search and solve the ambiguity in a three-dimensional coordinate space symmetrically about a central point. The central point of space-searching is selected according to the actual application, usually taking the location of defining the pulsar timing model. For the spacecrafts flying around the Earth or the Earth–Moon system, the center of mass of the Earth or the center of the Earth–Moon system can be selected as the central point of space-searching. For the space missions to investigate the planets in the solar system, the centers of mass of the planets can be selected as the central point of space-searching. For the spacecrafts lost in space, the last known orbit point can be selected as the central point of space-searching. In theory, any known position can be used as the central point of space-searching. The closer is the known position away from the real position of the spacecraft, the smaller the range of space-searching will be, and the faster the ambiguity resolution is.

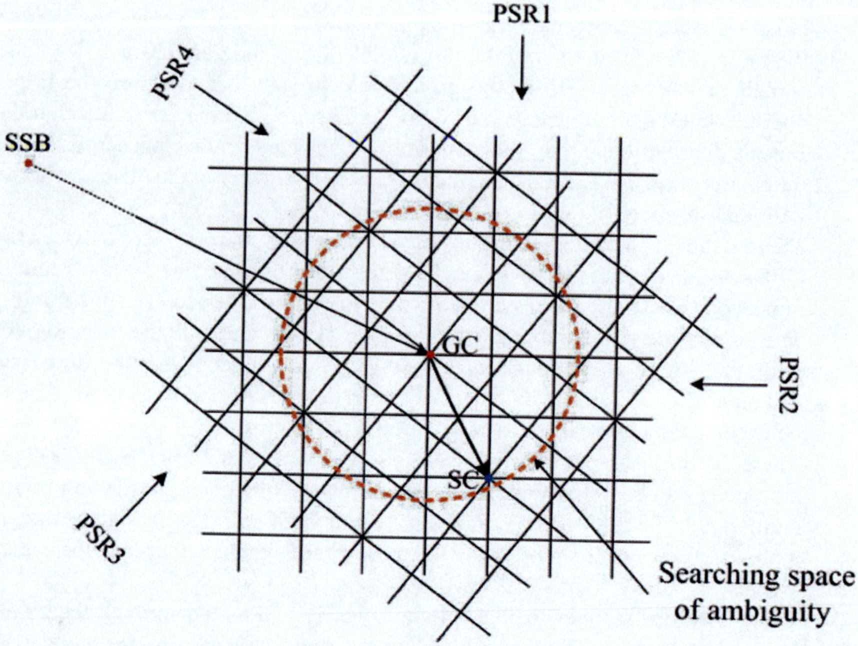

Fig. 5.11 Ambiguity space-searching demonstrated on two-dimensional plane

On a two-dimensional plane, the concept of the ambiguity space-searching with the center of mass of the Earth as the central point is demonstrated in Fig. 5.11, where the pulse phase arrival planes of four pulsars PSR1, PSR2, PSR3 and PSR4 are plotted; SSB stands for the center of mass of the solar system; GC stands for the center of mass of the Earth; SC stands for the position of spacecraft. When the pulsar timing model at the observational epoch is transformed from the SSB to GC, there is always only a candidate pulse phase group in the entire search space, i.e., the common intersection point of the four pulsars' pulse phase planes is the real spacecraft's position (SC).

For the autonomous navigation of spacecraft using X-ray pulsars, considering the long wavelength of a pulse cycle and the assistance with other navigation ways, as well as using the Kalman filtering and high-precision orbit propagation, it can be ensured that the approximate position error of the spacecraft is within the range of a pulse cycle, and both the measured pulse arrival-time and predicted one come from the same pulse cycle, so in general, there is not the problem of integer ambiguity. Only when the spacecrafts are in the extreme cases, such as the navigation system lack of other auxiliary navigation means and in initialization recovering, and the spacecrafts lost in space, and the distance between the approximate and real positions of the spacecraft is greater than the wavelength within a pulse cycle, there will be the problem of integer ambiguity.

5.6.5 Determination of Orbit and Time Parameters

The orbit and time parameters of spacecrafts can be determined by using the X-ray signals radiated from pulsars. According to the different basic observables and navigation ways, there are four types of measurement equations, including the equation of directly using distance, that of using pulse phase, that for absolute navigation, and that for relative navigation.

5.6.5.1 Measurement Equation of Directly Using Distance

Since the rotation periods of pulsars are extremely stable, the X-ray photons continuously radiated from the magnetic poles of pulsars can be gathered into the pulse signals as arriving at the spacecrafts. For a certain pulse signal, would the time when the pulsar emits the signals be known, the measurement equation can be established directly using the ranging observables, that is

$$\rho = \mathbf{n}_{SC} \cdot (\mathbf{D} - \mathbf{p}) + d_{\mathrm{Rel}}, \tag{5.109}$$

where $\rho = c(t_{SC} - t_T)$; c is the velocity of light; t_{SC} is the time when the X-ray pulse signal emitted from the pulsar arrives at the spacecraft under the timescale TCB; t_T is the time when the pulsar emits the X-ray pulse signal under the timescale TCB; $\mathbf{n}_{SC}$ is the direction vector from the spacecraft to the pulsar; $\mathbf{p}$ and $\mathbf{D}$ are, respectively, the position vectors of the spacecraft and pulsar in the BCRS; d_{Rel} is the correction of relativistic effect.

If the position error of pulsar, the time bias of pulsar emitting pulse signal, the residual of relativistic effect and other errors are collectively called measurement noise, and the spacecraft's clock bias and its position correction are considered, then there is the following formula.

$$\delta\rho_d = -\mathbf{n}_{SC} \cdot \delta\mathbf{p} - c\delta t_{SC} + \eta_d, \tag{5.110}$$

where $\delta\rho_d = c\left(\tilde{t}_{SC} - t_T\right) - \mathbf{n}_{SC} \cdot (\mathbf{D} - \tilde{\mathbf{p}}) - d_{\mathrm{Rel}}$; $\tilde{t}_{SC}$ and δt_{SC} are, respectively, the onboard clock reading time and its bias under the timescale TCB; $\tilde{\mathbf{p}}$ and $\delta\mathbf{p}$ are, respectively, the approximate position vector and its correction; η_d is the measurement noise.

Obviously, the basic observable of Eq. (5.110) is the modified distance $\delta\rho_d$ from the signal emitting source (the pulsar) to the user's position (the spacecraft), which is consistent with the GNSS measurement equation using the pseudorange. When more than four pulsars are observed at the same time, the position correction and clock bias parameters of the spacecraft can be directly estimated by using the method of least squares. Although the time when the pulse signals arrive at the spacecrafts can be measured, the pulsars belong to natural celestial bodies, and the time when the pulsars emit the pulse signals is unable to be known exactly. That is to say, the distances from the spacecrafts to the pulsars are unable to be measured accurately. Using the GNSS to provide navigation service for the spacecrafts, the transmitting time of the ranging code signals can be gotten from the navigation message, and the time delay within a ranging code period can be obtained by cross-correlation processing between the received ranging codes and the local codes, so the distance from the user spacecrafts to the GNSS satellites can be accurately measured. This is the natural discrepancy between the XPNAV and GNSS, and it is also the difficulty of XPNAV. Of course, the observables based on the stable pulse arrival-time have provided sufficient information for the spacecraft's orbit determination. Although the measurement equation of directly using the distance cannot be used for the spacecraft's navigation computing, it is helpful to introducing the concepts of the absolute navigation and relative navigation of the spacecrafts using X-ray pulsars.

5.6.5.2 Measurement Equation of Using Pulse Phase

The spacecrafts can autonomously navigate using the X-ray pulsars, and in fact, is to utilize the extremely stable periodic rotation characteristics of the pulsars, which are the most valuable navigation beacons. The stable pulse period signals create favorable conditions to detect and identify the specified pulse cycles. However, at any given epoch, it is difficult to detect the pulse peaks or the discernable feature points of a pulse profile, but only a point on the pulse profile is measured, which is the time of arrival of a photon, also the phase values within a pulse cycle. If the dimensionless number [0, 1] is used to represent a pulse cycle, the pulse phase measurement is the decimal fraction. Thereby, the measurement of pulse arrival-time with the pulse's peak as a reference point is actually that of pulse phase. When the number of integer cycles and the phase of a cycle have been measured, the distance from the pulsar to spacecraft can be computed easily. In practice, the measurement information for each pulse feature is not provided in the XPNAV, and hence the number of integer cycles at any given epoch is unable to be determined. In order to clearly explain the relationship between the pulse phase and distance measurement, a schematic diagram of measuring the pulse phase sequence is illustrated in Fig. 5.12, from which it is can

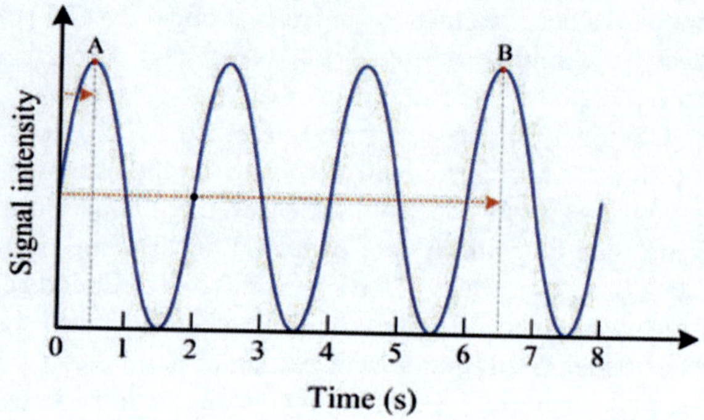

Fig. 5.12 Schematic diagram of pulse phase sequence measuring

be seen: the pulse period is 2 s; the pulse wavelength is $2c$ where c is the velocity of light; the distance and phase of point A are, respectively, $0.5c$ and 0.25; the distance and phase of point B are, respectively, $6.5c$ and 3.25.

For any given time, the total pulse cycle phase Φ can be expressed as the sum of the number N of integer cycles and a fraction ϕ of a cycle, and thus the measurement equation using the pulse phase is expressed as

$$\delta\rho_p = -\mathbf{n}_{SC} \cdot \delta\mathbf{p} - c\delta t_{SC} + \eta_p, \tag{5.111}$$

where $\delta\rho_p = \lambda\left(N + \tilde{\phi}\right) - \mathbf{n}_{SC} \cdot (\mathbf{D} - \tilde{\mathbf{p}}) - d_{\text{Rel}}$; λ is the wavelength of pulse signal; η_p is the phase measurement noise.

The pulse phase measurement is to measure the phase within a pulse cycle. When a pulsar is continuously tracked and observed, the number of integer cycles during the entire observation segment can be recorded, but that at the initial epoch is unknown. In the XPNAV, the integer-cycle number of the spacecraft relative to the pulsar is not solved, but that of the spacecraft relative to the SSB or a known point. Generally, it is considered that there is no error for solving the number of pulse cycles. If there is the error of one pulse cycle, it will result in the ranging error of hundreds or even thousands of kilometers. For the spacecraft's orbit determination, this is an intolerable gross error, which is detected and offset easily.

5.6.5.3 Measurement Equation for Absolute Navigation

The absolute navigation is to determine the three-dimensional position and velocity of the spacecraft in the BCRS, so as to obtain the relative relationship between spacecraft and its near celestial objects, control the spacecraft to avoid obstacles, and guide it to fly along the predetermined orbit. By comparing the time difference between the same pulse arriving at the spacecraft and at the SSB, the time delay of the spacecraft relative to the SSB along the line of sight of the pulsar can be gotten. That is to say, the ranging observables can be gotten. Without loss of generality, the simplified time delay model (5.106) is adopted here. For a given epoch t_k, using the detecting data from pulsar i, the measurement equation of the spacecraft relative to the SSB can be expressed as

$$\delta\rho_i = \mathbf{n}_i \cdot \delta\mathbf{r}_{SC,k} + c\delta t_k + \eta_i, \tag{5.112}$$

where

$$\begin{aligned}\delta\rho_i = {} & ct_{SSB} - ct_{SC} - \mathbf{n}_i \cdot \tilde{\mathbf{r}}_{SC,k} \\ & + \frac{1}{2D_0}\left[\left(\mathbf{n}_i \cdot \tilde{\mathbf{r}}_{SC,k}\right)^2 - r^2 + 2(\mathbf{n}_i \cdot \mathbf{b})\left(\mathbf{n}_i \cdot \tilde{\mathbf{r}}_{SC,k}\right) - 2\left(\mathbf{b} \cdot \tilde{\mathbf{r}}_{SC,k}\right)\right]\end{aligned}$$

$$+\frac{2\mu_s}{c^2}\ln\left|\frac{\mathbf{n}_i\cdot\tilde{\mathbf{r}}_{SC,k}+\mathbf{n}_i\cdot\mathbf{b}+\left\|\tilde{\mathbf{r}}_{SC,k}+\mathbf{b}\right\|}{\mathbf{n}_i\cdot\mathbf{b}+b}\right|;$$

$$\mathbf{n}_i=\left[\cos\alpha_i\cos\lambda_i\ \cos\alpha_i\sin\lambda_i\ \sin\alpha_i\right]^T;$$

$$\mathbf{r}_{SC,k}=\tilde{\mathbf{r}}_{SC,k}+\delta\mathbf{r}_{SC,k};$$

$\delta\mathbf{r}_{SC,k}$ is the correction of approximate position vector of the spacecraft at epoch t_k; δt_k is the onboard clock bias under the timescale TCB; η_i is the measurement noise of the pulse arrival-time; λ_i and α_i are, respectively, the right ascension and declination of pulsar i in the BCRS; $\mathbf{r}_{SC,k}$ and $\tilde{\mathbf{r}}_{SC,k}$ are, respectively, the real and approximate position vectors of the spacecraft relative to the SSB at epoch t_k.

It should be pointed out that only for the same pulse signal can the measured pulse arrival-time be compared with the predicted one, and then the difference of arrival-time obtained. If the distance difference between the approximate and real positions of the spacecraft is within a pulse cycle wavelength in the time transformation process, there will not be the problem of integer ambiguity, and thus it is ensured that the arrival-time of the same pulse is comparable. In the application of autonomous navigation of the spacecraft using the X-ray pulsars, the ground-based TT&C stations or other navigation ways can be used to assist the determination of system state parameters in the initial stage of navigation system. Taking the pulse arrival-time as the basic observables, the Kalman filter is designed to estimate the system state parameters. The high-precision orbit model is used to propagate for short-term orbit prediction, and hence the spacecraft's approximate position accuracy can be reached 1 km, while the pulse wavelength of millisecond pulsar is more than several hundred kilometers, so there is generally not the problem of integer ambiguity. In addition, the pulse phase is measurable within a pulse cycle. According to the relationship between the pulse phases and the ranging observables, the phase measurement equation can be established. In this way, the multiple phase measurements can be obtained within a pulse cycle, besides the phase measurement at the peak point (the measurement of arrival-time), so as to increase the redundant observations and improve the parameter-estimating accuracy.

5.6.5.4 Measurement Equation for Relative Navigation

The relative navigation is the process of determining the navigation parameters of the spacecraft relative to a reference object frame, which is a linear combination of range or phase measurements, also known as **differencing measurement**. Similar to the differencing measurement of the GNSS, there are generally three types of differencing linear-combinations: single difference, double difference and triple difference. The single difference, for the same pulsar, refers to the difference between the phase measured at the spacecraft and that predicted at the position of defining the pulsar timing model. The double difference refers to the difference between two single differences obtained by the spacecraft observing two pulsars at the same time. The triple difference refers to the difference between two double differences of observing more than two pulsars synchronously within different observation time. The common error terms, and the parameters which are difficult to be measured accurately, can be eliminated or weakened through the differencing measurements, so as to simplify the measurement equations and improve the system-state-estimated accuracy.

Usually, the pulsar timing model is defined by taking the SSB as the reference point, from which the single differencing measurement equation is obtained. This single differencing equation is the measurement equation for the absolute navigation. In order to simply the data processing, at any given epoch, the pulse arrival-time predicted by the pulsar timing model defined at the SSB can be transformed to any known reference positions, such as the centers of mass of the Earth, Moon and Mars, the Lagrange points between the Earth–Moon or the Earth–Sun, as well as another spacecraft or a base station. For two spacecrafts on which the detectors are installed, the X-ray photons radiated from the same pulsar are detected by the detectors, and the communication link between the two spacecrafts needs to be established for the differencing measurements to transmit the measurement data, which is the relative navigation mode in real sense. Choosing the SSB or the center of mass of a certain celestial body as the reference point, it is not required to transmit the measurement data for establishing the differencing measurement equations, which can be achieved only by calling the corresponding models and databases stored in the onboard computer.

5.6.6 Determination of Attitude Parameters

The attitude measurements of the spacecraft using X-ray pulsars are similar to the optical sensors, and their difference is that the X-ray photons are used to replace the visible light, resulting in the different signal processing algorithms. The pulse profile is the unique identifier of a pulsar. Once the pulse profile information is extracted, the one-to-one correspondence between the observed and the catalogued pulsars in the database will be established [29]. The fundamental problem of attitude determination of the spacecraft is to determine the attitude parameters of the spacecraft-body coordinate system (*b*-system) relative to the reference coordinate system (*r*-system), which is called **three-axis attitude determination**; or for a certain direction vector (unit vector) in the *b*-system, the process of determining its coordinates relative to the *r*-system is called **single axis attitude determination**. In this section, the three-axis attitude determination of the spacecrafts in the Earth orbits is mainly discussed, and for the deep-space probes, it can be processed similarly.

In the measurement coordinate system of the onboard detector, the measured direction vectors of pulsars are the basic observables of the spacecraft's attitude determination. The X-ray photons radiated from the pulsars are positioned and imaged with the mechanical collimators and by using optical focusing methods, and for the soft X-rays, the coded-aperture imaging technology can also be used. Supposed that

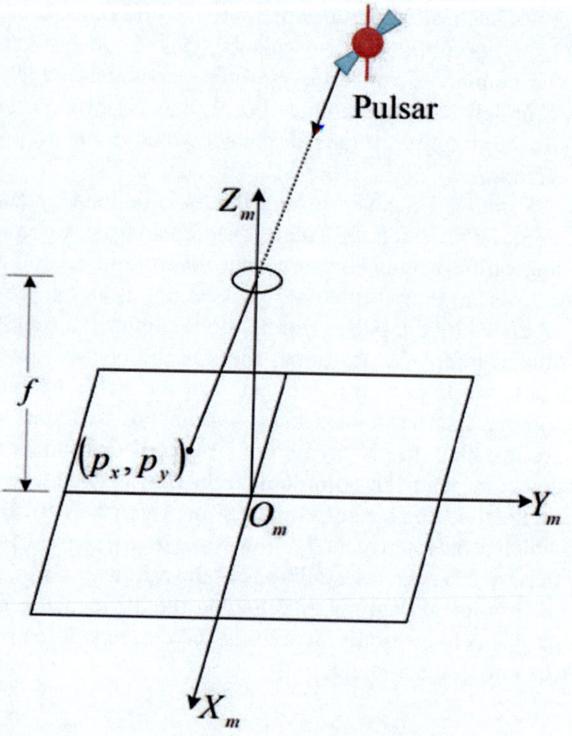

Fig. 5.13 Measurement coordinate system of the onboard X-ray detector

a detection array composed of the concentrating mirrors and their corresponding silicon drift detectors is installed on the spacecraft, the detector's measurement coordinate system is defined, as shown Fig. 5.13. Using the position coordinates (p_x, p_y) of the pulsar-imaging central point on the detector plane and the focal length f, the direction vector of the pulsar in the detector coordinate system is obtained easily. By the coordinate rotation transformation using the detector installation matrix, the pulsar's direction vector can be measured and expressed in the spacecraft-body coordinate system.

In the pulsar navigation database stored in the onboard computer, the angular positions of the X-ray pulsars used for navigation in BCRS have been accurately determined. That is to say, the direction vectors of the pulsars are known, also called reference direction vectors. During sufficient signal integration time, the measuring pulse profiles are extracted by the epoch folding for the X-ray photon sequences, and then through the cross-correlation processing of the measuring pulse profiles with the standard ones in the database, the observed pulsars can be identified. As a result, the one-to-one correspondences between the observed and the catalogued pulsars are established, and then the angular positions and relevant parameters of the navigation pulsars can be called up from the database. For the spacecrafts in the Earth orbits, considering that there is no kinematical rotation between the GCRS and BCRS, the reference direction vector of the pulsar in the spacecraft's orbit coordinate system can be easily obtained in the case of the spacecraft's orbit parameters known. When two pulsars are observed simultaneously on the spacecraft, and two direction vectors are obtained, the attitude parameters of the spacecraft can be estimated by the method of attitude determination using two vectors.

Using the X-ray detectors installed on the spacecraft to measure the direction vectors of pulsars, the spacecraft's attitudes in BCRS or in the orbit coordinate system can be determined with high precision, getting to the several arc-seconds, which is comparable to the level of current optical star sensors. Moreover, the attitude determination using the X-ray pulsars can avoid the complex star pattern recognition algorithms in the traditional celestial navigation. However, it takes time to detect the X-ray photons and extract the pulse profiles, so that there is a time lag for obtaining the attitude parameters. Furthermore, the measurement noise of the X-ray detector will also affect the attitude determination accuracy. The inertial attitude measurements using the gyroscopes have the advantages of complementary with the attitude measurements using the X-ray detectors. The inertial attitude has rapidly dynamic response; the random measurement error is small of the high-precision gyroscopes; the gyroscope drifts can be calibrated by the X-ray detector measurements. In practical application, the measurement data of the X-ray detectors and inertial gyroscopes is often combined to process. Thereby, within sampling interval, the inertial attitudes are measured by the gyroscopes to use for the spacecraft's navigation, and at sampling point of the X-ray detector unit, the measured direction vectors of the pulsars is used to correct the inertial attitudes for suppressing the accumulated error caused by the gyroscope drifts, so as to obtain highly accurate, continuous and stable attitude parameters of the spacecraft.

Supposed that the spacecraft's attitude measurement system is composed of the X-ray detector unit and the inertial attitude unit, in which the latter consists of three-axis strapdown rate integrating gyroscopes in the spacecraft-body coordinate system, the data processing flow of the combined attitude measurement of the spacecraft is shown in Fig. 5.14, where Δt_g is the sampling interval of the inertial attitude unit; Δt_d is the sampling interval of the X-detector unit; $\Delta\hat{\varepsilon}$ is the estimation of correction vector of the gyroscope drift; $\hat{\varepsilon}$ is the estimation of vector of the gyroscope drift; $\Delta\hat{\theta}$ is the estimation of correction vector of the spacecraft's attitude; $\hat{\mathbf{q}}$ is the estimation of the quaternion vector; $\hat{\omega}$ is the estimation of the rotational angular velocity of the spacecraft; $\mathbf{A}_{ob}$ is the outputted attitude matrix of the spacecraft. In order to improve the efficiency of data processing, a special data processor is configured in the X-ray detector unit, which is used for the X-ray photon measurement data acquiring, pulse profile folding, and pulsar's image extracting and positioning, so as to obtain the measuring direction vectors of the pulsars in the spacecraft-body coordinate system. The measurement data from the inertial attitude unit and the X-ray detector unit are sent to the onboard computer, respectively. The attitude determination process includes two parts: one is based on the data of inertial attitude unit, and the attitude estimation values of the spacecraft are computed by the attitude quaternion integral; another is to use the measurement data from the X-ray detectors to correct the attitude estimation error of the spacecraft and calibrate the gyroscope drifts. Finally, the

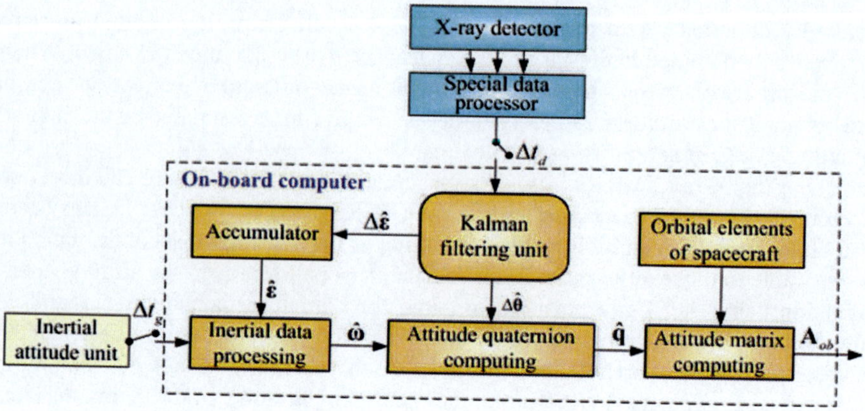

Fig. 5.14 Attitude determination flow of combining X-ray detectors with gyroscopes

attitude quaternion of the spacecraft in the inertial coordinate system is obtained, and then the spacecraft's attitude parameters in the orbit coordinate system can be gotten by the coordinate transformation.

Using the attitude measurement data from the different measurement systems, the Kalman filter is designed to process the combined data, and the attitude determination accuracy is improved significantly. The advantages of the attitude parameter estimation based on the combined measurement data are as follows:

(1) In each sampling period, the attitude measurement system can provide a group of attitude observables. The types of measurement data rely on the different sensors, and thus more redundancy data can be produced.
(2) The measurement data from the different sensors have different error characteristics. For example, the X-ray detector has high-frequency random noise, and the gyroscope has constant drift component, but the high-frequency noise of the gyroscope is low. So, the combination of the two devices will achieve the purpose of complementary advantage.
(3) The combined measurement data is processed by the Kalman filtering, which fully considers the spacecraft's attitude characteristics of kinematics and dynamics. Only if the high-frequency system process noise and measurement noise are suppressed, the high-precision attitude parameters can be estimated.

5.7 Design of Ground Test System

5.7.1 The Analysis of Necessity

It is well known that the X-ray photons radiated from pulsars are difficult to penetrate the dense atmosphere of the Earth, and they can only be detected in outer space. By constructing the test system of the XPNAV, the characteristics of the X-ray photon signals from the pulsars are simulated to test the function and performance of the X-ray detectors and verify the principle of the XPNAV system. Obviously, in the technical studies, space flight tests and engineering applications of the XPNAV, it is very necessary to construct the ground test system facilities, which can provide the closely realistic environment and detection platform for the pulsar's signal simulating, key technology testing, navigation principle verifying, and actual data processing.

(1) Constructing the ground test system is the inevitable requirement for simulating the pulsar's signals. When the electromagnetic wave passes through the Earth's atmosphere, it will be affected by the absorption and scattering of the atmosphere, so that its energy is attenuated. The effects that the atmosphere absorbs and scatters electromagnetic wave vary with wavelength. It is well known that there are only two electromagnetic radiation windows in the dense atmosphere of the Earth: one is the optical window, and another is the radio window. Generally, the atmospheric penetration rate of the visible light is more than 95%, while that of the radio wave is almost 100%. However, the X-rays emitted by the pulsars are a kind of high-energy electromagnetic wave, which is not covered by the atmospheric windows. When the X-rays propagate in the dense atmosphere of the Earth, they will be completely absorbed, so that the X-ray photon signals radiated from the pulsars cannot be received on the ground. So, in the vacuum environment of laboratory, constructing the ground test system, to closely realistically simulate the process of the X-ray photons radiated from the pulsars, is the precondition that the key technologies of the XPNAV are studied and verified on the ground.

(2) Constructing the ground test system is the core link to verify the key technologies of the XPNAV. The spacecraft's autonomous navigation based on the X-ray pulsars involves many key technologies, such as the X-ray detection, pulsar's database construction, time keeping, and large-scale navigation. Facing the requirements for developing the spacecraft's autonomous navigation, it is very necessary to build the ground test system and verify the key technologies of the XPNAV. In the environment of laboratory, the key technologies, such as the updating and calling of pulsar navigation data, high-precision transformation of arrival-time of the photons, X-ray photon detection, autonomous navigation algorithm and fault-tolerant processing, are tested and verified by using the ground test system, so as to demonstrate the feasibility of the XPNAV and provide the technical supports for the space flight tests.

(3) Constructing the ground test system is the important means to test the detectors of the XPNAV. The function of the detector system of the XPNAV is to detect the X-ray photons emitted by the pulsars, and accurately record the arrival-time of the photons using the onboard atomic clock. The X-ray detectors are made by using the properties of the interaction of the X-rays with the materials. Currently, there are many types of the X-ray detectors, such as the gas proportional counter, scintillation, micro-channel plate, silicon semiconductor, charge-coupled device, scanning-charge device and silicon drift device, which

have their own advantages and disadvantages in sensitivity, time resolution, spatial resolution, photon arrival-time measurement accuracy, working environment temperature, reliability, volume, weight and power consumption. For this reason, using the ground test system platform, the engineering applicability of the X-ray detectors developed by adopting the different technical approaches is tested and verified, to choose the types of X-ray detectors that can meet the requirements for the space flight test and actual engineering application of the XPNAV.

(4) Constructing the ground test system is the fundamental way to verify the system mechanism of the XPNAV. In general, the pulsars are the most stable clocks in nature, and with the X-ray photons radiated from the pulsars as the natural beacons, the navigation parameters like orbit, time and attitude can be determined with high precision by measuring the pulse arrival-time and pulsar's image information. The spacecraft's autonomous navigation based on the X-ray pulsars can be achieved by the onboard clock phase-locked loop, orbital parameter estimation loop and attitude parameter estimation loop. Before the space flight test of the XPNAV is carried out, the construction of the ground test system is the fundamental way to demonstrate the system mechanism of the XPNAV, laying the foundation for engineering application.

(5) Constructing the ground test system is the demonstration platform for the actual data processing of the XPNAV. The measurement data of the XPNAV are the arrival-time sequences of X-ray photons obtained by tracking the selected pulsars with the onboard detectors. When the measurement data are transmitted to the ground for processing, the pulse profiles can be extracted, the pulse arrival-time compared, the pulsar timing model and database established, the pulsar's period stability monitored and the system mechanism of the XPNAV demonstrated by post-processing. Moreover, the ground test system is also used to simulate the scene of the spacecraft's autonomous navigation using the X-ray pulsars, which can test the navigation algorithms, software and its interface relations, predict and reproduce the system faults of the XPNAV, and take the solving measures in time. Obviously, the ground test system has the function of data processing center of the XPNAV and provides the technical support and long-term maintenance for the space flight tests and applications of the XPNAV.

5.7.2 The Composition and Function

The ground test system of the XPNAV is the platform to simulate the environment of autonomous navigation for spacecrafts and test the prototype detectors. It is also the master control center for processing the measurement data of X-rays from the pulsars, with the test and verification capability of distribution, full-path, semi-physics and closed-loop. By simulating the characteristics of the X-ray photon signals radiated from the pulsars, as well as the space background noises, the key technologies, such as the X-ray photon detection, navigation database updating and calling, large-scale space-time transformation, time system keeping, autonomous navigation algorithm and fault-tolerant processing, are tested and verified. The ground test system is composed of four segments: central control, mathematical simulation, photon detection and evaluation demonstration, as shown in Fig. 5.15.

(1) The central control segment is the center of signal control, which is the top-level interfaces of external command execution, simulation parameter setting, control signal sending, data management, time-frequency synchronization and test task planning in the entire ground test system, with the functions of interface controlling, network monitoring, instruction broadcasting, parameter setting, network timing and system scheduling.
(2) The mathematical simulation segment is the center of data processing, which is the bottom components of the ground test system, including simulation data generation, interstellar chart preparation, large-scale space-time transformation and autonomous navigation computation, with the functions of orbital data generating, photon arrival-time simulating, navigation database constructing, large-scale and on-orbit dynamic effect simulating, pulse profile extracting and navigation parameter estimating.
(3) The photon detection segment is a physical signal unit, which is the hardware equipment of generating, detecting and processing the pulsar's X-ray signals in the ground test system, with the functions of real-time signal controlling, pulse

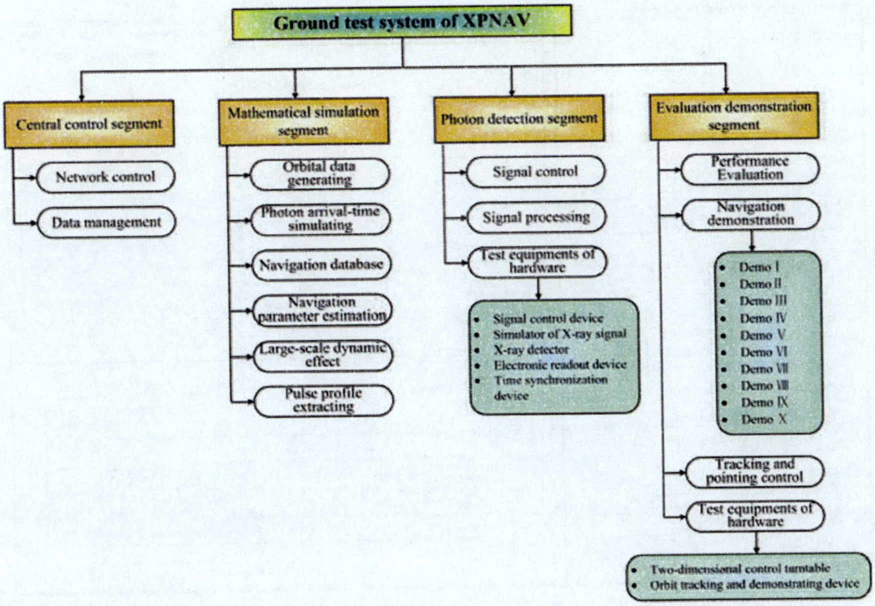

Fig. 5.15 Composition of ground test system for XPNAV

profile modulating, pulsar's X-ray signal generating, analog signal source calibrating, X-ray photon detecting, photon arrival-time measuring and processing, as well as remote data operating.

(4) The evaluation demonstration segment is the display window of the tested results, which is the display terminals of performance verification, data interpretation, anomaly diagnosis and orbit tracking demonstration in the ground test system, with the functions of navigation demonstrating, test evaluating, two-dimensional turntable tracking, result replaying and report printing.

5.7.3 Design of Overall Layout

According to the composition of the ground test system and the division of the four function segments, the overall layout design of the system is shown in Fig. 5.16, from which it can be seen that:

(1) The central control segment is composed of one network control computer, one data management computer and one workgroup switch, among which the network control computer is responsible for the software and hardware interface controlling, network operation status monitoring, system operational instruction sending, simulation parameter setting, test task planning and scheduling, and network timing; the data management computer is responsible

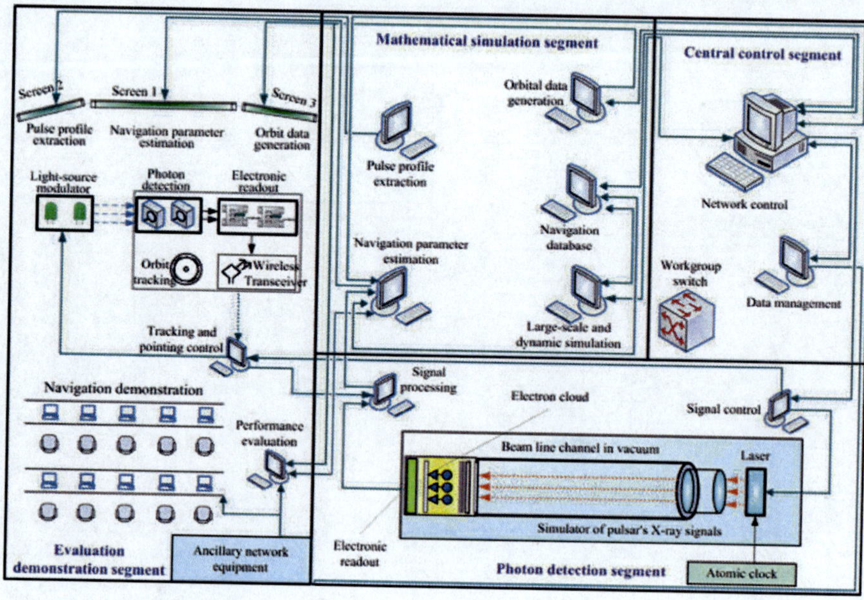

Fig. 5.16 Overall Layout of ground test system for XPNAV

for data classification, storage, query and extraction; the workgroup switch is responsible for network communication and data exchange.

(2) The mathematical simulation segment consists of six computers, respectively, used for orbital data generation, photon arrival-time simulation, navigation database construction, navigation parameter estimation, large-scale and dynamic effect simulation and pulse profile extraction. The orbital data generation computer is responsible for typical orbit and attitude data generation, two-dimensional and three-dimensional dynamic display, time and coordinate system transformation, and orbital parameter analysis. The photon arrival-time simulation computer, which is not shown in Fig. 5.16, is responsible for pulse phase computation and photon arrival-time sampling. The navigation database construction computer is responsible for establishing, updating and maintaining the pulsar's database, and the Sun's and planetary ephemeris parameters. The navigation parameter estimation computer is responsible for pulse profile delay estimating, phase tracking and filtering, orbit and attitude determining, as well as parameter estimation accuracy analyzing. The large-scale and dynamic effect computer is responsible for numerically simulating the large-scale delay effect of the solar system and the on-orbit dynamic effect of spacecrafts. The pulse profile extraction computer is responsible for photon arrival-time correcting, epoch folding and standard pulse profile getting.

(3) The photon detection segment is composed of one signal control computer, one signal processing computer, one signal control device, one X-ray signal simulation equipment, one X-ray detector, one set of electronic readout devices and one set of time synchronization devices. The signal control computer is responsible for digitizing the pulse profiles to output. The signal processing is responsible for processing the photon arrival-time sequences. The signal control device is responsible for generating the electric signals of the pulse profiles. The X-ray signal simulation equipment is responsible for simulating and generating the X-ray signals radiated from the pulsars. The X-ray detector is responsible for receiving the X-ray photons and converting them into electrical pulse signals. The electronic readout devices are responsible for measuring the photon arrival-time. The time synchronization devices are responsible for providing high-precision time-frequency signals.

(4) The evaluation demonstration segment is composed of one performance evaluation computer, ten navigation demonstration computers, one tracking and pointing control computer, one set of two-dimensional control turntable, one set of orbit tracking demonstration device and three sets of projection display equipment. The performance evaluation computer is responsible for the interpreting the test data and results, diagnosing the abnormal events and confirming the performance of ground test system. The navigation demonstration computers are responsible for displaying the navigation simulation data in real time, replaying the test results and printing the reports. The tracking and pointing control computer is responsible for sending the control commands for the two-dimensional turntable and the orbit tracking demonstration device. The two-dimensional control turntable is responsible for realizing the space

orientation of the detector by adjusting the pitch and yaw angles. The orbit tracking demonstration device is responsible for simulating the spacecraft's orbits, verifying the turntable pointing and tracking and detecting the single photon. The projection display equipment is responsible for displaying the test results on the large screens, such as the orbital data generation, pulse profile extraction and navigation parameter estimation.

5.7.4 Design of Operating Modes

There are six operating modes designed in the ground test system: pure mathematical simulation, laser detection simulation, short-range X-ray photon detection simulation, pure electronics access simulation, remote X-ray photon detection simulation and actual X-ray data processing.

(1) The pure mathematical simulation mode is with the mathematical simulation segment as the core, and under the unified scheduling of the network control computer, to test the correctness of orbital data generation, photon arrival-time simulation, navigation database updating and calling, large-scale space-time and dynamic effect simulation, pulse profile extraction, as well as autonomous navigation and fault-tolerant processing algorithm, and to quantitatively analyze and demonstrate the principle feasibility of the XPNAV.

(2) The laser detection simulation mode is on the basis of the pure mathematical simulation mode, and by connecting with the high-precision time synchronization devices, laser signal sources, photoelectric converters, two-dimensional control turntable and orbit tracking demonstration device, to test the hardware performance of light-source driven and modulated circuit, light intensity attenuation device, photoelectric converter, two-dimensional control turntable and orbit tracking demonstration device, so as to simulate and demonstrate the two-dimensional control turntable pointing and orbital-dynamic tracking process, and to carry out the closed-loop tests of the pulse signal detection and processing as well as the autonomous navigation process under the local coordinate system and large-scale space-time frame.

(3) The short-range X-ray photon detection simulating-mode is on the basis of the pure mathematical simulation mode, and by connecting with the high-precision time synchronization device, adjustable micro X-ray generator, and Silicon Drift Detector (SDD), to verify the X-ray photon intensity modulation technology, and to test the hardware performances of the X-ray generator and SDD, so as to achieve the simulation confirmation for the autonomous navigation process with distribution, full-path, semi-physics and short-range closed-loop.

(4) The electronic access simulation mode is on the basis of the pure mathematical simulation mode, and by connecting with the high-precision time synchronization device, signal control device and multi-channel electronic readout device, to test the performances of multi-channel pulse profile generators and electronic

readout devices, so as to achieve the simulation confirmation of multi-channel pulse signal control, signal processing and autonomous navigation algorithm.

(5) The remote X-ray photon detection simulating-mode is on the basis on the pure mathematical simulation mode, by connecting with the high-precision time synchronization device, to remotely control the modulated X-ray source, large-size beam line channel in vacuum, and large-area array detector through special fiber network, so as to test the functions and technical indicators of various types of detectors, and evaluate the autonomous navigation process.

(6) The actual X-ray data processing mode is under the unified scheduling of the network control computer, and by accessing the detected pulsar's X-ray data to the signal processing computer and cooperating with the computers of pulse profile extraction, navigation database, navigation parameter estimation, performance evaluation and navigation demonstration, to build and refine the X-ray pulsar database, so as to carry out the post-processing confirmation of the XPNAV, and finally achieve the construction of the unified space-ground time reference system.

5.7.5 *Numerical Experiments and Their Results*

On the basis of the built ground test system of the XPNAV, the numerical experiments can be carried out to verify the feasibility of algorithms and schemes of the XPNAV. There are two numerical experiments and their results mainly demonstrated in this section: one is the orbit and time parameter determination for navigation satellites; another is the attitude parameter determination for spacecrafts.

5.7.5.1 Experiment of Orbit and Time Parameter Determination

In the long-term and high-precision autonomous navigation application for spacecrafts, the accuracy of real-time orbit determination and time synchronization is required to be highest for the navigation satellites. In this way, the theories and methods of autonomous navigation of the navigation satellites based on the X-ray pulsars are completely applicable to other spacecrafts. If the X-ray pulsars can be used to realize the long-term and high-precision autonomous navigation for the navigation satellites, then navigation satellites are used for the ground users and LEO satellites to navigate; for the spacecrafts in the HEO, deep space and interstellar space, the X-ray pulsar can be directly used for navigation; and thus, the high-precision seamless navigation for the space flight missions from the ground to the near-earth space, deep space and interstellar space can be achieved [2, 30]. In this section, the numerical experiments of the XPNAV will be demonstrated with the navigation constellation Walker 24/6/1 as an example.

The **Walker constellation**, proposed by John Walker in 1970, is the uniform configuration one, where the orbital altitudes and inclinations are the same, the orbital planes are evenly arranged relative to the reference plane, and the satellites in the same orbital plane are evenly distributed. The typical Walker constellation is usually described by three parameters: total of satellites (T), number of orbital planes (P) and phase angle coefficient (F) of the corresponding satellites between the adjacent orbits, and then the Walker T/P/F is used to describe the formation of constellation, where the value range of parameter F is any integer between 0 and (P-1). In addition, one parameter of inclination (δ) of the orbital plane is added, and thus the constellation configuration can be determined completely. Therefore, sometimes the Walker constellation is also called **Walker-δ constellation**. The navigation constellation with the formation of Walker 24/6/1 is shown in Fig. 5.17, where the total number of satellites is 24, the number of orbital planes is 6 and the phase angle coefficient is 1. Moreover, there are 4 satellites evenly deployed in each orbit plane of constellation, and each satellite is with the orbital inclination of 55°, the orbital altitude of 21770.0 km, the orbital period of 13h3min and the repeat orbit of 11 cycles per 6 days.

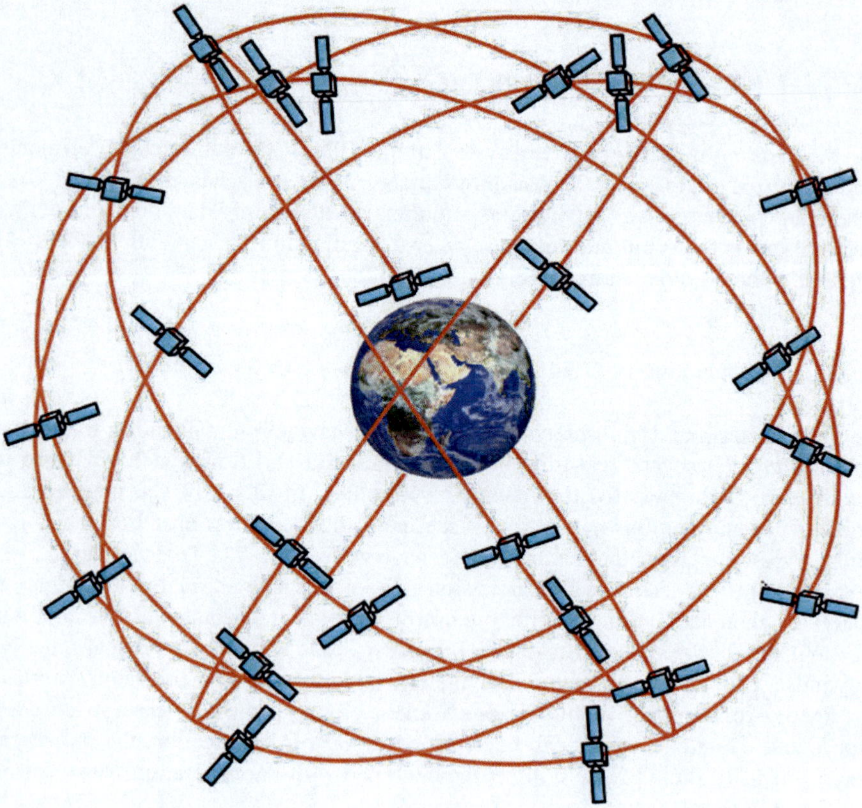

Fig. 5.17 Navigation constellation with configuration of Walker 24/6/1

In the numerical experiments that the orbit and time parameters of navigation satellites are determined by using the X-ray pulsars, two groups of numerical analysis test schemes are selected, for the application scenarios that only one detector is carried with each navigation satellite and only one pulsar is observed during each arc-segment. The conditions of numerical experiments are listed in Table 5.3, from which it can be seen that for four typical error sources of the XPNAV, the indicator's accuracies of scheme I are set to be one order of magnitude higher than that of scheme II.

According to the scheme I, three-dimensional orbit errors, radial orbit errors, clock synchronization errors and User Range Errors (URE) of the satellites are calculated, respectively, as shown in Figs. 5.18, 5.19, 5.20 and 5.21. Moreover, the RMS errors

Table 5.3 Numerical experiment conditions of orbit and time parameter determination of the navigation satellites based on X-ray pulsars

Number	Name of indicator	Value of indicator		Notes
		Scheme I	Scheme II	
1	Accuracy of Pulsar's angular position ($\times 10^{-3}$ arc-seconds)	0.1	1	There are four typical error sources for the XPNAV
2	Accuracy of pulsar timing model (microsecond)	0.1	1	
3	Accuracy of photon arrival-time measurement (microsecond)	0.1	1	
4	Accuracy of time transformation (microsecond)	0.1	1	
5	Pulse signal integral time (s)	1000		
6	Number of the pulsar tracked during each arc-segment	1		
7	Initial orbit error of satellites (m)	200		
8	Initial velocity error of satellites (m/s)	20		
9	Frequency-modulated random walk noise of clocks	4.0×10^{-34}		Only two kinds of noise processes are considered here, the frequency-modulated random walk noise process h_{-2} and the frequency-modulated white noise process h_0
10	Frequency-modulated white noise of clocks	2.0×10^{-22}		
11	Initial error of clocks (ns)	20		
12	Initial frequency drift error of clocks (Hz)	2×10^{-12}		
13	Time span of numerical experiment (day)	200		

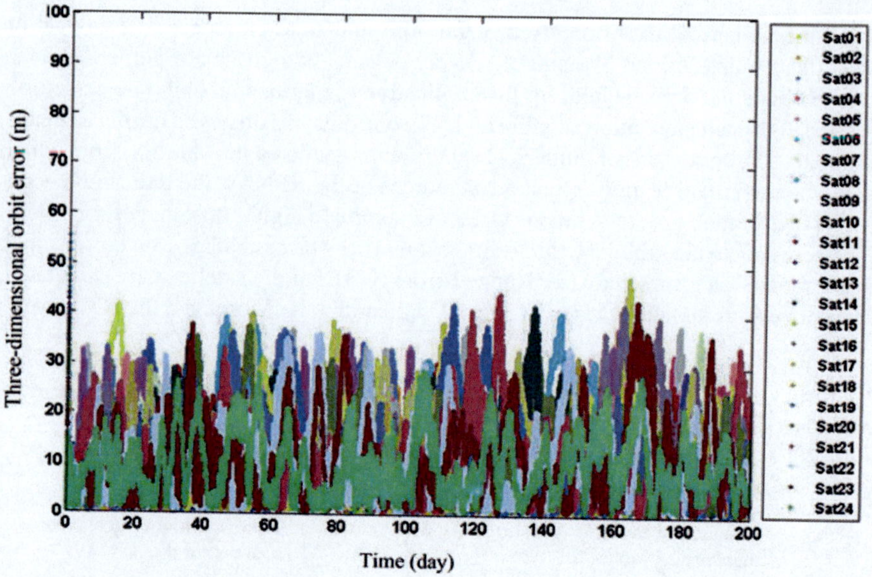

Fig. 5.18 Three-dimensional orbit errors for scheme I

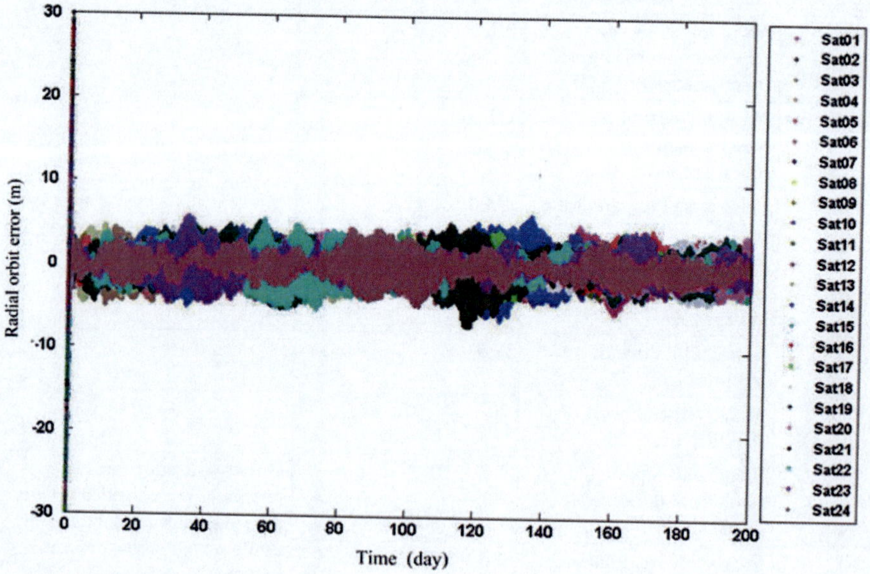

Fig. 5.19 Radial orbit errors for scheme I

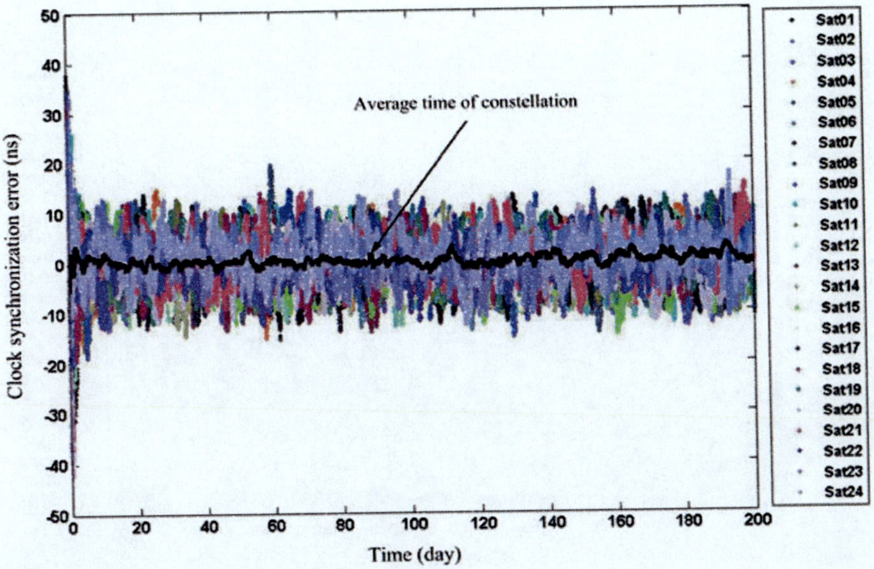

Fig. 5.20 Time synchronization errors for scheme I

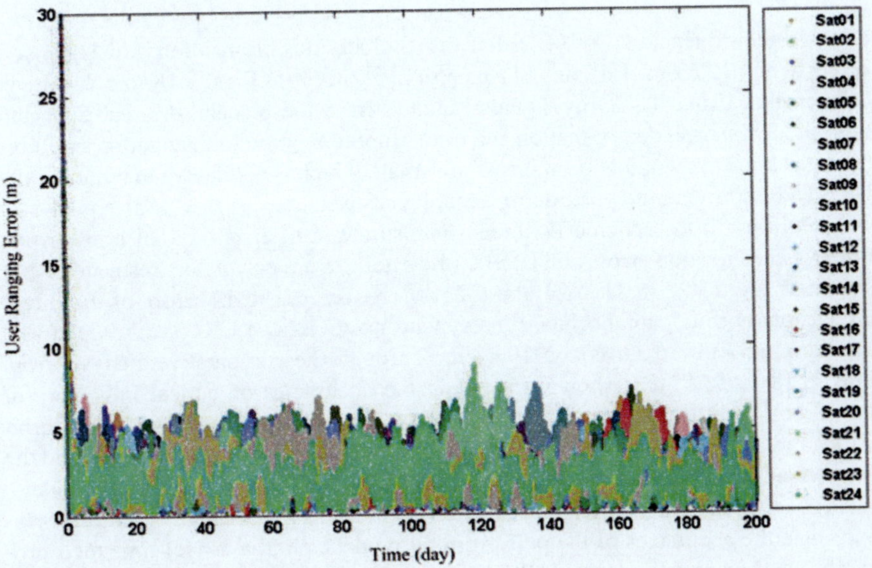

Fig. 5.21 User ranging errors for scheme I

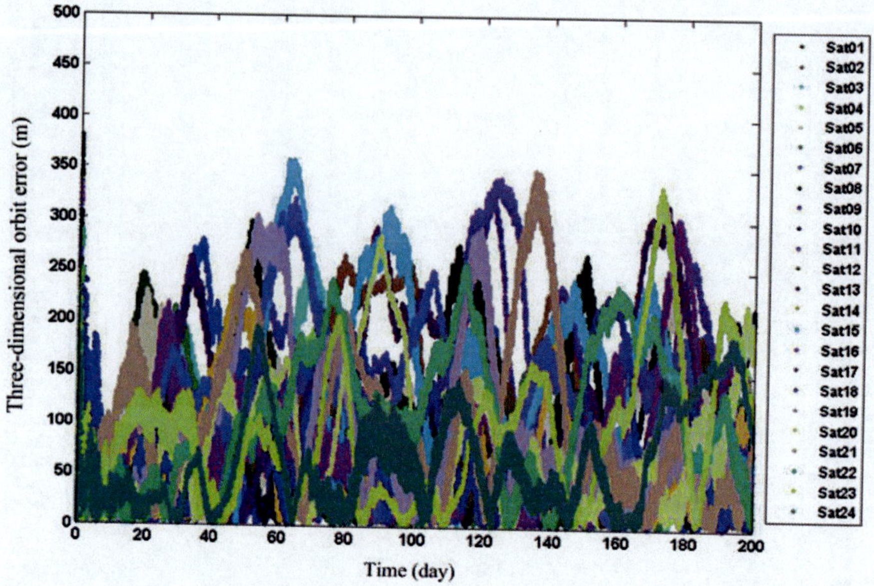

Fig. 5.22 Three-dimensional orbit errors for scheme II

of the three-dimensional orbit, radial orbit, clock synchronization and URE are, respectively, 12.69 m, 1.45 m, 4.17 ns and 2.59 m. From Figs. 5.18 to 5.21, it can be seen that using the X-ray signals radiated from the pulsars, the real-time and high-precision orbit determination and time synchronization for navigation satellites can be achieved; the radial orbit errors are small, which is conducive to reducing the URE and improving the positioning accuracy of users.

According to the scheme II, three-dimensional orbit errors, radial orbit errors, time synchronization errors and URE of the satellites are calculated, respectively, as shown in Figs. 5.22, 5.23, 5.24 and 5.25. Moreover, the RMS errors of the three-dimensional orbits, radial orbits, time synchronizations and URE are, respectively, 99.62 m, 5.57 m, 36.29 ns and 19.48 m. Through the comparative analysis from Figs. 5.18 to 5.25, it is shown that when the accuracies of typical indicators of the XPNAV are decreased by one order of magnitude, the three-dimensional orbit errors, radial orbit errors, time synchronization errors and URE will increase, and the corresponding accuracies also degrade approximately by one order of magnitude.

In conclusion, when the angular position accuracies of pulsars are 0.1 milliarcseconds, and the accuracies of the pulsar timing models, photon arrival-time measurements and time transformations all are 0.1 μs, the orbit and time parameter determination accuracies of navigation satellites based on the X-ray pulsars are, respectively, with the orders of 10 m and 1 ns. The X-ray pulsars provide an independent external space-time datum for the navigation constellations, which can determine the orbit and time parameters of navigation satellites with high precision, and there is not the whole rotation problem of navigation constellation [31, 32], so as to meet the

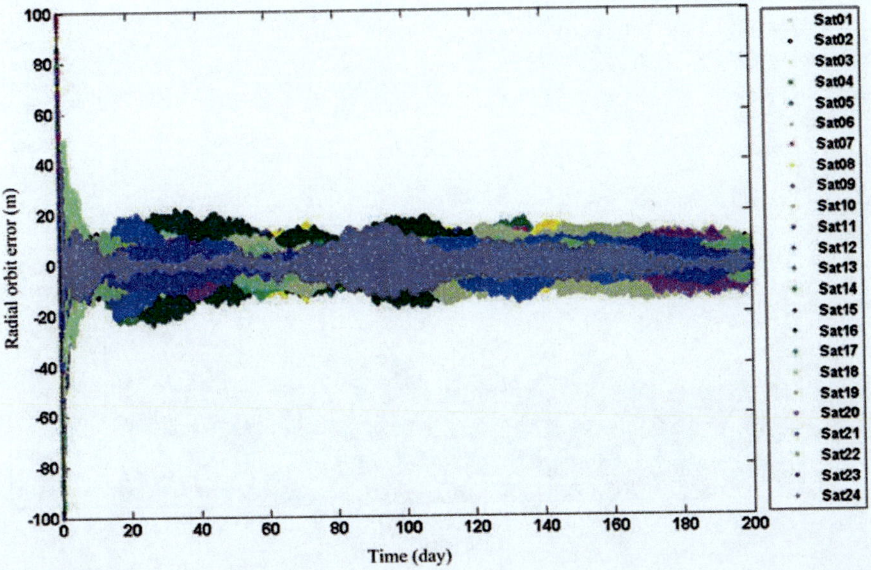

Fig. 5.23 Radial orbit errors for scheme II

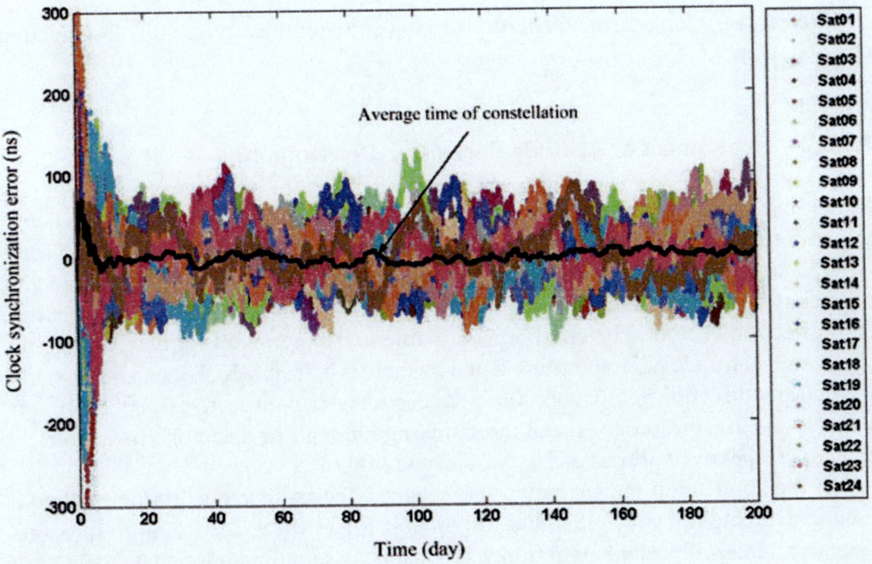

Fig. 5.24 Time synchronization errors for scheme II

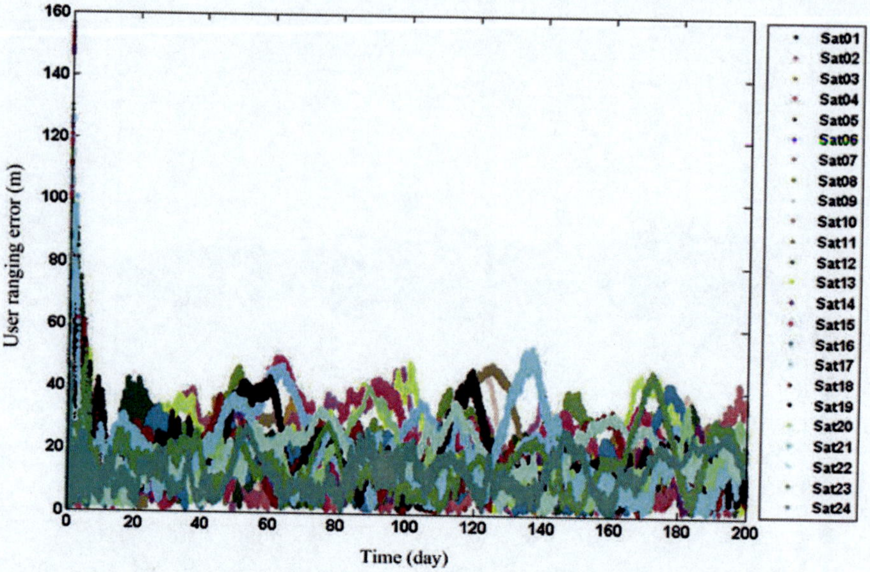

Fig. 5.25 User ranging errors for scheme II

requirements for long-term and high-precision autonomous navigation of navigation constellation.

5.7.5.2 Experiment of Attitude Parameter Determination

In the numerical experiment that the attitude parameters of spacecrafts are determined by using the X-ray pulsars, considering the requirement for double vector attitude measurements, two sets of detectors need to be carried with each spacecraft to track two X-ray pulsars for imaging observations at the same time, and measure their direction vectors. The conditions of numerical experiment for determining the spacecraft's attitude parameters are listed in Table 5.4, from which it can be seen that the pulsar's direction vector measurement accuracy are, respectively, set as 10 arc-seconds and 100 arc-seconds, and the sampling intervals of detector's measurement data are, respectively, set as 5, 10, 50, 100 and 500 s.

For the application scenario of any spacecraft, the attitude quaternion method is used to estimate the yaw, pitch and roll angles. Meanwhile, according to measurement accuracies of pulsar's direction vectors and the sampling intervals of detector's measurement data listed in Table 5.4, ten experimental examples are constructed. For the ten experimental examples, the attitude-estimated accuracies of the spacecraft are shown in Table 5.5, respectively. Due to the limitation of length in this section, only the attitude angle errors of a part of the experimental example are shown in Figs. 5.26, 5.27, 5.28 and 5.29, respectively. From the results of numerical experiments, it can be

Table 5.4 Numerical experiment Conditions of determining spacecraft's attitude parameters based on X-ray pulsars

Number	Name of indicator	Value of indicator	Notes
1	Accuracy of pulsar's angular position ($\times 10^{-3}$arc-second)	0.1	
2	Number of tracking pulsars at the same time	2	
3	Accuracy of measuring pulsar's direction vector (arc-second)	10, 100	Constructing ten experimental examples
4	Sampling interval of detector's data (s)	5, 10, 50, 100, 500	
5	Constant drift of gyroscope (°/h)	3	Only considering the constant drift and measurement random white noise of gyroscope
6	Measurement noise variance of gyroscope (°/h$^{1/2}$)	0.05	
7	Data updating frequency of gyroscope (Hz)	20	
8	Time span of numerical experiment (hour)	24	

Table 5.5 Estimation accuracies of attitude parameters of spacecraft based on X-ray pulsars

Accuracy of attitude angles / Direction vector accuracy & sampling interval		Yaw (as)	Pitch (as)	Roll (as)	Example number
10as	5 s	0.73	0.41	0.75	T01
	10 s	1.40	0.69	1.43	T02
	50 s	3.86	2.53	3.95	T03
	100 s	6.60	4.60	6.63	T04
	500 s	16.93	11.80	20.40	T05
100as	5 s	1.24	0.82	1.46	T06
	10 s	2.14	1.44	2.36	T07
	50 s	7.14	5.62	7.64	T08
	100 s	16.60	8.30	16.39	T09
	500 s	48.99	37.87	59.47	T10

shown that the lower is the measurement accuracies of the pulsar's direction vectors, the greater the attitude angle estimation errors; with the increase of the sampling interval of the detector data, the attitude estimation errors also increase gradually;

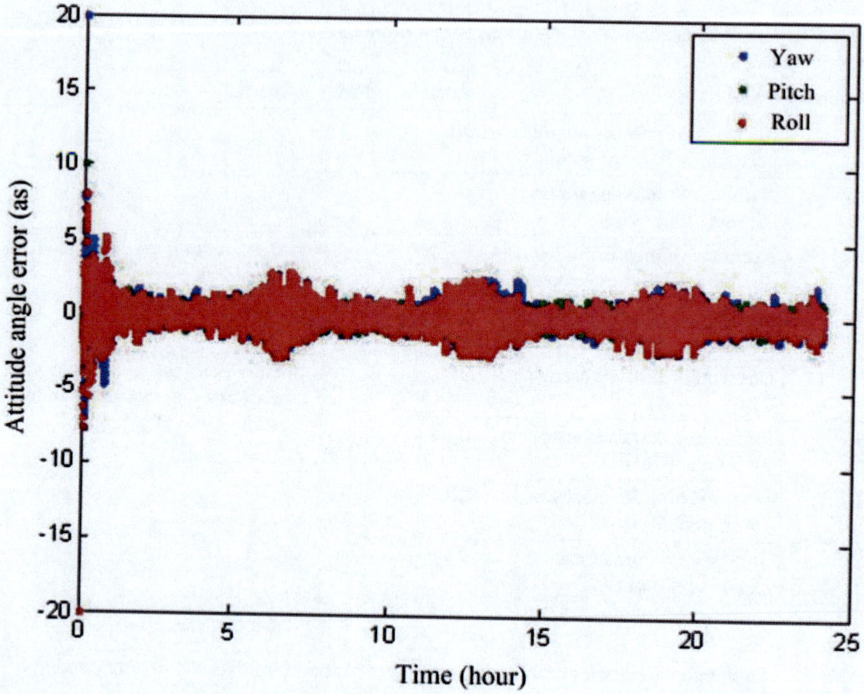

Fig. 5.26 Attitude angle errors for example T01

except for the experimental example T10 in which the pulsar's direction vector accuracy is 100 arc-seconds and the detector sampling interval is 500 s, the spacecraft's attitude angle estimation accuracies of the other nine examples are better than 21 arc-seconds. So, it is clearly demonstrated that the spacecraft's attitude parameters can be determined with high precision by using the X-ray pulsars.

5.8 The First Space Flight Test and Its Results

In November 2016, the world's first dedicated pulsar navigation test satellite, known as X-ray Pulsar-based Navigation-1 (XPNAV-1), was successfully launched by China, to conduct the first space flight test of the XPNAV [33, 34]. Up to now, the XPNAV-1 has normally operated in the orbit for 45 months, and a large amount of X-ray detection data has been obtained by this satellite, completing the scheduled scientific test missions as planned and achieving the expected test goals. The follow-up extended tests are been implementing with the XPNAV-1. In this section, the mission design of XPNAV-1 is briefly introduced first, and then the observation and

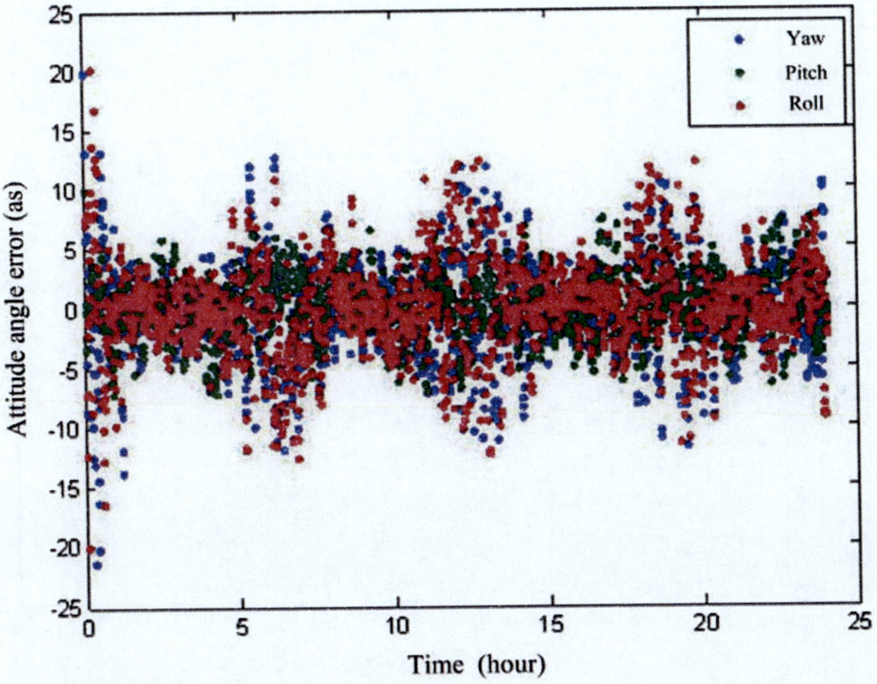

Fig. 5.27 Attitude angle errors for example T03

test results of XPNAV-1, including the observations of typical pulsars and X-ray binaries, the establishment of the Crab pulsar timing model and the orbit determination test using single pulsar, are discussed in detail [35].

5.8.1 Mission Design of XPNAV-1

The XPNAV-1 is a small satellite with a weight of 243 kg and consists of two parts, the platform and the payload. The platform is designed by adopting the idea of mechanical and electronic integration, composed of seven components, i.e., structure and mechanism, integrated electronics, navigation and control, power control, TT&C unit, thermal control system and GNSS antennas. The payload includes two types of X-ray detection equipment: one is the Time-resolved Soft X-ray Spectrometer (TSXS), and another is High Time resolution Photon Counter (HTPC). In principle, the TSXS is essentially a grazing-incidence focused X-ray detector with a multilayer nested optical system, with the time resolution of less than 1.5 μs, the effective detection area of 30 cm^2, the FOV of 15 arc-minutes, and the detecting energy spectrum range of 0.5–10 keV. All X-ray photons entering the FOV can be reflected

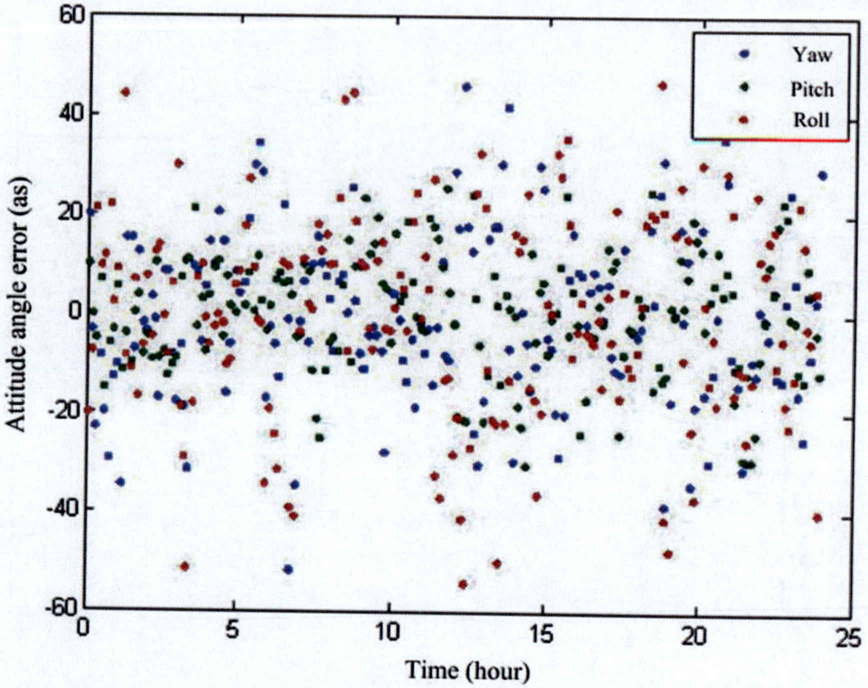

Fig. 5.28 Attitude angle errors for example T05

onto the SDD installed at the focal point of the multilayer nested optical system [36]. The HTPC is a collimated X-ray photon counter based on the MCP array, with the time resolution of 100 ns, the FOV of 2° and the effective detection area of 1200 cm^2. The structure of main body of the XPNAV-1 is a rectangular box with aluminum honeycomb panels, which is deployed in the orbit with a size of 2753 mm (*X*) × 1135 mm (*Y*) × 1785 mm (*Z*). The TSXS and HTPC are installed on the minus Z-axis and minus X-axis of the satellite, respectively. The structural shape and payload configuration of the XPNAV-1 are shown in Fig. 5.30.

In the mission design of the XPNAV-1, there are three scientific test objectives: the first is to test the performance of two types of X-ray detectors in the space environment, and to study the space background noise and detectors' response mechanism; the second is to detect the Crab pulsar or X-ray binaries and obtain the pulse profiles, so as to solve the problem of "seeing" pulsars with the X-ray detectors made in China; the third is to verify the mechanism of the XPNAV system in attempt, so as to demonstrate its principle feasibility.

In order to achieve the scientific test objectives, eight candidate observation objects were selected, including four Rotation Powered pulsars (RP) and four X-ray Binary stars (XB). The names, angular positions, pulse periods and photon fluxes of the eight candidate observation objects are, respectively, listed in Table 5.6, where the angular

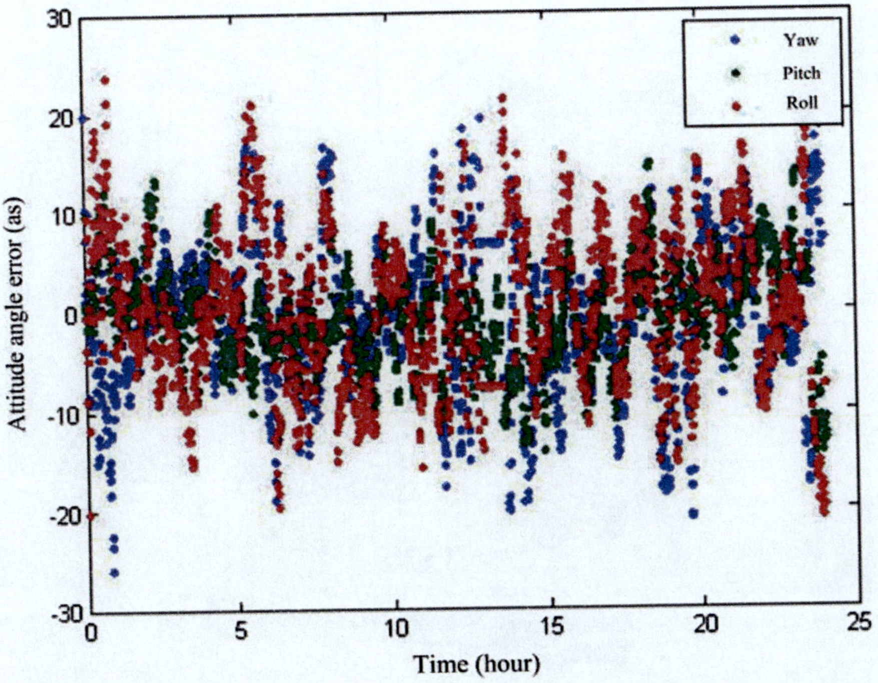

Fig. 5.29 Attitude angle errors for example T08

position parameters are the right ascensions and declinations in the ICRS at epoch J2000.0; the first four objects are the RP pulsars, and the last four objects are the XB stars; B0531+21 is the famous Crab pulsar, with the highest photon flux intensity in the discovered RP pulsars; B1617–155 is a close double star in the constellation Scorpius and is the first X-ray source other than the Sun discovered in 1962; J1744–28 is a type of X-ray transient source that was discovered by the bust and transient source experiment on the Compton Gamma-ray Observatory (CGRO or GRO) in 1995. Moreover, the distributions of the eight candidate observation objects on the celestial sphere are also shown in Fig. 5.31.

An attitude control mode with zero-momentum, magnetic control and three-axis stabilization is used to control the XPNAV-1's attitude changes. According to the requirements for the planed observation missions, the satellite can quickly point to any inertial space position, with the pointing accuracy of better than 2 arc-minutes, and adopts the control scheme of star sensor, gyroscope and momentum wheel to meet the requirements for high-precision pointing control in inertial space. Moreover, the satellite supports four observation modes: self-test mode, scanning observation mode, candidate object observation mode and arbitrary target-pointing mode. The self-test mode is used to test the noise level of X-ray detector equipment itself, known as the detector self-noise, when the TSXS lens is closed or the HTPC points

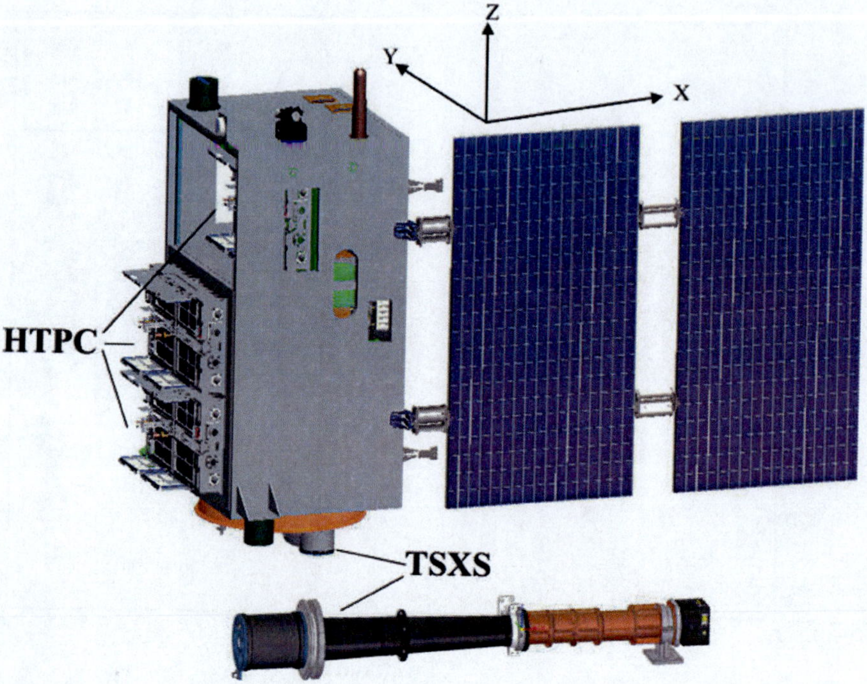

Fig. 5.30 Structural shape and payload configuration of XPNAV-1

Table 5.6 Relevant parameters of eight candidate observation objects

Number	PSR Number	Angular position at J2000.0 (deg)		Pulse period (ms)	Photon flux (ph/cm^2/s)	Notes
		Right ascension	Declination			
1	B0531+21	83.63303	22.01449	33.46	1.54×10^0	Crab, RP
2	B1509-58	228.48175	−59.13583	150.66	1.62×10^{-2}	RP
3	J1846-0258	281.60392	−2.97503	325.68	6.03×10^{-3}	RP
4	B0540-69	85.04668	−69.33171	50.50	5.15×10^{-3}	RP
5	B1617-155	244.979	−15.640	3.20	4.19×10^1	XB, Sco X-1
6	J1744-28	266.138	−28.741	467.05	3.80×10^1	XB, GRO
7	B1758-250	270.284	−25.079	3.00	3.74×10^0	XB
8	B1813-140	274.006	−14.036	3.30	2.10×10^0	XB

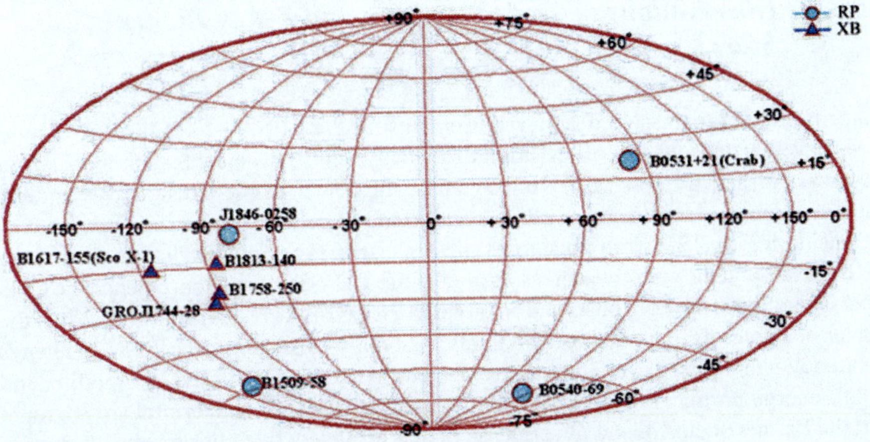

Fig. 5.31 Distribution of eight candidate observation objects on the celestial sphere

to the center of the Earth's disk. The scanning observation mode is used to evaluate the space background noise, through the scanning belt formed as the satellite moves along the orbit and the detector points to a sky area. The candidate object observation mode is a customized observation mode, by using which the X-ray detector TSXS or HTPC points to one of the eight candidate observation objects for carrying out the X-ray detections. The arbitrary target-pointing mode is similar to the third mode, but for the mode, the angular position parameters of the target celestial bodies are uploaded to the satellite through the ground TT&C stations at any time according to the actual observation plan, rather than those stored in the onboard memory in advance. Obviously, besides the eight candidate observation objects, the XPNAV-1 satellite can arbitrarily observe other X-ray sources, such as the center of the Milky Way system, and the supernova remnants.

At 7:42 Beijing time on November 10, 2016, the XPNAV-1 satellite was launched with the solid-fueled launch vehicle Long March 11 at Jiuquan Satellite Launch Center, China. After the launch vehicle took off ten minutes, the satellite was separated from it to enter into the predetermined orbit over the equator in the Indian Ocean. The XPNAV-1 orbit is designed as the Sun synchronization orbit with the local time of 6:00 AM at the descending node, the orbital altitude of 510 km and the inclination of 97.4°. After the in-orbit tests for a week, the XPNAV-1 satellite started to enter the stage of scientific test missions on November 18, 2016. From the observation data analysis, it is shown that the SNR of the observation data from the TSXS is higher than that of the observation data from the HTPC, so the observation and test results obtained by using the TSXS are mainly discussed and demonstrated in this section.

5.8.2 Observations of Typical Pulsars and X-Ray Binary Stars

According to the mission design requirements for the XPNAV-1 satellite, up to now, we have carried out the observations for many celestial objects, such as the four typical pulsars, the four X-ray binary stars, the center of the Galactic System and the supernova remnants, as well as conducted the relevant scientific experiments. Considering that the Crab pulsar has the highest X-ray photon radiation flux in the RP pulsars and the effective detection area of the TSXS is only 30 cm^2, hence the Crab pulsar becomes the core observation target in the entire XPNAV-1 mission. When the satellite passes through the South Atlantic Anomaly (SAA) area, the TSXS is always shut down in order to avoid damage to detector performance caused by high flux and high-energy particle radiation. When the satellite enters into the orbital arc-segment of the Earth's occlusion, no observation plan is arranged. In addition, when the angle between the observation target and the Sun is less than 45°, or the angle between the observation target and the Moon is less than 5°, the observation will be stopped to avoid a large amount of high-energy particles entering the detector. As a result, within an orbital period, there is only an observational arc-segment arranged, usually with the duration of 30–50 min.

In order to process the observed data from the XPNAV-1, a dedicated data processing software package has been developed under the actual X-ray data processing mode of the ground test system. The software package contains eleven software modules, which are responsible for original data unpacking, effective photon identifying, photon counting, energy-spectrum analysis, orbit propagation, photon arrival-time correction, pulsar database construction, pulse period searching, pulse profile extraction, pulsar timing parameter estimation, and orbit and time parameter determination, respectively [37]. By analyzing and processing the observation data, the observation results have been obtained for the candidate objects, such as the X-ray photon flux, energy-spectrum curve, photon arrival-time interval, pulse-period feature, pulse profile, pulsar timing model and navigation parameter estimation.

On November 20, 2016, the first batch of observation data from the XPNAV-1 was obtained by detecting the X-ray photons from the Crab pulsar with the detector TSXS, and composed of six observation arc-segments, with a total of 273 min. The results of photon counting per second for the six observation arc-segments are shown in Fig. 5.32. According to the effective detection area and detection efficiency of the TSXS, the average X-ray photon flux of the Crab pulsar, statistically calculated by using the first batch of observation data from the XPNAV-1, is 1.51 ph/cm^2/s, while the nominal value of X-ray photon flux of the Crab pulsar is 1.54 ph/cm^2/s generally. Apparently, both are almost same. Meanwhile, the energy-spectrum analysis results of the Crab pulsar within the energy band of 0.5–10 keV are shown in Fig. 5.33, where the horizontal axis is expressed with the common logarithm of energy. From Fig. 5.33, it can be seen that the highest X-ray photon flux radiated by the Crab pulsar occurs near the energy of 1 keV, and there is a peak value at energy of 10 keV, resulting from cumulatively counting the high-energy particles with the energy of

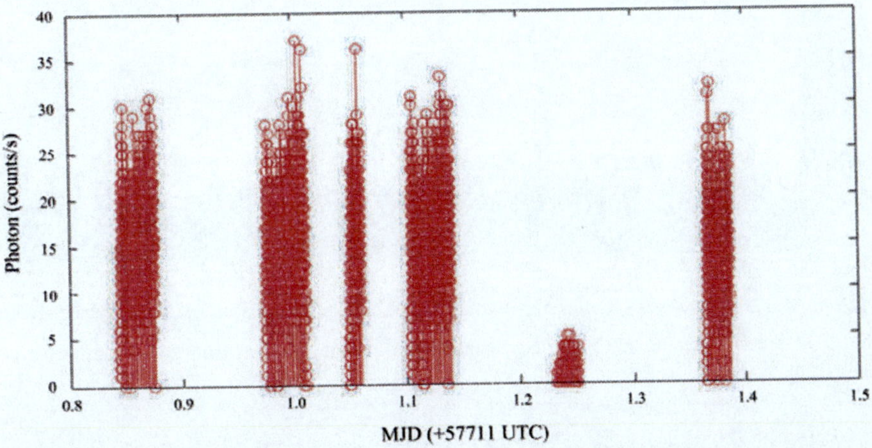

Fig. 5.32 X-ray photon counts of Crab pulsar observed by XPNAV-1 on November 20, 2016

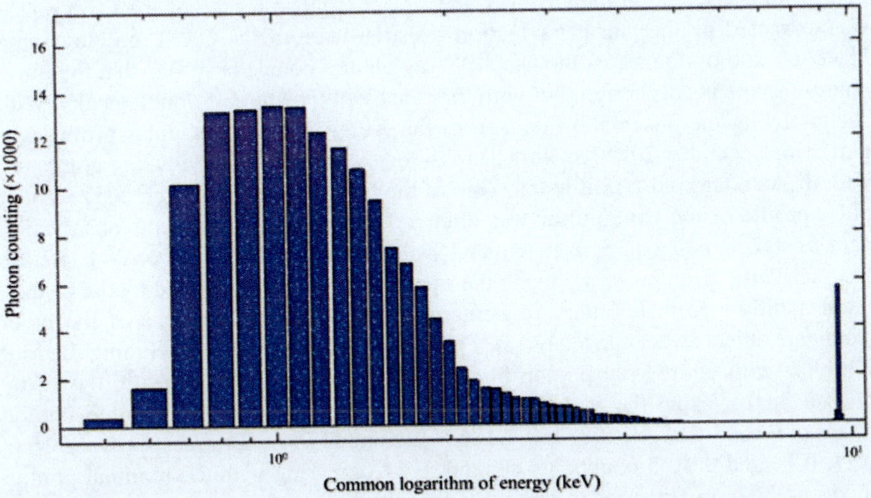

Fig. 5.33 Energy spectrum of Crab pulsar observed by XPNAV-1 on November 20, 2016

greater than 10 keV in the photon flux statistic. Therefore, in the data processing for the TSXS, only the X-ray photons within the energy spectrum range of 0.5–9 keV can be used usually.

Using the six arc-segments observation data of the Crab pulsar, the periodic feature of the Crab is searched in the frequency domain based on the method of Fast Fourier Transform (FFT), and thus it is found that there is a strong signal radiation of the Crab at the frequency 29.650879 Hz as shown in Fig. 5.34, whose corresponding period is 33.726 ms. At the corresponding observation epoch, the international reference

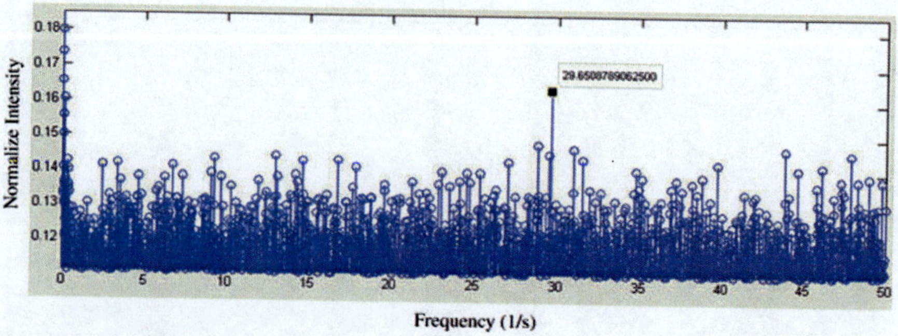

Fig. 5.34 Periodic feature of Crab pulsar searched in frequency domain using 273 min observation data from XPNAV-1

period of the Crab pulsar is 33.728 ms, and the difference between the two values is only 2 μs. That is to say, just at tenth day after the XPNAV-1 satellite was launched into the orbit, the Crab pulsar's periodicity was detected by using the first batch of 273 min observation data. In this way, the first pulse profile of the Crab pulsar was extracted by the pulsar navigation research team of the CAST on November 20, 2016, and of course, as the usable X-ray photon counts is so few that the pulse contour shape is very rough, but with the remarkable feature of double peaks. With accumulating the observation data from the XPNAV-1, a better pulse profile was gotten on December 20, 2016, with a cross-correlation degree of 99.995% compared with the standard pulse profile template of the Crab pulsar; by March 2017, a finer pulse profile of the Crab pulsar was obtained using 162 arc-segments observation data, as shown in Fig. 5.35, from which it is clearly shown that the XPNAV-1 satellite has "seen" the Crab pulsar by using the onboard detectors developed by the CAST.

In addition, from February to September 2017, the observations of the other candidate objects were carried out successively by the XPNAV-1, including the four XB stars with high X-ray photon fluxes and the three RP pulsars with low X-ray photon fluxes. From the statistical analysis, it is shown that the average photon fluxes of B1617–155, B1758–250, B1813–140 and J1744–28 are, respectively, 16.6, 0.21, 0.14 and 0.0155 counts per second. By Comparing with the nominal photon fluxes of the X-ray binaries in Table 5.6, it is shown that the higher is the fluxes, the better the coincidences between one another. It is resulted from the small effective detection area of the TSXS. The X-ray photon flux and energy spectrum of B1758–250 are shown in Fig. 5.36 and Fig. 5.37, respectively. For the observations of the three low flux RP pulsars, because the effective detection area of the TSXS is too small to extract the fine pulse profiles, only the analysis of their photon fluxes, energy spectrums and photon arrival-time intervals have been carried out currently.

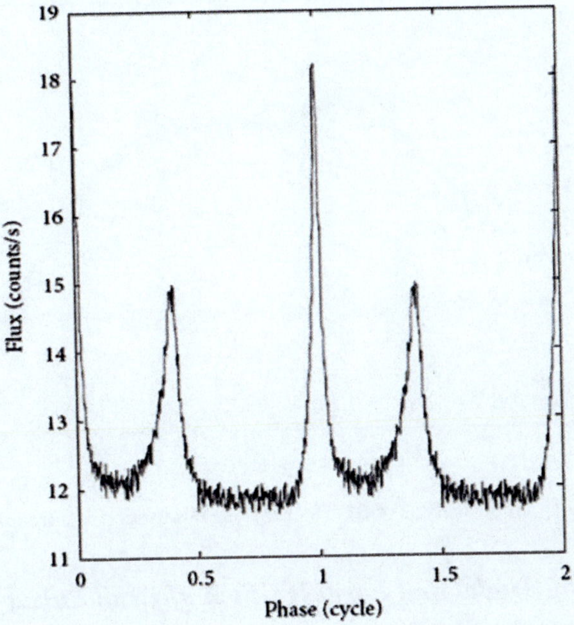

Fig. 5.35 A finer Crab profile obtained by using 162 arc-segments observation data from XPNAV-1

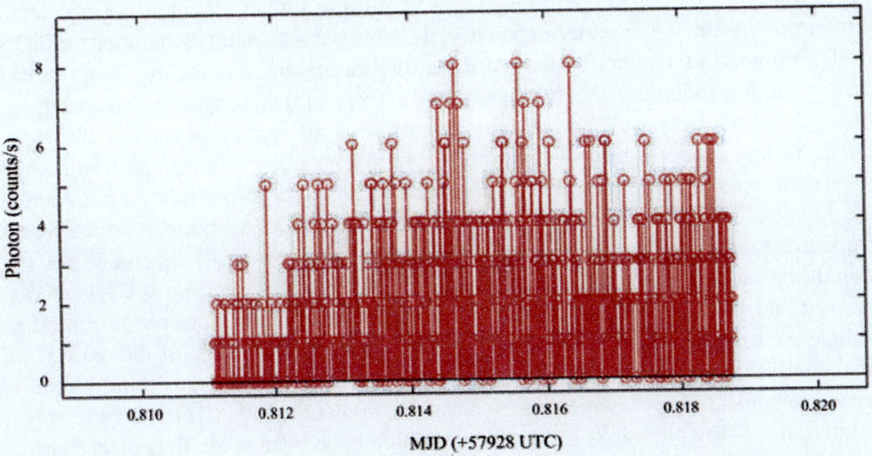

Fig. 5.36 X-ray photon flux analysis of B1758–250 using observation data from XPNAV-1

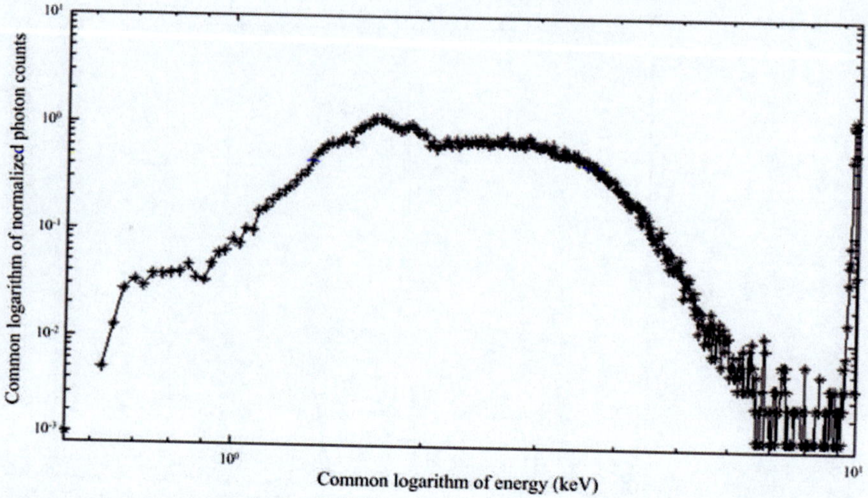

Fig. 5.37 Energy-spectrum analysis of B1758–250 using observation data from XPNAV-1

5.8.3 *Timing Model and Largest Glitch of Crab Pulsar*

Establishing the pulsar timing model is the core task of achieving the XPNAV. From Sect. 5.5.1, it is known that the pulsar timing model is actually a truncated expression of the Taylor series expansion of pulse phase, established under the TDB or TCB timescale. Generally, there are six rotation parameters for the X-ray pulsar timing model, including the frequency and its first-order, second-order and third-order derivatives, as well as the initial phase and reference epoch. For the radio pulsar timing model established by fitting long-term observation data, besides the six rotation parameters, the pulsar's angular position and proper motion parameters are included. However, the rotation parameters of the X-ray pulsars are different from those of the radio pulsars, and particularly there is large difference for the initial phase. Using the X-ray data of the Crab pulsar observed by the XPNAV-1, the rotation parameters obtained from the radio observation data are modified to reduce the timing residuals. In the BCRS, the corresponding time of the main peak of the pulse profile is usually called pulse TOA. In fact, the pulsar timing model is a precise "pulsar clock", which can accurately predict the pulse TOA. By the epoch folding, the X-ray photon TOA sequences which have been made the corrections of the timescale and time delay can be folded into a pulse cycle, and then a clear pulse profile shown. In the process of epoch folding, the prior rotation parameters of the Crab pulsar are gotten by extrapolating the Crab pulsar timing parameters from the monthly bulletin of the Jodrell Bank Observatory at the University of Manchester, the United Kingdom. As a result, there is a significant phase delay between the new pulse profile and the standard pulse profile. The phase delay can be calculated with the cross-correlation processing, and thus the zero-phase time estimation gotten.

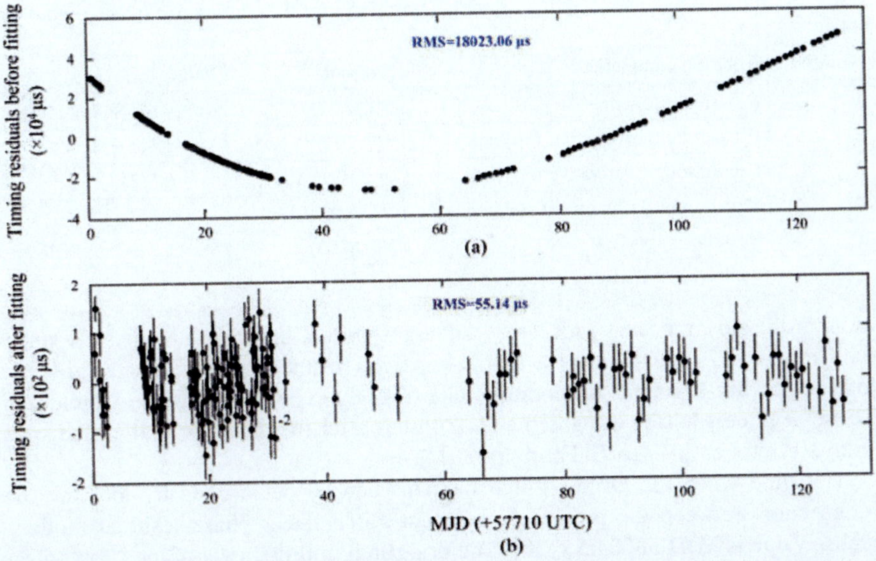

Fig. 5.38 Comparison between timing residual sequences before and after fitting for Crab pulsar timing model

In order to establish the Crab pulsar timing model, 179 arc-segments observation data from the XPNAV-1 are used, with a time span from November 20, 2016 to March 26, 2017. Correspondingly, there is the measured pulse TOAs with a total of 179 can be gotten under the TCB timescale, and compared with the predicted pulse TOAs, so as to obtain the timing residual sequences. The rotation parameters of the Crab pulsar are estimated by the method of weighted lest squares to reduce the timing residuals. The timing residual sequences of the Crab pulsar is shown in Fig. 5.38, where the sub-figure (a) is the timing residuals before fitting; the sub-figure (b) is the timing residuals after fitting; the small circle points denote the residual values; the short vertical lines denote the pulse TOA estimation uncertainties. It can be seen from Fig. 5.38 that there is obvious systematic error accumulation in the time residuals before fitting, with the RMS of 18023.06 μs, caused by using the prior rotation parameters with serious deviations; for the timing residuals after fitting, the systematic error accumulation has effectively been suppressed, so the RMS is only 55.14 μs. In other words, in the application of the XPNAV, using this Crab pulsar timing model will produce a ranging error of 16.5 km.

On November 8, 2017, the Crab pulsar suffered from the largest glitch in all of the recorded observations. After 10.7 h of the glitch occurring, the event that the pulse period of the Crab pulsar suddenly happened change was detected in processing the X-ray observation data from the XPNAV-1. For this reason, the observation plan of the XPNAV-1 was quickly adjusted to increase the observation arc-segments of the Crab pulsar from two arc-segments per day to six arc-segments, so as to frequently

Table 5.7 Rotation parameters estimated with 71-arc-segment data during one month before the glitch

Number	Rotation parameter	Symbol	Value
1	Rotation frequency	$\nu_0(1/s)$	29.6383989224(1)
2	First-order derivative of frequency	$\dot{\nu}(\times 10^{-10} 1/s^2)$	−3.686603(1)
3	Second-order derivative of frequency	$\ddot{\nu}(\times 10^{-21} 1/s^3)$	6.5(2)
4	Third-order derivative of frequency	$\dddot{\nu}(\times 10^{-27} 1/s^4)$	2.1(2)
5	Reference epoch	*RE*(MJD)	58011.729396056769

monitor the slow recovery process of the largest glitch. In order to intensively study the Crab pulsar's behavior in the soft X-ray band resulted from the largest glitch, a total of 455 arc-segments observation data from the XPNAV-1 was used, including 71 arc-segments before the glitch and 384 ones after the glitch, with the time span from 10 October 2017 to 12 February 2018.

The time when the largest glitch happens can be determined by locating the intersection between the pre-glitch and post-glitch linear phase residual models, whose value is MJD 58064.55 [38]. The pre-glitch timing model of the Crab pulsar is built by estimating the rotation parameters with the observation data of 71 arc-segments, and the rotation parameters are listed in Table 5.7. When glitch occurs for a pulsar, its rotation frequency will increase suddenly, and after the glitch, the pulsar's behavior can be described with one step component and one or more transient components. Therefore, the rotational frequency correction can be expressed as [39]

$$\Delta\nu(\mathrm{t}) = \Delta\nu_p + \Delta\dot{\nu}_p\, t + \frac{1}{2}\Delta\ddot{\nu}_p\, t^2 + \sum \Delta\nu_n \exp\left(-\frac{t}{\tau_n}\right), \tag{5.113}$$

where $\Delta\nu$ is the correction of the pulsar's rotation frequency; t is the elapsed time since the glitch occurs; $\Delta\nu_p$ and $\Delta\nu_n$ are, respectively, the permanent term and exponential decaying term of the step change of the rotation frequency at the glitch epoch; $\Delta\dot{\nu}_p$ and $\Delta\ddot{\nu}_p$ are, respectively, the first-order and second-order derivatives of the permanent term; τ_n is the transient decaying timescale.

Considering the difficulty of the rotation frequency estimation via the X-ray data, the rotation phase is usually used for the X-ray data processing. And thus, by integrating formula (5.113), the rotation phase correction is expressed as

$$\begin{aligned}\Delta\phi(t) &= -\int_0^{t-\Delta T_{GE}} \left[\Delta\nu_p + \Delta\dot{\nu}_p t + \frac{1}{2}\ddot{\nu}_p t + \sum \Delta\nu_n \exp\left(-\frac{t}{\tau_n}\right)\right] dt \\ &= -\Delta\nu_p(t-\Delta T_{GE}) - \frac{1}{2}\Delta\dot{\nu}_p(t-\Delta T_{GE})^2 - \frac{1}{6}\Delta\ddot{\nu}_p(t-\Delta T_{GE})^3 \\ &\quad - \sum \Delta\nu_n\tau_n\left[1 - \exp\left(-\frac{1}{\tau_n}(t-\Delta T_{GE})\right)\right],\end{aligned} \tag{5.114}$$

where ΔT_{GE} is the correction of the present epoch.

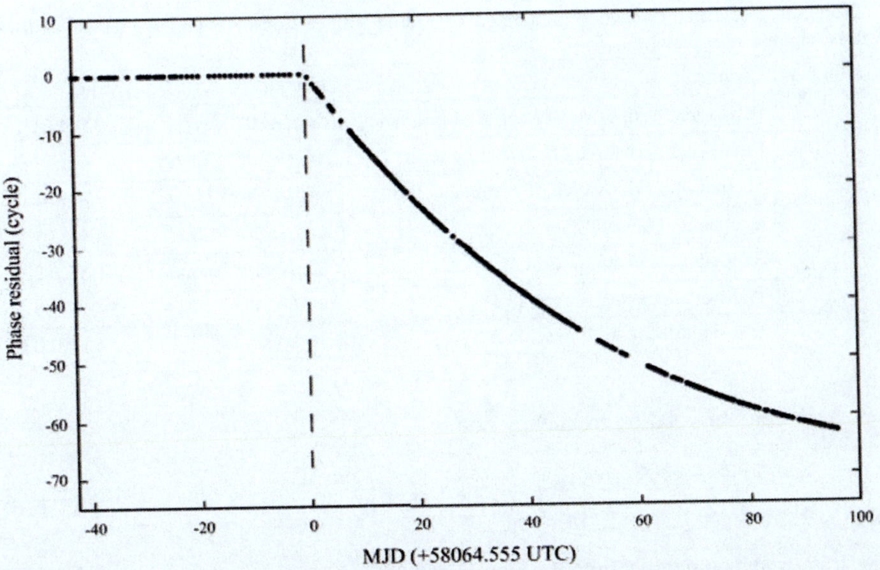

Fig. 5.39 Phase residuals without fitting the glitch

On the right side of formula (5.114), the minus sign at the beginning denotes the phase decreasing. The Levenberg–Marquardt method is used to solve the problem of the nonlinear least squares for formula 5.114. In the initial estimation, the glitch epoch is estimated as MJD 58064.55 $\pm$ 0.01 based on worse than the value MJD 58064.555 $\pm$ 0.005 estimated with the radio observation data [40]. With the glitch epoch MJD 58064.555 as the time zero point, and without fitting the glitch, the values of rotation parameters in Table 5.7 are directly used to calculate the phase residuals for the 455 observation arc-segments, as shown in Fig. 5.39, from which it can be seen that the phase residuals after the glitch are accumulated gradually because the Crab pulsar's rotation speeds up. Let $\Delta T_{GE} = 0.005$, which is the difference between the two glitch epochs from the X-ray and radio data, and the MJD 58064.555 is regarded as the fixed value. And thus, the optimal estimation results of the unknown parameters in formula (5.114) are listed in Table 5.8, using the 384 observation arc-segments [41].

From Table 5.8, it can be seen that there are two transient components of the rotation frequency change detected. One is the long transient component with the decaying time τ_1 of 38.6 days, corresponding to a positive exponential decaying part, i.e., $\Delta \nu_{n1} = 6.6 \times 10^{-6}$ Hz; another is the short transient component with the decaying time τ_2 of 2.37 days, corresponding to a negative exponential decaying part, i.e., $\Delta \nu_{n2} = -1.35 \times 10^{-6}$ Hz. The actual step change of the rotation frequency at the glitch epoch is

$$\Delta \nu_0 = \Delta \nu_p + \Delta \nu_{n1} + \Delta \nu_{n2} = (14.3 \pm 2.0) \times 10^{-6}\,\text{Hz}. \tag{5.115}$$

Table 5.8 Rotational correction parameters estimated with 384 observation arc-segments during three months after the glitch

Number	Parameter	Symbol	Value
1	Permanent term of step changes of frequency	$\Delta\nu_p(\times 10^{-6}\,\text{Hz})$	9.0(1.0)
2	First-order derivative of permanent term	$\Delta\dot{\nu}_p(\times 10^{-12}\,\text{Hz/s})$	−1.04(0.15)
3	Second-order derivative of permanent term	$\Delta\ddot{\nu}_p(\times 10^{-20}\,\text{Hz/s}^2)$	4.0(1.4)
4	Exponential decaying term 1	$\Delta\nu_{n1}(\times 10^{-6}\,\text{Hz})$	6.6(1.0)
5	Exponential decaying term 2	$\Delta\nu_{n2}(\times 10^{-6}\,\text{Hz})$	−1.35(0.02)
6	Transient decaying timescale 1	τ_1 (days)	38.6(3.4)
7	Transient decaying timescale 2	τ_2 (days)	2.37(0.05)

And then, the rotation frequency rate is

$$\frac{\Delta\nu_0}{\nu_0} = (0.48 \pm 0.09) \times 10^{-6}. \tag{5.116}$$

It is almost coincident with the results gotten from the radio observation data [40]. The time derivative of the rotation frequency at the glitch epoch can be expressed as

$$\Delta\dot{\nu}_0 = \Delta\dot{\nu}_p - \frac{\Delta\nu_{n1}}{\tau_1 \times 86400} - \frac{\Delta\nu_{n2}}{\tau_2 \times 86400} = (3.6 \pm 1.9) \times 10^{-12}\,\text{Hz/s}. \tag{5.117}$$

From the above results, it is shown that the glitch happened on November 8, 2017 is the largest for the Crab pulsar, and all of the rotation frequency rates of the glitches recorded before are usually less than 1×10^{-7}.

The fitted phase residuals before and after the glitch are shown in Fig. 5.40, where the dashed vertical line is the time zero point, set at the glitch epoch of MJD 58064.555; on the left side of the dashed line is the pre-glitch residuals obtained with the values in Table 5.7 as inputs, with the RMS of 76.73 μs; on the right side of the dashed line is the post-glitch residuals obtained with the values in Table 5.8 as inputs, with the RMS of 123.18 μs. Obviously, the Crab pulsar's behavior after the glitch can be described well by using formula (5.114) for the parameters fitting, which has important scientific significance for investigating the internal composition and property of the Crab pulsar.

5.8.4 Orbit Determination Experiment Using Single Pulsar

The Crab pulsar timing model has been built using the observation data from the XPNAV-1, and then how to use the observation data of single pulsar carries out the autonomous orbit determination experiment for the satellite? Due to the insufficiency of basic observables for navigation and the irregularity of measurement data

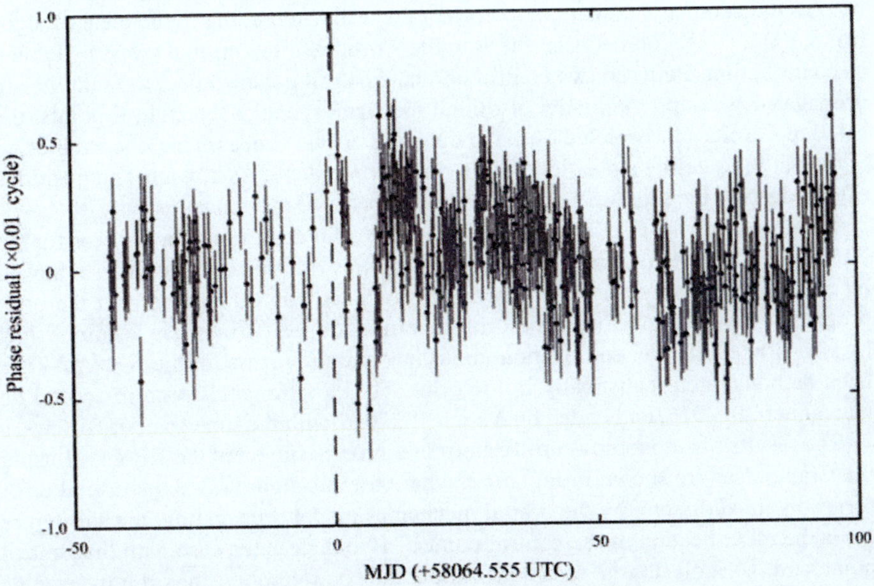

Fig. 5.40 The fitted phase residuals before and after the glitch

sequences, the autonomous orbit determination algorithm based on the extended Kalman filtering is often divergent. For this reason, an autonomous orbit determination algorithm based on normal plane constraints was proposed by the research team of the CAST [42], as shown in Fig. 5.41. The so-called **normal plane** refers to the plane perpendicular to the direction vector of the pulsar. The basic ideal of this algorithm is that the satellite's orbit must pass through the normal plane, which is

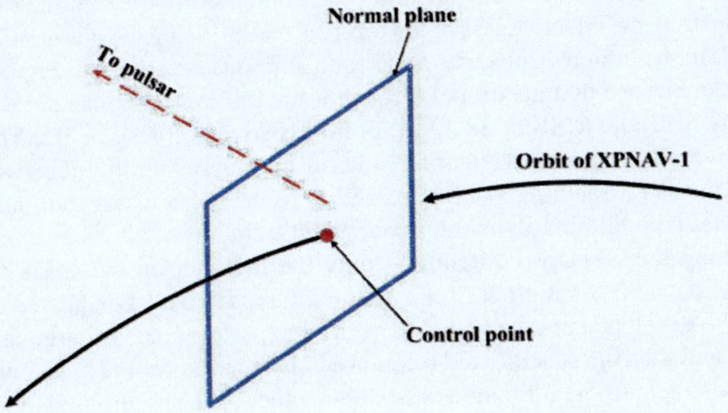

Fig. 5.41 Principle of orbit determination based on normal plane constraints

used as the geometric constraint to correct the orbit. According to the ideal of algorithm, the intersection between the satellite's orbit and the normal plane is defined as control point. Between two control points, the orbit parameters of the satellite are propagated by using the model of orbital mechanics, and at the control points, the satellite's orbits are corrected by using the pulse arrival-time ranging observables, to constrain the satellite's position within the normal plane. As a result, the positions of propagating the satellite's orbits can be corrected to the control points.

Supposed that the GNSS is unable to provide orbit determination services for the XPNAV-1 at a certain time, and then the satellite's orbit determination is achieved by completely relying on the X-ray signals radiated from the pulsar and the orbit propagation model. Although the orbit determination experiment is accomplished by post-processing, the entire autonomous navigation process for the XPNAV-1 has been demonstrated realistically in the orbit. The experimental scenario started on November 20, 2016, and ended on March 15, 2017 with the time span of 85 days.

The results of autonomous orbit determination experiment of the XPNAV-1 using the Crab pulsar are shown in Fig. 5.42, where the sub-figure (a) is the orbital error series obtained directly by the orbital mechanics model propagating; the sub-figure (b) is the identification of the control points, "1" indicates the case with the control point, and "0" indicates the case that without the control points, the orbit mechanics model is used for the orbital propagation; sub-figure (c) is the orbital error series obtained by using the Crab pulsar's X-ray observation data from the XPNAV-1.

From Fig. 5.42, it can be seen that the orbital errors obtained directly by using the satellite's orbit mechanics model for extrapolation have reached 3500 km during approximate 3 months; for the autonomous orbit determination of the XPNAV-1 satellite using the X-ray observation data of the Crab pulsar, although there are still the divergences of the orbital errors between two control points, the orbital errors at the control points can effectively be suppressed, so that the final cumulative maximum orbit error is only 820 km. Obviously, the accumulated orbit propagation errors of the XPNAV-1 have been effectively constrained and suppressed as the pulse arrival-time ranging observables of the Crab pulsar are added to modify the satellite's orbits. To further illustrate the effectiveness of the orbit correction using the observation data of single pulsar, the orbital error series at 25 control-points are shown in Fig. 5.43, from which it can be seen that the orbital error series at the control points are convergent completely, with the RMS of 38.4 km. From experimental results, it is shown that the effectiveness of pulsar navigation is confirmed by using the observation data from the XPNAV-1 satellite, and the feasibility of the pulsar navigation mechanism is demonstrated preliminarily by the space flight test.

Launching the XPNAV-1 satellite is only the first step in the space flight test program of pulsar navigation in China. In the future, a series of dedicated test satellites will be developed and launched to verify the performance of large-area X-ray detectors, and then the space-based reference constellation for the pulsar navigation will be constructed. The ultimate goals are to achieve the space-based autonomous operations and controls for navigation constellations, the autonomous navigations for deep-space vehicles and the construction of integrated space-ground pulsar time

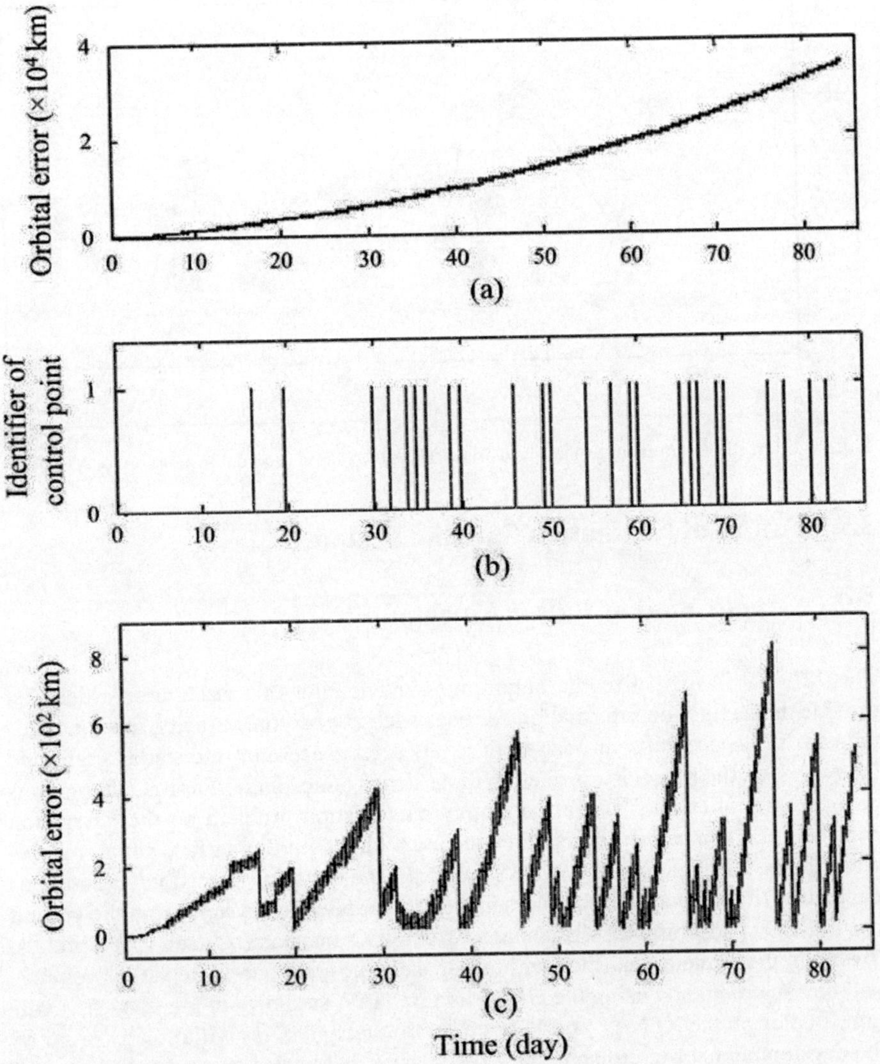

Fig. 5.42 Comparison between orbit propagation using orbital mechanics model and orbit determination using Crab pulsar, for XPNAV-1

system services, so as to speed up and haul the development of astronautic frontier technology.

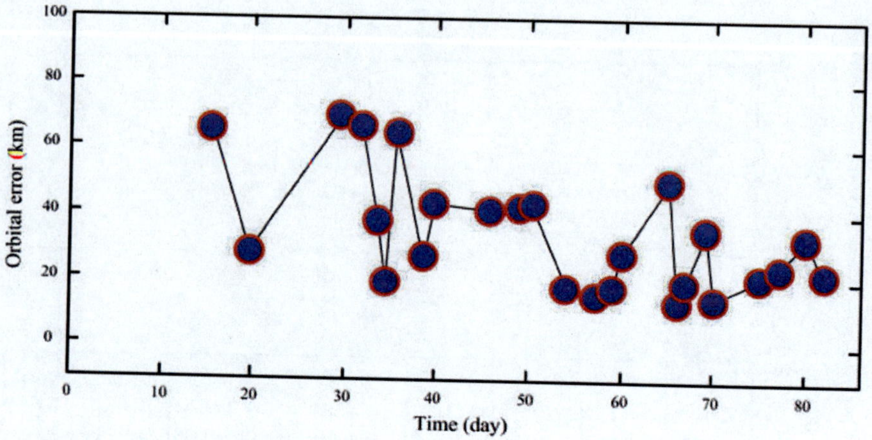

Fig. 5.43 Orbital errors at control points using Crab pulsar-based orbit determination for XPNAV-1

5.9 Future of Navigation Satellite System

5.9.1 *Application of Pulsar Navigation*

The XPNAV is to realize the autonomous navigation of spacecrafts, which can provide the navigation information services such as position, velocity, time and attitude for the spacecrafts in near-earth orbit, deep space and interstellar flight and the rovers on the planetary surface without dense atmosphere. For the autonomous navigation applications, if the autonomous navigation problem on the navigation constellations can be solved first, then the users on the Earth's surface and the spacecrafts in the LEO can use the GNSS for real-time and high-precision autonomous navigation. In this way, the seamless navigation application system from the ground users, LEO spacecrafts to MEO spacecrafts, HEO spacecrafts, deep space probes, rovers on the planetary surface without dense atmosphere and interstellar vehicles will be constructed by using the GNSS and XPNAV, as shown in Fig. 5.44. In addition, studies on the XPNAV and X-ray detection technologies will directly promote the developments of interdisciplinary fields, such as physics, astronomy, electronic science and technology, control science and engineering, aeronautic & astronautic science and technology, as well as nuclear science and technology. So, the application fields of the XPNAV mainly include the autonomous navigation of navigation constellation, deep-space exploration, interstellar flight, unified space-ground pulsar timing system and interdisciplinary research.

The first application field is to realize the long-term and high-precision autonomous navigation of navigation constellations. It is well known that the navigation satellite systems can proved all-weather, full-time, high-precision navigation information services for the users on the Earth's surface and the spacecrafts in the LEO, and have become an important space infrastructure, widely used in

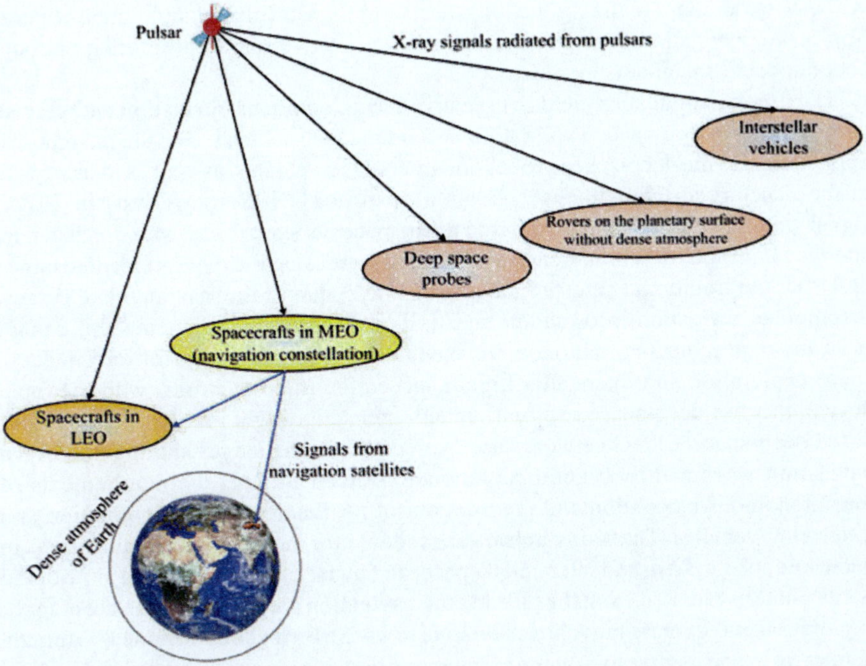

Fig. 5.44 Application system of XPNAV

various fields of national economy. At present, the application of the GNSS has become a necessary part of people's work and life. If the navigation constellation can achieve the autonomous navigation, the workload of ground TT&C system will be reduced effectively, the number of the deployed TT&C stations and the number of injecting the messages from the ground stations to the navigation satellites reduced, the construction and long-term operation maintenance cost of the navigation constellation reduced, the integrity of navigation information monitored in real time, and the dependence of the navigation constellation on ground TT&C system reduced, so as to enhance the autonomous survivability of the navigation satellite system. Currently, the navigation constellations themselves need frequent information support from the ground control system every day, so that they cannot operate independently for a long time in the case without the ground support. The autonomous navigation technology of navigation constellation based on the cross-link information has the problem that the whole rotation errors of the constellation accumulate with time, so that it is difficult for the constellation to run autonomously for a long time. The autonomous navigation of navigation satellites based on the X-ray pulsars has not the whole rotation of the constellation and can realize the autonomous operation of the constellation with high precision for a long time, which is in the line with the development trends of seamless and high-precision autonomous navigation application for the spacecrafts orbiting the Earth. Therefore, speeding up the development of the

XPNAV system and realizing the long-term and high-precision autonomous operation of the navigation constellation will play a positive role in promoting national economic and social development.

The second application field is to realize the autonomous navigation and precise control for the deep-space exploration and interstellar vehicles. For the navigations applications to the deep-space probes and interstellar vehicles, as well as to the rovers on the planetary surface without dense atmosphere, the GNSS is powerless; the TT&C signal strengths from the ground-based deep-space network decay with the distance, and the ranging error will increase by 4 km per increasing the measurement distance of 1 AU; the traditional celestial navigation ways, due to low navigation accuracy, incomplete navigation information and difficult technical implementation, cannot meet the requirements of autonomous navigation and precise control for the deep-space exploration and interstellar flight spacecrafts. In other words, with the rapid development of deep-space exploration and interstellar flight missions, the ground-based deep-space TT&C networks are facing serious challenges and the traditional navigation ways and their combinations are difficult to meet the requirements of the autonomous navigation and precise control for the deep-space exploration and interstellar vehicles. The X-ray pulsars are suitable for the navigation applications in the whole solar system and interstellar space, and by measuring the arrival-time of the X-ray photons radiated from the pulsars, the navigation parameters of the deep-space exploration and interstellar vehicles relative to the SSB can be determined. From the numerical experiment results in the previous section, it is shown that when the angular position accuracies of the pulsars are 0.1 milliarcseconds, and the accuracies of the pulsar timing model, time transformation model and pulse arrival-time measurement are 100 ns, respectively, the orbit determination and time parameters can be estimated, respectively, with the accuracy levels of 10 m and 1 ns, and the attitude estimation accuracy is better than 21 arc-seconds. So, using the X-ray signals radiated from the pulsars as natural beacons, the deep-space exploration and interstellar vehicles can achieve the autonomous navigation and precise control under the large-scale space-time reference frame.

The third application field is to construct the unified space-ground pulsar time system. The TAI is maintained together by the atomic clock groups distributed over the global time laboratories. The long-term instability of the TAI has been detected by using the pulsar timescale established by the ground radio pulsar observation network. The advantage of the pulsar timescale is with the long-term time stability of better than 10^{-19} s/s, which is higher than that of the most stable hydrogen atomic clock at present. By observing the X-ray signals radiated from the pulsars on the space-based platform and the radio signals radiated by the same pulsars with the ground-based radio telescopes, a long-term stable pulsar time reference system would be constructed to serve the national defense, scientific research and national economic development.

The fourth application field is to promote the interdisciplinary and basic science developments. The X-ray detections have always been the research fields of astrophysics, particle physics and astrometry, so the studies on the XPNAV and X-ray detection technology can not only be used in the field of spacecraft's autonomous navigation, but also directly promote the interdisciplinary researches, such as physics, astronomy, geophysics, mechanical engineering, information & communication engineering, control science and engineering, electronic science and technology, instrument science and technology, aeronautic & astronautic science and technology, surveying & mapping science and technology, nuclear science and technology, as well as computer science and technology. In addition, some basic research fields like the gravitational wave detection using the X-ray pulsars may also be carried out. It is conducive to accelerating the interdisciplinary integration researches, well developing and utilizing pulsar resources, improving the research ability and level of basic, cutting-edge and exploratory disciplines, and promoting the development of human civilization and the scientific and technological progress.

5.9.2 Generation Gap Between Navigation Satellite Systems

It is well known that a typical navigation satellite system consists of three segments, including space constellation, ground control and user terminal, where the space constellation segment is composed of tens of navigation satellites evenly distributed in the MEO, GEO or IGSO, providing more than four-signal coverage for global or regional users; the ground control segment is composed of the monitoring, major control and injection stations, which is responsible for tracking and monitoring the navigation satellites, compiling the control instructions and navigation messages, and uploading them to the navigation satellites regularly; the user terminal segment is composed of all kinds of user receivers on the Earth's surface and in the near-earth orbits, which is responsible for receiving and processing the signals from the navigation satellites to provide the position and time parameters for users. In general, the information transmission system between the navigation constellation segment and the ground control segment is referred to as **TT&C mechanism**, and the information transmission system between the navigation constellation segment and the user terminal segment is referred to as **navigation mechanism**. Obviously, the current navigation satellite system contains two aspects of the TT&C mechanism and the navigation mechanism, as shown in Fig. 5.45.

From the perspective of the development of navigation satellite system, the passive navigation mechanism based on the principle of the Doppler shifts is adopted for the First Generation Navigation Satellite Systems (1G-NAV), such as the Transit navigation system built in the United States in 1964, and the Tsiklon and Parus system built in Soviet Union in 1970 s. In addition, the active navigation mechanism based on the two-way timing and ranging is adopted for the 1G-NAV of China, i.e., the BeiDou navigation satellite test system built in 2000. At present, the passive navigation systems with one-way time ranging are adopted in the Second Generation Navigation Satellite Systems (2G-NAV), such as U.S. GPS system, Russian GLONASS system, China's BeiDou system and Europe's Galileo system. Obviously, the essential difference between the 1G-NAV and 2G-NAV is to use the different navigation mechanism,

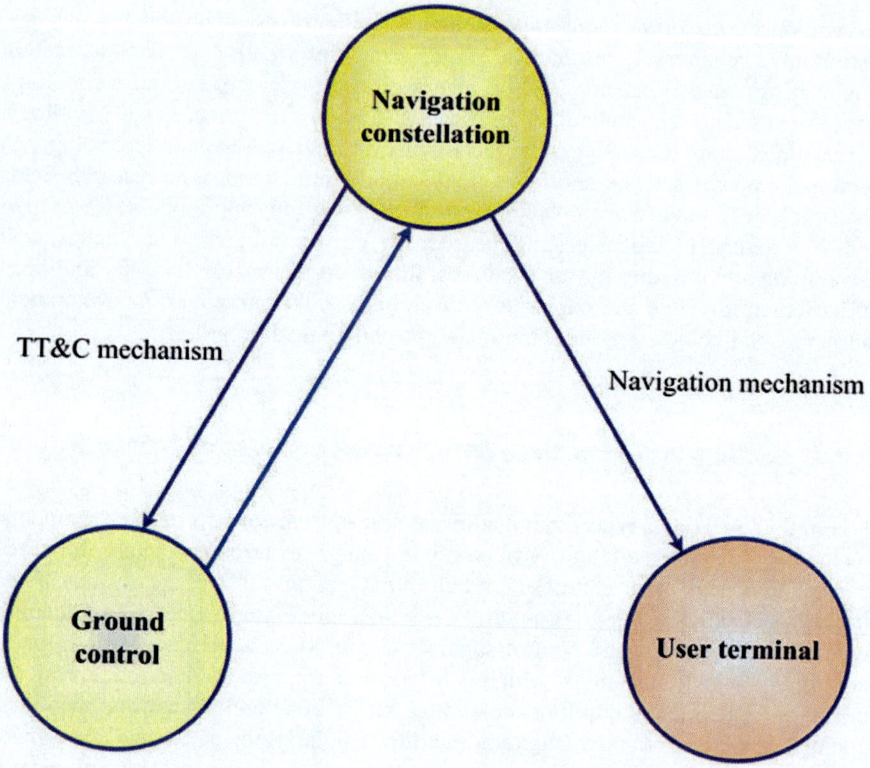

Fig. 5.45 Structure of navigation satellite system

which is called **generational gap**, which is considered as a sign that the "crossing-through-generation" occurs in the development of navigation satellite system. The so-called "**crossing-through-generation**" refers to the essential change of the mechanism of navigation satellite system, rather than some improvements of functions and performances of the system's components. In that way, is it necessary for the navigation satellite system to cross through generation and how is it to cross through generation? It is a major strategic issue facing the development of current navigation satellite system, and also the problem determining the development orientation of the Third-Generation Navigation Satellite System (3G-NAV).

From the prospective of the development of GPS system, there are seven kinds of GPS satellite series launched into the orbits since 1978, including Block I, II, IIA, IIR, IIR-M, IIF and III-A, as shown in Table 5.9, from which it can be seen that each GPS satellite series is compatible with the basic functions of the former series; the platform and payload technologies of the GPS satellites have been evolving constantly and their performance improving; the number of the launched GPS satellites has reached 75 as of June 2020. Although the performance of the space constellation, ground control and user terminal segments of GPS system has been improving for the past

Table 5.9 Technological evolutions of platform and payload of GPS satellites

GPS satellite series	Performance improvement	Number of satellite-launched
Block I	The C/A code, P1 code and navigation message are modulated at central frequency L1 (1575.42 MHz), and the P1 code and navigation message are modulated at central frequency L2 (1227.60 MHz). The Block I satellites are a basic type of GPS satellite series with the double frequencies	11
Block II	On the basis of the Block I satellites, the Nuclear Detection System (NDS) payload and the Selective Availability (AS) software are added in the Block II satellites; the military P codes are encrypted into P(Y) codes to carry out the Anti-Spoofing (A-S); the Block II satellites have the capacity of storing 14-day navigation messages. The NDS information is transmitted to the ground information analysis center for processing through frequency L3 (1381.05 MHz)	9
Block IIA	The Block IIA satellites belong to the advanced series of the Block II satellites, with the capacity of storing 180-day navigation messages. In the absence of the ground control system support, the satellites can autonomously broadcast the navigation messages for users positioning, but the positioning accuracy will gradually degrade with time	19
Block IIR	The Block IIR satellites are the replenishment series of the Block IIA satellites, which have the function of autonomous navigation by adding the cross-links for inter-satellite ranging and communicating. Moreover, the satellites have the capability of the on-orbit programmable and achieve the high-precision time keeping and seamless switching by improving the design of time-keeping system of the onboard rubidium clocks	13

(continued)

forty years, the system mechanism of GPS has not changed essentially. Apparently, the latest type of GPS satellite series, Block III-A, is the third generation of GPS satellite technology evolution, which does not show the new characteristics of the navigation system mechanism, and hence does not belong to the category of the 3G-NAV.

Table 5.9 (continued)

GPS satellite series	Performance improvement	Number of satellite-launched
Block IIR-M	The Block IIR-M satellites are the modernized series of the Block IIR satellites. The new military M-codes are added in the satellites, broadcasting them at the L1 and L2 frequencies. In addition, the civil ranging L2C codes are added at the L2 frequency. Compared with the P(Y) code, the M-code has stronger transmitting power, anti-jamming capability and security performance, and is conducive to direct acquisition	8
Block IIF	The Block IIF satellites are the follow-up series of the Block IIR satellites, in which the civil frequency L5 (1176.45 MHz) is added to ensure the safety in civil aviation application. Furthermore, the flexible software and hardware designs of the payload are adopted to enhance the cross-link data processing, inter-satellite communication, and high-speed uplink and downlink data transmission capabilities	12
Block III-A	The Block III satellites are the new generation GPS satellite series, with the design lifetime of 15 years, to meet higher demands of both military and civilian users. According to the different performance improvements, the GPS III satellites are divided into three types of A, B and C. The Block IIIA satellites deliver significant improvements over current GPS satellites, such as adding the new international civil signal (L1C) and increasing the M-code anti-jamming power with full Earth coverage for military users	3

It is imagined that the navigation satellite system serving for the near-earth space is copied to the deep-space exploration and interstellar flight applications, and the navigation constellations will be established at some feature positions in space, such as the Earth–Moon Lagrange points, the Moon and Mars, so as to provide the navigation information for the deep-space probes and interplanetary vehicles. On the one hand, the mechanism of the navigation satellite system has not changed substantially; on the other hand, the precise TT&C for the navigation constellation, as well as the long-term maintenance cost and economic benefits will become challenging problems. Therefore, constructing the dedicated navigation satellite system serving for deep-space exploration messions is also not the development orientation of the 3G-NAV system.

If the navigation constellation can operate autonomously for a long time, the TT&C mechanism of navigation satellite system will change essentially. So, the autonomous navigation will become the development orientation of the 3G-NAV system, which can effectively enhance the anti-jam and autonomous survival ability of the navigation satellite system and reduce the dependence of the system on the ground TT&C stations, with great engineering application value and strategic research significance.

5.9.3 Conception of Third-Generation Navigation Satellite System

The essential difference between the 1G-NAV and 2G-NAV is the difference of using the navigation mechanism, while the difference between the 2G-NAV and 3G-NAV is the difference of using the TT&C mechanism. So, the **3G-NAV** is defined as a kind of the navigation satellite system built by using the space-based TT&C mechanism, which can provide long-term autonomous operation service for the navigation constellation.

In the numerical analysis experiment in Sect. 5.7.5, the autonomous navigation of navigation constellation is achieved by installing the X-ray detectors on each navigation satellite platform to detect the X-ray photons radiated from the pulsars, measuring the pulse arrival-time and getting the pulsar's image information, as well as autonomously determining the orbit, time and attitude parameters of each satellite. However, the X-ray photon fluxes radiated from the pulsars are very so low and weak that the detector system with an array of large area is required to install on the platform of satellite for collecting the X-ray photons, while the detector system is often with a large volume, heavy weight and high power consumption, resulting in that the limited resource allocation of the satellite's payload is crowded out. From the perspective of the development of current X-ray detector technology, it is still difficult to develop the **navigation-type detectors** with high sensitivity, small size, light weight and low power consumption, which is suitable for the application of satellite's platform. For this reason, the constellation composed of several small or medium-sized satellites is constructed in the LEO, known as **reference constellation**, where the X-ray detector system is the only pay load of each satellite, with the effective detection area of $1-10\,m^2$, which is called **database-type detectors**. Taking the X-ray signals radiated from the pulsars as natural beacons, the satellites of the reference constellation autonomously determine the navigation parameters like orbit, time and attitude. Meanwhile, taking the reference constellation satellites as the TT&C stations with dynamically-accurately known positions, the orbit and time parameters of the GNSS satellites can be determined in real time by receiving the navigation signals emitted from them, and then the navigation messages are compiled and transmitted to the GNSS satellites through the inter-satellite links between the satellites in the reference constellation and those in the GNSS constellation. In this

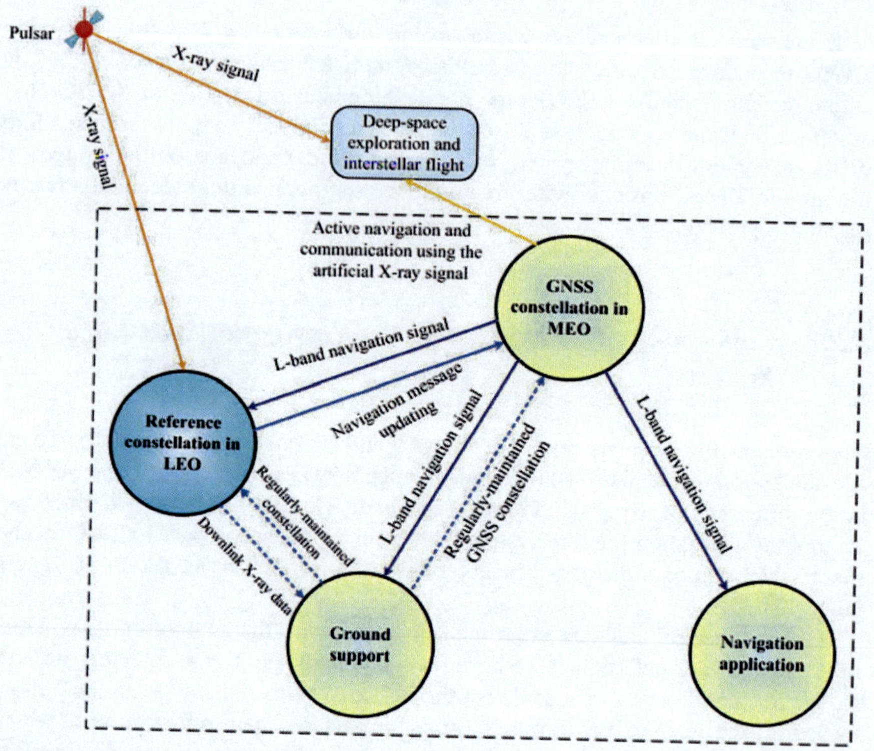

Fig. 5.46 Composition of 3G-NAV system

way, by using the LEO reference constellation based on the XPNAV, the GNSS constellation can be tracked, measured and controlled in real time, so as to achieve its space-based autonomous TT&C. Obviously, the 3G-NAV is mainly composed of four segments, including reference constellation, GNSS constellation, ground support and navigation application [43], as shown in Fig. 5.46.

(1) The reference constellation segment is built in the LEO, and its main tasks are to quickly construct the pulsar database used for the XPNAV and provide the space-time reference information for the GNSS constellation, so as to realize its space-based TT&C for the GNSS constellation, monitor the X-ray pulsars and maintain the pulsar database for a long time.

(2) The GNSS constellation segment consists of tens of the MEO satellites that can be broadcast the L-band navigation signals to the Earth and provide the global signal coverage with more than four satellites, so as to meet the navigation application requirements for the users on the Earth's surface and in the near-earth space. Furthermore, some MEO satellites can broadcast the artificial X-ray signals generated by the onboard small X-ray generators toward the deep space or interstellar space. The artificial X-ray signals can propagate in vacuum with almost no attenuation and high information transmission rate, to meet the

requirements for active navigation and high-baud-rate communication in the deep-space exploration and interstellar flight missions.

(3) The ground support segment is composed of a small number of the GNSS ground control stations and the radio observation stations, which are responsible for post-processing the orbit and time determination of the GNSS satellites to evaluate the GNSS performance; integrating the radio and X-ray pulsar data to build the fundamental database and provide the post-processing service; as well as regularly dumping and correcting the accumulated model errors of orbit determination and time synchronization data processing for the reference constellation satellites.

(4) The navigation application segment consists of the GNSS user terminals on the ground and in the near-earth space, and the X-ray detector terminals for the deep space and interstellar missions, so as to meet the requirements for the seamless autonomous navigation applications from the users on the ground to the spacecrafts in the near-earth orbits, the deep-space probes and interstellar vehicles, as well as the rovers on the planetary surface without dense atmosphere.

The 3G-NAV with the space-based TT&C as the core can realize the seamless signal coverage and navigation application from the ground to near-earth orbit, deep space and interstellar flight missions, with the technical indicators and performance features that cannot be achieved in the 2G-NAV system, as follow:

(1) There is the feature of autonomy. The 3G-NAV has the characteristics of autonomous navigation performance, such as the navigation information completeness, real-time operability, no emitting signals, no dependence on the ground stations and long-term automatic operation. Under the BCRF, the GNSS constellation can operate autonomously for more than 180 days. By using the reference constellation, the real-time orbit determination accuracies of the GNSS satellites are better than 10 m, and the time synchronization accuracies can reach the order of 1 ns.

(2) There is the feature of all space coverage. The third-generation navigation satellite and pulsar navigation system is suitable for the whole solar system and the interstellar space outside the solar system, meeting the requirements for the seamless navigations from the ground to the near-earth space, deep space and interstellar flight.

(3) There is the feature of the reconfigurable. The third-generation navigation satellite system will realize the constellation, software and hardware, and navigation signal reconfigurable. The constellation reconfigurable means that the GNSS satellites are designed by integrating the platform with the payload to get the compact structure and light weight, and then can be launched on demand and deployed dynamically to the corresponding station in the constellation, so as to maintain the constellation formation in time. The software and hardware reconfigurable means that the software is on-orbit programmable, and the hardware is designed by using the biological evolution algorithms, so as to realize the performance and function reconfigurable of the payload. The navigation signal

reconfigurable means that by using the X-ray signals radiated from the pulsars, the spacecrafts can autonomously navigate in the near-earth orbit, deep space and interstellar space; by using the artificial X-ray signals generated by a part of the GNSS satellites, the ultra deep space probes can carry out the active navigation and the high-baud-rate communication; by using the L-band signals from the GNSS satellites, the users on the ground and in the near-earth space can be positioned, navigated and timed; finally, the effects of the multiple signal reconfigurable connected with the seamless navigation will be achieved.

References

1. Shuai P, Chen SL et al (2007) Principle of X-ray Pulsar-based navigation. J Astronaut 28(6):1538–1543 (Chinese)
2. Buist PJ, Engelen S et al (2011) Overview of Pulsar navigation: past, present and future trends. Navigation 58(2):153–164
3. Downs GS (1974) Interplanetary Navigation Using Pulsating Radio Sources. NASA Tech Repo N74–34510:1–12
4. Shuai P, Chen SL et al (2006) X-ray Pulsar-based navigation technology and its application prospect. Aerosp China 10:27–32 (Chinese)
5. Shuai P, Chen SL et al (2007) Progress on X-ray Pulsar-based navigation technology. Chinese J Space Sci 27(2):169–176 (Chinese)
6. Chester TJ, Butman AS (1981) Navigation using X-ray Pulsar. TDA Progress Rep N81–27129:22–25
7. Wood KS (1993) Navigation studies utilizing the NRL-801 experiment and the ARGOS Satellite. In: Proceedings of international society of optical engineering (SPIE), Nashville, Tennessee, 105−116
8. Wood KS, Determan JR et al (2002) Using the unconventional stellar aspect (USA) experiment on ARGOS to determine atmospheric parameters by X-ray occultation. In: Proceedings of international society of optical engineering (SPIE), Nashville, Tennessee, 258−265
9. Wood KS (2004) The silicon X-ray imager (SIXI) on ISS express Pallet. In: The 42nd AIAA aerospace science meeting and exhibit, Reno, Nevada, Jan 2004, AIAA 2004-443:1−8
10. Hanson JE (1996) Principle of X-ray navigation. Doctoral Dissertation of Stanford University, CA, USA, Mar 1996
11. Sheikh SI, Pines DJ et al (2004) The use of X-ray Pulsar for spacecraft navigation. In: Proceedings of the 14th AAS/AIAA space flight mechanics conference, AAS 04-109, Maui, HI, Feb 2004:105-119
12. Sheikh SI (2005) The use of variable Celestial X-ray sources for spacecraft navigation. Dissertation for the degree of PhD, from the Graduate School of the University of Maryland, USA
13. Woodfork DW (2005) The use of X-ray pulsars for aiding GPS satellite orbit determination. Master's thesis of the US air force institute of technology, Wright-Patterson AFB OH School of Engineering and Management
14. Pines DJ (2004) ARPA/DARPA space programs. XNAV industry day, Aug 2004:1−15
15. Graven P, Collins J et al (2007) XNAV beyond the Moon. In: Proceedings of the 63rd annual meeting of the institute of navigation, 23−25 Apr 2007:423−431

16. Graven P, Collins J et al (2008) XNAV for deep space navigation. In: Proceedings of the 31st annual AAS guidance and control conference, American astronautical society, AAS 08-054, Feb 2008:1–16
17. Ray PS, Wood KS, Phlips BF (2006) Spacecraft navigation using X-ray Pulsars. NRL Rev Featur Res 95–102
18. Mitchell JM, Hassouneh MA et al (2015) SEXTANT-Station explorer for X-ray timing and navigation technology. In: Proceeding of the AIAA guidance, navigation and control conference, Jan 2015:1–16
19. Winterniz LMB, Hassouneh MA et al (2015) X-ray Pulsar navigation algorithms and tested for SEXTANT. In: Proceedings of IEEE aerospace conference, MT, US, Mar 2015: 1–13
20. Winterniz LMB, Mitchell JM et al (2016) SEXTANT X-ray Pulsar navigation demonstration: flight system and test results. In Proceedings of IEEE aerospace conference, MT, US, Mar 2016:1–11
21. Mitchell JW, Winternitz LB et al (2018) SEXTANT X-ray Pulsar navigation demonstration: initial on-orbit results, AAS 18–155, July 2018:1–12. https://ntrs.nasa.gov/search.jsp
22. Huang LW, Shuai P et al (2019) Pulsar-based Navigation results: data processing of the X-ray Pulsar Navigation-I Telescope. J Astronom Telescop Inst Syst 5(1)018003:1–8
23. Sala J, Urruela A et al (2004) Feasibility study for a spacecraft navigation system relying on pulsar timing information (Final Report). Adv Res Team of ESA 1–90. http://www.ESA.int/act
24. Shuai P, Li M et al (2009) X-ray Pulsar-based navigation: principle and methodology. China Astronautic Publishing House, Beijing (Chinese)
25. Taylor JH (1991) Millisecond Pulsars: nature's most stable clocks. In: Proceedings of the IEEE, Las Vegas, Nevada, pp 1054–1062
26. Matsakis DN, Taylor JH, Eubanks TM (1997) A statistics for describing pulsar and clock stability. Astron Astrophys 326:924–928
27. Hobbs G, Coles W et al (2012) Development of a pulsar-based timescale. Mon Not R Astron Soc 427(4):2780–2787
28. Emadzadeh AA, Speyer J (2011) Navigation space by X-ray pulsars. Springer Press, New York
29. Rad1 AM, Azari LV (2014) Determining attitude and position in deep space missions using X-ray Pulsars. Int J Astron Astrophys (4):628–648
30. Shuai P (2011) High-precision autonomous navigation technology: combined navigation system based on X-ray Pulsars. Space Int 7:44–51 (Chinese)
31. Shuai P, Li M et al (2007) A new approach to solve rotation issue of autonomous navigation constellation. In: Proceedings of the 58th international astronautical congress, Hyderabad, India, Sept 24–28. IAC-07-B7.1.08:1–10
32. Shuai P, Chen ZG, Qu GJ (2009) On Orbit mechanics of X-ray Pulsar-based navigation. Sci China (Volume E) 39(3):556–562 (Chinese)
33. Lin J, Singer PW (2016) China's deep spacecraft will be guided by X-ray Pulses from distant stars. Pop Sci 18:1–2
34. Zhang XY, Shuai P et al (2017) Mission overview and initial observation results of the X-ray Pulsar Navigation-I Satellite. Int J Aerosp Eng Hindawi, Article ID 8561830:1–7
35. Shuai P, Liu Q et al (2019) XPNAV-1 and its observation results. J Chinese Iner Technol 27(3):281–287 (Chinese)
36. Deng LL, Mei ZW et al (2017) The preliminary In-orbit observation results of Chinese first grazing incidence focusing X-ray Pulsar detector. In: Proceedings of the 68th international astronautical congress, Adelaide, Australia, 25–29 Sept 2017. IAC-17-A7.3.12:1–11
37. Huang LW, Shuai P et al (2018) A new explorer mission for soft X-ray timing-observation of the Crab Pulsar. Acta Astronaut 151:63–67
38. Huang LW, Zhang XY et al (2017) The largest Glitch of the Crab Pulsar detected at the X-ray band. Astron Telegr ATel #11025, 1–2. http://www.astronomerstelegram.org/?read=11025
39. Shemar SL, Lyne AG (1996) Observation of Pulsar Glitches. Mon Not R Astron Soc 282(2):677–690

40. Shaw B, Lyne AG, Stappers BW et al (2018) The Largest Glitch Observed in the Crab Pulsar. Mon Not R Astron Soc 478(3):3832–3840
41. Zhang XY, Shuai P et al (2018) X-ray observation of the 2017 November Glitch in the Crab Pulsar. Astrophys J 866:82–86
42. Huang LW, Shuai P et al (2019) Pulsar-based navigation results: data processing of the X-ray pulsar navigation-I telescope. J Astron Telesc Instrum Syst 01800, 35(1):1–8
43. Shuai P (2016) Pulsars: the lighthouse for space flights. National Defense Industry Press, Beijing (Chinese)

About the Author

Dr. Ping Shuai was born in Liupanshui City, Guizhou Province, China, in 1971. He graduated from Wuhan Technical University of Surveying and Mapping (now Wuhan University) in 1991. He received the M.S. degree in geodynamics and geotectonic physics from the Seismological Institute of Chinese Seismology Bureau in 1998, and the Ph.D. degree in Guidance and Navigation and Control from China Aerospace Science & Industry Corporation in 2003. He joined China Academy of Space Technology (CAST) in 2003, serving as a postdoctoral researcher. At present, he is employed by Qian-Xuesen Laboratory of CAST, serving as a professor and a chief researcher.

Dr. Shuai has ever majored in four subjects, including geodesy and surveying engineering, geodynamics and geotectonic physics, guidance and navigation and control, as well as aeronautic and astronautic science and technology. In the past 20 years, he is interested in many research fields, such as navigation satellite system, integrated navigation, navigation constellation design, pulsar navigation, astronautic technical history, space astronomy and cosmology. His space research activities mainly include two aspects, pulsar navigation and BeiDou navigation satellite system. He has been studying the pulsar navigation technology since 2005, and has served as a chief

P. Shuai, *Understanding Pulsars and Space Navigations*, Navigation: Science and Technology 11, https://doi.org/10.1007/978-981-16-1067-7

scientist for the project of the world's first dedicated pulsar navigation test satellite, X-ray Pulsar-based Navigation 1 (XPNAV-1), since 2013. The XPNAV-1 was launched successfully into orbit in November 2016, and he proposed the scientific missions and experimental programs of the satellite. As of October 2020, all scientific test objectives of the XPNAV-1 have been achieved, and the satellite still runs normally in the orbit, carrying out the extended tests. He has made the distinguished pioneering contributions to promote and demonstrate the first spaceflight test mission of the pulsar navigation technology, and establish a new space science research branch of autonomous navigation for spacecrafts by using X-ray pulsars as beacons. In addition, he has ever served as a senior engineer for the project of BeiDou system from 2003 to 2009. He has made the distinguished contributions to design the BeiDou navigation constellation. The regional BeiDou system was built in 2012, and the global BeiDou system was completed in 2020, officially providing the services for the worldwide users.

Index

P. Shuai, *Understanding Pulsars and Space Navigations*, Navigation: Science and Technology 11, https://doi.org/10.1007/978-981-16-1067-7

K

L

M

N

O

P

Q

R

S